Studies in Systems, Decision and Control

Volume 118

The series "Studies in Systems, Decision and Control" (SSDC) covers both new developments and advances, as well as the state of the art, in the various areas of broadly perceived systems, decision making and control- quickly, up to date and with a high quality. The intent is to cover the theory, applications, and perspectives on the state of the art and future developments relevant to systems, decision making, control, complex processes and related areas, as embedded in the fields of engineering, computer science, physics, economics, social and life sciences, as well as the paradigms and methodologies behind them. The series contains monographs, textbooks, lecture notes and edited volumes in systems, decision making and control spanning the areas of Cyber-Physical Systems, Autonomous Systems, Sensor Networks, Control Systems, Energy Systems, Automotive Systems, Biological Systems, Vehicular Networking and Connected Vehicles, Aerospace Systems, Automation, Manufacturing, Smart Grids, Nonlinear Systems, Power Systems, Robotics, Social Systems, Economic Systems and other. Of particular value to both the contributors and the readership are the short publication timeframe and the world-wide distribution and exposure which enable both a wide and rapid dissemination of research output.

More information about this series at http://www.springer.com/series/13304

Yongsheng Ding · Lei Chen
Kuangrong Hao

Bio-Inspired Collaborative Intelligent Control and Optimization

Springer

Yongsheng Ding
College of Information Science
and Technology
Donghua University
Shanghai, China

Kuangrong Hao
College of Information Science
and Technology
Donghua University
Shanghai, China

Lei Chen
College of Information Science
and Technology
Donghua University
Shanghai, China

Prof. Yongsheng Ding passed away on July 29, 2017.

ISSN 2198-4182 ISSN 2198-4190 (electronic)
Studies in Systems, Decision and Control
ISBN 978-981-10-6688-7 ISBN 978-981-10-6689-4 (eBook)
https://doi.org/10.1007/978-981-10-6689-4

Library of Congress Control Number: 2017954880

Printed on acid-free paper

This Springer imprint is published by Springer Nature
The registered company is Springer Nature Singapore Pte Ltd.
The registered company address is: 152 Beach Road, #21-01/04 Gateway East, Singapore 189721, Singapore

Preface

The industrial processes of high complexity are composed of various units, which are used in different working environments. The total productive processes change frequently with time, and the changes of the model are uncertain. In addition, the relevant industrial products demand of high quality and huge amount, which is one of the most challenging missions in the field of control engineering to build a reliable, efficient, and continuous control system.

The rapidly developing modern bioscience has been exploring the self-regulation mechanism from the natural organism featured by human body which is a perfect control system. The related studies expressed via mathematization, which not only enriched the study of biological sciences, but also provided a theoretical basis for bio-inspired tools and systems. And the self-regulation mechanism also became the source of inspiration to design novel intelligent control algorithms. The complex industrial processes are of high complexity, behavioral uncertainty, strict control, and high-quality requirements, which are similar to the organism. Therefore, bio-inspired intelligent control methods have potential utility in complex industrial processes. In this book, various intelligent coordination control systems and methods through mathematics have been designed, and based on the organism's (especially the human body) regulatory mechanism, such as the comprehensive regulation mechanism of neuroendocrine-immune system, the algorithms are applied to typical complex industrial production processes, the carbon fiber spinning process as an example.

The book is divided into nine chapters. Chapter 1 introduces the related background about the neuroendocrine-immune system and complex industrial production processes. Chapters 2–9 design various biological-inspired cooperative controls and optimization approaches in detail, and provide the related examples as well.

This work in the book was supported in part by the Key Project of the National Natural Science Foundation of China (no. 61134009), the National Natural Science Foundation of China (nos. 61473078, 61473077, 61503075), Cooperative research funds of the National Natural Science Funds Overseas and Hong Kong and Macao scholars (no. 61428302), National Key Research and Development Plan from

Ministry of Science and Technology (2016YFB0302701), Program for Changjiang Scholars from the Ministry of Education (2015–2019), International Collaborative Project of the Shanghai Committee of Science and Technology (no. 16510711100), and Shanghai Pujiang Program (no. 15PJ1400100). The research contents in this book, which elaborate and discuss systematically in many aspects of biological-inspired cooperative control theories and their applications, are multi-disciplinary cross connection with both theory and practice closely integrated.

Thanks to the research and work conditions provided by Donghua University and the persistent supports from our colleagues, which helped us accomplish the related research. Thanks to the intelligent system and network intelligence laboratory, since all the group members worked hard and cooperated with each other, we achieved great progress. I would like to acknowledge the generous contribution made by our Ph.D. students, *Bao Liu, Xiao Liang, Lei Zhang, Tong Wang, Tao Zhang, Nan Xu, Jiajia Chen, Fan Guo, Chuncai Xiao et al.*, and graduate students *Yi Wang, Sheng-Fang Dai, Lei Wang, Lezhi Wang et al.*, for their valuable research achievements.

Shanghai, China
June 2017

Yongsheng Ding
Lei Chen
Kuangrong Hao

Contents

Chapter 1
Related Background

The complex industrial process has the characteristics of high complexity, uncertainty, strict control requirement and product quality standard, and it is very closely to the characteristic of organism which is a perfect control system of human body. Through analysis of the collaborative regulation system of the organism, such as neuro-endocrine-immune (NEI) systems, whose structure models can be applied to design the intelligent controllers and optimization methods for industrial processes. Carbon fiber spinning is a typical complex industrial process, which needs various intelligent algorithms and optimization strategies based on bio-inspired algorithms to enhance the level of control system and realize the intelligent collaboration, so that the carbon fiber products can reach a certain quality.

1.1 Regulation Mechanism of Bio-Inspired Systems

1.1.1 Immune System

Immune system is a complex system, which is composed of many types of immune cells and adequate communication channels. The immune cells have the primitive function to detect the antigens and protect the body. The non-self ones are to be destroyed; the ones targeting at the self ones are to be negative selected. The superior ones detecting antigens will be memorized for quick second response. When the antigen is detected, the antibodies will be proliferated followed by hypermutation as variance method to produce diversity of antibodies. The ones aiming at the self body are soon tolerated by negative selection mechanism. After the elimination of antigen, the inferior antibodies are chosen to be eliminated to keep the balance of antibody population. This process is named "clonal selection". The various communication channels by various cytokines make the expression of complex immune function. Detection, destroy and cleaning of the invading antigens

Y. Ding et al., *Bio-Inspired Collaborative Intelligent Control and Optimization*,
Studies in Systems, Decision and Control 118,
https://doi.org/10.1007/978-981-10-6689-4_1

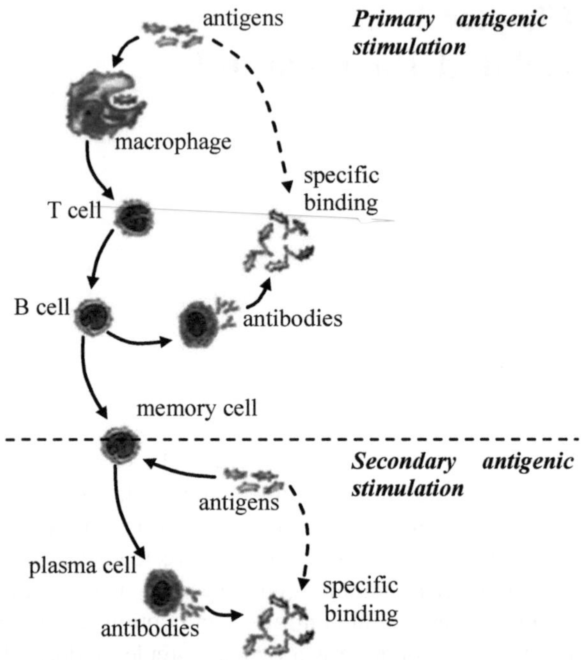

Fig. 1.1 Immunological response process

are all the results of the cooperation of different immune cells by cytokine networks. The cytokine networks provide negative or positive feedback for immune function. The immense discriminative ability of antibodies is also the result of combination mechanisms by cytokines.

The immune system is a system of biological structures and processes within an organism that protects against disease. A properly functioned immune system can detect a wide variety of agents, from viruses to parasitic worms, and distinguish them from the organism's own healthy tissue. Adaptive immune system, also known as the acquired immune system or as the specific immune system, is composed of highly specialized systemic cells and processes that eliminate or prevent pathogen growth. The adaptive immune system evolves in early vertebrates and allows for a stronger immune response as well as immunological memory, where each pathogen is "remembered" by a signature antigen [1]. The adaptive immune response is antigen-specific and requires the recognition of specific "non-self" antigens during a process call antigen presentation. Antigen specificity allows for the generation of responses that are tailored to specific pathogens or pathogen-infected cells. The ability to mount these tailored responses is maintained in the body by memory cells.

The detailed process of primary-secondary adaptive immunological responding is shown in Fig. 1.1. When pathogenic microorganisms such as bacteria or virus first invade into a human body, the macrophages will be first activated. They attack the microorganisms by swallowing them, and then dissolve them into pieces. These pieces will appear on the surface of the macrophages becoming antigens, and then present to T cells.

T cells can combine with the antigen as a key to its corresponding lock and then generate corresponding B cells. B cells will generate large amount of antibodies to destroy the pathogenic microorganisms by specific binding. Through such process, the immune system will defend our body successfully. Meanwhile, when immune cells are activated and begin to replicate, some of their offspring become long-lived memory cells. Throughout the lifetime of an individual, these memory cells remember each specific pathogen encountered and will mount a strong response if the pathogen is detected again. This process is adaptive because it occurs during the lifetime of an individual as an adaptation to infection with that pathogen and prepares the immune system for future challenges.

The major functions of the adaptive immune system can be summarized as:

(i) Immunological recognition. It is used to the recognition of specific non-self antigens during the process of antigen presentation.
(ii) Immunological response. It is the generation of responses which is tailored to maximally eliminate specific pathogens or pathogen-infected cells.
(iii) Immunological memory. With the function of immunological memory, each pathogen is remembered by a signature antibodies or T cell receptors. These memory cells can be called upon to quickly eliminate a pathogen should subsequent infections occur.

Artificial immune system is a computational intelligence paradigm inspired by the biological immune system, and has also been applied successfully to a variety of optimization problems [2–5]. Studies have shown that it possesses several attractive immune properties that allow evolution algorithms to avoid premature convergence and improve local search capability. In the widely studied immune algorithms, many results are about the clonal selection [6], negative selection [2] and immune network [7]. Another newly developed field is the study of cooperative model [8] in immune system. Immune system is very complex and expresses its immune function by cooperation of many immune entities. The cytokine network in immune system is an important substantial base for communication and cooperation among immune cells.

The immune system has the abilities of learning, memory, and recognition, and can memorize the invaded antigen and eliminate it fast, and will be more stable when the antigen invades again [9–11]. The immune system is a highly complicated physiological system, whose main function is to protect the body against the attack of external microorganisms. The immune system specifically recognizes and selectively eliminates foreign invaders by a process known as immune response. Its primary-secondary response mechanism is a process of reinforcement learning and has some special features of high efficiency, stableness, and self-adaptation. Since the later 1990s, the artificial immune system (AIS) inspired from the biological immune system has been developed quickly. As to immune-based intelligent controllers, several intelligent controllers based on the special feedback mechanism of the immune cell are introduced in [12–16], and several methods of regulating PID parameters dynamically to improve the control effectiveness by using immune algorithms are presented [17, 18]. However, the existing immune-based intelligent controllers have not considered the reinforcement learning mechanism of the immune system.

1.1.2 Neuroendocrine Regulation System

The biochemical systems in human body work cooperatively to establish a stable body environment which is critical to the human's live and behaviors. Among these systems, the neuroendocrine system plays a comparatively more important role in the regulation process of human life [19]. Figure 1.2 illustrates the basic structure of the neuroendocrine system and its working principles.

The neuroendocrine regulation system is a large feedback loop consisting of many organs. The core of the system is the cerebral cortex which takes action by secreting hormone to the hypothalamus. The hypothalamus sends instructions to two independent systems, namely, the neural system and endocrine system, respectively. For the neural system, the brain and spinal cord of human body act as the core and the millions of neurons with their varieties forms of connections act as the branches. The signals that the body receives, namely, the stimulations from the environment can be acquired by the neural system and then processed by the brain and spinal cord (in charge of signals of different importance). The instructions sent from the brain and spinal cord are finally transmitted to the corresponding organs to response to the stimulations. For the endocrine system, different types of glands are their actuators, which receive instruction and make regulation for the secretion of hormones. These two systems work independently, but there are some types of cooperation existed between them in order to achieve better control effect.

The structure of such biological system has some similarities with a common control system. The controller, the actuator, the target, and the feedback path in a control system can be replaced by the core (cerebral cortex, hypothalamus, neural system, and endocrine system), the organs and glands, the body, and the sensors, respectively. The principle of working independently with cooperation can also be regarded as the foundation of some decoupling regulation control schemes. Here the only problem left is to extend the original neuroendocrine system structure from two subsystems to three or more subsystems and this can be done by adding and tuning carefully more interactions among the subsystems, which requires deep analysis on the target control plant.

1.1.3 Neuroenodocrine-Immune System

Nervous system, endocrine system, and immune system are three main physiological regulation systems in body. Nervous system detects the inner and outside

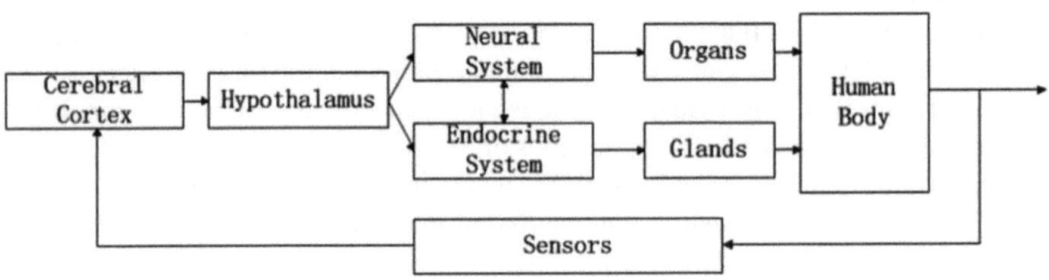

Fig. 1.2 The regulation mechanism of neuroendocrine system in human body

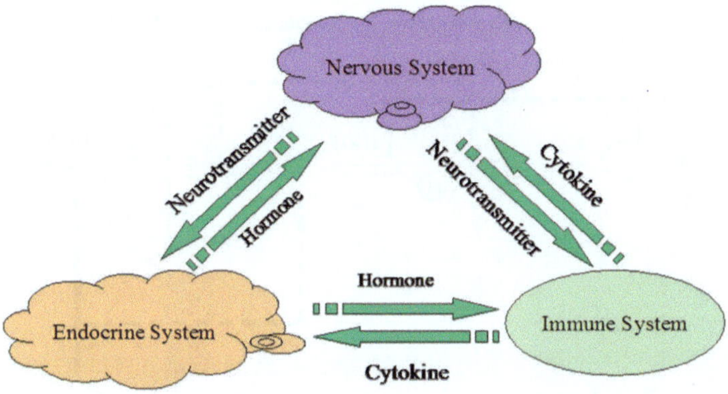

Fig. 1.3 The action relationship of NEI system

statement to regulation the function of all the apparatus in body; endocrine system secretes some kinds of hormones to adjust the speed of metabolism; and immune system is responsible for eliminating the invaded foreign antigen. These three physiological systems compose neuroenodocrine-immune (NEI) system, which is very complicated physiological system [20–22]. The nervous system can affect the endocrine system and the immune system via neurotransmitters, which is released in the nervous system. Endocrine system can simulate or inhibit the activation of nervous system and immune system through some hormones, e.g. GH, adrenalin. Immune system can modulate the physiological function of nervous and endocrine system via cytokines, which is secreted by immune cells. The interactions among them are bi-direction, reciprocal and complex [23, 24], as shown in Fig. 1.3.

1.1.3.1 Regulation Mechanism of Hormone

The regulation mechanism of hormone is similar to the closed-loop feedback control theory. Firstly, central nervous system makes hypothalamo secreting trophic hormone (TH), which stimulates the pituitary to secrete releasing-hormone (RH), and then RH stimulates the corresponded grand to secret hormone. Here, the TH and RH are generic terms of several kinds hormone, respectively. In the modulation system, there are at least four types of feedback, which include ultra-short loop feedback, short loop feedback, long loop feedback, and ultra-long loop feedback, as shown in Fig. 1.4.

The ultra-short loop feedback means the concentration of TH is fed back to hypothalamo, that of RH is fed back to pituitary, and that of hormone is fed back to grand. The short loop feedback means the concentration of TH is fed back to central nervous system, the concentration of RH is fed back to hypothalamo, and that of hormone is fed back to pituitary. The long loop feedback means that the convention of hormone is fed back to hypothalamo, and that of RH is fed back to central nervous system. The ultra-long loop feedback means the concentration of hormones

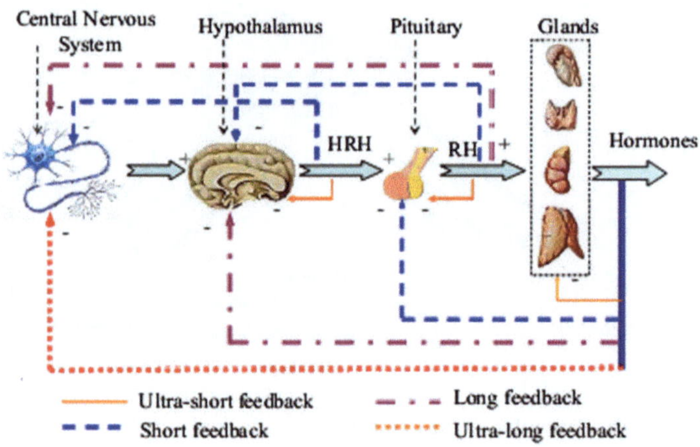

Fig. 1.4 Regulation mechanism of hormones

is fed back to central nervous system. Above these feedback are all achieved via cell-sensors. Through the complicated multiple feedback mechanism, the hormone concentration can be controlled stable easily.

1.1.3.2 Bi-Modulation Mechanism in NEI

Bi-modulation is the other special mechanism in NEI system. Neuroendocrine system may strengthen or weaken the activity of immune system through different hormones. For example, some hormones, i.e. adrenocorticotrophic hormone, androgenic hormones, and prostaglandin, can weaken the activity of immune system; while, some other hormones, i.e. thyroid hormone, growth hormone, prolactin, can strengthen it. This is the cooperatively of hormones.

The central nervous system lies in the highest level, which can harmoniously regulate the endocrine and immune system. When antigen invades the body, immune system not only tries to eliminate the antigen through the immune response, but also can cause the change of endocrine-immune system to adapt to the immune response. Thus, antigen can be easily eliminated quickly, and the homeostasis in body is kept.

1.2 The Application Background of the Complex Industrial Processes

Industrial control mainly includes the independent control of the production units and network control of production systems, and the intelligent applications include the development of intelligent algorithms, improvement of control strategies and construction of intelligent structures. Current development of the industrial intelligent control system is primarily focused on the followings: (1) the unit intelligent control system, such as fuzzy control, neurual network control and evolutionary algorithm control; (2) networked intelligent control system, such as embedded technolog and multi-agent system; (3) learning intelligent control system; (4) data-driven control system. In this chapter, a brief introduction of regulation mechanism of bio-inspired systems will be introduced, and the production process of carbon fiber as a typical example of complex industiral processes will be illustrated in details.

The fiber production is a common industrial process in material manufacturing. The basic concept for producing fiber is to shape the raw materials, liquid, colloid, or other forms, to tiny threads [25], and then they can be further used in other compound material productions, e.g. the weaving or molding process. As a complex process that requires many sections to cooperate with each other, the fiber production line is formed by a series of special equipments for different purposes. A classical instance usually includes sections such as raw material preparation, spinneret components, coagulation bath, one or more stretching units, fabricating, drying, and wire collection [26]. The raw material that passes through these sections may go with different phases and shapes. After extruded from the spinneret and emerged into the coagulation bath, the raw material becomes cold, dense and hard, which means it is gradually solidified and gained an original thread shape. With the stretching units, the threads are fully pulled so that it can be thinner but stronger. Then, the final production fibers are collected after some additional procedures conducted [27].

Carbon fiber is a kind of high-performance material with many advantages, such as high specific strength, high specific modulus, high-temperature resistance, corrosion resistance, fatigue resistance, creep resistance, electric conduction, heat transmission and a small thermal expansion coefficient. It is a novel breed of ideal engineering materials, hence it has great influence in aspects of military, industry, economy and so on. Due to its high strength, high break strength, high modulus, small diameter, low density, maximum crystallinity and low comonomer contents, polyacrylonitrile (PAN) based carbon fiber dominates the market and the optimization of its manufacture process receives increasing attention from scholars and researchers in recent years [28]. Carbon fibers are produced mainly from PAN, rayon, and pitch. Attributing to inherent structural composites, the PAN-based carbon fibers have maintained their predominance as engineering materials up to the present. It has been documented that 90% of the world's total carbon fibers used today are made from the PAN precursor [29]. Their high specific modulus and outstanding fatigue characteristics, combined with their stiffness and lower weight, make these fibers attractive for mass applications ranging from sporting goods to engineering components [30]. Producing a high quality carbon fiber is not an easy thing, since the technological process for carbon fiber is a complex nonlinear system which involves many process parameters that must be carefully optimized and adjusted. Here, the production process of carbon fiber as a typical example of complex industiral processes is illustrated in details.

Some literatures focused on analyzing the properties of carbon fibers by means of physical or chemical instruments. Liu et al. [31] referred to a surface treatment technique of electrochemical oxidation to improve interfacial bonding strength and tensile strength of carbon fibers. Wang et al. [32] investigated the chemical element potassium permanganate modification for carbon fibers during the heat treatment process by differential scanning calorimetry, infrared spectra, elemental analysis, and X-ray photoelectron spectroscopy. Rahman et al. [33] referred to the residence time of 3s as being the most suitable residence time for producing carbon fibers in a solvent-free coagulation process. Experimental data suggest that the Young's modulus of carbon fibers can reach a highest value of 2.55 Gpa. Liang et al. [34] used a bio-inspired intelligent cooperative controller to provide a plan for a stretching process for fiber production. Rennhofer et al. [35] investigated the structural change of carbon fibers with the use of an X-ray testing device at high temperatures under load. Belyaev et al. [36] investigated the kinetics of carbon fibers in oxidative stabilization by differential scanning calorimetry data. Chen et al. [37] proposed a hybrid model of genetic algorithm and improved particle swarm optimization to optimize the radial basis function neural network for real-time predicting of the carbon fiber manufacturing process.

According to all descriptions mentioned above, it is known that they have previously analyzed properties with the aid of different instruments, considering solely relationship between the productive parameters and the fiber properties [38]. This situation resulted for two main reasons, on the one hand, numerous researchers in materials science has different perspectives in the study of the productive process, and on the other hand, the technological process for carbon fiber is a nonlinear system, containing a lot of separate processes: polymerization, spinneret, coagulating baths, washing, stretching, applying oil, drying, pre-oxidation, carbonization, and more. These processes can be regarded as subsystems, and each subsystem has its own control parameters. These parameters affect and restrict the performance of the whole system directly, whereas they are affected not only by interrelation and coupling among subsystems but also by the external environment. Therefore, it is difficult to establish a precise mathematical model to represent linearly the relation between properties indices and productive parameters.

It is well known that before the fiber production line starts, the production parameters must be determined. If the fiber performance needed is changed, the corresponding production parameters must be changed too. If there is an approach to find the relationship between fiber performance and production parameters, the fiber production can be optimized. However, the fiber production line is a large-scale production system which has multiple processes, different modes and complex conditions, so it is difficult to complete the above task. Since 1960s, a large number of basic theory researches have been applied on the fiber production. The traditional optimization methods of this system involve controlling the production equipments, improving the production processes, and optimizing the fiber performance. But most of them are applied on the production controlling, such as the control of winding machines [39], coagulation bath [40], and stretching process [41]. In recent years, there are some researches in building mathematic models of production process, which use simulation technologies to find the accurate description and calculation for every step or a part of the fiber production, but not the whole process. Tan tried to find the relationship between diameter distributions to viscosity and elasticity of meltblown fibers [42]. Gou gave a two-dimensional model of dry spinning polymer fibers [43]. Lee gave a numerical reduction model of optical fibers [44]. Kadi gave a review of the influence from mechanical behavior [45]. Arafeh used neuro-fuzzy logic approach to model the material process [46].

1.2.1 The Basic Process of Carbon Fiber

Carbon fiber precursors refer to a kind of chemical fiber, which is made of specific high molecular compounds, such as acrylonitrile, and the production process includes stock preparation, spinning, post treatment, and so on. Common methods for producing carbon fiber precursor, includes wet, dry and dry-wet, etc., and wet spinning is the most classic method in the traditional spinning process. Wet spinning is mainly applied to the polymer which cannot be extruded as a filament, unless it was dissolved by solvent. The most important feature of this method is to immerse the entire spinneret in a configured bath. The production of carbon fiber is a long process which consists of a lot of critical technical processes, one of which is the spinning process. The optimization of carbon fiber spinning process has been studied intensively in recent years. A general approach is to look into the intrinsic relationship between the process parameters and fiber performance, and obtain optimum fiber properties by tuning the process parameters. Among the carbon fiber spinning processes, coagulation bath and drawing process are two important ones. Particularly, the drawing process comprises multiple drawing stages which form the protofilament, pre-oxidation drawing process, and carbonization drawing process.

Yang and Jaluria [47] considered maximum velocity lag, drawing-induced defect centers and draw tension as the quality measure for hollow optimal fiber. Based on these criteria they built a univariate model consisting of drawing temperature and feeding speed in the drawing process and solve the multivariable optimization problem via curve fitting. Bechtel et al. [48] analyzed the possible draw zones for the two-stage draw process and divided the physical equations into three stages: freespan, draw on a feed roller, and draw on a take-up roller. Then the authors took the temperature as a variation to simulate three phenomena: isothermal, heat the free span, and heat the roller. Gou and McHugh [43] built a two-dimensional model of the dynamics of polymer fiber's dry spinning based on the temperature, composition, and associated profiles, and suggested that the ratio of the Newtonian component viscosity was the most important parameter for spinline data. Makradi et al. [49] developed a temperature and strain-rate coupling cooperative model of polymeric fiber's postspinning draw process with variable fiber velocity, stress, strain rate, and temperature between the different rolls, and simulated the process with mass balance, momentum balance, energy equation, constitutive equation and the finite-element method. Mataram et al. [50] performed a tensile test to obtain the mechanical properties such as Fourier Transform Infrared Spectroscopy (FTIR) and Scanning Electron Microscope (SEM) of polyacrylonitrile (PAN)/silica composite fibers which fabricated by dry-jet wet spinning, showed an optimum Young's modulus at 5.94 GPa and tensile strength at 1.07 MPa at 1 wt% silica. Hobbs and Lesser [51] investigated the properties of poly fibers after two-stage drawing under high pressure CO_2, compared it with untreated fibers and solvent treated fibers. Chuo et al. [52] improved the motion equation based on the mass conservation principle, added time interval to impress evolution of fiber diameter when it was drawing, and suggested a close feedback loop to allow motors to dynamically tune the speed. Suman and Tandon [53] investigated the steady-state of multilayer fiber melting drawing, and analyzed linear stability of isothermal fiber drawing of Newtonian fluid, non-isothermal fiber drawing of Newtonian fluid: including heating only and simultaneous heating and cooling, isothermal fiber drawing of non-Newtonian fluid, respectively. Mabrouk [54] used a two-beam Pluta polarizing interference microscope to discuss three deformation mechanisms during low drawing which contain rubber model scheme, microcracks scheme and necking scheme. Mawardi and Pitchumani [55] used an analytical hyperbolic tangent function as a model to describe neck-down profile of optical fiber drawing process, and solved the optimization problem using the Nelder-Mead simplex method and a simulated annealing technique to achieve optimal process designs for target objectives from contour maps. They also considered the furnace wall temperature, cooling temperature, coefficients of the Arrhenius viscosity and diffusion models as uncertain parameters to guarantee robustness, and provided a valuable optical fiber drawing design tool. Liang et al. [41] proposed a novel bio-inspired multi-layered intelligent cooperative controller to control the rollers with proper stretching ratios during stretching process of fiber production.

The manufacture of polyacrylonitrile carbon fiber (PANCF) includes a series of processes with high fiber complexity [56, 57] such as polymerization, spun extrusion, coagulation, washing, stretching, pre-oxidation, carbonization and post treatment, etc. The aim for utilizing such processes is to create excellent carbon fibers with high strength and elastic modulus. Nevertheless, these processes with different work conditions are closely connected, and the devices applied for each process have various mechanical and control characteristics [58]. The nature of the raw materials (mainly the PAN) also vary within these processes and thus cannot be treated as stable during the whole production, and the coupling and lag existing in and between processes also bring challenge to the utilization of good control schemes.

The coagulation process of the PAN as-spun fiber in the water bath is one of the key processes during the whole carbon fiber production. The PAN solvent is firstly extruded out of the spinneret at a certain velocity, formed tiny streams and then submerged into a water bath slot that contains the coagulation solution of a certain concentration. The PAN streams there exchange components with the coagulation bath solution, and some components of the coagulation bath solution also infiltrate into the streams which is called double-diffusion [59, 60]. During this period, the streams start to coagulate and form the as-spun fibers that can be dragged to the

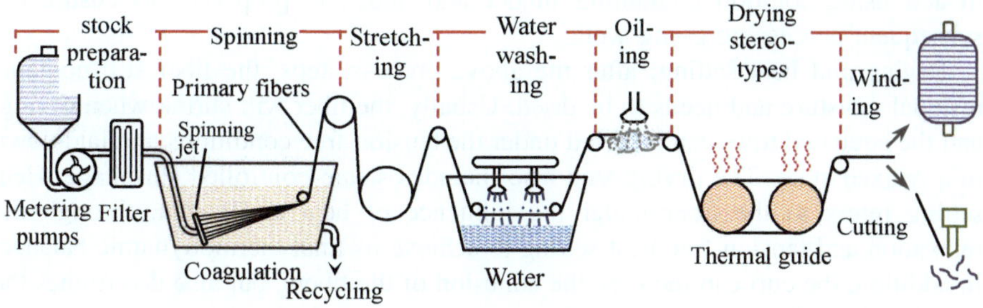

Fig. 1.5 The carbon fiber precursor production process

following process. The quality of the as-spun fiber, which is mostly determined by the parameters of the coagulation bath, has great impact on the final performance of the PANCF [58]. These parameters including temperature, concentration, and liquid-level of the bath have strong links which cannot be tuned independently and satisfactorily. Take the wet process as an example, a complete carbon fiber precursor process is shown in Fig. 1.5.

As can be seen from Fig. 1.5, the wet-process of carbon fiber precursor includes stock preparation, spinning, stretching, water washing, oiling, drying stereotypes, heat setting, winding, etc. Each link is briefly described as follows:

Stock preparation: the stock solution preparation includes polymerization and spinning preparation. Polymerization refers to the spinning of raw materials (polymers) dissolved in a certain solvent to get spinnable solution, known as spinning dope, which can also be obtained directly from homogeneous solution polymerization. Polymerization involves both swelling and dissolution processes, and it will take a long time. The spinning solution must be mixed, filtered and defoamed before spinning to remove the gel pieces, bubbles and impurities.

Spinning: The prepared spinning solution is sent to a spinning machine through circulating pipes, metered by a pump, passed through a filter into a spinneret and then extruded from spinneret holes, the spinneret holes are distributed in the spinneret, to form a fine stream of the raw liquid and to enter the coagulation bath. A spinneret has a large number of spinneret holes, each spinneret spun out a monofilament, while these monofilaments form the final product together, that is the multifilament.

Stretching: it is an integrated process with elastic deformation and plastic deformation. Solidification of the nascent fibers needs to be stretched to improve the fiber material and mechanical properties. The strecting process can be done in one step or multiple steps, usually in order to upgrade the fiber, multi-level stretching process is applied.

Water washing and oiling: in the process of washing, the washing equipment can remove residual solvents, salt and other substances, and the fiber is in tension for washing in the traditional washing process. In the process of oiling, the fiber is treated using solution containing binder and antistatic properties to ensure the subsequent processing going well.

Drying and heat setting: after the above process steps, the fiber still contains residual moisture and needs to be dried. Usually, the fiber will shrink when drying, and the stretched tows can be dried under the tension-free conditions and laid down in a relaxed state. The drying step also includes steam-controlled shrinkage. Heat setting refers to the fiber under the influence of heat and steam, through the relaxation and tension-free heat setting to achieve its final thermodynamic balance. In addition, the curl can increase the adhesion of the fibers, but also determines the visual and tactile properties of fibers.

Winding or cutting: eventually, the chemical fibers can be formed as filament products, and also be cut as staple fibers.

The fiber stretching process plays the key role in the process of fiber production and its effects is measured by the stretching ratio. The stretching ratio is determined by the relative speed of the winding roller, and it has impact on the performance of the final fiber filament and production directly. The fiber production process is one of the complicated production processes in industrial production process [61]. These complicated systems have some of the same characteristics, such as the diversity production environment, the precise measuring instruments, and the expensive production costs. In this type system, on the one hand, a minor variation occurred in one section of the whole line can cause huge change for the final product, and on the other hand, these minor variations cannot be avoided in the industrial production process.

Fiber production process is a typical industrial process in material manufacturing. The fiber production line is formed by a series of special equipment for different purposes, such as the manufacturing or weaving process. In the fiber productive process, the basic purpose for producing fiber is to shape the raw materials, liquid, colloid to tiny threads [62]. A general fiber production process consists of seven sections, namely raw material polymerization, spinneret components, coagulation bath, stretching units, fabricating, drying, and wire collection [63]. As a complex process, it requires many sections to cooperate with each other. After extruded from the spinneret and emerged into the coagulation bath, which means it is gradually solidified and obtained an original silk shape, the raw material becomes dense and hard. Furthermore, after conducting some other procedures, such as oiling, drying and wire collection, the production fibers are completed [27]. In the fiber productive process, the stretching process is an important part, which makes the fibers gradually stretch in the roller using a plurality of different rotational speeds of the rotating motor at different stretch ratios. The variety of the roller speeds causes the changing of the fiber molecular chains structure, rearranges the inherent molecular structure, and then increases strength and toughness of the final fiber products [62]. Generally speaking, the stretching process consists of several stretch stages, and each stretch unit has to achieve different values of elongation. In addition, the number of the stretch stages can affect the quality of the final fiber products [63]. During the stretching process, winding roller is the basic equipment in a stretching unit, which is shaped like a cylinder with different diameter. Fibers that are full of plasticity are wound on two or more rollers one by one, and can be lengthened when the rollers rotate at a specific series of speed [64]. The stretching ratio is defined as the speed proportion between one roller and its neighborhood, which is an important parameter in the control process of quality of the fiber product.

As to the control of the coagulation bath, a traditional scheme is to establish independent control units with conventional proportional-integral-derivative (PID) controllers for the variables, respectively [65]. The closed-loop control model can therefore be established to each individual process several times for there are many variables to be controlled in the coagulation bath. Another practice is the cascade control system for some important parameters which are selected from the whole variables set while keeping others omitted [25, 66]. In some special cases (e.g., the educational and research scenarios), even the open-loop scheme can be applied for each variable by making the values of variables stable with the aid of auxiliary devices, e.g., a heating devices to regulate the temperature manually [67]. For the PANCF production, all the control strategies above do not take the interrelation and coupling among variables into consideration, which cannot guarantee the persistent stabilization of the coagulation bath. Consequently, the PAN precursor fibers produced by such systems can not enjoy a high quality. Another side of such issue is that some schemes or algorithms dealing with plant decoupling have been proposed but not applied to the PAN production process yet due to its special characteristics and complex decoupling relations among variables [68, 69].

Recently, some researchers are gradually focusing on the regulation principles of human body and try to apply such principles to the industrial control domain [70–72]. The control of the coagulation bath of PANCF can be regarded as a representative of complex control toward large-scale systems, which has the potentiality to utilize the bio-inspired control schemes.

1.2.2 Process and Equipment of Raw Silk Spinning

Wet spinning process is one of the spinning processes for various chemical fibers including PANCF. It includes a series of complex processes such as polymerization, coagulation and drawing [40, 56, 57]. The coagulation process of the PANCF in the water coagulation bath is one of the most important processes during the whole wet spinning process. A structure of the coagulation bath, which is commonly used in industry, is depicted as in Fig. 1.6.

The PAN solution, which is a soluble derivative formed by dissolving the PAN in DMSO solvent, is firstly extruded out of the spinneret at a certain velocity. Then the extruded solution jets are submerged into the coagulation solution which is a composite of water and DMSO with a certain concentration. When the PAN jets are coagulating in the bath, the components are exchanging between the PAN jets and the bath solution. This exchanging process is called double-diffusion because in this process the DMSO inside the PAN jets diffuses into the bath solution while the

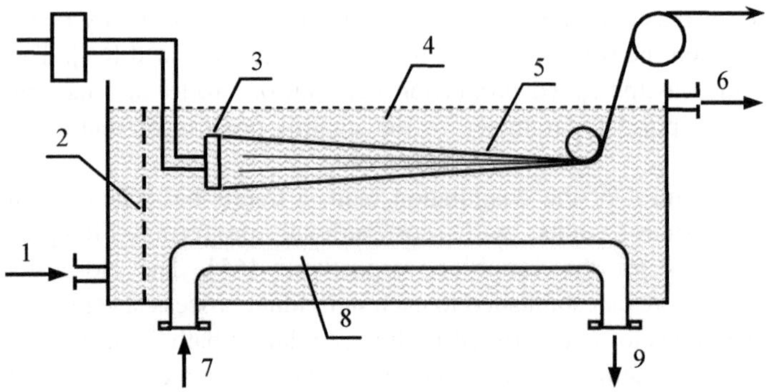

Fig. 1.6 Coagulation bath of wet spinning. **1** water-adding intake; **2** baffle plate; **3** spinneret; **4** coagulation solution; **5** PAN threads; **6** overflow vent; **7** vapor entrance; **8** heat exchange tube; **9** vaporexit

water molecule diffuses into the jets. With the coagulation degree deepened, the jets start to solidified and form the as-spun fibers which can be dragged to the following process.

The quality of the as-spun fiber is mostly determined by the technical parameters of the coagulation bath and it has a great influence on the final performance of the PANCF. These parameters include the bath's temperature, concentration, size, and so on. Amongst all the technical parameters, temperature of the coagulation bath is the most crucial one. Relative to other parameters, temperature has an obvious impact on the diffusion rate in the coagulation process. The coagulation speed is very sensitive to the change on temperature, if the coagulation temperature is high, the diffusion speed will be fast, so that the coagulation will be quick and the compactness will be better; but if the diffusion speed is too fast, a clear skin-core structure will appear and it will cause homogeneity of the fiber. If the temperature is low, the homogeneity and quality of the fiber is better, but if the coagulation bath is not long enough, the solidification of the fiber's core will not be thorough, and this will cause more broken filament during the follow-up spinning processes. Thus, the priority of the coagulation control system is about the temperature control.

The coagulation process of PANCF production is regarded as a key one because it is where the PAN as-spun fibers first generated [73]. The quality of the as-spun fiber has great impact on the final performance of the PANCF [58]. A traditional control scheme for the coagulation bath is to use the conventional PID controllers for each variable respectively [65]. Another scheme is to design a cascade control system for some important parameters [74]. An intelligent cooperative decoupling controller based on the neuroendocrine regulation principle is proposed and applied to the coagulation bath [40].

1.2.3 *Multi-Level Stretching Process*

The rapid development of the manufacturing industry has resulted in a rapidly growing demand for high tensile, light-weight and high modulus PANCFs [57]. PAN filament has been well known as the perfect precursors for preparing high-performance carbon fibers after a set of primary processing steps including thermal stabilization treatment, carbonization and stretching process. Particularly, the stretching process plays a crucial role in improving the overall properties of entire production line, which mainly consists of a cold-stretched process and a hot-stretched process [56, 58]. The mechanical properties of PANCFs become superior only after they undergo the stretching process. This raises a challenging demand for innovative stretching control techniques, which are able to not only smoothly and accurately track the preset stretching ratios but also to achieve a cooperative regulation between two processes, thereby realizing the desired tension of fiber precursors in the presence of various disturbances [75].

1.2.3.1 Stretching Process

The stretching process is one of the key processes in the production of polymer fibers. The main target of stretching process is to contribute threads of certain diameters so that better internal structure can be acquired (that is useful to gain higher strength and larger toughness) for the following processes [25]. In the stretching process, fibers full of plasticity are wound on two or more rollers one by one, and can be lengthened when the rollers rotate at a specific series of speeds [64]. In practice, the stretching process is often organizing into several small parts in different working environments because these environments can exert different influence on fibers, e.g. in hot air, water or high pressure vapor. Each part includes several rollers, usually two or three as a group. This group is called “a stretching unit”. A typical stretching process may include one or two stretching units. In a unit, the stretch ratio is defined as the speed proportion between one roller and its neighborhood, which is an important parameter in controlling the quality of the fiber products. Figure 1.7 shows the basic structure and equipments used for a stretching process in fiber production. In this figure, two stretching units and a bath between them are illustrated. Note that the fibers in the bath may not always be stretched, so the stretching unit referred in this section does not include those in the bath.

Figure 1.8 shows the structure of multi-level stretching system with several consecutive driving motors. ***LS*** denotes the tension sensor mounted on each stretching zone for measuring precursor tension. The velocities of the driving motors are controlled in decentralized inner-loops to implement the re-allocated stretching ratios together. Moreover, the input voltages are determined by control allocation algorithm for online self-tuning of the stretching ratios and the steady convergence of tension at the same time [40, 61, 76, 77].

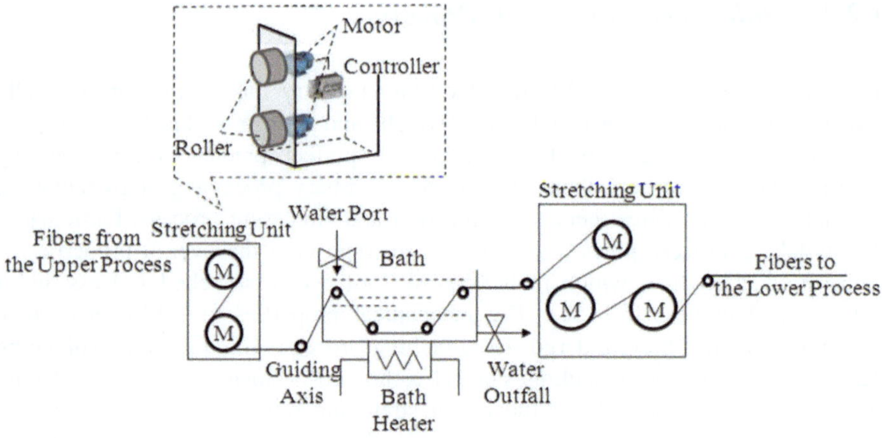

Fig. 1.7 The stretching process in fiber production

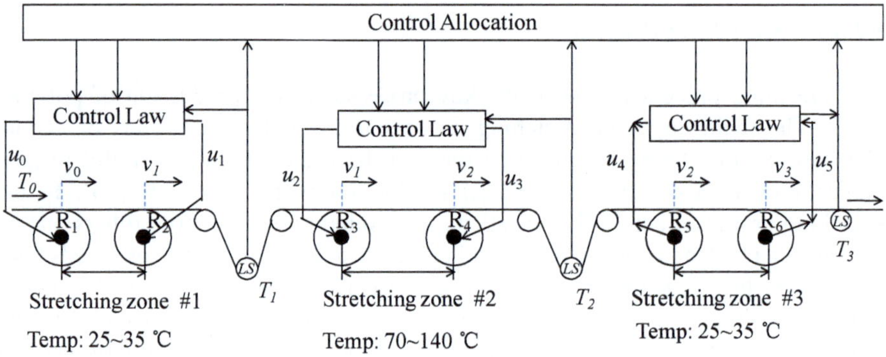

Fig. 1.8 The structure of multi-level stretching system

For this system, it is important to regulate the stretching ratios among the multi-level sub-processes throughout the production line. It is deeply further involved into final tension stable dynamics, which determines the performance of industrial products. For example, when the stretching ratio of the first level stretching process changes, it will influence the next one, for example, the driving motor R3 should keep the same speed with that of the driving motor R2 as shown in Fig. 1.8.

The stretching ratios are regulated through the speed difference between neighboring coupled rollers driven by the DC motors [40, 78]. The control allocation is applied to regulate the input voltages of these motors for the redundant stretching ratios. There are several situations existed in speed regulation, for example, a certain velocity difference between two motors (i.e. roller #2 and roller #3) causes stretching operation in relevant zone, but the rollers in non-stretching zones should maintain the same speed. Therefore, the regulation of one stretching zone inevitable affects the next one, which refers to a unidirectional coupling effect and should be eliminated in the actual system. Similarly, the tension of one sub-process affects the neighboring process which is the main controlled variable.

In order to solve the above problem, distributed control strategy and decoupled control allocation algorithm among these inner-loops should be operated at the same time [75, 79, 80]. That is to say, on the one hand, in order to operate decoupling tensions of these sub-processes, the inner-loops are firstly used with the purpose of regulating the response of each stretching zone separately and rapidly; on the other hand, an optimal control allocation is used to eliminate the coupling effects synergistically. In a word, for studying the relationship between distributed control strategy and online optimization, the primary goal is to figure out that the dynamical optimization can drive the velocities of two successive rollers for attaining appropriate stretching ratios. Considering the ideal dynamics of tension responses, the velocities of the rollers can be determined within corresponding constraints. The combination of these stretching ratios is finally determined for satisfying the tension requirement. As such, we can design the control law and achieve optimal stretching ratio in each sub-process. Minimizing the coupling effects on two consecutive driving motors and fluctuation of tension responses at the same time is a critical and indispensable procedure for ensuring the quality of the finished products.

The PANCF precursors are made through the multi-stage stretching process in different situations, such as cold-stretched and hot-stretched processes. In the cold-stretched process, the mechanical properties such as modulus and tensile strength are enhanced. The filaments are not yet in perfect alignment with respect to the molecular orientation and crystallinity. Therefore, the high active molecules reaction in precursors, along with an appropriate stretching ratio in hot-stretched process, are believed to contribute to the more effective treatment for attaining high performance PAN fibers [56, 58]. The schematic flow of stretching process in carbon fiber production is as shown in Fig. 1.9.

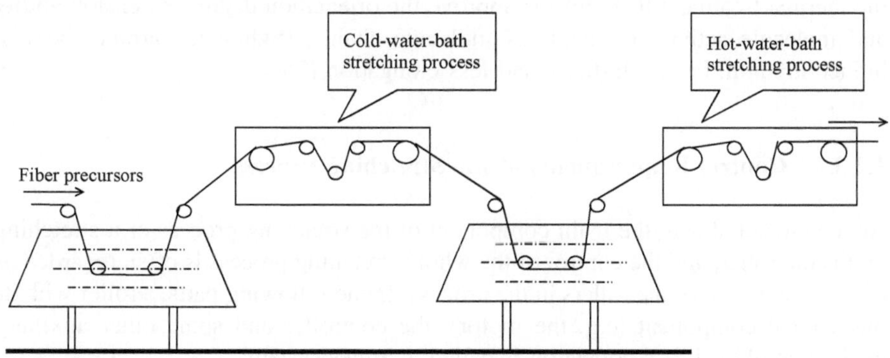

Fig. 1.9 The stretching process on carbon fiber production

Table 1.1 Tension and stretch ratio at different measuring points during preoxidation process

Test sample	Cold-stretched process		Hot-stretched process	
	Tension/cN	Stretch ratio	Tension/cN	Stretch ratio
1	320	1.65	257	1.25
2	335	5.38	325	6.17
3	238	0.769	272	1.16
4	190	0.57	221	0.85

After the molten polymer changes into filaments in coagulating bath, filaments are uniaxially stretched into the solidified filaments. In different stretching situations, the aggregated state microstructure and orientation degree of macromolecule chains of as-spun PAN fibers become sufficiently stretched and regularly oriented along with fiber axis, which is designed to produce filaments with a desired strength. At the same time, the increase of the number of molecule chains, which can hold more external tension, leads to the improvement of the fracture strength [57, 58].

The tensile strength and the breaking elongation ratio of the precursors depend on the combination of individual stretching ratios of sub-processes in different environments. Table 1.1 shows the tension values of the same test sample in one experiment ranging from 320 to 335 cN with varying stretching ratios in the cold-water-stretched process and the almost same situation in hot-water-stretched processes. On the contrary, another consequence is obtained by comparing the values in hot-stretched process, where the tension value of precursors changes in the range of 257 to 325 cN depending on various stretch ratio of 1.16 to 6.17 at the different test samples. In addition, we can observe from table that plastic deformation in as-spun fibers increases as the temperature of fiber precursors becomes higher.

Table 1.1 also shows that the properties of PAN as-spun fibers heavily depend on the real-time adaptive allocation of the stretching ratios in different units because different stretching processes have different influences on the orientation degree and the physical relaxation or shrinkage of PAN molecule chains. This is mainly due to the thermal treatment that aims to improve the orientation degree, extension ability and molecule activity of the PAN molecule chains, leading to carbon fibers of higher strength, better elasticity and less elongation [75].

1.2.3.2 Control Requirements of the Stretching Process

As mentioned above, the main component of the stretching process or a stretching unit is the roller, and the control of the whole stretching process is often regarded as equivalent as that of the rollers in the process. In the following parts, a roller with its mechanical component (e.g. the motor), the controller and some other auxiliary devices combined as a collection is called a "roller system".

A typical type of roller system is called independent roller system, usually consisting of a motor, a metal cylinder, and an electronic controller. The motor used for roller can be of any kind available for speed regulation, such as direct current (DC) motors, alternative current (AC) motors, and servo motors. The cylinder is the only exposed part (namely, can be seen by the operator) of roller system on which the fibers to be stretched are winding, and a stretching unit of such type may include several cylinders of different diameters. The controller is the kernel of the whole roller system. It takes responsibility of regulating the speed of the cylinder it connects according to the instructions from the upper commanders in the production line, e.g. the industrial process computers. Many types of controllers ranging from the most conventional PID controller, fuzzy controller to the updated intelligent controllers can be applied in the roller system. In the systematic view, these roller controllers can also be regarded as the lowest supervising level in the whole production control system. To meet the requirements of these controllers, a central control system is usually established to unitarily manage these low-level controllers and thus distribute appropriate speed ratio to each stretching unit by setting the speed of each roller in a unit.

Another type of roller system is called interlocked roller system. In this system, all the rollers in a stretching unit (or even in the whole stretching process) are driven by one single motor (called a "driving roller" that dominates other rollers). Other rollers are connected to the driving roller by some mechanical parts, e.g. gears. So the speeds of all the rollers but the driving one are regulated by the driving roller in uniform, and the transmission ratios between gears of different sizes are the stretching ratio of the stretching units.

Both of the roller systems above are the most common types in practical fiber production. However, there are still some disadvantages existed in these systems and cannot be well treated at present. For the independent roller system, it requires a central control system taking charge of all the roller controllers, which is easy for unified management and full of agility but leaving potentially instability. If one roller needs to be tuned, the speed variation information must be spread to all the other roller controllers through the central command system, and the whole stretching process takes time to move to a new balance status. This regulation may result in more defective fiber products. For the interlock system, all the rollers are running according to the given precise speed ratios, and the speed errors are only relative to the mechanical errors of its moving parts, e.g. the gears. However, the connections of gears may change thoroughly to fit the varying stretching ratio requirements. This kind of roller system keeps comparatively stable at the cost of agility.

In coagulating process, the polymer material is extruded from the spinneret hole and formed the precursor in the form of filaments. Therefore, for the different concentration and temperature of polymeric melt, the extruding rate, thickness, modulus and also cross-sectional area of filaments may be changed, which regarded as the disturbances to the wholly system, yield the deterioration of pre-defined PI gains and further the unacceptable tensile force of fibers. Moreover, the PI parameters of inner velocity loop and the outer tension loop are difficult to determine and need a lot of experiments. Due to these special situations, especially the variable elastic modulus of as-spun fibers attached onto the rollers, the best combination of the P and I gains of these inner-loops in the normal conditions may be deteriorated and could not adapt to new circumstance anymore. With more than one desired control objections and two distributed control effectors to generate the appropriate solutions for these inner-loops, control modes and combination of both control surfaces are of great importance and should be carefully selected for the tension control performance. In this case, comparing with traditional methods, by means of optimization, namely, the compensation of controller gains for individually tension and velocity error corrections, the systematic disturbances can be eliminated easily and precisely.

Therefore, during this stretching process, two aspects are important for precursor properties:

Control performance: As mentioned above, the main difficulty that results in inaccurate tension tracking and high sensitivity to disturbances can mainly be attributed to the nonlinear speed coupling between these neighboring driving motors. In most related controller design, the ratios in different units (subsystems) are specified separately in advance. For achieving high performance precursors, more efficient control strategies must be designed.

The stretching mechanism: The tension of the precursors after the second driving roller should be maintained at a normal and desired value, which is the most important requirement in production, because the deviation from it will introduce fluctuations in elongations at break of filaments, and may decrease the crystallization and orientation of fibers. In order to guarantee the quality of PANCF, the precursor tension value should be maintained at a desired value during operation.

We take into account the following two properties.

Tension tracking: Once the stress value of filament tension is set to a desired value in order to regulate both mechanical properties and surface roughness, stretching controllers should interact with each other to maintain the desired tension value in different conditions.

Disturbance rejection: Since the stretching network is expected to be robust to disturbances, the system should be able to detect abnormal conditions with the help of multi-layered feedback loops within the network and then reallocate the stretching ratios accordingly to these stretching units to maintain the desired tension.

1.3 Conclusions

In this chapter, the regulation mechanisms of bio-inspired systems are introduced, such as immune system, neuroendocrine system and neuroenodocrine-immune system. The application background of the complex industiral processes is the production process of carbon fiber, after introducing the basic process of carbon fiber, raw silk spinning and multi-level stretching processes are also descripted in details.

References

1. Pancer, Z., Cooper, M.D.: The evolution of adaptive immunity. Annu. Rev. Immunol. **24**(1), 497–518 (2006)
2. Cao, X., Qiao, H., Xu, Y.: Negative selection based immune optimization. Adv. Eng. Softw. **38**(10), 649–656 (2007)
3. Hou, T.H., Su, C.H., Chang, H.Z.: Using neural networks and immune algorithms to find the optimal parameters for an IC wire bonding process. Expert Syst. Appl. Int. J. **34**(1), 427–436 (2008)
4. Polat, K., Güneş, S.: Principles component analysis, fuzzy weighting pre-processing and artificial immune recognition system based diagnostic system for diagnosis of lung cancer. Expert Syst. Appl. **34**(1), 214–221 (2008)
5. Zhang, Z.: Immune optimization algorithm for constrained nonlinear multiobjective optimization problems. Appl. Soft Comput. **7**(3), 840–857 (2007)
6. Dong, W., Shi, G., Zhang, L.: Immune memory clonal selection algorithms for designing stack filters. Neurocomputing **70**(4–6), 777–784 (2007)
7. Freschi, Fabio, Repetto, Maurizio: VIS: An artificial immune network for multi-objective optimization[J]. Eng. Optim. **38**(8), 975–996 (2006)
8. Toma, N., Endo, S., Yamada, K.: A study of a parallelized immune coevolutionary algorithm for division-of-labor problems[J]. Artif. Life Rob. **9**(2), 76–80 (2005)
9. Dasgupta, D.: Artificial Immune Systems and Their Applications. Springer (1999)
10. Leandro, N.D.C., Jonathan, T.: Artificial Immune System: A New Computational Intelligence Approach. Springer (2002)
11. Cai, M.-Y.: Medical Immunology. Scientific Publishing House, Beijing, China (2002)
12. Ding, Y.-S., Ren, L.-H.: Fuzzy self-tuning immune feedback controller for tissue hyperthermia. In: Proceedings of The Ninth IEEE International Conference, Fuzzy IEEE 2000, San Antonio, Texas, USA, May 7–10, vol. 1, pp. 534–538 (2000)
13. Ding, Y.-S., Ren, L.-H.: A novel fuzzy self-tuning immune feedback control system. Control and Decis. (in Chinese). **15**(4), 443–446 (2000)
14. Qi, Z.-Q., Song, S.-M.: A novel immune feedback control algorithm and its applications. In: Proceedings of Genetic and Evolutionary Computation Conference–GECCO 2004, Seattle, WA, USA, June 26–30, pp. 318–320 (2004)
15. Motohiro, K., Minoru, S., Kazuhiko, T.: Adaptive learning method of network controller using an immune feedback. In: Proceedings of the 1999 IEEE/ASME International Conference Advanced Intelligent Mechatronics, Atlanta, USA, September 19–23, pp. 641–646 (1999)
16. Takashi, K., Yamada, T.: Application of an immune feedback mechanism to control system. JSME Int. J., Series C. **41**(2), 184–191 (1998)
17. Kim, D.H.: PID controller tuning of a boiler control system using immune algorithm typed neural network. In: Proceedings of Computational Science-ICCS 2004: 4th International Conference, Krakow, Poland, June 6–9, pp. 695–698 (2004)
18. Kim, D.H., Park, J.I.: Intelligent PID control by immune algorithms based fuzzy rule auto-tuning. Lect. Notes Comput. Sci. **2715**, 474–482 (2003)
19. Mitra, S., Hayashi, Y.: Bioinformatics with soft computing. IEEE Trans. Syst., Man, Cybern. C, Appl. Rev. **36**(5), 616–635 (2006)

20. Keenan, D.M., Veldhuis, J.D.: A biomathematical model of time-delayed feedback in the human male hypothalamic-pituitary-Leydig cell axis. Am. J. Physiol. **275**, 157–176 (1998)
21. Farhy, L.S.: Modeling of Oscillations of Endocrine Networks with Feedback. Methods Enzymol. **384**, 54–81 (2004)
22. Vargas, P., Moioli1, R.L., Castro, N.D., et al.: Artificial homeostatic system: A novel approach. ECAL and LNAI. **3630**, 754–764 (2005)
23. Mester, D., Bräysy, O., Dullaert, W.: A multi-parametric evolution strategies algorithm for vehicle routing problems. Expert Syst. Appl. **32**(2), 508–517 (2007)
24. Niu, Y.-L., Wang, J.-M.: Vehicle route algorithm of logistics distribution. J. Traffic Transp. Eng. **6**(2), 83–87 (2006)
25. Carrol, J.R., Givens, M.P.: Design elements of the modern spinning control system. In Proc. IEEE Annu. Text. Fiber Film Ind. Tech. Conf., 1–2 (1994)
26. Bazbouz, M.B., Stylios, G.K.: Novel mechanism for spinning continuous twisted composite nanofiber yarns. Eur. Polym. J. **44**(1), 1–12 (2008)
27. Ismail, F., Rahman, M.A., Mustafa, A., Masuura, T.: The effect of processing conditions on a polyacrylonitrile fiber produced by a solvent-free coagulation process. Mater. Sci. Eng. **485** (1–2), 251–257 (2008)
28. Rahaman, M.S.A., Ismail, A.F., Mustafa, A.: A review of heat treatment on polyacrylonitrile fiber. Polym. Degrad. Stab. **92**(8), 1421–1432 (2007)
29. Yusof, N., Ismail, A.F.: Post spinning and pyrolysis processes of polyacrylonitrile (PAN)-based carbon fiber and activated carbon fiber: A review. J. Anal. Appl. Pyrol. **93**, 1–13 (2012)
30. Chand, S.: Review carbon fibers for composites. J. Mater. Sci. **35**, 1303–1313 (2000)
31. Liu, J., Tian, Y., Chen, Y., Liang, J., Zhang, L., Fong, H.: A surface treatment technique of electrochemical oxidation to simultaneously improve the interfacial bonding strength and the tensile strength of PAN-based carbon fibers. Mater. Chem. Phys. **122**(2–3), 548–555 (2010)
32. Wang, Y., Yin, W.: Chemical modification for PAN fibers during heat-treatment process. Phys. Procedia **18**, 202–205 (2011)
33. Rahman, M.A., Ismail, A.F., Mustafa, A.: The effect of residence time on the physical characteristics of PAN-based fibers produced using a solvent-free coagulation process. Mater. Sci. Eng. **448**(1–2), 275–280 (2007)
34. Liang, X., Ding, Y.S., Ren, L.H., Hao, K.R., Wang, H.P., Chen, J.: J: A bio-inspired multi-layered intelligent cooperative controller for stretching process of fiber production. IEEE Trans. Syst. Man Cybern. C. **42**, 367–377 (2012)
35. Rennhofer, H., Loidl, D., Puchegger, S.: Peterlik, H: Structural development of PAN-based carbon fibers studied by in situ X-ray scattering at high temperatures under load. Carbon **48**, 964–971 (2010)
36. Belyaev, S.S., Arkhangelsky, I.V., Makarenko, I.V.: Nonisothermal kinetic analysis of oxidative stabilization processes in PAN fibers. Thermochim. Acta **507–508**, 9–14 (2010)
37. Chen, J.J.; Ding, Y.S.; Hao, K.R: The bidirectional optimization of carbon fiber production by neural network with a GA-IPSO hybrid algorithm. Math. Probl. Eng. 1–16 (2013)
38. Sugimoto, Y., Shioya, M., Yamamoto, K., Sakurai, S.: Relationship between axial compression strength and longitudinal microvoid size for PAN-based carbon fibers. Carbon **50**, 2860–2869 (2012)
39. Skinner Micheal, L., Roser Ronald, R., Samowitz Kirk, J.: Computer control of fiber glass filament winding machines. IEEE Trans. Ind. Appl. **21**(4), 1057–1063 (1985)
40. Ding, Y.-S., Liang, X., Hao, K.-R., Wang, H.-P.: An intelligent cooperative decoupling controller for coagulation bath in polyacrylonitrile carbon fiber production. IEEE Trans. Control Syst. Technol. **21**(2), 467–479 (2013)
41. Liang, X., Ding, Y.-S., Ren, L.-H., Hao, K.-R.: A bioinspired multilayered intelligent cooperative controller for stretching process of fiber production. IEEE Trans. Syst. Man Cybern. Part C Appl. Rev. **42**(3), 367–377 (2012)
42. Tan, D.H., Zhou, C.-F., Ellison, C.J., Kumar, S., Macosko, C.W., Bates, F.S.: Meltblown fibers: Influence of viscosity and elasticity on diameter distribution. J. Nonnewton. Fluid Mech. **165**(15–16), 892–900 (2010)

43. Gou, Z.-M., McHugh, A.J.: Two-dimensional modeling of dry spinning of polymer fibers. J. Non-Newtonian Fluid Mech. **118**(2–3, 5), 121–136 (2004)
44. Lee, K.-M., Wei, Z.-Y., Zhou, Z.: Modeling by numerical reduction of modes for multivariable control of an optical-fiber draw process. IEEE Trans. Autom. Sci. Eng. **3**(1), 119–130 (2006)
45. Kadi, H.E.: Modeling the mechanical behavior of fiber-reinforced polymeric composite materials using artificial neural networks-A review. Compos. Struct. **73**(1), 1–23 (2006)
46. Arafeh, L., Singh, H., Putatunda, S.K.: A neuro fuzzy logic approach to material processing. IEEE Trans. Syst. Man Cybern. Part C Appl. Rev. **29**(3), 362–370 (1999)
47. Yang, J., Jaluria, Y.: Feasibility and optimization of the hollow optical fiber drawing processs. Int. J. Heat Mass Transf. **52**(17–18), 4108–4116 (2009)
48. Bechtel, S.E., Vohra, S., Jacob, K.I.: Modeling of a two-stage draw process. Polym. **42**(5), 2045–2059 (2001)
49. Makradi, A., Cox, C.L., Ahzi, S., et al.: Postspinning draw of polymeric fibers: Multiscale micromechanical model for a solid polymer under finite deformation and strain-induced crystallization. J. Appl. Polym. Sci. **100**(3), 2259–2266 (2006)
50. Mataram, A., Ismail, A.F., Mahmod, D.S.A., et al.: Characterization and mechanical properties of polyacrylonitrile/silica composite fibers prepared via dry-jet wet spinning process. Mater. Lett. **64**(17), 1875–1878 (2010)
51. Hobbs, T., Lesser, A.J.: Preparation of high performance poly (ethylene terephthalate) fibers: two-stage drawing using high pressure CO 2. Polym. **41**(16), 6223–6230 (2000)
52. Chuo, S.M., Wan, M.H., Wang, L.A., et al.: Multistage Modified Fiber Drawing Process and Related Diameter Measuring System. J. Lightwave Technol. **27**(27), 2983–2988 (2009)
53. Suman, B., Tandon, P.: Fluid flow stability analysis of multilayer fiber drawing. Chem. Eng. Sci. **65**(20), 5537–5549 (2010)
54. Mabrouk, M.A.: Experimental investigation of deformation mechanisms during low drawing of polymeric fibres. Polym. Testing **21**(6), 653–658 (2002)
55. Mawardi, A., Pitchumani, R.: Optical Fiber Drawing Process Model Using an Analytical Neck-Down Profile. IEEE Photonics J. **2**(4), 620–629 (2010)
56. Wang, S., Chen, Z.-H., Ma, W.-J., Ma, Q.-S.: Influence of heat treatment on physicalchemical properties of PAN-based carbon fiber. Ceram. Int. **32**(3), 291–295 (2006)
57. Sedghi, A., Farsani, R.E., Shokufar, A.: The effect of commercial polyacrylonitrile fibers characterizations on the produced carbon fibers properties. J. Mater. Process. Tech. **198**(1–3), 60–67 (2008)
58. Tan, L.-J., Chen, H.-F., Pan, D., Pan, N.: Investigating the spin ability in the dry-jet wet spinning of PAN precursor fiber. J. Appl. Polym. Sci. **110**, 1997–2000 (2008)
59. Ismail, F., Rahman, M.A., Mustafa, A., Matsuura, T.: The effect of processing conditions on a polyacrylonitrile fiber produced using a solvent-free coagulation process. Mater. Sci. Eng. A **485**(1–2), 251–257 (2008)
60. Nain, S., Sitti, M., Jacobson, A., Kowalewski, T., Amon, C.: Dry spinning based spinneret based tunable engineered parameters (STEP) technique for controlled and aligned deposition of polymeric nanofibers. Macromol. Rapid Commun. **30**, 1406–1412 (2009)
61. Liang, X., Ding, Y.S., Ren, L.H., Hao, K.R.: Jin, Y.L: Data-Driven cooperative intelligent controller based on the endocrine regulation mechanism. IEEE Trans. Control Syst. Technol. **22**(1), 94–101 (2014)
62. Carroll, J.R., Givens, M.P., Piefer, R.: Design elements of the modem spinning control system. IEEE Annual. Textile Fiber Film Industrial Technology Conference. 4–5 (1994)
63. Bazbouz, M.B., Stylios, G.K.: Novel mechanism for spinning continuous twisted composite nanofiber yarns. Eur. Polym. J. **44**(1), 1–12 (2008)
64. Jeffrey, K.C.F.: An entire strategy for control of a calendar roller system. Part III: Intelligent settling time-optimal control. Text. Res. J. **78**(1), 81–87 (2008)
65. Stoker, M., White, A.S.: Mechatronic cine-film copying using transputer control. Mechatronics **10**(7), 773–807 (2000)

66. Karaman, M., Batur, C.: Draw resonance control for polymer fiber spinning process. In Proc. Amer. Cont. Conf. 2155–2159 (1998)
67. Ghosh, A., Das, S.K.: Open-loop decoupling of MIMO plants. IEEE Trans. Autom. Control **54**(8), 1977–1981 (2009)
68. Wu, M., Yan, J., She, J.-H., Cao, W.-H.: Intelligent decoupling control of gas collection process of multiple asymmetric coke ovens. IEEE Trans. Ind. Electron. **56**(7), 2782–2792 (2009)
69. Liu, H.-B., Li, S.-Y., Chai, T.-Y.: Intelligent decoupling control of power plant main steam pressure and power output. Electr. Power Energy Syst. **25**(10), 809–819 (2003)
70. Farhy, L.S., Straume, M., Johnson, M.L., Kovatchev, B., Veldhuis, J.D.: A construct of interactive feedback control of the GH axis in the male. Amer. J. Physiol. Reg. Integr. Comput. Physiol. **281**(1), 38–51 (2001)
71. Keenan, D.M., Licinio, J., Veldhuis, J.D.: A feedback-controlled ensemble model of the stress-responsive hypothalamo-pituitary-adrenal axis. Proc. Nat. Acad. Sci. **98**(7), 4028–4033 (2001)
72. Mitra, S., Hayashi, Y.: Bioinformatics with soft computing. IEEE Trans. Syst. Man, Cybern. C, Appl. Rev. **36**(5), 616–635 (2006)
73. Sedghi, A., Farsani, R.E., Shokufar, A.: The effect of commercial polyacrylonitrile fibers characterizations on the produced carbon fibersproperties. J. Mater. Process. Tech. **198**(1–3), 60–67 (2008)
74. Labinaz, G., Bayoumi, M.M., Rudie, K.: Viable cascade control and application to a batch polymerization process. IEEE Trans. Control Syst. Technol. **8**(3), 396–407 (2000)
75. Chen, J.C., Harrison, I.R.: Modification of polyacrylonitrile (PAN) carbon fiber precursor via post-spinning plasticization and stretching in dimethyl formamide (DMF). Carbon **40**(1), 25–45 (2002)
76. Ji, M., Egerstedt, M.: Distributed coordination control of multiagent systems while preserving connectedness. IEEE Trans. Rob. **23**(4), 693–703 (2007)
77. Zhang, H.W., Lewis, F.L.: Adaptive cooperative tracking control of higher-order nonlinear systems with unknown dynamics. Automatica. **48**(7), 1432–1439 (2012)
78. Hu, Y.F., Ding, Y.S., Hao, K.R., et al.: An immune orthogonal learning particle swarm optimization algorithm for routing recovery of wireless sensor networks with mobile sink. Int. J. Syst. Sci. **45**(3), 337–350 (2014)
79. Pagilla, P.R., Dwivedula, R.V.: Decentralized control of web processing lines. IEEE Trans. Control Syst. Technol. **15**(1), 106–117 (2007)
80. Zhang, T.P., Xia, X.N.: Decentralized adaptive fuzzy output feedback control of stochastic nonlinear large-scale systems with dynamic uncertainties. Inf. Sci. **315**(1), 17–38 (2015)

Chapter 2
Human Body Based Intelligent Cooperative Decoupling Controllers

Decoupling control, a vital approach in control engineering, is one of the most critical and effective methods to solve complex coupling system. In this chapter, we discuss the developments of some effective decoupling control methods for the coupling phenomenon among complex systems. Firstly, a bio-inspired decoupling controller (BDC) from the bi-regulation principle of the growth hormone (GH) in neuroendocrine system is proposed [1]. Secondly, an intelligent bi-cooperative decoupling controller (IBCDC) inspired from the modulation mechanism of internal environment in body is proposed [2]. Finally, an intelligent cooperative decoupling controller (ICDC) based on the neuroendocrine regulation principle of human body is proposed [3].

2.1 Introduction

Coupling is also a very popular phenomenon in multiple-input-multiple-output (MIMO) systems. For example, during the industrial process, many control loops are designed for a control plant to guarantee the product quality. Usually, there exists coupling influence among these loops. If the influence is not eliminated successfully, these loops will not be auto-controlled or controlled stably. As such, it will affect the stability of the process and the product quality. Hence, it is of great importance to eliminate the coupling influence. Decoupling control is one of the most critical and effective methods to solve this problem. And the essentiality of the decoupling control is to reduce or eliminate the coupling influence among various loops via control compensation based on the mathematical model or the process mechanism. The conventional decoupling control methods include diagonal matrix

Y. Ding et al., *Bio-Inspired Collaborative Intelligent Control and Optimization*, Studies in Systems, Decision and Control 118,
https://doi.org/10.1007/978-981-10-6689-4_2

decoupling, identification matrix decoupling, feed forward compensation decoupling [4, 5], and pole assignment decoupling [6, 7]. However, the existing methods may be a little complicated, or lack very satisfactory decoupling effectiveness. And, most of them are based on the mathematical models, which is more or less impractical for the industrial control.

As we know, bio-systems are highly stable with harmonious regulation and accurate control, and have been imitated to develop some novel intelligent control methods for complex systems. Since the early 1990s, many intelligent decoupling control technologies have been developed, such as fuzzy decoupling [8–11], neural networks decoupling [12–14], self-adaptive decoupling [15], genetic algorithm-based decoupling [16–18], and kernel decoupling [19]. Some researchers focus on the regulation principles of human body and try to apply such principles to the industrial control domain [20–22]. Under the control of the nervous system, a number of human physiological regulation systems build a highly complex and stable regulation network which maintains physiological homeostasis through the cooperative regulation mechanisms in human body [23–25]. Besides from the above intelligent control methods, some novel ones are recently developed from neuroendocrine (NE) system [26, 27]. NE system involves nervous system and endocrine system, which are two important physiological modulation systems in body. Endocrine is responsible for secrete some hormones and modulate the secreting process via complex feedback mechanism. Nervous system also regulates the hormone-secreting process by neurotransmitters secreted in it. As thus, the hormone concentration in body can be controlled stable and quickly [28, 29]. NE system is a complex coupling one, and several simple decoupling architectures are introduced to simulate the process of the hormone release in the medical field [28–32].

2.2 Intelligent Decoupling Control System Inspired from Regulation of the Growth Hormone

In this section, we present a bio-inspired decoupling controller (BDC) from the bi-regulation principle of the growth hormone (GH) in neuroendocrine system, and provide its decoupling algorithm for complex coupling system. The BDC consists of an identification unit and two or more control units. The identification unit can obtain the mathematical model via the step-response method and correct the model online. The control unit is composed of a control module, a coupling error prediction (CEP) module, an inverse control (INC) module, and an output module. The CEP module can predict the coupling error between loops, while the INC module can calculate the corresponding decoupling output. The control units communicate with each other to exchange the control information and to adjust the actuators harmoniously. As thus, the coupling influences among the various control loops can be reduced or eliminated. In order to examine the decoupling control effectiveness of the BDC, we apply it to the methanol-water distillation column model [33, 34] that is a two-input-two-output (TITO) strong coupling system. The simulation results demonstrate that the BDC can completely eliminate the coupling influence with better control performance and adaptability. Compared with other decoupling control algorithms, the BDC can be implemented more easily and practically.

2.2.1 *Bi-Regulation Principle of the Growth Hormone*

GH is one of vital endocrine hormone secreted by pituitary, which is indispensable to the body growth. The secretion of GH is regulated by the GH-releasing hormone (GHRH) and the GH-release-inhibiting hormone (SRIF). Both GHRH and SRIF are secreted by hypothalamus. GHRH stimulates the synthesis and release of GH, while SRIF antagonizes the secretion of GH. The concentration of GH in blood circulation system or interstitial fluid is fed back to neurons of GHRH and SRIF in body via the cell sensors to modulate their release. When the concentration of GHRH exceeds that of SRIF, the concentration of GH will increase, and vice versa.

In addition, there is an interaction between the regulating neural unit of GHRH and that of SRIF. The increase of GHRH concentration may also cause the increase of SRIF, which may inhibit the secretion of GHRH in turn. Similarly, the increase of SRIF concentration may cause the increase of GHRH, which may inhibit the secretion of SRIF in turn. In addition, the concentration of GH is also fed back to the central nervous system, which includes spinal cord, brain, white matter, gray matter, and so on. Then the central nervous systems can further modulate the GHRH and SRIF regulation process in hypothalamus. Through the harmonious regulation of GHRH and SRIF, the secretion of GH can be controlled accurately and stably [31, 32]. The regulating loop of GH is as shown in Fig. 2.1 [1].

2.2.2 *Design of the Bio-Inspired Decoupling Controller*

2.2.2.1 The Structure of the BDC

According to Fig. 2.1, we design the BDC for the decoupling control system as shown in Fig. 2.2 [1]. The regulating neural unit of GHRH and that of SRIF are corresponding to the control unit A and unit B with the output u_1 and u_2, respectively. The central nervous system is corresponding to the identification unit. The pituitary is corresponding to the process plant. The concentration of the GH in body is corresponding to the output variables y_1 and y_2.

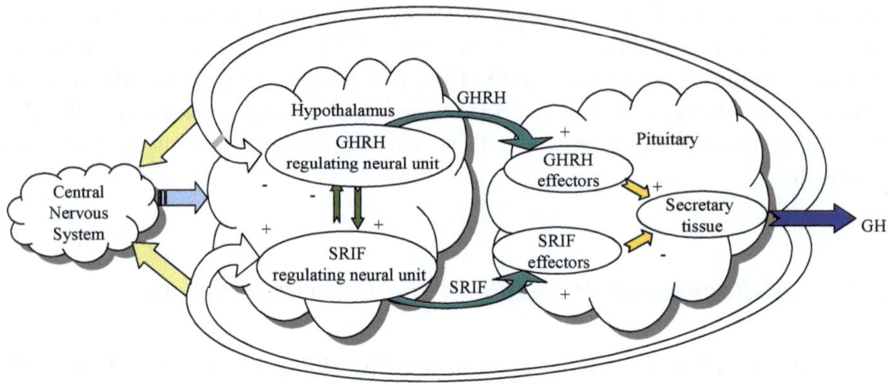

Fig. 2.1 The regulating loop of the GH

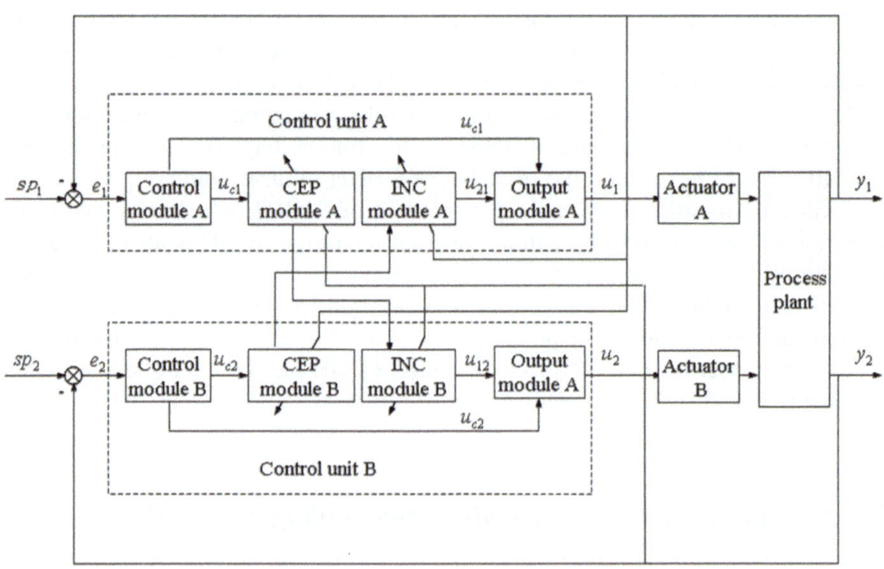

Fig. 2.2 The structure of the BDC

The corresponding relationship between the bi-regulation of the GH and the structure of the BDC is as shown in Table 2.1 [1].

The BDC consists of an identification unit and two or more control units. The identification unit can obtain the mathematical model via the step-response method and correct the model online. The control unit is composed of a control module, a coupling error prediction module (CEP), an inverse control module (INC), and an output module. The CEP can predict the coupling error between two control loops,

Table 2.1 The relationship between the Bi-regulation of the GH and the structure of the BDC

No.	Bi-regulation of the GH	Structure of the BDC
1	Regulating neural unit of GHRH	Control unit A
2	Regulating neural unit of SRIF	Control unit B
3	GHRH	u_1
4	SRIF	u_2
5	Central nervous system	Identification unit
6	Pituitary	Process plant
7	Concentration of GH	Output variables y_1 and y_2

while the INC can calculate the corresponding decoupling output. Based on the decoupling algorithm explained in the next section, the control units communicate with each other to exchange the control information and to adjust the actuators harmoniously. As thus, the coupling influences among the various control loops can be reduced or eliminated.

The mathematical description for the BDC is as follows. For convenience, we denote the transfer function of the decoupling unit as $D(s)$, including CEP, INC and output module, and that of the conventional control unit and the process plant as $C(s)$ and $G(s)$, respectively. As thus, the transfer function $W(s)$ of the closed loop control system can be denote as,

$$W(s) = \frac{C(s)D(s)G(s)}{1 + C(s)D(s)G(s)}$$

2.2.2.2 The Decoupling Control Algorithm

In order to describe the decoupling control algorithm conveniently, we take a TITO linear coupling plant as an example. For a TITO linear coupling plant

$$\begin{bmatrix} y_1(s) \\ y_2(s) \end{bmatrix} = \begin{bmatrix} G_{11}(s), G_{21}(s) \\ G_{12}(s), G_{22}(s) \end{bmatrix} \begin{bmatrix} u_1(s) \\ u_2(s) \end{bmatrix}$$

where $G_{ij}(s)$ (i, j = 1, 2) is the transfer function between the i-th input and the j-th output; $y_1(s)$ and $y_2(s)$ are the process output variables; and $u_1(s)$ and $u_2(s)$ are the outputs of the controllers. If $G_{ij}(s)$ is known, we can predict the coupling influence from the input u_i to the output variables y_j (i, j = 1, 2, and $i \neq j$), and then take a measure to reduce the influence.

We define the following variables for the BDC. uc_1 and uc_2 are the outputs of the control modules A and B, respectively, u_{12} is the decoupling signal of the INC B, and u_{21} is the decoupling signal of the INC A.

The output of the control unit A is

$$u_1 = u_{c1} + u_{21}$$

and that of the control unit B is

$$u_2 = u_{c2} + u_{12}$$

According to the predicted coupling error, the corresponding compensation output can be obtained via the inverse control model. In the following, we introduce the decoupling algorithm by taking the decoupling process from $u_1(s)$ to $y_2(s)$ as an example.

When y_1 is regulated, y_2 should be kept unchanging in order to eliminate the coupling influence from u_1 to y_2. The coupling influence $\Delta de_{12}(s)$ from $u_1(s)$ to $y_2(s)$ with the dynamic changing output $\Delta u_1(s)$ can be predicted based on the following equation,

$$\Delta de_{12}(s) = G_{12}(s)\Delta u_1(s) \tag{2.1}$$

To keep y_2 unchanged, we should let $\Delta u_{21}(s) = 0$ and $\Delta u_1(s) = \Delta u_{c1}(s)$. To eliminate the coupling influence, we employ the inverse model control, noted as $IC_{12}(s)$, and make the following equation true,

$$\Delta de_{12}(s) + \Delta u_{c1}(s)F_{12}(s)IC_{12}(s)G_{22}(s) = 0 \tag{2.2}$$

where, $F_{12}(s)$ is the transfer function of the CEP A. It is well-known that the coupling transfer function $G_{12}(s)$ in most process systems may be described as a first-order plant with time delay,

$$G_{12}(s) = \frac{K_{12}e^{-\tau_{12}s}}{T_{12}s + 1} = G'_{12}(s)e^{-\tau_{12}s} \tag{2.3}$$

To predict the coupling influence, we make $F_{12}(s)$ equal to $G'_{12}(s)$. Then, from Eqs. (2.1)–(2.3), we have

$$IC_{12}(s) = -\frac{e^{-\tau_{12}s}}{G_{22}(s)}.$$

Obviously, $IC_{12}(s)$ is an inverse model of $G_{22}(s)$ with a time delay.
Similarly, $G_{22}(s)$ can be noted as

$$G_{22}(s) = \frac{K_{22}e^{-\tau_{22}s}}{T_{22}s + 1}$$

Then, we get

$$IC_{12}(s) = -\left(\frac{1}{K_{22}} + \frac{T_{22}}{K_{22}}s\right)e^{(\tau_{22}-\tau_{12})s} \tag{2.4}$$

Compare with the conventional PID control algorithm, we can find that $IC_{12}(s)$ in Eq. (2.4) is a PD controller with a differentiation/lead or lag element. In fact, τ_{22} should be less than τ_{12}, which means $IC_{12}(s)$ is a controller with lag element that is easily implemented. If τ_{22} is larger than τ_{12}, $IC_{12}(s)$ is a PD controller with a differentiation/lead element, which is difficult or impossible to be implemented in computer.

Finally, we can get the decoupling control signal from the control unit A to B,

$$\Delta u_{12}(s) = F_{12}(s)IC_{12}(s)\Delta u_{c1}(s).$$

In the same way, for regulating y_2, we can get the decoupling compensation equation from $u_2(s)$ to $y_1(s)$ as follows,

$$IC_{21}(s) = -\left(\frac{1}{K_{11}} + \frac{T_{11}}{K_{11}}s\right)e^{(\tau_{11}-\tau_{21})s}.$$

The decoupling signal from the control unit B to A is

$$\Delta u_{21}(s) = F_{21}(s)IC_{21}(s)\Delta u_{c2}(s)$$

where,

$$F_{21}(s) = \frac{K_{21}}{T_{21}s+1}$$

As such, we can obtain the dynamic decoupling output of the control unit A,

$$\Delta u_1(s) = \Delta u_{c1}(s) + F_{21}(s)IC_{21}(s)\Delta u_{c2}(s) \tag{2.5}$$

and that of the control unit B,

$$\Delta u_2(s) = \Delta u_{c2}(s) + F_{12}(s)IC_{12}(s)\Delta u_{c1}(s) \tag{2.6}$$

Then, we can get the transfer function of the decoupling unit as follows,

$$D(s) = \begin{bmatrix} 1 & F_{21}(s)IC_{21}(s) \\ F_{12}(s)IC_{12}(s) & 1 \end{bmatrix}$$

2.2.2.3 Mathematical Model Identification

If the mathematical model of the process plant is available, Eqs. (2.5) and (2.6) can be derived from it easily. However, the mathematical model is not always available in some control systems. In this case, the step-response identification is employed to obtain the approximate transfer matrix $G(s)$ online. In the BDC, we design a harmonious adjusting method for the control units A and B by exchanging the control information between them. When the BDC is taken into effect for the first time, the control unit A sends "HOLD" signal to the control unit B, then the output of the conventional control module B will be unchanged. At same time, the control unit A outputs a step signal. When both control loops become stable, $G_{11}(s)$ and $G_{12}(s)$ can be obtained by using the step-response identification. In the same way, the control unit B sends "HOLD" signal to the control unit A and outputs a step signal, then $G_{22}(s)$ and $G_{21}(s)$ can be obtained. As thus, the mathematical model can be obtained quickly and easily.

In practice, the values of the plant parameters are always changing, so the mathematical model should be corrected online. There are some methods to correct the model, such as neural networks, fuzzy logic, and least squares method. In the BDC, the identification unit is to perform this task. To implement it more easily, we use the least squares method to correct the mathematical model. For the least squares method, the input variables include control output signals u_1, u_2, and the process variables y_1, y_2. The identified results are used to adjust the INC and CFP models.

2.2.2.4 Extension to MIMO Systems

The BDC and its decoupling control algorithm for TITO systems can be easily extended to MIMO systems. For a MIMO system,

$$\begin{bmatrix} y_1(s) \\ \vdots \\ y_n(s) \end{bmatrix} = \begin{bmatrix} G_{11}(s) & G_{21}(s) & \cdots & G_{n1}(s) \\ & \vdots & \vdots & \\ G_{1n}(s) & G_{2n}(s) & \cdots & G_{nn}(s) \end{bmatrix} \begin{bmatrix} u_1(s) \\ \vdots \\ u_n(s) \end{bmatrix} \tag{2.7}$$

where $y_i(s)$ (i = 1, 2, …, n) are the output variables; $G_{ij}(s)$ (i, j = 1, 2, …, n) are the transfer functions between the i-th input and the j-th output; and $u_i(s)$ (i = 1, 2, …, n) are the outputs of the BDC. According to the proposed decoupling algorithm, while regulating the state variables $y_i(s)$, all the values of the dynamics influences of $y_1, y_2 \cdots, y_{i-1}, y_{i+1}, \cdots, y_n$ should be equal to zero. Based on the derived algorithm, the coupling influence can be eliminated by adjusting n actuators harmoniously. The corresponding dynamic decoupling output of every control unit can be derived in the similar way as the decoupling control algorithm for TITO systems. For example, if the dynamical changing output of conventional control model of i-

th control unit is Δu_{ci}. Suppose that the dynamic changes of the other control units are equal to zero. For the first control unit, we have

$$\Delta u_{i1} G_{11}(s) + \Delta u_i G_{i1}(s) = 0$$

where, Δu_{i1} is the decoupling output of first control unit for regulating $y_i(s)$, and.

$$\Delta u_i = \Delta u_{ci}.$$

Then we can get,

$$\Delta u_{i1} = \frac{-G_{i1}(s)}{G_{11}(s)} \Delta u_i$$

In the similar way, we can get the decoupling output of *j*-th control unit for regulating $y_i(s)$,

$$\Delta u_{ij} = \frac{-G_{ij}(s)}{G_{jj}(s)} \Delta u_i$$

We note that the number of control variables need not be equal to that of output variables.

Moreover, if the mathematical model of a MIMO system is not available, we can also obtain the approximate transfer matrix $G(s)$ via the step-response identification, and correct the mathematical model online by applying the least squares method. Then, the decoupling equations can be obtained from the identified mathematical model.

2.2.2.5 Simulation Results

In order to examine the decoupling control effectiveness of the BDC, we apply it to the methanol-water distillation column model that is a two-input-two-output (TITO) strong coupling system [42, 43]. The Wood and Berry's FOPDT transfer function model is considered:

$$\begin{bmatrix} y_1(s) \\ y_2(s) \end{bmatrix} = \begin{bmatrix} \frac{12.8e^{-s}}{16.7s+1} & \frac{-18.9e^{-3s}}{21.0s+1} \\ \frac{6.6e^{-7s}}{10.9s+1} & \frac{-19.4e^{-3s}}{14.4s+1} \end{bmatrix} \begin{bmatrix} u_1(s) \\ u_2(s) \end{bmatrix} + \begin{bmatrix} \frac{3.8e^{-8s}}{14.9s+1} \\ \frac{4.9e^{-s}}{13.2s+1} \end{bmatrix} F(s) \tag{2.8}$$

where, y_1 is the mole fraction of the methanol at the top; y_2 is the mole fraction of the methanol at the bottom; u_1 is the reflux flow rate; u_2 is the steam flow rate; and $F(s)$ is the feed flow rate. The model (2.8) is for a methanol-water distillation column, which is a coupling TITO system with large time delay. There are strong coupling influences between the mole fraction of the methanol at the top and that at the bottom. The steady state values of the methanol composition at the top and at the bottom are 96.25 mol% and 0.5 mol%, respectively. The conventional control scheme is that y_1 and $y_2$is controlled through adjusting $u_1$and u_2, respectively.

Following, we apply the BDC to the model (2.8) with ignoring the dynamical changing rate of feed flow $F(s)$ in Fig. 2.2. We choose the control unit A for y_1, and the control unit B for y_2. We first identify the mathematical model via step-response identification method. The identified mathematical model is as follows:

$$\begin{bmatrix} y_1(s) \\ y_2(s) \end{bmatrix} = \begin{bmatrix} \frac{12.6e^{-s}}{16.9s+1} & \frac{-18.7e^{-3s}}{21.1s+1} \\ \frac{6.5e^{-7s}}{10.9s+1} & \frac{-19.6e^{-3s}}{14.5s+1} \end{bmatrix} \begin{bmatrix} u_1(s) \\ u_2(s) \end{bmatrix}$$

Based on the decoupling algorithm discussed in Sect. 2.2.3, we obtain the dynamic decoupling output to control y_1,

$$\Delta u_1(s) = \Delta u_{c1}(s) + \frac{1.48}{21.1s+1}(16.9s+1)e^{-2s}\Delta u_{c2}(s)$$

and that to control y_2,

$$\Delta u_2(s) = \Delta u_{c2}(s) + \frac{0.33}{10.9s+1}(14.5s+1)e^{-4s}\Delta u_{c1}(s)$$

Using the BDC, we can obtain satisfactory control performance for the coupling TITO system (2.8), as shown in Fig. 2.3 [1].

Compare the control effectiveness of the BDC with that of the conventional control scheme. Here, the conventional PID control algorithm is adopted in the control units A and B. To make the comparison be fair, the control parameters of the PID controller are same with those of control units A and B, as shown in Table 2.2 [1]. In order to illustrate the contrast effectiveness of the influence on y_2 with the changing of y_1, we change the set points of y_1 at the 300-th, 600-th, 1000-th, and 1500-th minute in Fig. 2.3(a). From Fig. 2.3(a), we can see that the influence of y_1 on y_2 is very large in PID control scheme, while y_2 can be kept unchanging in the BDC scheme. Also, we examine the influence on control performance with the changing of the plant parameters. In Fig. 2.3(a), the plant parameter Kc_{12} is changed from 6.5 to 10.2 and T_{12} is changed from 21.1 to 18.5 at 800-th minute. The BDC can eliminate the coupling influences successfully with online identification. At the same time, the contrast effectiveness of the influence on y_1 with the changing of y_2 is as shown in Fig. 2.3(b). The plant parameter Kc_{12} is changed from −18.9 to −15.6 and T_{21} is changed from 10.9 to 11.5 at 1000-th minute. The set points of y_2 at the 1200-th and 1500-th minute are changed. From Fig. 2.3(b), we can see that y_2 influences on y_1 badly in the PID scheme, but y_1 is kept unchanging in the BDC scheme. Also, we can see that even with the plant parameters changing, the BDC can eliminate the coupling influences successfully, as shown in Table 2.3 [1].

The simulation results demonstrate that the BDC can almost eliminate the coupling influence, and has better control performance and adaptability than that of

Fig. 2.3 The performance comparison between the BDC and PID control scheme

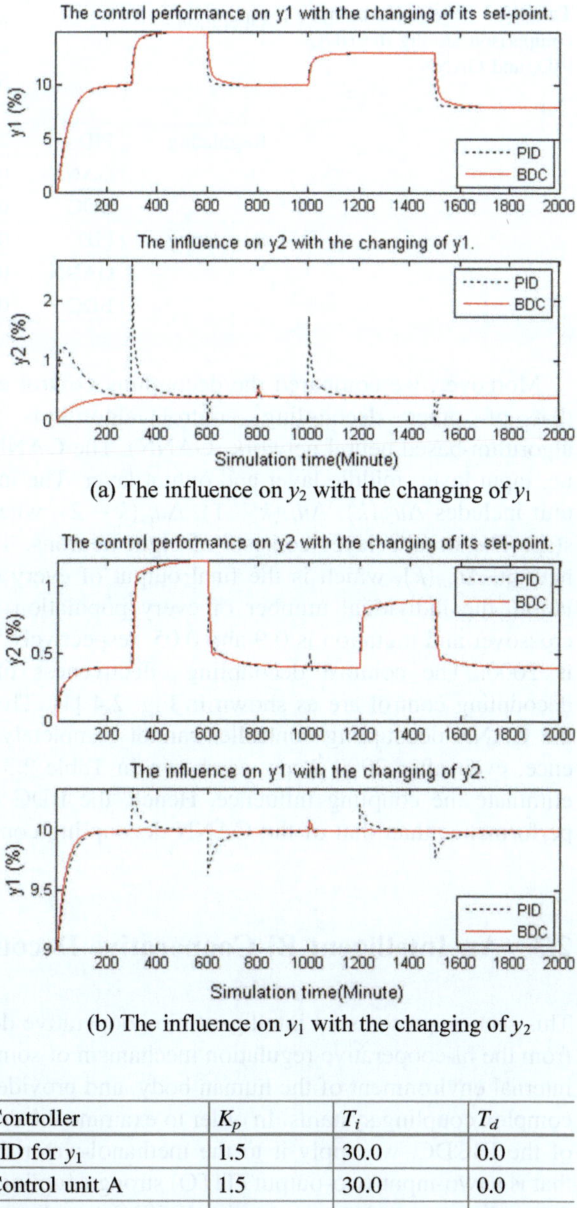

(a) The influence on y_2 with the changing of y_1

(b) The influence on y_1 with the changing of y_2

Table 2.2 The values of the parameters in the BDC and those of the conventional PID controller

Controller	K_p	T_i	T_d
PID for y_1	1.5	30.0	0.0
Control unit A	1.5	30.0	0.0
PID for y_2	0.5	50.0	0.0
Control unit B	0.5	50.0	0.0

the conventional control scheme. Also, the BDC has better robustness under the changing of the plant parameters.

Table 2.3 The performance comparison among the BDC, PID, and GANN

Operations		Maximum coupling influences (absolute value)	
		Set-point change	Parameter change
Regulating y_1	PID	2.23	1.81
	GANN	0.78	0.4
	BDC	0.01	0.01
Regulating y_2	PID	0.41	0.26
	GANN	0.36	0.34
	BDC	0.01	0.01

Moreover, we compared the decoupling control effectiveness of the BDC with that of other decoupling control algorithms, for example, the genetic algorithm-based neural network (GANN). The GANN is composed of three layers, i.e. input layer, middle layer and output layer. The input layer of every decoupling unit includes $\Delta u_{ci}(k)$, $\Delta u_{ci}(k-1)$, $\Delta u_{ci}(k-2)$, where $i = 1, 2$, and k is the time step. The middle layer consists of eight neurons. The output layer includes one neuron, $\Delta u_{di}(k)$, which is the final output of every decoupling unit. In the simulation, the individual number of every population is 50, and the probability of crossover and mutation is 0.9 and 0.05, respectively. The number of iterative steps is 2000. The contrast decoupling effectiveness of the BDC and the GANN decoupling control are as shown in Fig. 2.4 [1]. The simulation results show that the GANN decoupling controller cannot completely eliminate the coupling influence, even after 2000 steps, as shown in Table 2.3 [1] but the BDC can almost eliminate the coupling influence. Hence, the BDC has better decoupling control performance than that of the GANN decoupling control scheme.

2.3 An Intelligent Bi-Cooperative Decoupling Controller

This section presents an intelligent bio-cooperative decoupling controller (IBCDC) from the bi-cooperative regulation mechanism of some physiological indexes in the internal environment of the human body, and provides its decoupling algorithm for complex coupling systems. In order to examine the decoupling control effectiveness of the IBCDC, we apply it to the methanol-water distillation column model [33] that is a two-input-two-output (TITO) strong coupling system. Compared with other decoupling control methods, the IBCDC can be implemented more easily and practically with better control performance and adaptability. The decoupling control algorithm can also be easily generalized to the MIMO coupling control systems. The study not only favors the design of intelligent decoupling controllers for complex coupling systems, but also benefits the future studies of coordination mechanism of physiological systems or biological systems.

Fig. 2.4 The decoupling performance comparison between the BDC and the GANN decoupling control scheme

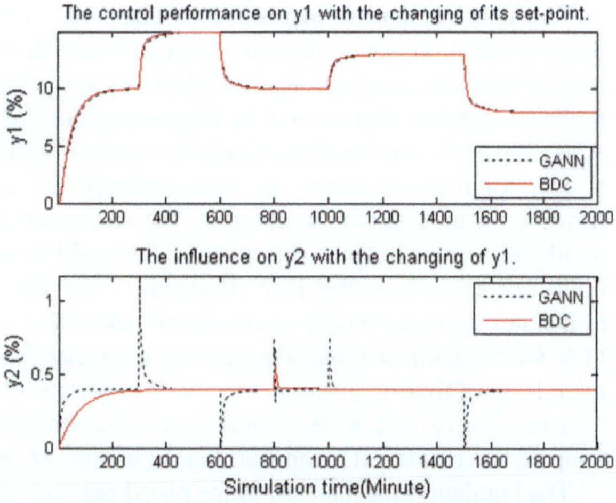

(a) The influence on y_2 with the changing of y_1

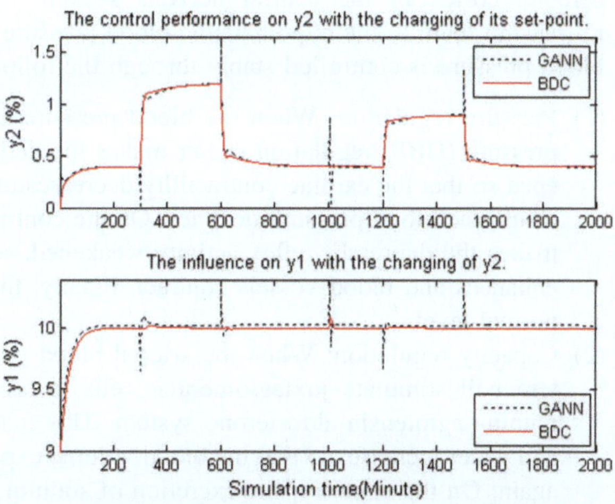

(b) The influence on y_1 with the changing of y_2

2.3.1 *Bi-Cooperative Regulation of Internal Environment in Body*

Under the control of the nervous system, a number of human physiological regulation systems build a highly complex and stable regulation network, which maintains physiological homeostasis through the cooperative regulation mechanisms in the body. Among them, the body temperature (BT) regulation system and the blood pressure (BP) regulation system belong to a typical multi-coordination control system [30, 35, 36].

The core regulation institution of the body temperature is the hypothalamic thermoregulation center. This center coordinates the heat-production organs and the heat-dissipation tissues to maintain the body temperature at a stable level, according to the temperature signals sent by the temperature sensor cells from the whole body. When the body temperature arises, the nervous system modulates the surface pore to open for sweat secretion, and makes skin blood vessels expand to decrease blood flow for promoting skin cooling. At the same time, the hypothalamic temperature regulation center reduces the release of thyroid hormones and lowers basal metabolic rate to reduce the heat production. On the contrary, the nervous system modulates the surface pore to close and makes skin vessel contracts to reduce blood flow for reducing the skin cooling. At the same time, the hypothalamus increases the release of thyroid hormones to improve the basal metabolic rate and increase the heat release, or makes the muscles increase tension and even tremble to achieve temperature equilibrium through the regulation of the nervous system.

The regulation mechanism of the blood pressure system in the body is that blood pressure sensor cells distribute the whole-body transmit pressure signals to the cerebral cortex of the central nervous system. Then, the conditioned signals re-transmit them to the hypothalamic blood-pressure regulation center. Finally, the blood pressure is controlled stably through the following three modulation ways.

(1) Pressure regulation: When the blood pressure arises, the hypothalamic blood pressure (HBP) regulation center makes the depressor reflex activity strengthened so that the cardiac contractility decreases and blood vessels expand. As a result, the blood pressure declines. On the contrary, the HBP regulation center makes the depressor reflex activity weakened, so that the cardiac contractility enhances and blood vessels contract. Finally, the blood pressure ascends to a normal level.
(2) Capacity regulation: When the arterial blood pressure drops, the nervous system will stimulate juxtaglomerular cells to secret therennin and actives the rennin-angiotensin-aldosterone system. This makes the absorption of sodium and water increase, so that the blood volume expands and blood pressure arises again. On the contrary, the excretion of sodium and water increases to reduce the blood volume. As a result, the blood pressure declines to a normal level quickly.
(3) Body fluid regulation: Adrenaline, norepinephrine, angiotensin, etc. in blood and tissue have a role in contracting the blood vessel, which may make the blood pressure ascend. However, bradykinin, prostaglandin E, atrial natriuretic peptide, etc. have a strong vasodilative effect, which can make the blood pressure down.

During the regulation process of the body temperature and the blood pressure, when the blood flow increases or decreases for regulating the body temperature, the blood pressure also changes. In the same way, with the increase or decrease of the blood flow for the regulation of blood pressure, the body temperature also alters.

Therefore, there are serious coupling effects between the body temperature regulation system and the blood pressure regulation system.

During the regulation process of the above two systems, not only the temperature signal and the blood pressure signal are rapidly fed back to the cerebral cortex, but also the secretions of central hormone concentration information from hypothalamus temperature and blood pressure control centers are timely fed back there. Under the regulation of the senior nervous system, the body temperature and the blood pressure control systems not only adjust their own parameters to provide a steady change, but also are able to quickly eliminate the coupling influence between them. The coordinated approaches of the blood flow and the blood vessel contraction and dilation are the main ways to eliminate the coupling influence. The above bi-cooperative regulation mechanism can be abstracted to be the control structure as shown in Fig. 2.5. In Fig. 2.5, BT, BP, HT, and HBP is body temperature, blood pressure, hypothalamus temperature, and hypothalamus blood pressure, respectively.

2.3.2 *Design of the Bi-Cooperative Decoupling Controller*

If the regulation processes of multiple physiological indexes are parallel studied in the human bio-information regulation network, we can find a special bi-cooperative mechanism for complex decoupling control, for example, the ascend and descend bi-cooperative modulation mechanism in the body temperature and the blood pressure regulation systems. In order to design a novel decoupling controller, we make the multiple physiological indexes corresponding to the MIMO process variables, the central nervous system to the supper control unit, the hypothalamus

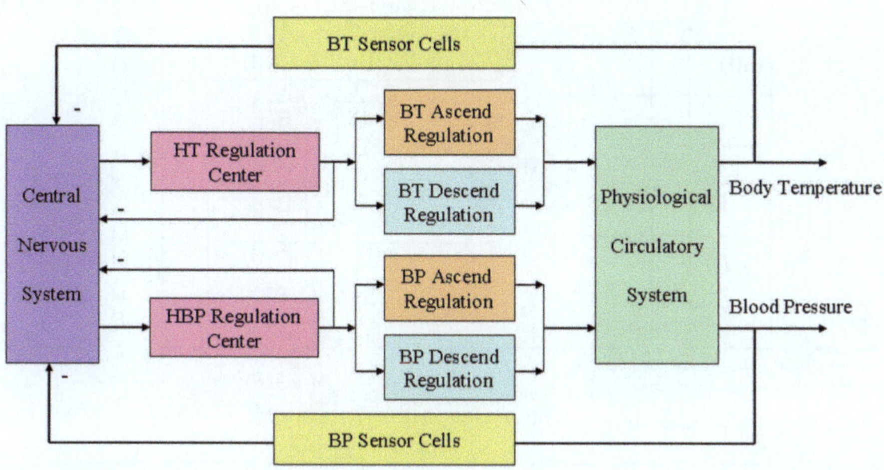

Fig. 2.5 The regulation structure of a physiological system

and the pituitary to the control unit and the decoupling control unit, respectively. Following, we will discuss the bio-inspired IBCDC architecture and its decoupling compensation algorithm.

2.3.2.1 Bio-Inspired Architecture of the IBCDC

Inspired from the mutual cooperative regulation mechanism in the body temperature system and the blood pressure system, a novel cooperative decoupling control architecture and its algorithm can be designed according to Fig. 2.5.

The IBCDC consists of a coordination control unit (CCU), a model identification unit (MIU), a decoupling control evaluation unit (DCEU), control unit 1 and control unit 2. The DCEU includes a reinforcing decoupling control block (RDCB) and a suppression decoupling control block (SDCB). Each control unit or block is designed according to different physiological organs or systems, it includes an original control block (OCB) and a decoupling compensation block. As shown in Fig. 2.6, the CCU and the MIU are designed corresponding to the cerebral cortex; the DCEU and two control units are designed corresponding to the hypothalamus temperature and the blood pressure regulation centers, respectively; the OCB and the DCB are corresponding to hypothalamus and hormone secretion glands.

Inspired from the multi-level complicated feedback mechanism of the physiological regulation systems, we design a feedback network of control information.

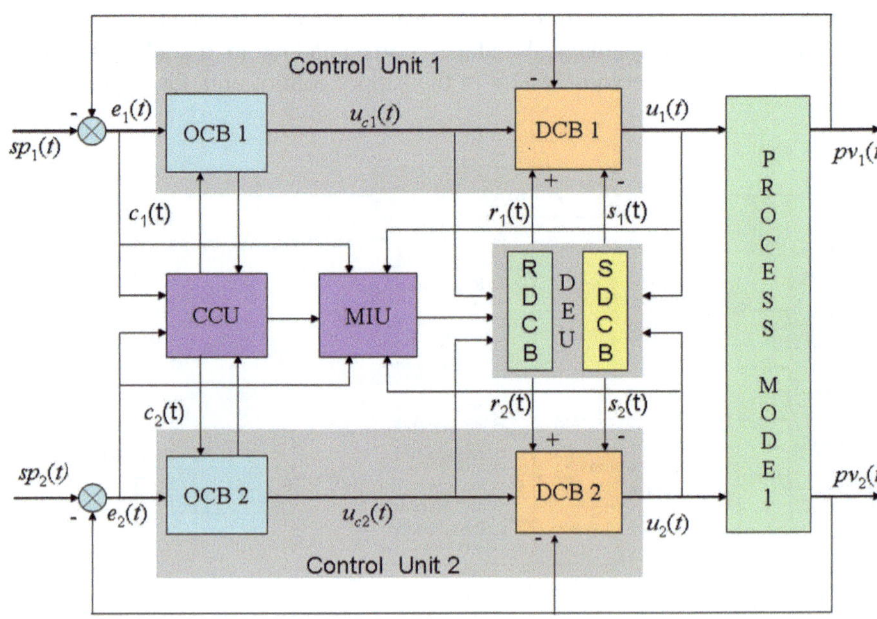

Fig. 2.6 The architecture of the IBCDC

The process variables are not only fed back to the OCB, but also to the DCB; the output signals of the OCBs are fed back to the CCU separately; and all the control output and the decoupling compensation signals are fed back to the DCEU for the effectiveness evaluation of the decoupling control.

In Fig. 2.6, $u_{c1}(t), u_{c2}(t)$ denotes the original control signals of RCB 1 and RCB 2, respectively; $\Delta u_{c1}(t)$ may be calculated according to one control algorithm, such as PID control algorithm. $u_1(t), u_2(t)$ denotes the output of DCB 1 and DCB 2, respectively; $c_1(t)$, $c_2(t)$ denotes the coordination signal sent from the CCU for control unit 1 and control unit 2, respectively; $e_1(t)$, $e_2(t)$ denotes the control error of two loops, respectively. $r_i(t)$, $s_i(t)$, $(i = 1,\ 2)$ denotes the reinforcing and the suppression regulation signals of the decoupling compensation, respectively.

Under the coordination of the CCU, all the control units or blocks communicate with each other to exchange the control or decoupling control information. Through adjusting the actuators harmoniously, the coupling influences among different control loops can be reduced or eliminated in time.

2.3.2.2 General Decoupling Control Algorithm

For convenience to introduce the principle of the IBCDC, we define $G_{c1}(s)$, $G_{c2}(s)$ as the transfer function of RCB 1 and RCB 2, respectively; $\Delta d_1(t)$, $\Delta d_2(t)$ as the dynamic decoupling compensation for the two loops, respectively. The output of the control unit 1 is $\Delta u_1(t) = \Delta u_{c1}(t) + \Delta d_1(t)$, and that of the control unit 2 is $\Delta u_2(t) = \Delta u_{c2}(t) + \Delta d_2(t)$.

For a two-input-two-output (TITO) linear discrete coupling plant

$$\begin{bmatrix} y_1(z) \\ y_2(z) \end{bmatrix} = \begin{bmatrix} G_{11}(z), G_{21}(z) \\ G_{12}(z), G_{22}(z) \end{bmatrix} \begin{bmatrix} u_1(z) \\ u_2(z) \end{bmatrix} \tag{2.9}$$

where $G_{ij}(z), (i,\ j = 1,\ 2)$ is the transfer function between the i-th input and the j-th output; $y_1(z)$ and $y_2(z)$ are the process output variables; and $u_1(z)$ and $u_2(z)$ is the output of control unit 1 and control unit 2, respectively.

If $G_{ij}(z)$ is obtained, we can predict the coupling influence from the input $u_i(z)$ to the output variable $y_j(z)$ $(i, j = 1, 2, \text{and } i \neq j)$, then take a measure to reduce the influence. In the following, we take the decoupling process from $u_1(z)$ to $y_2(z)$ as an example to introduce the decoupling algorithm.

When $y_1(z)$ is regulated, $y_2(z)$ should be kept unchanged, i.e. $\Delta y_2(z) = 0$, to eliminate the coupling influence from $u_1(z)$ to $y_2(z)$. If the output of control unit 1 is $\Delta u_1(z) = \Delta u_{c1}(z)$, the coupling influence from $u_1(z)$ to $y_2(z)$ can be predicted according to the following equation,

$$\Delta c_{12}(z) = G_{12}(z)\Delta u_{c1}(z) \tag{2.10}$$

where, $\Delta c_{12}(z)$ is the coupling influence from $u_1(z)$ to $y_2(z)$ with the dynamic changing output $\Delta u_1(z)$.

Based on the cooperative process narrated above, control unit 2 is in decoupling control status at this time, and the output of its original control block is equal to zero, i.e. $\Delta u_{c2}(z) = 0$. The output of control unit 2 is just the decoupling compensation output of DCB 2, i.e. $\Delta u_2(z) = \Delta d_2(z)$.

To eliminate the coupling influence, the following equation should be true,

$$\Delta c_{12}(z) + \Delta d_2(z) G_{22}(z) = \Delta y_2(z) = 0 \tag{2.11}$$

Take Eq. (2.10) into Eq. (2.11), the decoupling compensation of control unit 2 can be gotten as

$$\Delta d_2(z) = -\frac{G_{12}(z)}{G_{22}(z)} \cdot \Delta u_{c1}(z) \tag{2.12}$$

Similarly, when $y_2(z)$ is regulated, $y_1(z)$ must be kept unchanged to eliminate the coupling influence from $u_2(z)$ to $y_1(z)$, through the decoupling compensation $\Delta d_1(z)$ of DCB 1. The decoupling compensation $\Delta d_1(z)$ can be gotten as follows,

$$\Delta d_1(z) = -\frac{G_{21}(z)}{G_{11}(z)} \cdot \Delta u_{c2}(z) \tag{2.13}$$

As most plants in the industrial process can be approximately represented as a first order with time delay, i.e.

$$G_{ij}(z) = \frac{K_{ij}}{T_{ij}} \frac{z^{-\tau_{ij}/Ts}}{1 - exp(-T_s/T_{ij}) z^{-1}} \tag{2.14}$$

where K_{ij}, T_{ij}, τ_{ij} are the open-loop gain, time constant, and delay time, respectively. Substituting Eq. (2.14) into Eq. (2.12), we can get the decoupling compensation of DEB 2,

$$\Delta d_2(z) = -\Delta u_{c1}(z) \cdot K_{12}^* \cdot \left(\frac{1 - exp(-T_s/T_{22}) z^{-1}}{1 - exp(-T_s/T_{12}) z^{-1}} \right) \cdot z^{-(\tau_{12} - \tau_{22})/Ts} \tag{2.15}$$

where $K_{12}^* = K_{12}/K_{22} \cdot T_{22}/T_{12}$.

In the same way, we can get the decoupling compensation of DEB 1,

$$\Delta d_1(z) = -\Delta u_{c2}(z) \cdot K_{21}^* \cdot \left(\frac{1 - exp(-T_s/T_{11}) z^{-1}}{1 - exp(-T_s/T_{21}) z^{-1}} \right) \cdot z^{-(\tau_{21} - \tau_{11})/T_s} \tag{2.16}$$

where $K_{21}^{*} = K_{21}/K_{11} \cdot T_{11}/T_{21}$.

Consider the difference expression of Eq. (2.15) as follows,

$$\begin{aligned} \Delta d_2(k_2) &= exp(-T_s/T_{12}) \cdot \Delta d_{12}(k_{12}-1) - K_{12}^{*} \cdot \\ &\quad (\Delta u_{c1}(k_{12}-l_{12}) - exp(-T_s/T_{22}) \cdot \Delta u_{c1}(k_{12}-l_{12}-1)) \end{aligned} \tag{2.17}$$

where k_2 is the discrete step of control unit 1, and l_{12} is the delay step of the decoupling compensation output, $l_{12} = (\tau_{12} - \tau_{22})/T_s$.

Similarly, we can get the difference expression of Eq. (2.16) as follows,

$$\begin{aligned} \Delta d_1(k_1) &= exp(-T_s/T_{21}) \cdot \Delta d_{21}(k_{21}-1) - K_{21}^{*} \cdot \\ &\quad (\Delta u_{c2}(k_{21}-l_{21}) - exp(-T_s/T_{11}) \cdot \Delta u_{c2}(k_{21}-l_{21}-1)) \end{aligned} \tag{2.18}$$

where k_1 is the discrete step of control unit 2, and l_{21} is the delay step of the decoupling compensation output, $l_{21} = (\tau_{21} - \tau_{11})/T_s$.

(1) If $\tau_{ij} = 0$ or $\tau_{12} = \tau_{22}$, $\tau_{21} = \tau_{11}$, the pure delay time has no effect on the decoupling compensation, i.e. $l_{12} = 0$, $l_{21} = 0$.
(2) If $\tau_{12} > \tau_{22}$ or $\tau_{21} > \tau_{11}$, the affection of the pure delay time should be counted, and $l_{12} > 0$, $l_{21} > 0$.
(3) If $\tau_{12} < \tau_{22}$ or $\tau_{21} < \tau_{11}$, the pure delay time changes to the lead time, which makes it impossible to implicate the decoupling algorithm. On this condition, we let $l_{12} = 0$ and $l_{21} = 0$.

From Eq. (2.17), we can get the following general decoupling compensation expression of DEB 1,

$$\Delta d_1(k_1) = \alpha_{21} \cdot \Delta d_{21}(k_{21}-1) - \beta_{21} \cdot \Delta u_{c2}(k_{21}-l_{21}) + \gamma_{21} \cdot \Delta u_{c2}(k_{21}-l_{21}-1) \tag{2.19}$$

where $\alpha_{21} = exp(-T_s/T_{21})$, $\beta_{21} = K_{21}/K_{11} \cdot T_{11}/T_{21}$, $\gamma_{21} = \beta_{21} \cdot exp(-T_s/T_{11})$.

In the same way, we can also get the general decoupling compensation expression of DEB 2,

$$\Delta d_2(k_2) = \alpha_{12} \cdot \Delta d_{12}(k_{12}-1) - \beta_{12} \cdot \Delta u_{c1}(k_{12}-l_{12}) + \gamma_{12} \cdot \Delta u_{c1}(k_{12}-l_{12}-1) \tag{2.20}$$

where $\alpha_{12} = exp(-T_s/T_{12})$, $\beta_{12} = K_{12}/K_{22} \cdot T_{22}/T_{12}$, $\gamma_{12} = \beta_{12} \cdot exp(-T_s/T_{22})$.

As a result, the discrete control expressions of control unit 1 and control unit 2 are as follows,

$$u_1(k_1) = \begin{cases} u_1(k_1 - 1) + \Delta u_{c1}(k_1), & (k_1 = 2k + 1) \\ u_1(k_1 - 1) + \Delta d_1(k_1), & (k_1 = 2k) \end{cases} \tag{2.21}$$

$$u_2(k_2) = \begin{cases} u_2(k_2 - 1) + \Delta u_{c2}(k_2), & (k_2 = 2k) \\ u_2(k_2 - 1) + \Delta d_2(k_2), & (k_2 = 2k + 1) \end{cases} \tag{2.22}$$

If the accurate mathematical model of a process plant is obtained, the decoupling compensation parameters α_{12}, β_{12}, γ_{12}, α_{21}, β_{21}, γ_{21} may be calculated easily. However, in most cases, it is impossible to get the accurate mathematical model of the process plant. Here, the step-response identification method is employed to obtain the approximate transfer matrix $G_{ij}(z)$ online. When the IBCDC is taken into effect for the first time, the CCU sends "*HOLD*" signal to control unit 2, then the output of control unit 2 will be unchanged. At the same time, control unit 1 outputs a step signal. When both control loops become stable, $G_{11}(s)$ and $G_{12}(s)$ can be obtained by using the step-response identification. Similarly, CCU sends "*HOLD*" signal to control unit 1and outputs a step signal, then $G_{21}(s)$ and $G_{22}(s)$ will be obtained. Thus, the mathematical model can be obtained quickly and easily, and the parameters of α_{12}, β_{12}, γ_{12}, α_{21}, β_{21}, γ_{21} can be calculated approximately.

2.3.2.3 Bi-Cooperative Decoupling Control Algorithm

As the mathematics models used in Eqs. (2.19) and (2.20) are approximate models, and if only the decoupling compensation algorithms shown in Eqs. (2.19) and (2.20) are used, the decoupling control effectiveness may not be satisfactory. To improve the decoupling control performance, we design the reinforcing decoupling control block (RDCB) and the suppression decoupling control block (SDCB) in the DEU.

(1) Evaluation for the decoupling control effectiveness

To improve the decoupling control algorithm, it is important to evaluate the decoupling control effectiveness. Obviously, the ideal goal of the decoupling control is to make the coupling influence or the error minimal during every control period. Considering the pure delay time in the process, the current decoupling control result may depend on the decoupling compensation during the time period $(k - l_{ii})^{th}$, $(i = 1, 2)$, where k is the discrete control steps and l_{ii} is the delay steps or time of control plant $G_{ii}(z)$, $l_{ii} = \tau_{ii}/T_s$. As thus, we can evaluate the decoupling result according to the following steps.

(1) Judge the future change direction of the decoupling control error, according to the decoupling compensation $\Delta d_1(z)$ *or* $\Delta d_2(z)$ and the open loop gain k_{ii} of the corresponding transfer function $G_{ii}(z)$. When the decoupling compensation $\Delta d_1(z)$ *or* $\Delta d_2(z)$ is the output, the future change direction of the decoupling control may be predicted based on the following equation,

$$dir_{ij}(k) = sign(\Delta d_j(k)) \cdot sign(k_{jj}) \tag{2.23}$$

where $dir_{ij}(k)$ indicates the change direction. When $dir_{ij}(k) > 0$, the corresponding decoupling control error $e_j(k + l_{jj}), (j = 1,\ 2)$ should descend, i.e. $\Delta e_j(k + l_{jj}) < 0$. Otherwise, $\Delta e_j(k + l_{jj}) > 0$.

2. Evaluate the decoupling control effect based on $dir_{ij}(k)$ and its corresponding decoupling control error $e_j(k + l_{jj})$. According to the mathematics model, l_{jj} can be obtained. Thus, the current decoupling control effect may be evaluated based on the current decoupling control error $e_j(k)$ and its corresponding decoupling compensation $\Delta d_j(k - l_{jj})$ and the predicted change direction $dir_{ij}(k - l_{jj})$. As thus, we can estimate the decoupling control error according to the following criterion.

   (i) If $sign(dir_{ij}(k - l_{jj})) \otimes sign(\Delta e_j(k)) = 0$, the direction of the corresponding decoupling compensation is consistent with the change trend of the decoupling control error. If the decoupling control error is not equal to zero, the compensation should be reinforced according to the current control error.

   (ii) If $sign(dir_{ij}(k - l_{jj})) \otimes sign(\Delta e_j(k)) = 1$, the decoupling compensation direction is not right, or the decoupling compensation may overshoot. The compensation output should be suppressed according to the current control error.

According to the evaluation result of the decoupling compensation, the original decoupling control output, as shown in Eqs. (2.19) and (2.20), may be strengthened or weakened.

$$Eva_{ij}(k) = \begin{cases} 1 & (sign(dir_{ij}(k - l_{jj})) \otimes sign(\Delta e_j(k)) = 0) \\ -1 & (sign(dir_{ij}(k - l_{jj})) \otimes sign(\Delta e_j(k)) = 1) \end{cases} \tag{2.24}$$

(2) Algorithm of the RDCU and the SDCU

According to the original decoupling control algorithm in Eqs. (2.19) and (2.20), we design the improved decoupling control algorithm with the reinforcing or suppression function Eqs. (2.25) and (2.26) as follows,

$$\Delta d_1(k_1) = \{\alpha_{21} \cdot \Delta d_{21}(k_1 - 1) - \beta_{21} \cdot \Delta u_{c2}(k_1 - l_{21}) + \gamma_{21} \cdot \Delta u_{c2}(k_1 - l_{21} - 1)\} \cdot (1 + \eta_1(k_1)) \tag{2.25}$$

$$\Delta d_2(k_2) = \{\alpha_{12} \cdot \Delta d_{12}(k_2 - 1) - \beta_{12} \cdot \Delta u_{c1}(k_2 - l_{12}) + \gamma_{12} \cdot \Delta u_{c1}(k_2 - l_{12} - 1)\} \cdot (1 + \eta_2(k_2)) \tag{2.26}$$

where $\eta_1(k_1)$, $\eta_2(k_2)$ is the strengthen and weaken factor, respectively. If the original decoupling compensation needs to be reinforced, the factors $\eta_1(k_1)$, $\eta_2(k_2)$ should be larger than zero, i.e. $\eta_1(k_1) > 0$, $\eta_2(k_2) > 0$. Otherwise, the factors should be smaller than zero, i.e. $\eta_1(k_1) < 0$, $\eta_2(k_2) < 0$. Thus, when $\eta_1(k_1) > 0$, $\eta_2(k_2) > 0$, the expression $(1+\eta_1(k_1))$, $(1+\eta_2(k_2))$ are larger than 1. So the decoupling compensation, as shown in Eqs. (2.25) and (2.26), can be strengthened. Otherwise, if $0 < (1+\eta_1(k_1)) < 1$ and $0 < (1+\eta_2(k_2)) < 1$, then the decoupling compensation is weakened.

Inspired from the hormone modulation law in endocrine system [37], we make the decoupling control error $e_j(k)$ correspond with the stimulation signal of hormone, and $\eta_1(k_1)$, $\eta_2(k_2)$ with the secreting concentration of hormone in the body. Thus, we design the regulation algorithm as follows,

$$\begin{cases} \Delta\eta_1(k_1) = Eva_{21}(k) \cdot \frac{abs(e_1(k))}{\alpha_{21} + exp(abs(e_1(k)))} \\ \eta_1(k_1) = \eta_1(k_1 - 1) + \Delta\eta_1(k_1) \end{cases} \tag{2.27}$$

$$\begin{cases} \Delta\eta_2(k_2) = Eva_{12}(k) \cdot \frac{abs(e_2(k))}{\alpha_{12} + exp(abs(e_2(k)))} \\ \eta_2(k_2) = \eta_2(k_2 - 1) + \Delta\eta_2(k_2) \end{cases} \tag{2.28}$$

where $-1 < \eta_1(k_1) < 1, -1 < \eta_2(k_2) < 1$; α_{12}, α_{21} are real factors and may be set according to the real-time decoupling effect.

(3) Feedback of decoupling error

The corrections of reinforcement or suppression of the decoupling compensation algorithms in Eqs. (2.27) and (2.28), are delayed and a little rough. Because their evaluations are according to the $\Delta d_j(k - l_{ii})$ and the current control error $e_j(k)$. Therefore, the improved decoupling compensation should further consider the direct compensation according to the current decoupling error $e_j(k)$ in the IBCDC. Here, we employ the conventional proportional action to the designed decoupling algorithm. The final decoupling compensation is as follows,

$$\begin{aligned} \Delta d_1(k_1) = &\{\alpha_{21} \cdot \Delta d_{21}(k_1 - 1) - \beta_{21} \cdot \Delta u_{c2}(k_1 - l_{21}) + \gamma_{21} \cdot \Delta u_{c2}(k_1 - l_{21} - 1)\} \cdot (1 + \eta_1(k_1)) \\ &+ sign(k_{11}) \cdot K_{c1} \cdot (e_1(k) - e_1(k - 1)) \end{aligned} \tag{2.29}$$

$$\begin{aligned} \Delta d_2(k_2) = &\{\alpha_{12} \cdot \Delta d_{12}(k_2 - 1) - \beta_{12} \cdot \Delta u_{c1}(k_2 - l_{12}) + \gamma_{12} \cdot \Delta u_{c1}(k_2 - l_{12} - 1)\} \cdot (1 + \eta_2(k_2)) \\ &+ sign(k_{22}) \cdot K_{c2} \cdot (e_2(k) - e_2(k - 1)) \end{aligned} \tag{2.30}$$

where K_{c1}, K_{c2} are the supplement proportional gains, and their value should be very small.

(4) The cooperative logical control

In order to eliminate the coupling influence, all controllers in a MIMO control system should work cooperatively. In most cases, each control unit not only modulates regularly the control error, but also makes compensation for the coupling influence at the same time. In such case, the decouple control algorithm may be very complicated or difficult to be realized. In order to make the decoupling algorithm be easer or simpler, the IBCDC is designed as follows. In the IBCDC, all the control units cooperatively work during different periods. To eliminate the control error and the coupling influence stably, the CCU can coordinate the work status of control unit 1 and control unit 2, according to the real-time control error information. The detail working sequence of the IBCDC is as shown in Fig. 2.7. In Fig. 2.7, control unit 1 and control unit 2, including the RCB and the DCB, always work in different status during every control period. In other words, when one control unit works normally during a period, the other must keep in decoupling control status at the same time. In Fig. 2.7, $C_1(t), C_2(t)$ is the cooperative logical control signal of control unit 1 and control unit 2 sent from the CCU, and $ST_1(t), ST_2(t)$ indicates the working status of control unit 1 and control unit 2, respectively. The detail working process is explained as follows.

(1) If $C_1(t) = 0$, control unit 1 is in the decoupling status, and control unit 2 is in the normal work status. At this time, the real-time output of RCB 1 is equal to zero, i.e. $\Delta u_{c1}(t) = 0$, and the decoupling compensation output of DCB 1 is $\Delta d_{21}(t)$. As thus, the final output of control unit 1 is the compensation output of DCB 1, i.e. $\Delta u_1(t) = \Delta d_{21}(t)$. Otherwise, if $C_1(t) = 1$, control unit 1 is in normal control status, and the compensation output is equal to zero, i.e. $\Delta d_{21}(t) = 0$. As thus, the final output of control unit 1 is the original control output of OCB 1, i.e. $\Delta u_1(t) = \Delta u_{c1}(t)$.
(2) Similarly, if $C_2(t) = 0$, control unit 2 is in decoupling control status. The original output of OCB 2 is equal to zero, i.e. $\Delta u_{c2}(t) = 0$. The output of DCB 2 is the decoupling compensation $\Delta d_{21}(t)$ for control loop 2. The final output

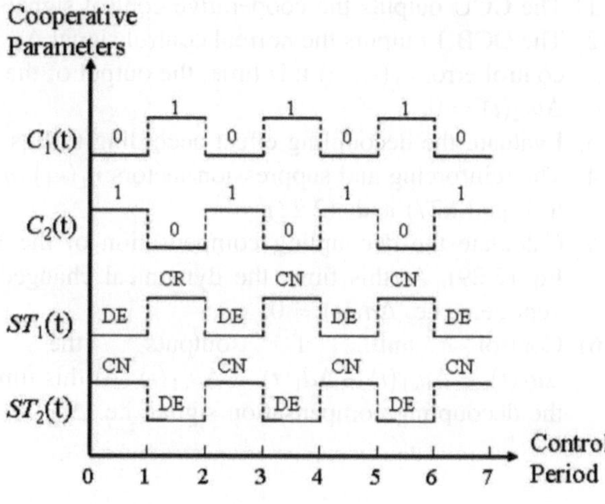

Fig. 2.7 Phase sequence of the IBCDC

of control unit 2 is $\Delta u_2(t) = \Delta d_{c12}(t)$. Otherwise, if $c_2(t) = 1$, control unit 2 is in normal control status, and its output is the original control output of RCB 2, i.e. $\Delta u_2(t) = \Delta u_{c2}(t)$.

With the cooperation of the CCU, control unit 1 and control unit 2 work cooperatively in asynchronous status. During the even control periods, control unit 1 is in normal control status, while control unit 2 is in decoupling control status. Similarly, during the odd control periods, control unit 2 is in normal control status, while control unit 1 is in decoupling control status, as shown in Fig. 2.7, where "CR" and "DE" indicates "control" and "decoupling", respectively.

(5) Working process of the IBCDC

Before the IBCDC takes into effect, the following pre-tasks must be carried out.

1. Identify the mathematics model $G_{ij}(z), (i,j = 1,2)$ using the step response method.
2. Calculate the decoupling compensation factors α_{12}, β_{12}, γ_{12}, α_{21}, β_{21}, γ_{21} in Eqs. (2.19), (2.20), (2.25), and (2.26).
3. Adjust the reinforcement and suppression parameters α_{12}, α_{21} in Eqs. (2.27) and (2.28).
4. Set the proportional parameters K_{c1}, K_{c2} in Eqs. (2.28) and (2.29).

When the above work is finished, the IBCDC will take into effect. During every control period, the CCU needs to coordinate the work status of control unit 1 and control unit 2 according to the working sequence mentioned in Fig. 2.7. In fact, control unit 1 and control unit 2 work in dual working status during the even or odd control periods. Next, we take the process during the odd periods as an example to introduce the detail working process as follows.

During the odd periods, control unit 1 works in normal control status while control unit 2 is in decoupling status. The detail working process during the odd periods is,

1. The CCU outputs the cooperative control signals $C_1(t) = 1, C_2(t) = 0$.
2. The OCB 1 outputs the normal control signal $\Delta u_{c1}(t)$ according to the real-time control error $e_1(t)$. At this time, the output of the OCB 2 keeps unchanged, i.e. $\Delta u_{c2}(t) = 0$.
3. Evaluate the decoupling effect according to Eqs. (2.27) and (2.24).
4. The reinforcing and suppression factors $\eta_1(k_1)$, $\eta_2(k_2)$ are calculated according to Eqs. (2.27) and (2.28).
5. Calculate the decoupling compensation of the DEB 2, $\Delta d_2(t)$, according to Eq. (2.29). At this time, the dynamical changed decoupling compensation is kept zero, i.e. $\Delta d_1(t) = 0$.

(6) Control unit 1 outputs the final control signal $\Delta u_1(t) = \Delta u_{c1}(t) + \Delta d_1(t) = \Delta u_{c1}(t)$; At this time, control unit 2 only outputs the decoupling compensation signal, i.e. $\Delta u_2(t) = \Delta u_{c2}(t) + \Delta d_2(t) = \Delta d_2(t)$.

Fig. 2.8 The work process during the odd control period

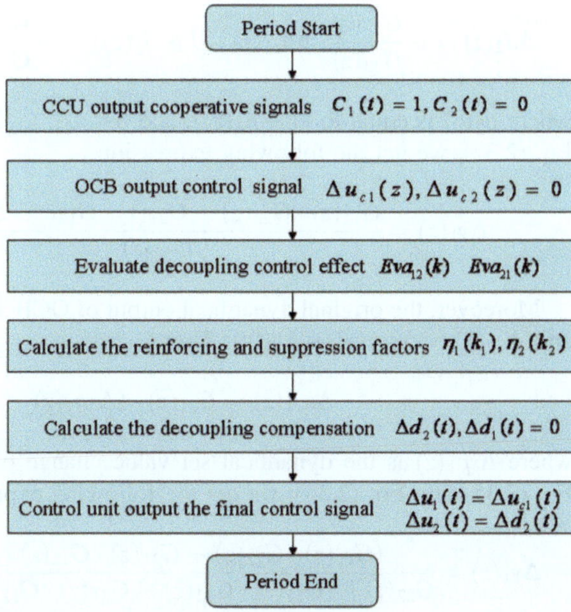

The detail working process (**Algorithm 2.1**) during the odd periods is illustrated in Fig. 2.8. The process during the even periods is similar to that during the odd periods.

2.3.2.4 Stability Analysis on the IBCDC

First, we discuss the stability of the IBCDC during even periods. From Eq. (2.9) we can get the dynamical mathematical model as follows,

$$\begin{bmatrix} \Delta y_1(z) \\ \Delta y_2(z) \end{bmatrix} = \begin{bmatrix} G_{11}(z), G_{21}(z) \\ G_{12}(z), G_{22}(z) \end{bmatrix} \begin{bmatrix} \Delta u_1(z) \\ \Delta u_2(z) \end{bmatrix} \tag{2.31}$$

During the even periods, the original control output of RCB 2 is equal to zero, i.e. $\Delta u_{c2}(z)=0$. According to work principle of the IBCDC, the output of control unit 1 and control unit 2 can be obtained, i.e. $\Delta u_1(z)=\Delta u_{c1}(z),\ \ \Delta u_2(z)=\Delta d_2(z)$. As thus, the following expression holds,

$$\Delta y_1(z) = G_{11}(z)\Delta u_{c1}(z) + G_{21}(z)\Delta d_2(z) \tag{2.32}$$

According to Eqs. (2.12) and (2.30), we get the final expression of $\Delta d_2(z)$ as follows (ignore the proportional action term),

$$\Delta d_2(z) = -\frac{G_{12}(z)}{G_{22}(z)} \cdot \Delta u_{c1}(z) \cdot (1 + \eta_2(z)) = -\frac{G_{12}(z)}{G_{22}(z)} \cdot \Delta u_{c1}(z) \cdot \mu_2(z) \quad (2.33)$$

where $\mu_2(z)$ is equal to $(1 + \eta_2(z))$, and $0 < \mu_2(z) < 2$. Substituting Eq. (2.33) into Eq. (2.32), we get the following expression,

$$\Delta y_1(z) = \frac{G_{11}(z) \cdot G_{22}(z) - G_{21}(z) \cdot G_{12}(z) \cdot \Delta\mu_2(z)}{G_{22}(z)} \cdot \Delta u_{c1}(z) \quad (2.34)$$

Moreover, the original dynamical output of OCB 1 is $\Delta u_{c1}(z) = G_{c1}(z) \cdot \Delta e_1(z)$, and $\Delta e_1(z) = \Delta sp_1(z) - \Delta y_1(z)$. Thus,

$$\Delta u_{c1}(z) = G_{c1}(z) \cdot (\Delta sp_1(z) - \Delta y_1(z)) \quad (2.35)$$

where $\Delta sp_1(z)$ is the dynamical set value change of control loop 1. Substituting Eq. (2.35) into Eq. (2.34), we get the following expression,

$$\Delta y_1(z) = \frac{(G_{11}(z) \cdot G_{22}(z) - G_{21}(z) \cdot G_{12}(z) \cdot \mu_2(z))G_{c1}(z)}{G_{22}(z) + (G_{11}(z) \cdot G_{22}(z) - G_{21}(z) \cdot G_{12}(z) \cdot \mu_2(z))G_{c1}(z)} \cdot \Delta sp_1(z) \quad (2.36)$$

According to **Algorithm 2.1** in Fig. 2.8, the dynamical change output of $y_2(z)$ is equal to zero at this time, i.e. $\Delta y_2(z) = 0$.

In the similar way, during the odd periods the dynamical output of loop 2 is obtained as follows,

$$\Delta y_2(z) = \frac{(G_{11}(z) \cdot G_{22}(z) - G_{21}(z) \cdot G_{12}(z) \cdot \mu_1(z))G_{c2}(z)}{G_{11}(z) + (G_{11}(z) \cdot G_{22}(z) - G_{21}(z) \cdot G_{12}(z) \cdot \mu_1(z))G_{c2}(z)} \cdot \Delta sp_2(z) \quad (2.37)$$

where $\Delta sp_2(z)$ is the dynamical set value change of loop 2, where $\mu_1(z)$ is equal to $(1 + \eta_1(z))$ (where $0 < \mu_1(z) < 2$) and the dynamical output of loop 1 is equal to zero, i.e. $\Delta y_1(z) = 0$.

From Eqs. (2.36) and (2.37), the stability of the IBCDC during the odd and even periods depends on the characteristic roots of its close loop characteristic equation,

$$G_{22}(z) + (G_{11}(z)G_{22}(z) - G_{21}(z)G_{12}(z)\mu_2(z))G_{c1}(z) = 0$$

and

$$G_{11}(z) + (G_{11}(z)G_{22}(z) - G_{21}(z)G_{12}(z)\mu_1(z))G_{c2}(z) = 0.$$

Obviously, it is only when its characteristic root lies in the unit circle that the decoupling control system is stable. As $\mu_1(z)$ and $\mu_2(z)$ are both larger than 0 and smaller than 2, i.e. $0<\mu_1(z)<2$ and $0<\mu_2(z)<2$, the stability of the control system is mainly determined by the transfer functions of control unit 1 and control unit 2, i.e. $G_{c1}(z), G_{c2}(z)$, and the process plant model $G_{ij}(z), (i,j=1,2)$.

Extension to MIMO Systems

The IBCDC and the decoupling control algorithm for TITO systems can be easily extended to MIMO systems. For a MIMO system,

$$\begin{bmatrix} y_1(z) \\ \vdots \\ y_n(z) \end{bmatrix} = \begin{bmatrix} G_{11}(z) & G_{21}(z) & \cdots & G_{n1}(z) \\ & \vdots & \vdots & \\ G_{1n}(z) & G_{2n}(z) & \cdots & G_{nn}(z) \end{bmatrix} \begin{bmatrix} u_1(z) \\ \vdots \\ u_n(z) \end{bmatrix} \tag{2.38}$$

where $y_i(z)$ (i = 1, 2, …, n) is the output variable; $G_{ij}(z)$ (i, j = 1, 2, …, n) is the discrete transfer function between the i-th input and the j-th output; $u_i(z)$ (i = 1, 2, …, n) is the output of the controller. For the decoupling algorithm mentioned above, when regulating the state variable $y_i(z)$, we should ensure that all the values of the dynamics influence of $y_1, y_2 \cdots, y_{i-1}, y_{i+1}, \cdots, y_n$ are equal to zero. Based on the derived algorithm, the coupling influence can be eliminated by adjusting n actuators harmoniously. The corresponding dynamic decoupling output of every control unit can be derived in the similar way. Moreover, if the mathematical model of the MIMO system is not available, we can also obtain the approximate transfer matrix $G(z)$ via the step-response identification. Then, the decoupling equations can be obtained from the identified mathematical model.

2.3.2.5 Simulation Results

In order to examine the decoupling control effectiveness of the IBCDC, we consider the Wood and Berry's FOPDT transfer function model of a methanol-water distillation column [35]:

$$\begin{bmatrix} y_1(s) \\ y_2(s) \end{bmatrix} = \begin{bmatrix} \frac{12.8e^{-s}}{16.7s+1} & \frac{-18.9e^{-3s}}{21.0s+1} \\ \frac{6.6e^{-7s}}{10.9s+1} & \frac{-19.4e^{-3s}}{14.4s+1} \end{bmatrix} \begin{bmatrix} u_1(s) \\ u_2(s) \end{bmatrix} + \begin{bmatrix} \frac{3.8e^{-8s}}{14.9s+1} \\ \frac{4.9e^{-s}}{13.2s+1} \end{bmatrix} F(s) \tag{2.39}$$

where, y_1 is the mole fraction of methanol at the tops; y_2 is the mole fraction of methanol at the bottoms; u_1 is the reflux flow rate; u_2 is the steam flow rate; and $F(s)$ is the feed flow rate. The system (2.39) is a coupling TITO system with large time delay. The conventional control scheme is that y_1 and y_2 is controlled using PID controller to adjust u_1 and u_2, respectively. However, there is a violent coupling influence between the control loops.

Following, we control the system (2.39) by the IBCDC in Fig. 2.6, and ignore the feed flow rate $F(s)$. We choose the control unit 1 for y_1, and the control unit 2 for y_2. We simulate the actual process field using the mathematical model, and identify the mathematical model via step-response identification method. The identified mathematical model is as follows:

$$\begin{bmatrix} y_1(s) \\ y_2(s) \end{bmatrix} = \begin{bmatrix} \frac{12.6e^{-s}}{16.9s+1} & \frac{-18.7e^{-3s}}{21.1s+1} \\ \frac{6.5e^{-7s}}{10.9s+1} & \frac{-19.6e^{-3s}}{14.5s+1} \end{bmatrix} \begin{bmatrix} u_1(s) \\ u_2(s) \end{bmatrix}$$

Based on the decoupling algorithm discussed in Sect. 2.1.3 and the selected sample time T_s is 0.5 s, the dynamic decoupling output to control y_1 is

$$\begin{cases} \Delta d_{21}(k_{21}) = \{0.9535 \cdot \Delta d_{21}(k_{21}-1) + 1.1742 \cdot \Delta u_{c2}(k_{21}-4) + 0.8981 \cdot \Delta u_{c2}(k_{21}-5)\} \cdot (1+\eta_{21}) \\ \eta_{21} = Eva_{21}(k) \cdot \frac{abs(e_1(k))}{\alpha_{21} + exp(abs(e_1(k)))} \\ Eva_{21}(k) = \begin{cases} 1 & (sign(dir_{21}(k-l_{11})) \otimes sign(\Delta e_1(k)) = 0) \\ -1 & (sign(dir_{21}(k-l_{11})) \otimes sign(\Delta e_1(k)) = 1) \end{cases} \end{cases} \tag{2.40}$$

where $\alpha_{21} = 0.4$. In the same way, we can also get the dynamic decoupling compensation to control y_2 is,

$$\begin{cases} \Delta d_{12}(k_{12}) = \{0.9123 \cdot \Delta d_{12}(k_{12}-1) + 0.4494 \cdot \Delta u_{c1}(k_{12}-8) - 0.41 \cdot \Delta u_{c1}(k_{12}-9)\} \cdot (1+\eta_{12}) \\ \eta_{12} = Eva_{12}(k) \cdot \frac{abs(e_2(k))}{\alpha_{12} + exp(abs(e_2(k)))} \\ Eva_{12}(k) = \begin{cases} 1 & (sign(dir_{12}(k-l_{22})) \otimes sign(\Delta e_2(k)) = 0) \\ -1 & (sign(dir_{12}(k-l_{22})) \otimes sign(\Delta e_2(k)) = 1) \end{cases} \end{cases} \tag{2.41}$$

where $\alpha_{12} = 0.5$.

Using the IBCDC, we can obtain the satisfactory control performance for the coupling TITO system with large time delay, as shown in Figs. 2.9 and 2.10.

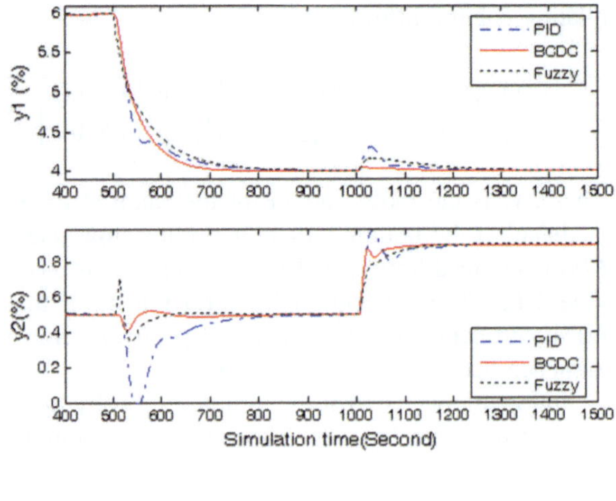

Fig. 2.9 Comparison of decoupling control effect among the IBCDC, PID and Fuzzy control

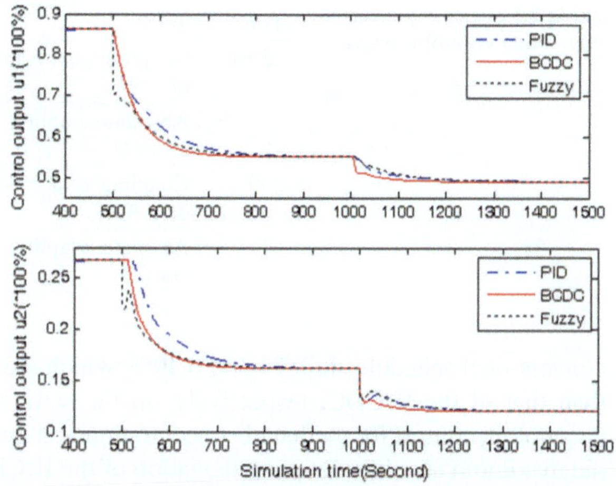

Fig. 2.10 Comparison of control output among the IBCDC, PID and Fuzzy control

We also compare the control effectiveness of the IBCDC with that of the conventional control scheme and fuzzy control algorithm. Here, the conventional PID control algorithm is adopted in control unit 1 and control unit 2. The fuzzy control algorithm is used in the decoupling control unit, and the dynamical change output of original control in every control unit and the dynamical control error of the other loop are selected as fuzzy control variables. In the fuzzy controller, the domains of input and output are both selected as [−1, 1], and triangle function is chosen as fuzzy membership function. In order to conveniently compare the decoupling control effectiveness of different control algorithms, the control parameters of two control units of the IBCDC and the main control unit of the fuzzy controller are set to be same, as shown in Table 2.4.

The comparison of decoupling control effectiveness among the IBCDC, the fuzzy decoupling control and the conventional control is as shown in Fig. 2.9. In Fig. 2.9, when the system is stable, the set value of methanol concentration in the tower top is changed from 6 to 4% at the 250-th second, and the set value of methanol concentration in the tower bottom is changed from 0.5 to 0.9% at the 500-th second. From the coupling effectiveness of the methanol concentration at the tower bottom, we can find that the maximum decoupling error of the IBCDC is 0.093%, while the largest absolute coupling deviations of the fuzzy decoupling control and the

Table 2.4 Decoupling control parameters

Control algorithm	Control parameters	K_p	T_i	T_d
IBCDC	Control unit 1	0.45	5.5	0.08
	Control unit 2	0.35	10.3	0.06
Fuzzy	Control unit 1	0.45	5.5	0.08
	Control unit 2	0.35	10.3	0.06
PID	Loop 1	0.45	5.5	0.08
	Loop 2	0.35	10.3	0.06

Table 2.5 Coupling error in decoupling control process

Time	Term	IBCDC	Fuzzy	PID
250 s	Coupling peak-value of y_2/%	0.407	0.71	0.01
	Absolute coupling error/%	0.093	0.21	0.49
500 s	Coupling peak-value of y_1/%	4.04	4.16	4.32
	Absolute coupling error/%	0.04	0.16	0.32

conventional schedule are 0.21 and 0.49%, which are 2.3 times and 5.3 times larger than that of the IBCDC, respectively. In the same way, from the process of the decoupling control for methanol concentration at the tower top at the 500-th second, the maximum absolute coupling deviation of the IBCDC is 0.04%, while the largest absolute coupling deviations of the fuzzy decoupling control and the conventional control are 0.16 and 0.32%, which are 4 times and 8 times than that of the IBCDC at this time, respectively. The detail data is as shown in Table 2.5.

The output comparison of the IBCDC, the fuzzy controller and the conventional controller is as shown in Fig. 2.10. From the figure, we can see that the output of the IBCDC is more stable than that of the other two controllers. Consequently, the IBCDC has much better decoupling control effectiveness than the fuzzy decoupling controller and the conventional controller, according to Figs. 2.9 and 2.10.

2.4 A Neuroendocrine Regulation Based Intelligent Cooperative Decoupling Controller

In this section, an intelligent cooperative decoupling controller (ICDC) and its control scheme are presented based on the neuroendocrine regulation principle of human body, and applied to the coagulation bath of PANCF production. Compared with other decoupling control methods, the ICDC can be implemented more easily and practically but provides better control performance and adaptability. Also, it is the first to discuss the decoupling control method for the control in the coagulation bath of PANCF production, and can be easily extended to other phases of the production line.

2.4.1 Neuroendocrine Regulation Principle in Human Body

The biochemical systems in human body work cooperatively to establish a stable body environment which is critical to the human's live and behaviors. Among these systems, the neuroendocrine system plays a comparatively more important role in the regulation process of human life [22]. Figure 2.11 illustrates the basic structure of the neuroendocrine system and its working principles.

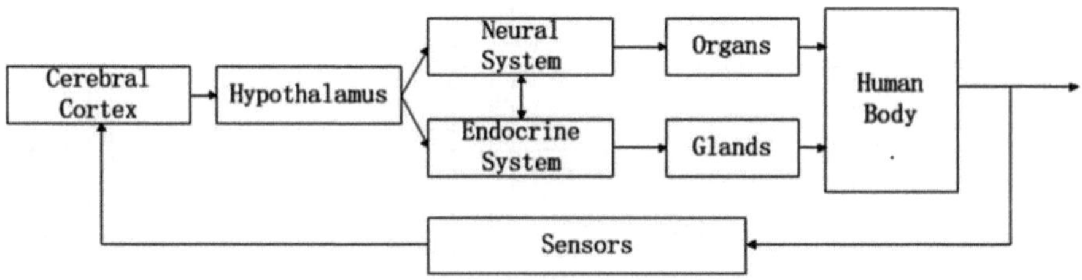

Fig. 2.11 The regulation mechanism of neuroendocrine system in human body

The neuroendocrine regulation system is a large feedback loop consisting of many organs. The core of the system is the cerebral cortex which takes action by secreting hormone to the hypothalamus. The hypothalamus sends instructions to two independent systems, namely, the neural system and endocrine system, respectively. For the neural system, the brain and spinal cord of human body act as the core and the millions of neurons with their varieties forms of connections act as the branches. The signals that the body receives, namely, the stimulations from the environment can be acquired by the neural system and then processed by the brain and spinal cord (in charge of signals of different importance). The instructions sent from the brain and spinal cord are finally transmitted to the corresponding organs to response to the stimulations. For the endocrine system, different types of glands are their actuators, which receive instruction and make regulation for the secretion of hormones. These two systems work independently, but there are some types of cooperation existed between them in order to achieve better control effect.

The structure of such biological system has some similarities with a common control system. The controller, the actuator, the target, and the feedback path in a control system can be replaced by the core (cerebral cortex, hypothalamus, neural system, and endocrine system), the organs and glands, the body, and the sensors, respectively. The principle of working independently with cooperation can also be regarded as the foundation of some decoupling regulation control schemes. Here the only problem left is to extend the origin neuroendocrine system structure from two subsystems to three or more subsystems and this can be done by adding and tuning carefully more interactions among the subsystems, which requires deep analysis on the target control plant.

2.4.2 Design of the ICDC

2.4.2.1 System Structure

Based on the principle of neuroendocrine regulation in human body, the ICDC controller is designed. Figure 2.12 shows the overall system structure of the controller. The whole system consists of four parts, the control center unit (CCU), the

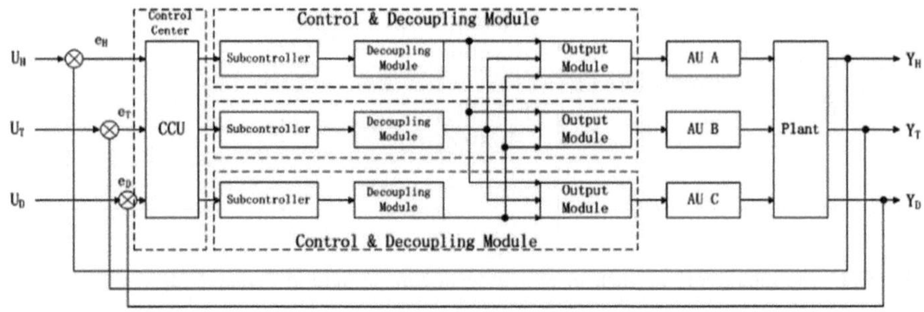

Fig. 2.12 The system structure of the ICDC

control and decoupling unit (CDU), the actuator unit (AU), and the industrial plant to be controlled. These four parts work together according to the principle of the organs in the neuroendocrine regulation system. The CCU includes a control algorithm switching module to determining whether the neuroendocrine decoupling algorithm should be added according to the dynamic plant error, and whether the parameters of the applied CDUs should be tuned. The number of the CDUs in the system is determined by the number of the variables to be controlled (here three CDUs are equipped for the liquid-level, the temperature, and the concentration, respectively). Each CDU consists of a subcontroller module, a decoupling module and an output module. The CDUs take effect independently, but the information of the target plant can be exchanged among them, so the cooperative control scheme can be realized. The AU receives signals from the CDU it connects to and sends them to the plant.

2.4.2.2 The Algorithm of the ICDC Scheme

(1) The PID-DC scheme as a foundation

The whole ICDC scheme consists of two components, namely, a PID decoupling controller (PID-DC) component and a neuroendocrine decoupling component. The PID-DC component is responsible for the basic regulation which also acts as a foundation of the following neuroendocrine decoupling component.

The control plant shown in Fig. 2.12 can be written as follow.

$$\begin{bmatrix} H(s) \\ T(s) \\ D(s) \end{bmatrix} = \frac{1}{Ss+\frac{1}{R}} \begin{bmatrix} e^{-\tau_1 s} & e^{-\tau_2 s} & e^{-\tau_3 s} \\ \frac{T_{DMSO}-T_0}{h_0} e^{-\tau_1 s} & \frac{T_{H_2O,H}-T_0}{h_0} e^{-\tau_2 s} & \frac{T_{H_2O,L}-T_0}{h_0} e^{-\tau_3 s} \\ \frac{D_{DMSO}-D_0}{h_0} e^{-\tau_3 s} & \frac{D_{H_2O,H}-D_0}{h_0} e & \frac{T_{H_2O,L}-D_0}{h_0} e^{-\tau_3 s} \end{bmatrix} \begin{bmatrix} U_H(s) \\ U_T(s) \\ U_D(s) \end{bmatrix} \tag{2.42}$$

where the H, T, D are the control outputs, a_{ij} are the corresponding coefficients, U_H, U_D and U_T are the control signals generated by three PID controllers of the PID-DC scheme.

For this plant shown in Eq. (2.42), a state space transformation is applied so that the interrelationship among variables can be expressed by their internal states and therefore observed clearly. Take the first vector H in Eq. (2.42) as an example, it can be achieved by the expression

$$H = H_H + H_T + H_D = (a_{11}U_H + a_{12}U_T + a_{13}U_D) \tag{2.43}$$

where H_H, H_T and H_B are the components of H contributed by the three control signals U_H, U_D and U_T, respectively. It can be observed that the coefficients a_{11}, a_{12} and a_{13} all have the form of the Laplace s-formula, compared with Eq. (2.42). Take the first component H_H for further transformation and its state space expression can be written as

$$\begin{cases} \dot{V}_{11} = A_{11}V_{11} + B_{11}U_H \\ H_H = C_{11}V_{11} \end{cases} \tag{2.44}$$

where V_{11} is the state matrix selected in the subsystem represented by Eq. (2.42), A_{11} is the state matrix, B_{11} is the input matrix, and C_{11} is the output matrix. For a practical industrial system with discrete sampling, Eq. (2.44) can be written as

$$\begin{cases} \Delta V_{11,\mathrm{k}} = F_{11}V_{11,k-1} + G_{11}U_{H,k} \\ H_{H,k} = E_{11} + V_{11,k} \\ \quad = E_{11}(V_{11,k-1} + \Delta V_{11,k}) \\ \quad = H_{H,k-1} + E_{11}\Delta V_{11,k}, \; k = 1, 2, \ldots \end{cases} \tag{2.45}$$

where $V_{11,k}$ is the state matrix at the k-th sampling period and $\Delta V_{11,k}$ is its increment. F_{11}, G_{11} and E_{11} are the discrete forms of the matrices A_{11}, B_{11} and C_{11}, respectively. $U_{H,k}$ is the control signal at the k-th sampling period and $H_{H,k}$ is the corresponding output. It is clear that the value of H_H at every sampling period, $H_{H,k}$, can be calculated according to Eq. (2.45) as

$$H_{H,k} = H_{H,k-1} + E_{11}(F_{11}V_{11,k-1} + G_{11}U_{H,k}) \tag{2.46}$$

Note that $U_{H,k}$ is the dynamic output of the PID controller equipped for regulating the variable H in the PID-DC scheme. The value of the variable $H_{H,k}$ can be acquired in Eq. (2.46). Similarly, the other two components of H can also be acquired with the approach above and Eq. (2.43) at the k-th sampling period can be rewritten as

$$\begin{aligned} H_k &= H_{H,k} + H_{T,k} + H_{D,k} \\ &= [H_{H,k-1} + E_{11}(F_{11}V_{11,k-1} + G_{11}U_{H,k})] \\ &\quad + [H_{T,k-1} + E_{12}(F_{12}V_{12,k-1} + G_{12}U_{T,k})] \\ &\quad + [H_{D,k-1} + E_{13}(F_{13}V_{13,k-1} + G_{13}U_{D,k})] \end{aligned} \tag{2.47}$$

where E, F and G are the same as those of their ancestors before transformation as shown in Eq. (2.43). The dynamic value of the other two variables in the system represented by Eq. (2.42) can also be calculated with the procedures above. With the PID scheme, the control signals generated by the controllers for the liquid height, temperature and the density at the $k + 1$-th sampling period can therefore be generated as

$$\begin{cases} U_{H,k+1} = PID(\Delta H_{k+1}, k+1) \\ U_{T,k+1} = PID(\Delta T_{k+1}, k+1) \\ U_{D,k+1} = PID(\Delta D_{k+1}, k+1) \end{cases} \tag{2.48}$$

where

$$\begin{cases} \Delta H_{k+1} = H_k - H_{k-1} \\ \Delta T_{k+1} = T_k - T_{k-1} \\ \Delta D_{k+1} = D_k - D_{k-1} \end{cases} \tag{2.49}$$

and

$$\begin{cases} PID(\Delta H_{k+1}, k+1) = K_{P,H} \times (\Delta H_{k+1} - \Delta H_k) \\ \qquad + K_{I,H} \times \Delta H_{k+1} \\ \qquad + K_{D,H} \times (\Delta H_{k+1} - 2\Delta H_k + \Delta H_{k-1}) \\ PID(\Delta T_{k+1}, k+1) = K_{P,T} \times (\Delta T_{k+1} - \Delta T_k) \\ \qquad + K_{I,T} \times \Delta T_{k+1} \\ \qquad + K_{D,T} \times (\Delta T_{k+1} - 2\Delta T_k + \Delta T_{k-1}) \\ PID(\Delta D_{k+1}, k+1) = K_{P,D} \times (\Delta D_{k+1} - \Delta D_k) \\ \qquad + K_{I,D} \times \Delta D_{k+1} \\ \qquad + K_{D,D} \times (\Delta D_{k+1} - 2\Delta D_k + \Delta D_{k-1}) \end{cases} \tag{2.50}$$

Equation (2.50) can therefore be applied to the plant at the next period. As such, an initial closed-loop control model is established with the PID-DC scheme. This scheme can be used as an instance for comparison with the one embedded with the proposed ICDC scheme which will be discussed next.

(2) The cooperative decoupling in the ICDC

Suppose T, D and H are the variables of the temperature, the concentration, and the liquid-level of a PANCF coagulation respectively, and the coagulation bath model with a MIMO linear expression in Eq. (2.42) under control can be written as

$$\begin{bmatrix} \dot{T} \\ \dot{D} \\ \dot{H} \end{bmatrix} = \begin{bmatrix} a_{11} & a_{12} & a_{13} \\ a_{21} & a_{22} & a_{23} \\ a_{31} & a_{32} & a_{33} \end{bmatrix} \begin{bmatrix} T \\ D \\ H \end{bmatrix} + \begin{bmatrix} b_{11} & b_{12} & b_{13} \\ b_{21} & b_{22} & b_{23} \\ b_{31} & b_{32} & b_{33} \end{bmatrix} \begin{bmatrix} u_1 \\ u_2 \\ u_3 \end{bmatrix} = A \begin{bmatrix} T \\ D \\ H \end{bmatrix} + B \begin{bmatrix} u_1 \\ u_2 \\ u_3 \end{bmatrix} \tag{2.51}$$

where A, B are the matrices of linear coefficients representing by $a_{ij}, b_{ij}(i = 1, 2, 3; j = 1, 2, 3)$, $u_i(i = 1, 2, 3)$ are the output values of control units for each variable, respectively. There are three variables coupling with each other in this model, so the control signal for each variable is therefore composed by three different parts coming from three different control units.

For a dynamic model of the bath, suppose the set point of the temperature is changed, and the corresponding control unit, namely, the CDU for temperature should be regulated so that the actual temperature follows the change of the set point. According to the principle of the controller above, the control signal sent from the CDU for temperature will be also sent to the other two controllers, so the actual signal that each CDU sends to its controlled variable should consist of three parts. One part is the signal received from its own controller and the other two from the other controllers. The combination of the control signals is as follows

$$\begin{cases} u_1 = u_{11} + u_{21} + u_{31} \\ u_2 = u_{12} + u_{22} + u_{32} \\ u_3 = u_{13} + u_{23} + u_{33} \end{cases} \tag{2.52}$$

where $u_{ij}(i = 1, 2, 3; j = 1, 2, 3)$ is the control signals distributed from the i-th CDU that controls one of the three variables (the temperature, the concentration, and the liquid-level) to the j-th CDU, respectively.

On the other hand, the change of temperature should not affect the liquid-level and the concentration of the coagulation bath, or the effects should be kept within a strict range. So the following requirements should be met

$$\begin{cases} \dot{D} = 0 \\ \dot{H} = 0 \end{cases} \tag{2.53}$$

namely,

$$\begin{cases} a_{21}T + a_{22}D + a_{23}H + b_{21}u_1 + b_{22}u_2 + b_{23}u_3 = 0 \\ a_{31}T + a_{32}D + a_{33}H + b_{31}u_1 + b_{32}u_2 + b_{33}u_3 = 0 \end{cases} \tag{2.54}$$

Since the liquid-level and the concentration of the bath are supposed to be kept unchanged, $a_{22}D + a_{23}H$ and $a_{32}D + a_{33}H$ can be regarded as constant, so

$$\begin{cases} a_{21}T + b_{21}u_1 + b_{22}u_2 + b_{23}u_3 = -(a_{22}D + a_{23}H) = C_1 \\ a_{31}T + b_{31}u_1 + b_{32}u_2 + b_{33}u_3 = -(a_{32}D + a_{33}H) = C_2 \end{cases} \tag{2.55}$$

From Eqs. (2.54) and (2.55), it could be acquired that

$$\begin{cases} a_{21}T + b_{21}(u_{11} + u_{21} + u_{31}) + b_{22}(u_{12} + u_{22} + u_{32}) + b_{23}(u_{13} + u_{23} + u_{33}) = C_1 \\ a_{31}T + b_{31}(u_{11} + u_{21} + u_{31}) + b_{32}(u_{12} + u_{22} + u_{32}) + b_{33}(u_{13} + u_{23} + u_{33}) = C_2 \end{cases} \tag{2.56}$$

All the discussions above are based on a continuous system. For a practical system, these control models should be discrete, and the increments of components in Eq. (2.56) should be considered in replace of their full values as below

$$\begin{cases} a_{21}\Delta T + b_{21}\Delta u_{11} + b_{22}\Delta u_{12} + b_{23}\Delta u_{13} = 0 \\ a_{31}\Delta T + b_{31}\Delta u_{11} + b_{32}\Delta u_{12} + b_{33}\Delta u_{13} = 0 \end{cases} \tag{2.57}$$

which can be solved according to the Cramer principle. The values of Δu_{12} and Δu_{13} are the control signals that distribute by the CDU for temperature to the CDUs for liquid-level and concentration to satisfy the requirement that the liquid-level and the concentration should be kept constant. With $\Delta T = \Delta e_T(k) = e_T(k) - e_T(k-1)$, Eq. (2.57) can be solved as

$$\begin{cases} \Delta u_{12} = \frac{a_{31}b_{23} - a_{21}b_{33}}{b_{22}b_{33} - b_{23}b_{32}}\Delta e_T(k) + \frac{b_{23}b_{31} - b_{21}b_{33}}{b_{22}b_{33} - b_{23}b_{32}}\Delta u_{11} \\ \Delta u_{13} = \frac{a_{21}b_{32} - a_{31}b_{22}}{b_{22}b_{33} - b_{23}b_{32}}\Delta e_T(k) + \frac{b_{21}b_{32} - b_{22}b_{31}}{b_{22}b_{33} - b_{23}b_{32}}\Delta u_{11} \end{cases} \tag{2.58}$$

which is the expression of the decoupling signals transferred by the CDU for temperature to the CDUs for liquid-level and concentration. To make the structure of Eq. (2.58) more explicit, the polynomial form of it can be changed to a matrix form as follows

$$\begin{cases} \Delta u_{12} = \dfrac{-\begin{vmatrix} a_{21} & b_{23} \\ a_{31} & b_{33} \end{vmatrix}}{\begin{vmatrix} b_{22} & b_{23} \\ b_{32} & b_{33} \end{vmatrix}}\Delta e_T(k) + \dfrac{-\begin{vmatrix} b_{21} & b_{23} \\ b_{31} & b_{33} \end{vmatrix}}{\begin{vmatrix} b_{22} & b_{23} \\ b_{32} & b_{33} \end{vmatrix}}\Delta u_{11} \\ \Delta u_{13} = \dfrac{-\begin{vmatrix} a_{21} & b_{22} \\ a_{31} & b_{32} \end{vmatrix}}{\begin{vmatrix} b_{22} & b_{23} \\ b_{32} & b_{33} \end{vmatrix}}\Delta e_T(k) + \dfrac{-\begin{vmatrix} b_{21} & b_{22} \\ b_{31} & b_{32} \end{vmatrix}}{\begin{vmatrix} b_{22} & b_{23} \\ b_{32} & b_{33} \end{vmatrix}}\Delta u_{11} \end{cases} \tag{2.59}$$

Note that each denominator in Eq. (2.59) is identical and let $K_1 = -\begin{vmatrix} b_{22} & b_{23} \\ b_{32} & b_{33} \end{vmatrix}$

Eq. (2.59) can be rewritten as

$$\begin{cases} \Delta u_{12} = \dfrac{\begin{vmatrix} a_{21} & b_{23} \\ a_{31} & b_{33} \end{vmatrix}}{K_1} \Delta e_T(k) + \dfrac{\begin{vmatrix} b_{21} & b_{23} \\ b_{31} & b_{33} \end{vmatrix}}{K_1} \Delta u_{11} \\ \Delta u_{13} = \dfrac{\begin{vmatrix} a_{21} & b_{22} \\ a_{31} & b_{32} \end{vmatrix}}{K_1} \Delta e_T(k) + \dfrac{\begin{vmatrix} b_{21} & b_{22} \\ b_{31} & b_{32} \end{vmatrix}}{K_1} \Delta u_{11} \end{cases} \quad (2.60)$$

Take the system model in Eq. (2.43) into consideration, it is concluded that the expressions of the incremental decoupling signals can be achieved only by making some modifications on certain rows and columns in the system model without complicated calculations. So the incremental decoupling signals distributed by the CDU for concentration to the other CDUs with the changing of concentration D can be written as

$$\begin{cases} \Delta u_{21} = \dfrac{\begin{vmatrix} a_{12} & b_{13} \\ a_{32} & b_{33} \end{vmatrix}}{K_2} \Delta e_D(k) + \dfrac{\begin{vmatrix} b_{12} & b_{13} \\ b_{32} & b_{33} \end{vmatrix}}{K_2} \Delta u_{22} \\ \Delta u_{23} = \dfrac{\begin{vmatrix} a_{12} & b_{11} \\ a_{32} & b_{31} \end{vmatrix}}{K_2} \Delta e_D(k) + \dfrac{\begin{vmatrix} b_{12} & b_{11} \\ b_{32} & b_{31} \end{vmatrix}}{K_2} \Delta u_{22} \end{cases} \quad (2.61)$$

where $K_2 = -\begin{vmatrix} b_{11} & b_{13} \\ b_{31} & b_{33} \end{vmatrix}$. Similarly, the incremental decoupling signals with the changing of liquid-level H can be written as

$$\begin{cases} \Delta u_{31} = \dfrac{\begin{vmatrix} a_{13} & b_{12} \\ a_{23} & b_{22} \end{vmatrix}}{K_3} \Delta e_H(k) + \dfrac{\begin{vmatrix} b_{13} & b_{12} \\ b_{23} & b_{22} \end{vmatrix}}{K_3} \Delta u_{33} \\ \Delta u_{32} = \dfrac{\begin{vmatrix} a_{13} & b_{11} \\ a_{23} & b_{21} \end{vmatrix}}{K_3} \Delta e_H(k) + \dfrac{\begin{vmatrix} b_{13} & b_{11} \\ b_{23} & b_{21} \end{vmatrix}}{K_3} \Delta u_{33} \end{cases} \quad (2.62)$$

where $K_3 = -\begin{vmatrix} b_{11} & b_{12} \\ b_{21} & b_{22} \end{vmatrix}$.

The Eqs. (2.60)–(2.62) are all based on some rank-2 determinants which do not need much calculation, and the corresponding parameters such as $a_{ij}, b_{ij}, K_i (i = 1,\ 2,\ 3; j = 1,\ 2,\ 3)$ are all fixed when the whole production line is established in practice. Note that the procedures above only provide an increment for the control signal generated by the CDU, which is unlike that of the PID-DC that can generate an independent control signal. So the cooperative decoupling algorithm inspired by the neuroendocrine principle follows the former conventional PID-DC part and acts as an enhancement. Compared to the conventional PID-DC scheme, the proposed ICDC scheme takes the decoupling as its primary control objective, and the aim of the cooperative decoupling algorithm is to restrict the coupling effects to minimum, so it is reasonable to evaluate that the proposed ICDC should acquire better performance than the PID-DC. On the other hand, the Eqs. (2.60)–(2.62) are not just independent operations but can be regarded as operators embedded in other ordinary control algorithms to take effect, so the original algorithms considering less on decoupling can be changed to those with intelligent decoupling features.

In some special cases, the discussion above may not be applied properly, e.g. some of the components among K_1, K_2 and K_3 are zero. It means that some variables to be controlled have proportional relations which result in the failure of using the Cramer principle. As a result, the MIMO system degrades to a two-input-two-output (TITO) or even single-input-single-output (SISO) system. For the control system, only the disproportionate variables should be controlled by the cooperative decoupling algorithm referred above, on which the control signals for the other variables can be generated by simply multiplying the corresponding proportional factors.

(3) The overall procedures of the ICDC

The overall procedure of the proposed ICDC is illustrated as the flow diagram in Fig. 2.13. As mentioned above, it consists of two main components, namely, the conventional PID-DC as a foundation and the cooperative decoupling process inspired by the neuroendocrine principle as an enhancement, and a controller parameter switching process is implemented by the CCU before the decoupling module proceeds. The aim of realizing PID-DC in the ICDC is to provide a foundation to the whole controller so that the plant can be tuned to an approximately stable status, and the introduction of the neuroendocrine decoupling module is to make the process with more accuracy and proficiency. A CCU with a parameter reconfiguring scheme makes such decision by collecting the dynamic errors of variables under control and then determining whether the error is large enough so that the parameters of the controller should be reconfigured. A CCU without such a mechanism, however, is also acceptable, which means the function of such CCU is just collecting errors and transferring them to the CDUs. The strategy on how and when the neuroendocrine decoupling algorithm should take effect also depends on the instructions of the CCU, and the dynamic error is still the main criteria for selecting different decoupling strategies. With these procedures, the final control signal can be generated by

$$u_{final} = \begin{cases} u_{conv} + \Delta u_{icdc}, & if \quad e_k \geq e_{threshold} \\ u_{conv}, & if \quad e_k < e_{threshold} \end{cases} \tag{2.63}$$

where u_{final} is the final control signal possibly synthesized by the output of PID-DC component and the cooperative decoupling component, or only that of the PID-DC component. u_{conv} is the output of the conventional PID-DC component, Δu_{icdc} is the output of the cooperative decoupling component (the symbol Δ as a prefix means it is an increment), e_k is the real-time plant error and $e_{threshold}$ is the predefined switching threshold.

With Eq. (2.63) the CCU can make a selection between an individual PID-DC scheme and an integrated PID-DC and cooperative decoupling scheme. As a result, the rapidness and accuracy of the whole control system can be guaranteed by

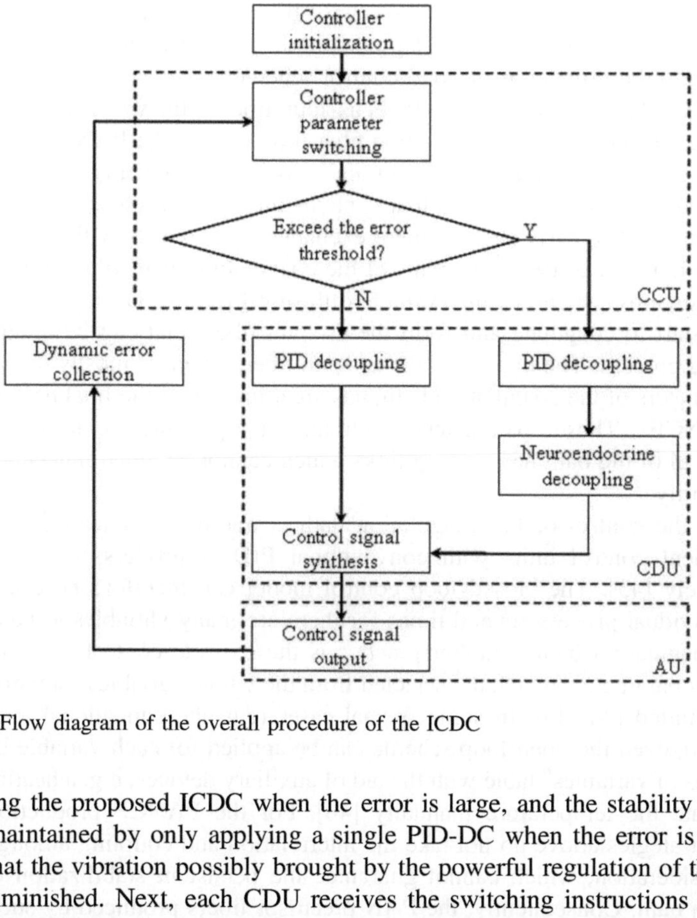

Fig. 2.13 Flow diagram of the overall procedure of the ICDC

introducing the proposed ICDC when the error is large, and the stability of it can also be maintained by only applying a single PID-DC when the error is small or zero so that the vibration possibly brought by the powerful regulation of the ICDC can be diminished. Next, each CDU receives the switching instructions from the CCU and generates the corresponding control signals, then sends it to the plant it takes responsibility for. The dynamic output of the plant are collected by sensors and then transferred to the CCU and each CDU, which guides them to make switching judgment or generate decoupling and control signals.

2.4.3 *Verification and Results*

The manufacture of PANCF includes a series of processes with high complexity such as polymerization, spun extrusion, coagulation, washing, stretching, pre-oxidation, carbonization and post treatment, etc. [38, 39]. The aim for utilizing such processes is to create excellent carbon fibers with high strength and elastic modulus. Nevertheless, these processes with different work conditions are closely connected, and the devices applied for each process have various mechanical and control characteristics [40]. The nature of the raw materials (mainly the PAN) also vary within these processes and thus cannot be treated as stable during the whole production, and the coupling and lag existing in and between processes also bring challenge to the utilization of good control schemes.

The coagulation process of the PAN as-spun fiber in the water bath is one of the key processes during the whole carbon fiber production. The PAN solvent is firstly extruded out of the spinneret at a certain velocity, formed tiny streams and then submerged into a water bath slot that contains the coagulation solution of a certain concentration. The PAN streams there exchange components with the coagulation bath solution, and some components of the coagulation bath solution also infiltrate into the streams which is called double-diffusion [41, 42]. During this period, the streams start to coagulate and form the as-spun fibers that can be dragged to the following process. The quality of the as-spun fiber, which is mostly determined by the parameters of the coagulation bath, has great impact on the final performance of the PANCF. These parameters including temperature, concentration, and liquid-level of the bath have strong links which cannot be tuned independently and satisfactorily.

As to the control of the coagulation bath, a traditional scheme is to establish independent control units with conventional PID controllers for the variables, respectively [43]. The closed loop control model can therefore be established to each individual process several times for there are many variables to be controlled in the coagulation bath. Another practice is the cascade control system for some important parameters which are selected from the whole variables set while keeping others omitted [44, 45]. In some special cases (e.g. the educational and research scenarios), even the open-loop scheme can be applied for each variable by making the values of variables stable with the aid of auxiliary devices, e.g. a heating devices to regulate the temperature manually [46]. For the PANCF production, all the control strategies above do not take the interrelation and coupling among variables into consideration, which cannot guarantee the persistent stabilization of the coagulation bath. Consequently, the PAN precursor fibers produced by such systems can not enjoy a high quality. Another side of such issue is that some schemes or algorithms dealing with plant decoupling have been proposed but not applied to the PAN production process yet due to its special characteristics and complex decoupling relations among variables [47].

2.4.3.1 Modeling of the Coagulation Bath in PANCF Production

The coagulation system in the whole PANCF production line consists of two slots, a bath solution preparation slot and a coagulation slot. The main ingredients in the coagulation solution are dimenthylsulfoxide (DMSO) and water (other trace components can be ignored). The coagulation solution is firstly made in the preparation slot by mixing three different kinds of solutions (including cold water, hot water, and high concentration DMSO solution) coming from three independent ports on the preparation slot, and then transferred to the coagulation slot for formal use. Through such two-step process, the fluctuation in the solution mixing can be eliminated,

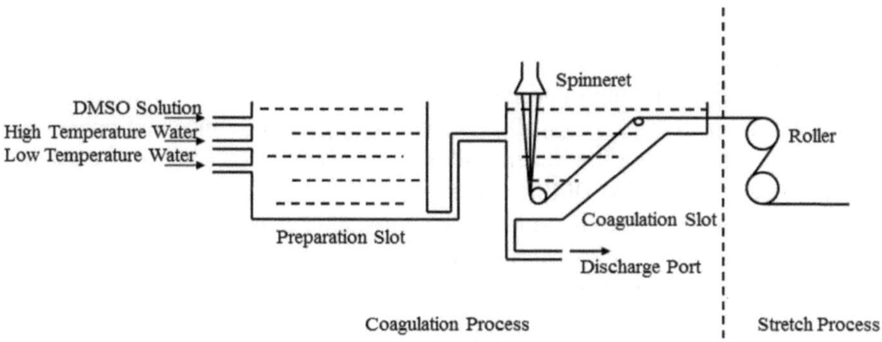

Fig. 2.14 The coagulation bath in PANCF production

which is beneficial to the stability of the coagulation. The shape of the preparation slot is supposed to be a standard cube. The coagulation slot is equipped with a discharge port, from which some solution can be discharged when the volume of the coagulation solution is more than required, and the discharged solution can be recycled for coagulation again. Figure 2.14 shows the main structure of the coagulation bath slot and the ports on it. There are many variables to be controlled in the coagulation bath of the PANCF production, and the liquid-level, the temperature, and the concentration of the bath play essential roles. However, these main variables often take effect together, namely, coupling with each other. It is the main challenge for the design of a control system for the coagulation bath. Meanwhile, other minor parameters may also affect the overall status of the bath. To simplify the problem discussed, three main variables of the coagulation bath referred above are selected as the main targets, and then its mathematic model is established.

(1) Regulation of the liquid-level

The relation between the liquid-level of the coagulation bath and the volume of solution components coming from each entrance port can be written as

$$S\frac{dh}{dt} + K\frac{h}{R} = V_{DMSO} + V_{H_2O,H} + V_{H_2O,L} \tag{2.64}$$

where S denotes the area of the slot bottom; h is the real-time liquid-level; R is the hydraulic resistance of the discharge port; K is the conversion factor (usually fixed to 1); V_{DMSO}, $V_{H_2O,H}$, and $V_{H_2O,L}$ are the volumes of the high concentration DMSO solution, the hot water and the cool water, respectively.

The stable state of the coagulation system can be expressed by the following equation

$$Sh_0\frac{dD}{dt} = V_{DMSO}(D_{DMSO} - D) + V_{H_2O,H}(D_{H_2O,H} - D) + V_{H_2O,L}(D_{H_2O,L} - D) \tag{2.65}$$

where h_0 denotes the liquid-level at the stable state; $V_{0.DMSO}$, $V_{0.H_2O,H}$, and $V_{0.H_2O,L}$ are the volumes of the high concentration DMSO solution, the hot water, and the cool water, respectively; D_{DMSO}, $D_{H_2O,H}$, and $D_{H_2O,L}$ are the concentration values of the DMSO solution, the hot water, and the cool water, respectively (the solute in either kind of water is also DMSO, so $D_{H_2O,H} \approx 0$ and $D_{H_2O,L} \approx 0$). D_0 is the steady value of the concentration of the bath with a dynamic changing value labeled ΔD_0. The Laplace transformation can be utilized to Eq. (2.65) and get the regulation model for the liquid-level of the coagulation bath as follows

$$H(s) = \frac{V_{DMSO}(s) + V_{H_2O,H}(s) + V_{H_2O,L}(s)}{Ss + \frac{K}{R}} \tag{2.66}$$

(2) Regulation of the temperature

Similar to the model for the liquid-level, the relation between the temperature and other variables of the coagulation bath are given by the equation as below

$$Sh_0 \frac{dT}{dt} = V'_{DMSO}(T_{DMSO} - T) + V'_{H_2O,H}(T_{H_2O,H} - T) + V'_{H_2O,L}(T_{H_2O,L} - T) \tag{2.67}$$

where

$$\begin{aligned} V'_{DMSO} &= V_{0.DMSO} + \Delta V_{0.DMSO} \\ V'_{H_2O,H} &= V_{0.H_2O,H} + \Delta V_{0.H_2O,H} \\ V'_{H_2O,L} &= V_{0.H_2O,L} + \Delta V_{0.H_2O,L} \\ T &= T_0 + \Delta T \end{aligned} \tag{2.68}$$

Here T_{DMSO}, $T_{H_2O,H}$, and $T_{H_2O,L}$ are temperatures of the high concentration DMSO solution, the hot water and the cool water, respectively; T_0 is the started-state value of the bath temperature with ΔT_0 as its transient change. As the coagulation bath becomes stable, its liquid-level keeps unchanged, so does the temperature. Here

$$\begin{gathered} \tfrac{h_0}{R} = V_{DMSO} + V_{H2O,H} + V_{H2O,L} \\ V_{0.DMSO}(T_{DMSO} - T_0) + V_{0.H_2O,H}(T_{H_2O} - T_0) + V_{0.H_2O,L}(T_{H_2O} - T_0) = 0 \end{gathered} \tag{2.69}$$

where the specific heat capacities of each kind of solution has been reduced as a common fraction. Take Eqs. (2.68) and (2.69) into consideration and then the regulation model for the temperature of the coagulation bath can be acquired as follows

$$T(s) = \frac{V_{DMSO}(s)(T_{DMSO} - T_0) + V_{H_2O,H}(s)(T_{H2O,H} - T_0) + V_{H_2O,L}(s)(T_{H2O,L} - T_0)}{Sh_0 s + \frac{h_0}{R}} \tag{2.70}$$

(3) Regulation of the concentration

The principle for deducting the temperature model of the coagulation bath can be followed here to achieve the concentration model, in which the main idea is keeping the balance among different solutes. At the steady-state point of the dynamic regulating process for the coagulation bath, the DMSO also becomes stable, namely, the concentration of the bath solution is balanced (with minor fluctuations dynamically). Note that the double-diffusion process between the as-spun fibers and the bath solution keeps on running and can certainly bring extra DMSO. But such process has also reached its balance point when the system becomes stable. So the DMSO coming from the fibers to the bath is as much as that coming from the bath to the as-spun fibers, and the DMSO brought by such process can therefore be ignored. Here

$$V_{0.DMSO}(D_{DMSO} - D_0) + V_{0.H_2O,H}(D_{H_2O,H} - D_0) + V_{0.H_2O,L}(D_{H_2O,L} - D_0) = 0 \tag{2.71}$$

Referring to the deduction for the temperature above, the mathematical model for regulating the concentration of the coagulation bath can be generalized as follows

$$D(s) = \frac{V_{DMSO}(S)(D_{DMSO} - D_0) + V_{H_2O,H}(D_{H_2O,H} - D_0) + V_{H_2O,L}(D_{H_2O,L} - D_0)}{Sh_0 s + \frac{h_0}{R}} \tag{2.72}$$

where D_0 denotes the set point of the concentration determined by the production specification.

(4) Comprehensive regulation model

In the sections above, the three main variables of the coagulation bath, the liquid-level, the temperature, and the concentration, have been described by a series of mathematical models (2.66), (2.70) and (2.72), respectively. So a comprehensive model of the coagulation bath for further regulation may have the form as below

$$\begin{bmatrix} H(s) \\ T(s) \\ D(s) \end{bmatrix} = \frac{1}{Ss + \frac{1}{R}} \begin{bmatrix} 1 & 1 & 1 \\ \frac{T_{DMSO} - T_0}{h_0} & \frac{T_{H_2O,H} - T_0}{h_0} & \frac{T_{H_2O,L} - T_0}{h_0} \\ \frac{D_{DMSO} - D_0}{h_0} & \frac{D_{H_2O,H} - D_0}{h_0} & \frac{D_{H_2O,L} - D_0}{h_0} \end{bmatrix} \begin{bmatrix} U_H(s) \\ U_T(s) \\ U_D(s) \end{bmatrix} \tag{2.73}$$

However, many processes in the practical manufacturing line require extra time to response for the regulations to take effect which results in some modifications on the model above (by adding some lag coefficients), so

$$\begin{bmatrix} H(s) \\ T(s) \\ D(s) \end{bmatrix} = \frac{1}{Ss + \frac{1}{R}} \begin{bmatrix} e^{-\tau_1 s} & e^{-\tau_2 s} & e^{-\tau_3 s} \\ \frac{T_{DMSO} - T_0}{h_0} e^{-\tau_1 s} & \frac{T_{H_2O,H} - T_0}{h_0} e^{-\tau_2 s} & \frac{T_{H_2O,L} - T_0}{h_0} e^{-\tau_3 s} \\ \frac{D_{DMSO} - D_0}{h_0} e^{-\tau_1 s} & \frac{D_{H_2O,H} - D_0}{h_0} e^{-\tau_2 s} & \frac{D_{H_2O,L} - D_0}{h_0} e^{-\tau_3 s} \end{bmatrix} \begin{bmatrix} U_H(s) \\ U_T(s) \\ U_D(s) \end{bmatrix} \tag{2.74}$$

where τ_1, τ_2 and τ_3 are the lag coefficients in the regulation of the liquid-level, the temperature, and the concentration, respectively.

The coagulation bath can be regarded as a MIMO system with the coupling effects of the liquid-level, the temperature, and the solution concentration. According to the principle of the neuroendocrine cooperative regulation in the human body, the ICDC can be designed to meet the requirement of perfectly controlling the bath status and eliminating the coupling among variables.

A coagulation bath model as shown in Fig. 2.14 is applied here to verify the effectiveness of the ICDC above. The coagulation bath contains three entrance ports that are responsible for separately transmitting different solutions (hot/cold water and high concentration DMSO solution) and an exit port that connects directly to the entrance port of the bath slot. The proposed ICDC can be applied to conduct regulation to the bath slot. Here, Eq. (2.42) is extended by Eq. (2.74), and its parameters (the parameters of the preparation slot with the initial conditions of the coagulation system) are list in Table 2.6.

Note that the concentrations of the water, both hot and cold, are set to 0.001 instead of pure zero due to the actual circumstance that the DMSO exists as one kind of impurities in the water.

The control model for the coagulation bath with the actual system parameters in Table 1 substituted can be written as

Table 2.6 Slot Parameters and initial conditions of the coagulation system

Slot parameters		Initial conditions	
Parameter	Value	Parameter	Value
S/m^2	0.15	$T_{DMSO}/°\text{C}$	30
R	0.1	$T_{H_2O,H}/°\text{C}$	65
h_0/m	1	$T_{H_2O,L}/°\text{C}$	10
τ_1/s	1	$T_0/°\text{C}$	15
τ_2/s	2	$D_{DMSO}/\%$	80
τ_3/s	5	$D_{H_2O,H}/\%$	0.001
		$D_{H_2O,L}/\%$	0.001
		$D_0/\%$	0.65

$$\begin{bmatrix} H(s) \\ T(s) \\ D(s) \end{bmatrix} = \frac{1}{0.15s+1} \begin{bmatrix} e^{-s} & e^{-2s} & e^{-5s} \\ 150e^{-s} & 500e^{-2s} & -50e^{-5s} \\ 1.5e^{-s} & -6.49e^{-5s} & -6.49e^{-5s} \end{bmatrix} \begin{bmatrix} U_H(s) \\ U_T(s) \\ U_D(s) \end{bmatrix} \quad (2.74)$$

where the lag coefficients of the temperature, the liquid-level, and the concentration are 1 s, 2 s and 5 s, respectively.

The ICDC is then applied to this target with conventional PID algorithm as the main control scheme of each CDU, but the parameters of the PID algorithm are fixed or dynamically changed according to the instructions of the CCU, and these two plans with their performances are simulated separately in the following experiments. The initial set point for the temperature is 15 °C. For the liquid-level and concentration, the set points are 0.1 m and 65%, respectively. A conventional PID-DC scheme mentioned in Sect. 2.3 for the coagulation bath is applied for comparison, and the initial PID parameters and the error thresholds for scheme switching applied in both the PID-DC and the ICDC schemes are listed in Table 2.7 (Both the schemes include three controllers, so each of them has three sets of PID parameters.). All the simulation processes are carried on with MATLAB R2009a on a workstation computer with a dual-core processor and 2 GB memory.

As the coagulation bath is a complex coupling object, the simulations should reach some requirements as follows.

(1) The other two variables should not be influenced when one variable is regulated independently.
(2) Each variable should not be influenced by other variables when these variables are regulated together.

Table 2.7 Initial PID parameters in the PID-DC and the ICDC schemes

	PID-DC	ICDC	$e_{threshold}$
Liquid-level			
P	0.6	0.6	$10\% \times H_{ref}$ [a]
I	5×10^{-2}	5×10^{-2}	
D	5×10^{-2}	5×10^{-2}	
Temperature			
P	10^{-3}	10^{-3}	$10\% \times T_{ref}$ [b]
I	5×10^{-4}	5×10^{-4}	
D	0	0	
Concentration			
P	5×10^{-2}	5×10^{-2}	$10\% \times D_{ref}$ [c]
I	10^{-2}	10^{-2}	
D	0	0	

Simulation with Fixed Parameters

To reach the goals below, a series of simulations consisting of two phases are designed to demonstrate the performance of the proposed ICDC, and the parameters for the controllers in both the PID-DC and the ICDC are fixed, as shown in Table 2.7.

a, *b*, *c*: The symbol H_{ref}, H_{ref} and H_{ref} are the references of the liquid-level, the temperature and the concentration, respectively. Note that there are several possible reference values for a single variant for simulation.

Phase I: Individual Regulation

Given that the total simulation time occupies 100 sampling periods (here one sampling period in simulation can be regarded as one second in practice, and a same configuration is also applied in the Phase II below). The system is firstly regulated from all-zero to a stable status with the initial set points of the three variables, then the set points of the liquid-level, the temperature and the concentration are changed to another value at the 20-th, 40-th and 60-th sampling period, respectively. That is to say, the simulation of this phase should be carried out three times and the variables should be changed each time one by one, while the others keep unchanged. The set point of temperature is supposed to change to 30 °C, and for the liquid-level and concentration the corresponding values are 0.2 m and 75%. The purpose of Phase I is to verify the effectiveness of the proposed ICDC against the variation occurred on each individual variable and the ability of the unaffected variables for not being disturbed by the deliberately changed variable.

Phase II: Integrated Regulation

Given the total simulation time occupies 100 sampling periods. As in Phase I, the system is also stable when the regulation begins. Then the set points of the liquid-level, the temperature and the concentration are changed to another value at the 20-th, 40-th and 60-th sampling period, respectively. At the 20-th sampling period, the set point of the liquid-level is firstly changed to 0.2 m, then followed by the changes for the temperature (to 30 °C) and the concentration (to 75%) at the 40-th and the 60-th sampling period. These steps in this phase are to verify the ability of the coagulation system with the proposed ICDC for dealing with concurrent regulations of all the variables.

The simulations above are carried out by applying large modifications to the set points of variables so that the effects of the proposed ICDC can be shown explicitly. However, such large changes cannot be tolerated in the practical production process. On most occasions, the status of the coagulation bath may only encounter tiny fluctuations, so the controller applied should be less sensitive to these fluctuations to maintain the overall state of the coagulation bath stable. Thus, the simulation for this condition should also be conducted to emulate the practical production. At the last part of the simulation, some small modifications are made to the set point of each variable, namely, changing the liquid-level from 0.1 m up to 0.12 m, the temperature from 15 °C up to 17 °C, and the concentration from 65% down to 60%.

At the 80-th sampling period, all the variables are tuned to their origin values, namely, 0.1 m for the liquid-level, 15 °C for the temperature, and 65% for the concentration. The aim of doing so is to verify whether the system has the ability to return to its normal working status after a period of fluctuation.

Figure 2.15 shows the comparison between the PID-DC and the ICDC when an individual variable is regulated while the others are kept stable (Phase I). Figure 2.16 shows the performance of the two types of controllers for the integrated regulation (Phase II) and Fig. 2.17 is the performance of the ICDC against the PID-DC with practical modifications. Table 2.8 shows comparison of the performance between the ICDC and the PID-DC. The percent overshoot and the settling time at the beginning of the whole control process are calculated to evaluate the swiftness of the two types of decoupling controllers. The overshoot and settling time when the disturbance (means the set point modifications here) occurs are also collected to evaluate the robustness of the controllers. The simulation results show that the PID-DC is very sensitive toward the changes of the set points. The value of one variable to be controlled fluctuates much when another variable is changing, which means the decoupling between the two variables cannot be eliminated satisfactorily. Although the variables controlled by the PID-DC can finally reach to their set points, the regulating time elapsed is too long to be accepted in practice. Such phenomenon does not appear when the proposed ICDC is applied. The fluctuation of one variable can be restricted to a small scale when another variable is modified. Besides, the ICDC can guarantee the smooth regulation without overshoot, which is essential to the PANCF production process with high precision. That's because the parameters of the ICDC change dynamically according to the instruction of the CCU. Such mechanism can help the status of the product line switch from one state to another one smoothly without much fluctuation, which is benefit to reduce the amount of substandard PANCF.

2.4.3.2 Simulation with Variable Parameters

The simulation with variable controller parameters is conducted here to test the function of the CCU and its performance in the ICDC. Two schemes are available for the CCU, namely, the Scheme A for selecting controller parameters from a predefined set with multiple parameter configurations and the Scheme B for making them fluctuating with the dynamic response according to some rules. For Scheme A, three groups of controller parameters are predefined and take actions according to the dynamic error intensity. Table 2.9 provides the parameters for each level of controller parameters. For Scheme B, an original group of parameters is provided to the controllers, which can fluctuate with the dynamic error intensity by a certain percentage, and the details of such scheme are provided in Table 2.10. The switching strategies for the parameter tuning are determined according to the nature of the coagulation bath as the plant, and the changes on parameters, if necessary, should be built on a smooth and incremental basis. The conditions fortuning the parameters, namely, the ranges of error in Table 2.10 are determined by the

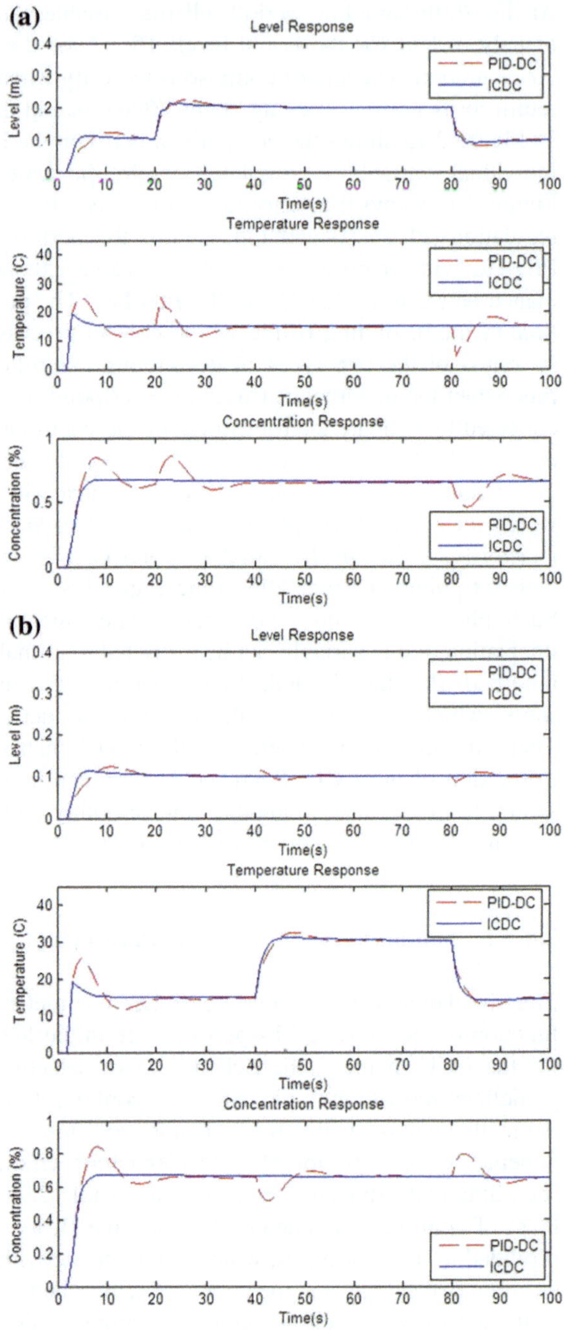

Fig. 2.15 Performance of the PID-DC and the ICDC for individual regulation of variables. **a** Effect of liquid level change on coagulation bath system. **b** Effect of temperature change on coagulation bath system. **c** Effect of concentration change on coagulation bath system

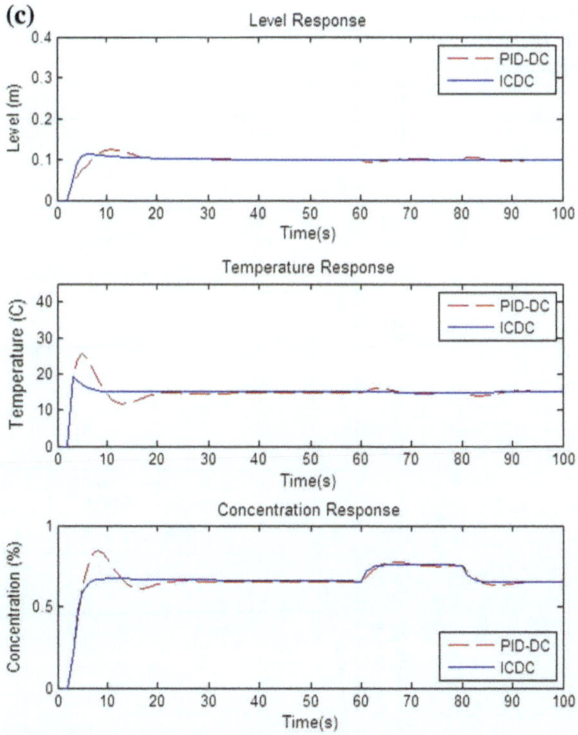

Fig. 2.15 (continued)

experimental simulations. It should be noted that all the parameters cannot be set below zero or it would go against the laws for the physical devices. The simulation procedures under the practical circumstance mentioned in Sect. 4.1 are carried out, and the system responses and the corresponding switching procedures of controller parameters are then recorded. The PID-DC is also taken as a comparison and its parameters are changed according to the rules listed in Table 2.10. Figures 2.18 and 2.19 are the responses of the coagulation bath model with the PID-DC and the ICDC enhanced by Scheme A and Scheme B, respectively. Figures 2.20 and 2.21 represent the switching processes of the controller sets (for Scheme A) or parameters (for Scheme B), respectively. For the PID-DC with Scheme A, the responses of liquid level, temperature and concentration all fluctuate at the beginning as the results of rapid changing of parameter sets, which can be observed in Fig. 2.18. The regulating time is therefore extended accordingly. Similar phenomenon can also be noticed in the system with the ICDC, but only happens during the regulation of temperature. Taken the parameters shown in Table 2.8 into consideration, it can be deducted that a selection of smaller parameters applied to the controller of temperature are more sensitive against changes, and the target variables, consequently, would vibrate more than those with larger parameters. It also can be observed that

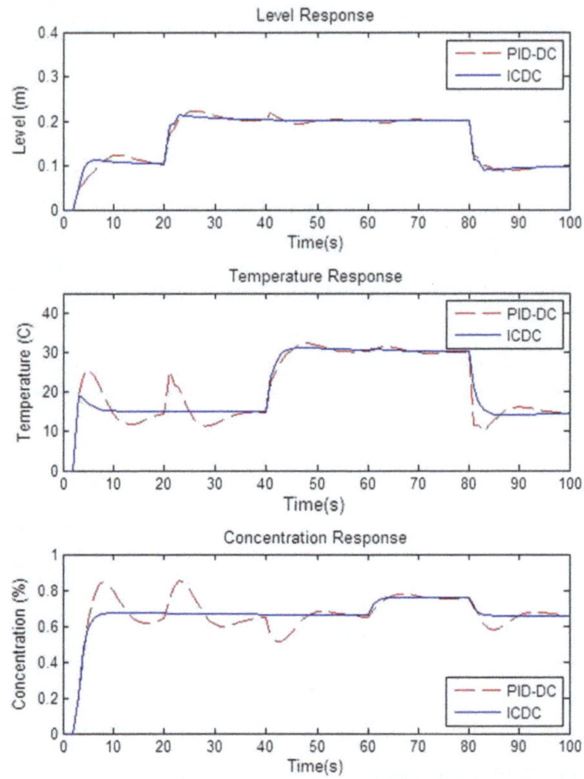

Fig. 2.16 Performance of the PID-DC and the ICDC for comprehensive regulation of variables

the switching process of the ICDC with Scheme A is simpler than that of the PID-DC which means the enhancement brought by the neuroendocrine-inspired decoupling scheme has the ability to compensate the performance for a controller that may not be tuned well by other schemes. For the control system with Scheme B, the corresponding responses are much more like those with the fixed parameters, as shown in Fig. 2.17, and the fluctuations are smaller than those with Scheme A-embedded plans. But the overshoot of some variables, e.g. the temperature and concentration, can still not be neglected (e.g., the value of temperature at 4 s in Fig. 2.19 is over 20, while its correspondence in Fig. 2.17 is not more than 20), and the changes of the references may also lead to minor vibrations. Such simulation results demonstrate that the ICDC with dynamic parameters has better performance than the conventional PID-DC schemes.

$$a : e_{percent} = \frac{(actual\,output - reference)}{reference}.$$

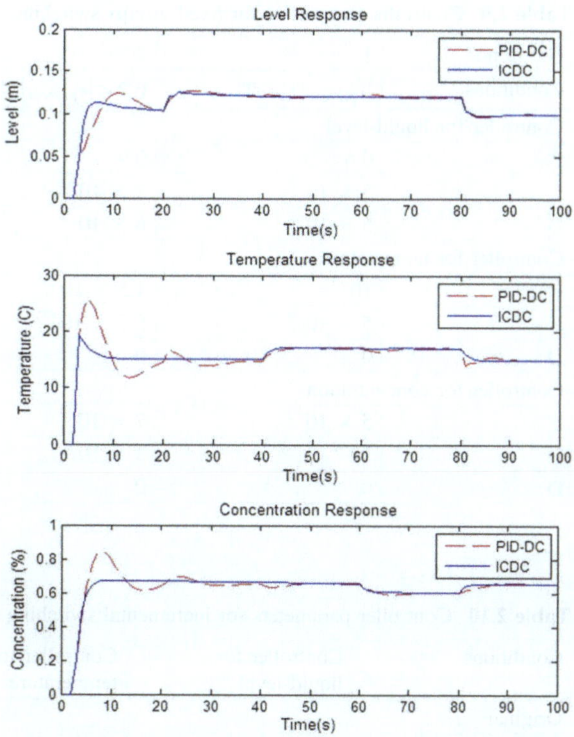

Fig. 2.17 Performance of the PID-DC and the ICDC for comprehensive regulation of variables

Table 2.8 Numerical performance of the ICDC against the PID-DC with practical parameters applied

	Liquid-level		Temperature		Concentration	
	PID-DC	ICDC	PID-DC	ICDC	PID-DC	ICDC
Initial overshoot/%	39.2	9.8	78.2	38.5	35.1	0
Settling time/s	27	15	31	7	27	6
Disturbance overshoot/%	1.9	0	13.7	0	10.4	0
Recover time/s	15	0	14	0	22	0

2.5 Conclusions

In this chapter, we presented the bio-intelligent cooperative decoupling controller BDC, IBCDC and ICDC for complex coupling systems. The BDC is based on the bi-regulation principle of the GH, the IBCDC is inspired from the modulation mechanism of internal environment in body, and the ICDC is inspired by the neuroendocrine regulation principle of human body.

Table 2.9 Controller parameters for fixed groups switching

Group no.	1	2	3
Conditions	$\lvert e_{percent} \rvert \leq 0.2$[a]	$0.2 < \lvert e_{percent} \rvert \leq 0.5$	$\lvert e_{percent} \rvert > 0.5$
Controller for liquid-level			
P	0.6	0.9	1.1
I	5×10^{-2}	7×10^{-2}	9×10^{-2}
D	5×10^{-2}	6×10^{-2}	7×10^{-2}
Controller for temperature			
P	10^{-3}	1.5×10^{-3}	2×10^{-3}
I	5×10^{-4}	5.5×10^{-5}	6×10^{-5}
D	0	0	0
Controller for concentration			
P	5×10^{-2}	7×10^{-2}	9×10^{-2}
I	10^{-2}	3×10^{-2}	5×10^{-2}
D	0	0	0

Table 2.10 Controller parameters for incremental switching

Conditions	Controller for liquid-level	Controller for temperature	Controller for concentration
Original			
P	0.6	10^{-3}	5×10^{-2}
I	5×10^{-2}	5×10^{-4}	10^{-2}
D	5×10^{-2}	0	0
$\lvert e_{percent} \rvert \leq 0.2$			
P	No operation		
I			
D			
$0.2 < \lvert e_{percent} \rvert \leq 0.5$			
P	Decrease by 5% based on the values of the last sampling period		
I			
D			
$\lvert e_{percent} \rvert > 0.5$			
P	Increase by 5% based on the values of the last sampling period		
I			
D			

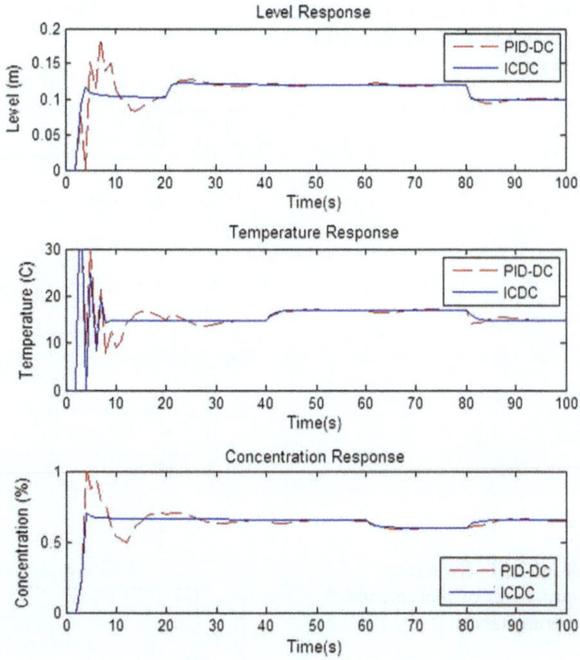

Fig. 2.18 Responses with fixed parameter groups switching

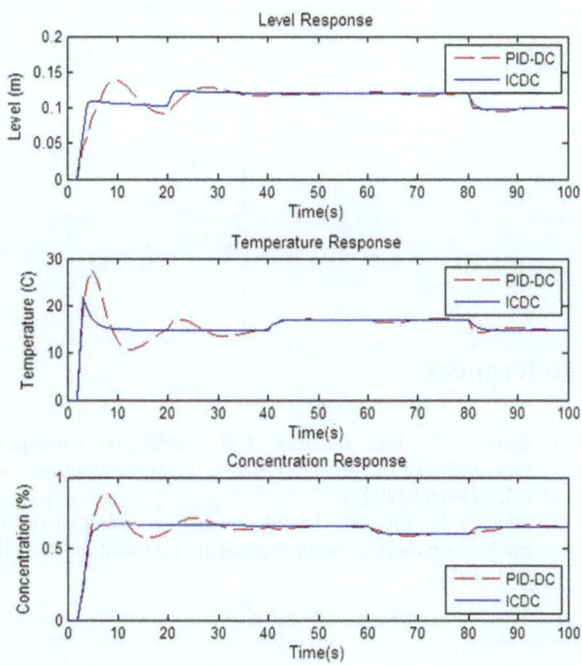

Fig. 2.19 Responses with parameter incremental switching

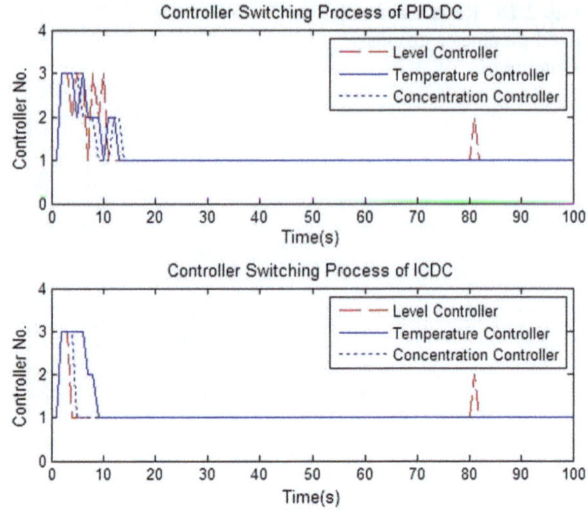

Fig. 2.20 Switching processes of fixed parameter groups

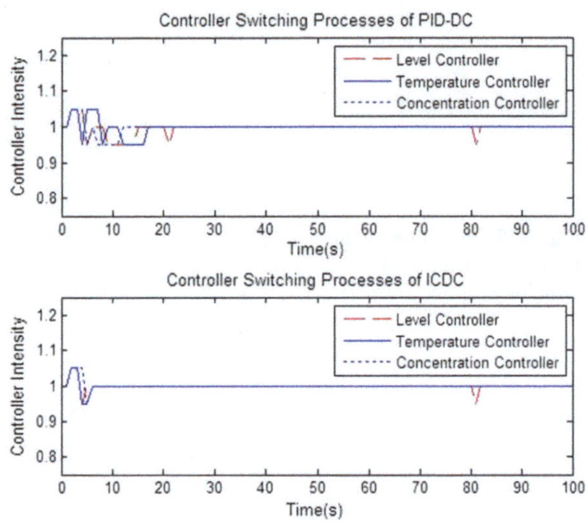

Fig. 2.21 Switching processes of incremental parameters

References

1. Ding, Y.S., Liu, B., Ren, L.H.: Intelligent decoupling control system inspired from modulation of the growth hormone in neuroendocrine system. Dynam. Cont. Dis. Ser. B. **14** (5), 679–693 (2007)
2. Ding, Y.S., Liu, B.: An intelligent bi-cooperative decoupling control approach based on modulation mechanism of internal environment in body. IEEE. T. Contr. Syst. T. **19**(3), 692–698 (2011)

3. Ding, Y.S., Liang, X., Hao, K.R., Wang, H.P.: An intelligent cooperative decoupling controller for coagulation bath in polyacrylonitrile carbon fiber production. IEEE. T. Contr. Syst. T. **21**(2), 467–479 (2013)
4. Zheng, J.C., Guo, G.X., Wang, Y.Y: Feedforward decoupling control design for dual-actuator system in hard disk drives. IEEE. T. Mag. **40**(4), 2080–2082 (2004)
5. Wu, X.J., Jiang, J.G., Dai, P., Zuo, D.S.: Full digital control and application of high power synchronous motor drive with dual stator winding fed by cycloconverter. Paper present at the 5th International Conference on Power Electronics and Drive Systems (PEDS). **2**, 1194–1199 (2003)
6. Hauksdottir, A.S., Ierapetritou, M.: Simultaneous decoupling and pole placement without canceling invariant zeros, pp. 25–27. Paper present at the Amer. Contr. Conf. Conf. Arlington, USA (2001)
7. Deng, Z.L., Huang, X.R.: Multivariable decoupling pole assignment self-tuning feedforward controller. Paper present at IEE. P.—Contr. Theor. Ap. **138**, 85–88 (1991)
8. Lin, M.C., Mon, Y.J.: Decoupling control by hierarchical fuzzy sliding-mode controller. IEEE. T. Contr. Syst. T. **13**(4), 593–598 (2005)
9. Song, Z., Sukthankar, P., Chen, Y.Q., Gu, J.: Progressive fuzzy fusion control of two coupled inverted pendulum. Paper present at IEEE International Symposium on Computational Intelligence in Robotics and Automation (CIRA). **3**, 1457–1462 (2003)
10. Liu, H.B., Li, S.Y., Chai, T.Y.: Intelligent decoupling control of power plant main steam pressure and power output. Int. J. Elec. Power. **25**, 809–819 (2003)
11. Saravanan, S., Kher, S.: Fuzzy control of multivariable process by modified error decoupling. Isa. T. **41**(4), 437–444 (2002)
12. Lin, S.T.: Tsai, H.C: Impedance control with on-line neural network compensator for dual-arm robots. J. Intell. Robot. Syst. **18**(1), 87–104 (1997)
13. Ma, Z., Jutan, A.: Control of a pressure tank system using a decoupling control algorithm with a neural network adaptive scheme. IEE. P.—Contr. Theor. Ap. **150**(4), 389–400 (2003)
14. Shu, H.L., Guo, X.C.: Decoupling control of multivariable time-varying systems based on PID neural network. Paper present at the 5th Asian Control Conference (ACC), Melbourne, Australia. 20–23 (2004)
15. Warwick, K., Zhu, Q.M., Ma, Z.: A hyperstable neural network for the modeling and control of nonlinear systems. Sadhana. **25**(2), 169–180 (2000)
16. Zhang, J., Chung, H.S.H., Lo, W.L., et al.: Implementation of a decoupled optimization technique for design of switching regulators using genetic algorithms. IEEE. T. Power. Electr. **16**(6), 752–763 (2001)
17. Hazzab, A., Bousserhane, I.K., Kamli, M.: Design of a fuzzy sliding mode controller by genetic algorithms for induction machine speed control. Int. J. Emerg. Electr. Power Syst. **1** (2), 1–17 (2004)
18. Hou, Y.B.: A decoupling control method with improving genetic algorithm. Paper present at 2002 International Conference on Machine Learning and Cybernetics (ICMLC). **4**, 2112–2115 (2002)
19. Quan, Y., Yang, J.: Optimal decoupling control system using kernel method. J. Syst. Eng. Electron. **15**(3), 364–370 (2004)
20. Farhy, L.S., Straume, M., Johnson, M.L., Kovatchev, B., Veldhuis, J.D.: A con- struct of interactive feedback control of the GH axis in the male. Am. J. Physiol. - Reg. I. **281**(1), 38–51 (2001)
21. Keenan, D.M., Licinio, J., Veldhuis, J.D.: A feedback-controlled ensemble model of the stress-responsive hypothalamo-pituitary-adrenal axis. P. Natl. Acad. Sci. **98**(7), 4028–4033 (2001)
22. S. Mitra, S., Hayashi, Y.: Bioinformatics with soft computing. IEEE. T. Syst. Man. Cy. C. **36** (5), 616–635 (2006)
23. Fu, Y., Chai, T.Y.: Intelligent decoupling control of nonlinear multivariable systems and its application to a wind tunnel system. IEEE.T. Contr. Syst. T. **17**(6), 1376–1384 (2009)

24. Liang, J., Wang, Z., Liu, Y., Liu, X.: Robust synchronization of an array of coupled stochastic discrete-time delayed neural networks. IEEE. T. Neural. Networ. **19**(11), 1910–1921 (2008)
25. Chunghtai, S.S., Wang, H.: A high-integrity multivariable robust control with application to a process rig. IEEE. T. Contr. Syst. T. **15**(4), 775–785 (2007)
26. Liu, B., Ding, Y.S.: A novel intelligent controller inspired from the regulation mechanism of testosterone. J. Shanghai Jiaotong University (in press)
27. Liu, B., Ren, L.H., Ding, Y.S.: A novel intelligent controller based on modulation of neuroendocrine system. International Symposium on Neural Networks, vol.3, 119–124. Springer, Berlin Heidelberg (2005)
28. Dan, G., Lall, S.B.: Neuroendocrine modulation of immune system. Indian. J. Pharmacol. **30**, 129–140 (1998)
29. Timmis, J., Neal, M.: Artificial homeostasis: integrating biologically inspired computing. http://www.cs.kent.ac.uk/pubs/2003/1586/content.pdf(2004). Access (2004)
30. Keenan, D.M., Licinio, J., Veldhuis, J.D.: A feedback-controlled ensemble model of the stress-responsive hypothalamo-pituitary-adrenal axis. P. Natl. Acad. Sci. USA **98**, 4028–4033 (2001)
31. Farhy, L.S., Straume, M., Johnson, M.L., et al.: A construct of interactive control of the GH axis in the male. Am. J. Physiol. Reg. I. **281**(1), 38–51 (2001)
32. Farhy, L.S.: Modeling of oscillations of endocrine networks with feedback. Method. Enzymol. **384**, 54–81 (2004)
33. Lee, M., Park, S.: Process control using a neural network combined with the conventional PID controllers. T. Contr. Automat. Syst. Eng. **2**(3), 196–200 (2000)
34. Loquasto, F., and Seborg, D.E.: Model predictive controller monitoring based on pattern classification and PCA. Paper present at Amer. Contr. Conf. Conf. **3**, 1968–1973 (2003)
35. Ackermann, U.: Regulation of arterial blood pressure. Surgery. **22**(5), 120a–120f (2004)
36. Campbell, I.: Body temperature and its regulation. Anaesth. Intens. Care. Med. **9**(6), 259–263 (2008)
37. Fanga, B., Kelkara, A.G., Joshib, S.M., Pota, H.R.: Modelling, system identification, and control of acoustic-structure dynamics in 3-Denclosures. Control. Eng. Pract. **12**, 989–1004 (2004)
38. Wang, S., Chen, Z.H., Ma, W.J., Ma, Q.S.: Influence of heat treatment on physical chemical properties of PAN-based carbon fiber. Ceram. Int. **32**(3), 291–295 (2006)
39. Sedghi, A., Farsani, R.E., Shokufar, A.: The effect of commercial polyacrylonitrile fibers characterizations on the produced carbon fibers properties. J. Mater. Process. Tech. **198**(1–3), 60–67 (2008)
40. Tan, L.J., Chen, H.F., Pan, D., Pan, N.: Investigating the spin ability in the dry-jet wet spinning of PAN precursor fiber. J. Appl. Polym. Sci. **110**, 1997–2000 (2008)
41. Ismail, F., Rahman, M.A., Mustafa, A., Matsuura, T.: The effect of processing conditions on a polyacrylonitrile fiber produced using a solvent-free coagulation process. Mater. Sci. Eng., A **485**(1–2), 251–257 (2008)
42. Nain, S., Sitti, M., Jacobson, A., Kowalewski, T., Amon, C.: Dry spinning based spinneret based tunable engineered parameters (STEP) technique for controlled and aligned deposition of polymeric nanofibers. Macromol. Rapid. Comm. **30**, 1406–1412 (2009)
43. Stoker, M., White, A.S.: Mechatronic cine-film copying using transputer control. Mechatronics **10**(7), 773–807 (2000)
44. Karaman, M., Batur, C.: Draw resonance control for polymer fiber spinning process. Paper present at Amer. Contr. Conf. Conf, 2155–2159, (1998)
45. Carrol, J.R., Givens, M.P.: Design elements of the modern spinning control system. Paper present at Textile, Fiber & Film Industry Technical Conference, IEEE, 1–12 (1994)
46. Ghosh, A., Das, S.K.: Open-loop decoupling of MIMO plants. IEEE. T. Automat. Contr. **54** (8), 1977–1981 (2009)
47. Wu, M., Yan, J., She, J.H., Cao, W.H.: Intelligent decoupling control of gas collection process of multiple asymmetric coke ovens. IEEE. T. Ind. Electron. **56**(7), 2782–2792 (2009)

Chapter 3
Bio-Inspired Collaborative Controllers for Multi-Level Systems

In this chapter, three bio-inspired collaborative controls are proposed for multi-level systems. Firstly, a novel bio-inspired multi-layered intelligent cooperative controller (BMLICC) is proposed [1]. Secondly, a novel cytokine-network-based stretching network (CNSN) is proposed to achieve the desired control performance in large production lines [2]. Finally, a novel immune-based self–adaptive collaborative control allocation (ICCA) method is proposed for multi-level stretching processes in large industrial production lines [3].

3.1 Introduction

The industrial process is one of the large-scale complex systems created by mankind. One distinct characteristic of it is that it has many sections linking with each other. This feature is more explicit in some production lines that work non-stop for a long time, e.g. those applied in chemical, material, and pharmaceutical industries. The complexity of equipments, the variant forms of raw materials, and the elegancy of the corresponding automatic mechanisms merge together and therefore form a complete production line that is very sophisticated and full of challenges.

In large-scale industrial production lines, the multiple sub-processes are complicated with non-linear dynamics, and many control variables are coupled with each other, for example, the multi-level stretching processes in large chemical fiber production lines [4]. In such multi-level process, allocation of the stretching ratio at each level determines each stable tension condition on precursor, which is the primary reason for the final fiber break or fold. In normal conditions of industrial environment, the stretching ratios are often determined empirically and pre-fixed offline by using preliminary experimental studies. Consequently, from the control point of view, during the system operation, the stretching ratios are interrupted and varied with the coupling effects of tensions in neighboring levels, which is the primary reason leading to the fluctuation in tension response [5, 6].

Y. Ding et al., *Bio-Inspired Collaborative Intelligent Control and Optimization*, Studies in Systems, Decision and Control 118,
https://doi.org/10.1007/978-981-10-6689-4_3

The control and elimination of vibrations occurred in the stretching and winding processes with a long transmission way implemented by several independent machines has raised interest of some researchers. It can be divided into two categories by different aspects they focus on, namely, the roller system itself that mainly consists of several motors and the corresponding designing of controllers or control schemes. To the roller system itself, Kuo and Fang analyzed the overall system structure of the calendar roller system and proposed its dynamic mathematical model, added with the settling time-optimal control approach using an intelligent control scheme with expert system in their following work [7]. Pagilla et al. built dynamic models for the cascaded roller systems in the winding procedure of the web processing lines, and then designed a decentralized controller [8]. Shin and Kwon [9] took the tension existing among the motors in the winding systems as a main factor to be dealt with. On the other hand, the coupling of motor motions is another problem that is critical to the system performance, and some decoupling approaches have been introduced. Li and Rahn [10] tried to isolate the vibrations of motors to their own scopes by appropriately modeling. Abjadi et al. [11] proposed a sliding-mode controller to implement decoupling. Yerashunas et al. [12] effectively decreased the sensitiveness of web between rollers by introducing a transverse vibration model incorporating viscoelasticity. It should be noted here that the multi-motor system with an X-Y layout is also a branch of such area, and some intelligent methods such as fuzzy algorithms and neural networks have been applied [13]. To the controller design of the multi-motor system, several approaches including conventional PID algorithm (with its variances), fuzzy logic, sliding-mode controller and robust controller have been proposed [14–17]. Besides of these conventional strategies, some intelligent control schemes have been applied to solve problems like intelligent decoupling of variables [18]. The neural networks are also applied to form control schemes with a parallel controller structure, such as Cao et al. [19], Dinh and Afzulpurkar [20], and Lin et al. [21]. Claveau et al. Gueaieb et al. [22] applied the decentralized adaptive hybrid intelligent control scheme to implement simultaneous position and force control in multiple cooperative manipulator systems. There are also some literatures focusing on the random uncertainties existing in the industrial processes and some modern control thoughts, e.g. the H-infinity controller have been applied to build a reliable system against possible disturbances [23–25].

To address the weaknesses of the PID controllers, more advanced control schemes, such as decentralized control, multi-layered feedback control, and H∞ robust control, sliding-mode control have been proposed. A major remaining issue in these methods is the requirement for an accurate mathematical model of the controlled plant. To control the processing lines of fibers, Pagilla et al. [8] analyzed the dynamic model by taking into account the time-varying nature of the roll inertia and radius, and proposed a decentralized controller [26, 27]. Although these schemes have successfully been used to solve some specific problems existing in interconnected processes, they mostly rely on an accurate mathematical model of the plant. On the other hand, some ad hoc control strategies are designed, which cannot be easily extended to other large scale systems [28]. In addition to the conventional decoupling strategies discussed above, intelligent cooperative strategies have been proposed. Ding and Liu proposed a bi-cooperative decoupling control approach to achieve better multi-variable performance [29], in which a more general and easy implementing method suited for all these sorts of reconfigurable problems is proposed. Liang et al. proposed an intelligent cooperative controller based on the regulation network in human body for coordinated and synchronous regulation in stretching system [5, 30]. There is also research work that focuses on adaptive cooperative strategies. For example, Rashed et al. suggested an adaptive control scheme in speed regulation of voltage-fed induction multi-motor and Gren et al. reported an adaptive cooperative control of mobile sensor networks in a distributed environment [31]. All of these papers provide a new thought for combination of online optimization and distributed control. Therefore, control allocation of the stretching ratios opens up a new avenue for cooperative control of multi-stage processes involved in large scale production lines [32, 33]. The central idea of this approach is to map the alterable desired virtual inputs given the condition of output fiber performance onto real-time set-points of the controllers. The purpose of this strategy is to build a structure including interconnections of all stages in resource, information sharing and control deployment optimizations according to the corresponding control allocation algorithms. The control allocation algorithm here refers to an algorithm that allocates desired stretching ratios to each controller according to the feedback of the fiber tension value and the change of control. Control allocation deals with the problem of unalterable control demands for distributed multi-level sub-processes, leading to a restricted performance, which changes the pre-fixed inner-loops conditions and improves the optimization oriented operations. A control allocation problem in the stretching process can be formulated as an optimization problem for the final fiber quality, where the objective is typically to search for the best combination of control efforts (or power) subject to actuator constrains and other operational constraints [34]. Conventional methods for solving the reconfigurable issue include the unconstrained least-square method, linear programming and quadratic programming [35]. Most recently, intelligent optimization methods inspired from nature have been developed to address the weaknesses in traditional control allocation algorithms to achieve faster and more accurate global search [36].

For the traditional control methods including PID and robust control, the accurate model is necessary [37–39]. In these methods, the dynamic model is introduced for the behavior of pliable material tension variations in machinery. Differential equations are used to describe the relationship between tension of elastic webs and velocity of each roller based on the general laws of physics [40–42]. An improved dynamic model for the wholly elastic material is proposed for distinguishing among multiple sub-processes or different tension zones [37, 43], then a decentralized controller is designed based on this proposed model [40, 44].

For the aforementioned methods, due to the lacking of the proper stretching ratios in all sub-processes, the decentralized control also may not lead to a collaborative regulation of these sub-units [38, 45, 46]. To solve this type of problem, the gain-tuning control strategies are addressed based on the mathematical model and some assumed conditions with rarely varying systematic parameters [6, 47, 48], and the gain-tuning control is designed to treat the dynamics without mathematical model. Furthermore, the model-free gain-tuning method is proposed in order to adapt with the varying conditions to the non-linear parameters. The critical issue is how to systematically and simultaneously operate the online tuning for relevant driving actuators. The adaptive control is proposed to solve the coupling effects [42, 49]. Using the gain-tuning controllers for the speed online setting of two neighboring permanent magnet synchronous driving motors, the control of desired stretching ratio in each stretching zone and the tension response are attainable [47, 49, 50]. In addition, Su and Wu proposed an output feedback control for solving the Markovian jump scalar nonlinear system [4, 44, 51]. The interconnected control problem was also addressed by decentralized adaptive fuzzy output feedback control to deal with dynamic uncertainties. It is the most important problem for the decentralized control structure to attain the optimal solution of online gain-tuning. For achieving the gain-tuning in large scale systems, an improvement of the transient collaborative performance has been addressed by control allocation [47]. Moreover, control allocation for online optimization is proposed to combine the gain-tuning controller for the stretching ratios reallocation [37, 52, 53]. Using the dynamical programming for approximating the real-time sub-models, the manipulations of the driving motor velocities within a feasible range are used as manipulated control variables [51, 54], and then the dynamical programming is used to reversing optimization for minimizing the tracking error from reference sub-models [35, 55–57]. Decentralized adaptive fuzzy control for a class of large-scale MIMO nonlinear systems is proposed and its application is expanded to industrial systems [8, 58, 59]. Based on a self-tuning controller capable of handling with error tracking, the motor speeds can be regulated by online regulation according to the fiber overall tension requirement aforementioned. The reconfigurable control is also one of those control strategies to solve the online regulations [60].

3.2 Multi-Layered Intelligent Cooperative Controllers

In this section, a novel bio-inspired intelligent cooperative controller is proposed to provide a plan for controlling the interlinked rollers by forming them to unified stretching units. The basic structure of the typical stretching units is firstly analyzed and some of its characteristics useful for controller design are then generalized. Based on the multi-layer regulation networks in human body, a bio-inspired networked controller structure is established.

3.2.1 A Bio-Inspired Multi-Layered Intelligent Cooperative Controller

3.2.1.1 Multi-Layered Regulation in Neuroendocrine System

The neuroendocrine system in human body is a complex system that plays a fundamental role in regulating life activities. It mainly includes two parts, namely, the nervous system and the endocrine system. Most of the regulation processes conducted by these two systems can be seen as negative close-loop feedback processes, and the neuroendocrine system takes action by combining and then organizing them into different layers. Each layer has its own function and in-charge scope, and messages can be transferred among them so that the whole system can act as integrity. In general, the neuroendocrine system can be separated into four layers, the ultra-short feedback layer, the short feedback layer, the long feedback layer, and the ultra-long feedback layer whose hierarchical structure is shown in Fig. 3.1. The ultra-short feedback means the hormone released by a certain gland or organ is directly feedback to its source and changes its status. The short feedback happens between two endocrine glands that have supervision relations. In the short feedback cycle, the administrative gland secretes hormones to regulate the function of the subordinate one while the secretion of the subordinate one also has influence back to its supervisor. The long feedback exists among three or more glands that messages are fed back from one gland not to its previous gland but to its further commanding glands. The ultra-long feedback is established on the basis of the interactions between nervous system and endocrine system. The nervous system which initially aims for responding stimulations can therefore acquires a new function that monitors the endocrine system, and this can form a balance and integrity between body feelings, movements and metabolism. In that case, mistakes made by either of the two systems can be quickly observed and corrected by another one, which enhances the stability of the whole body.

A more detailed example is taken here to illustrate the working procedure of such multi-layered regulation networks, see also in Fig. 3.1. The thyrotropin releasing hormone (TRH) released by the hypothalamus in human brain regulates the hormone secreting level of pituitary. The TRH also comes back to the hypothalamus itself to make a self regulation (namely, the ultra-short feedback).

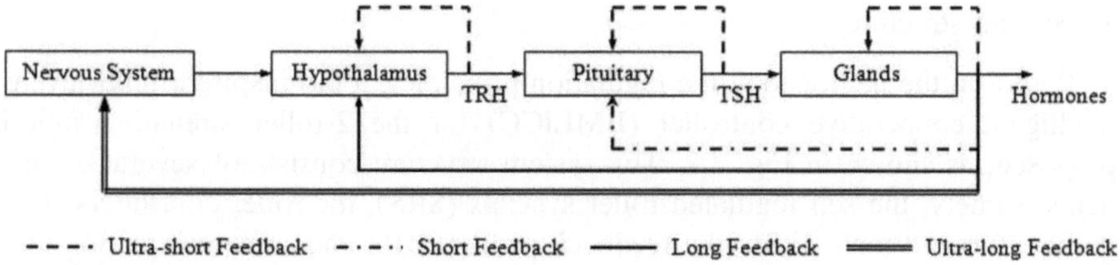

Fig. 3.1 Multi-layered regulation network of the neuroendocrine system

On one hand, the feed forward path from the hypothalamus to the pituitary is a part of the short feedback so the hypothalamus can be seen as a commander and the pituitary is its subordinate member. On the other hand, the hormone released by the pituitary called thyroid stimulating hormone (TSH) is used for modulating the thyroid, and the concentration of TSH has influence back onto the hypothalamus through short feedback path. Such relation can also be found between the pituitary and the thyroid. The hormone that releases from the thyroid (called thyroxine) can also stimulate the hypothalamus in order to control the releasing speed of TRH (the long feedback). And the hypothalamus (or say, the whole endocrine system) is under control of the nervous system and receives instructions from it. Conversely, the hormone secreting level of endocrine glands is fed back and helps to enhance or suppress the performance of nervous system. In this way, an integrated neuendocrine system with four layers is established.

In the neuroendocrine system with multi-layered feedbacks, usually, the regulation of a single gland or organ needs more than one feed forward/feedback loops around it to build a complex mechanism. Different loops have different glands as the controller to play their specific roles. The glands in the outer loops deal with the stability of the whole system and those in the inner loops care for the agility. It is a guarantee that the glands to be controlled can quickly response to the stimulations and variations of the body, and meanwhile keeps its working status stable toward the possible disturbance. The characteristics of the multi-layered neuroendocrine system can be summarized as below: (1) The multi-layered neuroendocrine system has several feedback loops and several glands or organs. The messages transmit among these glands or organs in a fixed, forward direction and the feedback messages are received by all the glands or organs simultaneously. (2) Different layers of the neuroendocrine system take different responsibilities. The outer layers deal with the stability while the inner loops care for the agility. Therefore, the inspiration of designing a new type of intelligent cooperative controller is based on such characteristics. It should consist of two essential parts as below: (1) A multi-layered structure to fulfill regulations with both stability and agility. (2) An appropriate series of regulation algorithms for the controllers with high precision and short response time.

3.2.1.2 The Multi-Layered Intelligent Cooperative Controller

1. System structure

Based on the neuroendocrine regulation principles, a bio-inspired multi-layered intelligent cooperative controller (BMLICC) for the 2-roller stretching unit is proposed, as shown in Fig. 3.2. The system structure consists of several components, namely, the self-regulated roller systems (SRS), the roller controllers (RC), the feedforward path (FFP), the feedback paths (FBP), and some conversion units

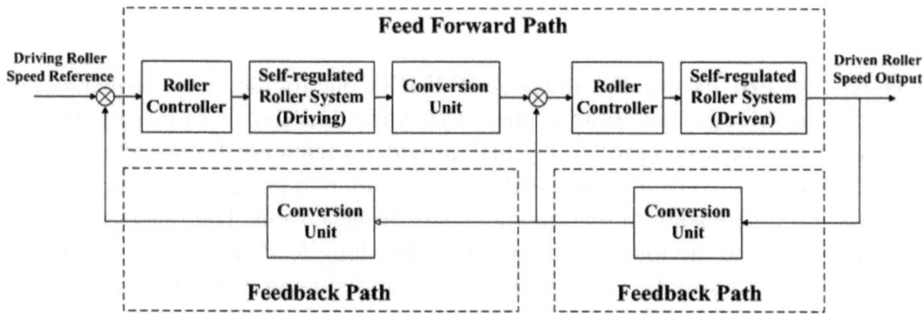

Fig. 3.2 The Bio-Inspired intelligent cooperative controller

(CU) on the FFP and FBP. Note that all these devices and the structures they form can be representatives of different layers in the original endocrine regulation system. But not every layer has an accurate correspondence in the proposed controller due to the limited complexity of the 2-roller stretching unit. The flexibility and integrity of the endocrine regulation system with its multi-layered structure can guarantee that the layers can be added or removed for plants with different complexities and characteristics.

Self-regulated roller system (SRS)

The SRS is established to ensure each roller has the ability to keep itself stable and minimize its speed variation to other components as much as possible. The SRS is a classical closed-loop feedback motor system, e.g. a double closed-loop system for DC motor. Compared to the neuroendocrine system, the SRS represents the ultra-short feedback layer which can keep itself (namely, the glands) stable.

Roller controller (RC)

The RC is located in front of each SRS. It receives error signals and generates instructions for the SRS. Some ordinary controllers having been widely used in industry can be applied as RC, e.g. PID controller and fuzzy controller.

Feedforward path (FFP)

The FFP is responsible for transmitting control signals from the driving roller to the driven roller. With the aid of FFP, the driven roller can keep connection with the driving roller, thus adjust its speed according to that of the driving roller. Note that there is a CU located on the FFP.

Feedback path (FBP)

Two types of FBP exist in the proposed BMLICC, one for the driving roller and the other for the driven roller. Both of them transmit speed signals of the driven roller backward. For the driving roller, the FBP provides a reference for speed adjustment of itself. For the driven roller, the FBP is another loop to prevent the possible speed variation from spreading to other components of the system. The functions of such FBP are similar to the long-feedback (even ultra-long feedback which depends on the complexity of specific target plants) layers in the neuroendocrine system. Like the FFP, each FBP has its own CU.

Conversion unit (CU)

The CU plays an essential role in the proposed controller. The aim of the CU is to convert signals from one type to another and make possible adjustment. On one hand, the output signal of the SRS is usually the speed of the roller (representing by a counting number or a voltage proportional to the motor speed). However, the input signal of the RC is voltage on most occasions. So the conversion between the speed signals and the voltage should be done by the CU beforehand. On the other hand, the relations between the input signals of the RC and the output signals of the SRS are not all linear, which requires to be converted by a non-linear converter with certain forms of compensation. Such characteristics should also be embodied by the CU. The CUs in the proposed system are located on the signal transmitting paths, namely, on the FFP and the FBP.

2. Cooperative regulation algorithm

The regulation algorithm of the BMLICC consists of the control algorithm of the RC and the transformation procedure of the CU, as shown in Fig. 3.3. The static configuration and basic working procedure of the proposed controller are illustrated in this figure. As mentioned above, each RC can be one of the commonly used controllers. The transformation procedure of the CU will be further discussed. It is clear that two roller systems are connected by the signal conversion sections (namely, the CU) to exchange their information while they are also able to regulate their own working status with the help of the closed-loop feedback speed regulation systems equipped for each roller system. Note that a system status identification section is included in the driving roller system to detect whether the whole system encounters vibration or external disturbance. This function is implemented by comparing the converted signal from the driving roller with the speed reference and the predefined stretch ratio.

Figure 3.4 shows the regulation process against the vibration or external disturbance. The disturbance is firstly detected by the system status identification section of the driving roller system, and then used as input of the RC in front of the driving roller to generate corresponding control instruction. With the changing

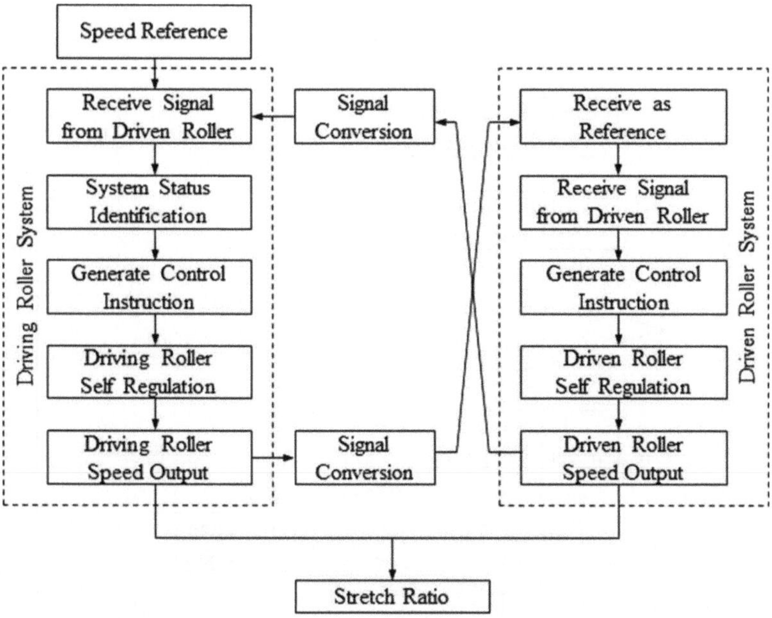

Fig. 3.3 The regulation algorithm of the BMLICC

instruction, the speed of the driving roller is tuned, trying to eliminate the vibration within the driving roller system. If the vibration cannot be completely eliminated (e.g. the vibration is not instantaneous but lasting for a long time.), it is then transferred to the driven roller system. The driven roller is therefore tuned to follow the change of the driving roller so that the stretch ratio is maintained. The stretch ratio is the main quality factor of a stretching unit, and such mechanism is a guarantee for keeping the stretch ratio unchanged among possible disturbance rather than fixing the speed of rollers at a static point.

As to the CU, two types of signal transmitting paths, the FFP and FBP, are equipped with the CU. So the algorithms of the CU have two corresponding types.

CU on the FFP

The CU on the FFP receives the real-time speed signals from the driving roller and then sends it to the controller of the driven roller. The conversion formula of such CU can be written as

$$\begin{aligned} C_{U,\,out} &= \frac{C_{V,\,in}}{K} + \sigma(C_{U,\,out}, C_{V,\,in}) \\ \sigma(C_{U,\,out},\ C_{V,\,in}) &\begin{cases} = 0 & (linear) \\ \neq 0 & (nonlinear) \end{cases} \end{aligned}, \tag{3.1}$$

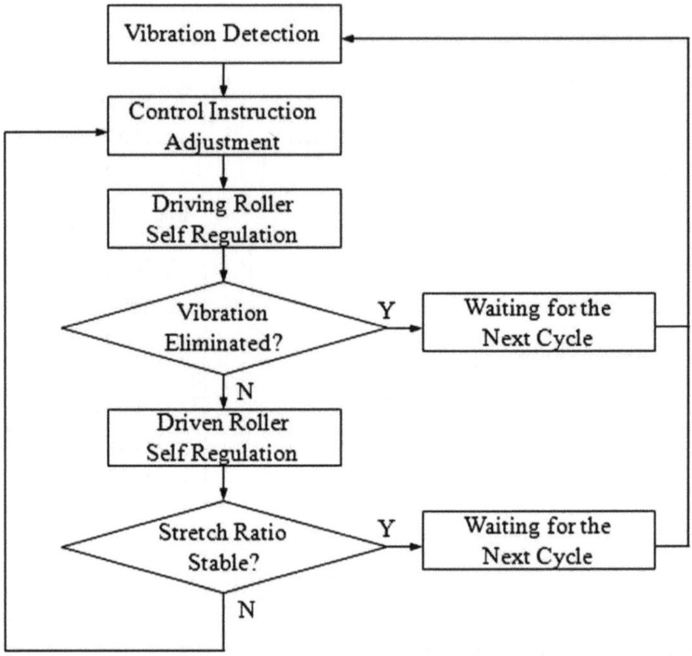

Fig. 3.4 Vibration elimination process

where $C_{V,\,in}$ represents the speed signals (in rpm), $C_{U,\,out}$ represents the speed set voltage of the driven roller (in volt), K is the conversion efficient, and $\sigma(C_{U,\,out}, C_{V,\,in})$ is a function related to the input and output of the CU. In Eq. (3.1), the relation between the speed set voltage and the corresponding roller speed can be embodied by the efficient K. K also includes the stretching ratio between the driving roller and the driven roller. The function $\sigma(C_{U,\,out}, C_{V,\,in})$ is used to make non-linear conversion and conduct compensations if the speed-voltage relation is not linear. The value of this function remains zero if the speed-voltage relation is linear.

CU on the FBP

In the proposed system, two CUs are equipped on two FBPs, respectively. Similar to the CU on the FFP, the CUs on the FBPs also take charge in converting the speed signals to voltages. Therefore, Eq. (3.1) can also be applied without much trouble. However, the stretching system sometimes needs to do some unordinary adjustments that require good dynamic performances, e.g. a large-range speed regulation in a short time, an inhibition to the possible disturbance or a precision compensation. So the CU in these cases should act like a normal controller, which needs to be designed following the rules of designing a controller. One example of such CUs can be written as

$$\begin{aligned} C_{out} &= C_{in} \times K \\ K &= K_s \times K_p[e_{in}(t) + \tfrac{1}{T_i}\int_0^t e_{in}(t)dt + T_d\tfrac{de_{in}(t)}{dt}], \end{aligned} \tag{3.2}$$

where C_{in} and C_{out} are the input and the corresponding converted output of the CU, K is the conversion efficient and K_s is the proportional efficient representing the speed-voltage matching and stretching ratio (like the K of the CU on the FFP). K_p, T_i and T_d are the proportional efficient, the integral time constant and the derivative time constant, respectively. $e_{in}(t)$ is the input error of the CU. Note that $e_{in}(t)$ is not calculated by comparing the input of the CU with some certain fixed value but by comparing the input at the current sampling time with that in the last sampling time.

3. A typical stretching control system

The actual stretching process in practice usually includes several stretching units to meet the complex requirements of fiber quality. These stretching units are linked with each other, supervised by a command central. The command central can be an industrial process computer (IPC) or other platforms. All the stretching units communicate with the command central through the input and output devices, e.g. cables, ports and cards. Figure 3.5 shows a basic model of an IPC controlled stretching process. It is a hardware-implemented closed-loop feedback system, which therefore provides a natural foundation for applying the BMLICC with convenience. The whole stretching process is divided into several smaller units, and each of them consists of two sets of rollers, namely, a driving one and a driven one.

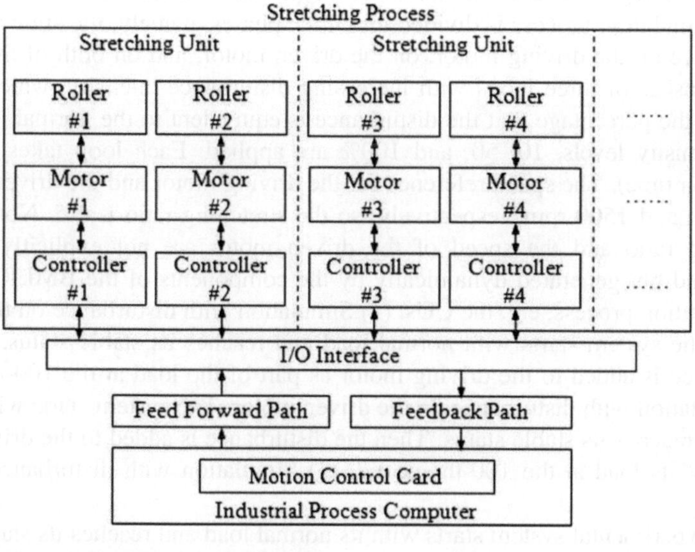

Fig. 3.5 A typical stretching control system

Their own drivers and controllers are also included in the unit and therefore form complete closed-loop speed regulating systems. The stretching units exchange information with the IPC through the I/O interface, which represents the feed forward path or the feedback path of the BMLICC. The IPC takes charge of receiving working information, making necessary conversion and distributing control instructions to the units, so it plays as the RC and the CU.

3.2.2 Simulation and Experimental Results

3.2.2.1 Computer-Based Simulation Results

An experimental stretching unit consisting of two DC motors is applied here as a test platform to verify the effectiveness of the proposed BMLICC. Although the most widely used type of motor for fiber production is the induction motor (AC motor) with inverter, the frequency transformation done by the inverter in the stretching unit brings extra vibrations, which is also the reason of applying DC motor instead of the AC one. According to the sequence in which the fiber is winded, two DC motors in this stretching unit take the roles of the driving motor and the driven motor, respectively. Both motors are equipped with their independent double closed-loop speed regulation systems. An adjustable disturbance is added to each motor together in the form of extra physical load which would result in the speed changes of the corresponding motor. This phenomenon usually occurs during the practical manufacturing process due to the mechanical friction and abrasion, which is one of the main results that cause the degradation of fiber quality. So it can be regarded as a simulation of the practical circumstance.

The simulation process is divided into three phases, namely, the simulation with disturbance on the driving motor, on the driven motor, and on both of them. Each phase consists of three loops with increasing disturbance intensity, which is measured by the percentage that the disturbance is equivalent to the normal load. Here three intensity levels, 10, 50, and 100% are applied. Each loop takes 250 s (in simulation time). The speed reference for the driving motor and the driven motor is 1000 rpm and 1500 rpm, respectively, so the stretching ratio is 1.5. Note that the stretching ratio and the speed of the driven motor are not explicitly assigned beforehand but generated dynamically by the components of the BMLICC during the simulation process, e.g. the CUs. (1) Simulation with disturbance on the driving motor. The system starts with normal load and reaches its stable status. Then the disturbance is added to the driving motor as part of the load at the 100-th second. (2) Simulation with disturbance on the driven motor. The system starts with normal load and reaches its stable status. Then the disturbance is added to the driven motor as part of its load at the 100-th second. (3) Simulation with disturbance on both motors.

The experimental system starts with its normal load and reaches its stable status. Then the disturbance is added to both the driving motor and the driven motor at the

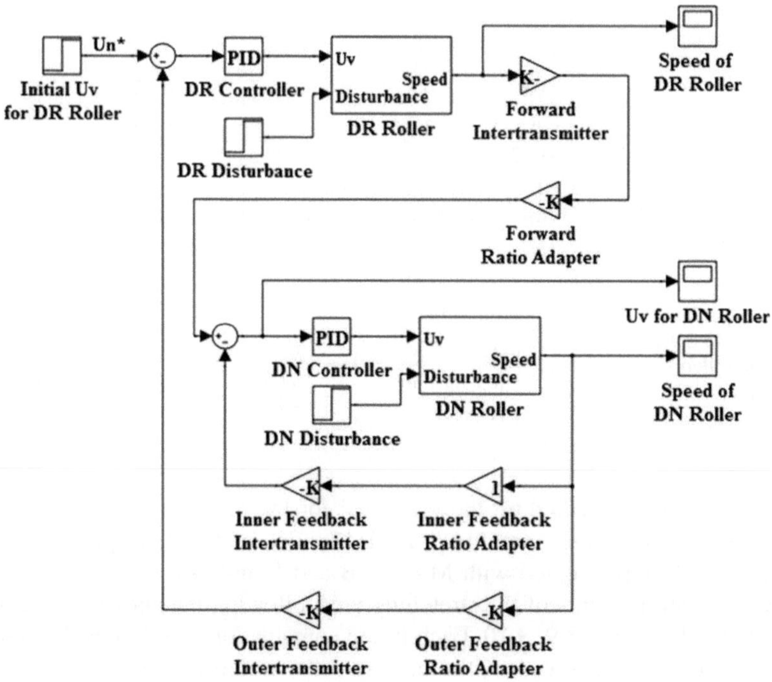

Fig. 3.6 Schematic of the stretching system

100-th second. The structure of the whole stretching system can be represented by Fig. 3.2. Figure 3.6 shows the schematic of the whole stretching system with the proposed control scheme, and the internal structure of the self-regulated roller system is provided in Fig. 3.7. (The acronym "DR" and "DN" in the figure refer to the devices for the driving roller and the driven roller, respectively.) The parameters

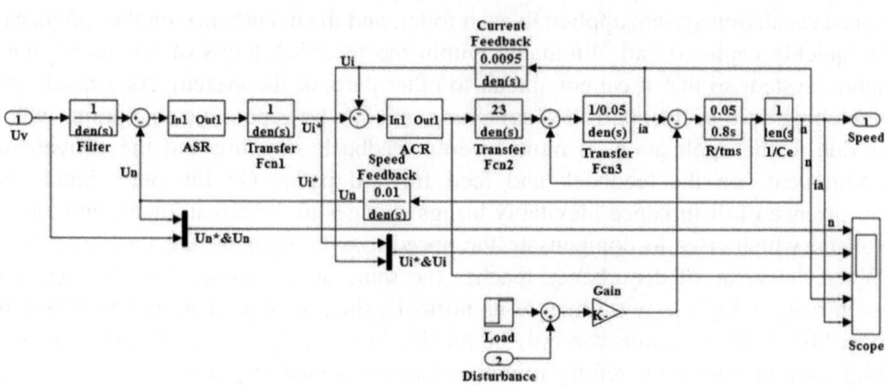

Fig. 3.7 Schematic of the self-regulated roller system

Table 3.1 Parameters of the self-regulated rollers for simulation

Item	K_S	$R(\Omega)$	T_m(s)	T_s(s)	T_l(s)	C_e [a]
Value	23	0.05	0.8	0.0017	0.025	0.185
Item	T_{on}(s)	T_{oi}(s)	α	β	Load[b]	U_n^*(V)
Value	0.01	0.002	0.01	0.0095	500	10

[a] The electromotive force (EMF) coefficient of DC motors
[b] The load of the rollers are measured in the form of additional rotating speed of the corresponding roller (RPM)

Table 3.2 Parameters of the roller controllers for simulation

Item	K_P	K_I	K_D
DR controller	0.08	0.15	0.15
DN controller	2.5	0.35	0.11

for the roller system including those applied for the motors of the rollers and the corresponding controllers are listed in Tables 3.1 and 3.2, respectively. These systems are all implemented with MATLAB and Simulink.

The simulation results of the stretching unit following the above procedures are illustrated in Figs. 3.8, 3.9, 3.10. Each figure contains four subfigures that show the armature currents and speed of the driving roller and driven roller with different levels of load, respectively. The armature currents of the rollers are recorded as a part of response because the disturbance is attached to the system in the form of extra load, which directly affects the current of the affected roller. The stability of speed of the rollers is the main objective that the stretching unit with the BMLICC should firstly reach, and the ratio of the stretching unit can be therefore calculated by comparing the speeds of the driving roller and the driven roller. It can be observed from these figures that the stretching process with the proposed BMLICC can maintain the speeds of its motors at their expected values, and the disturbance of different levels attached to either of them (even both) does not bring distinct changes to the roller speed. Such advantage can be attributed to the independent speed regulation system applied to each roller, and the disturbance on the rollers can be quickly captured and eliminated within the feedback loops of the speed regulation system so that it cannot spread to other parts of the system. As a result, the stretching ratio of the controlled stretching unit can be kept to a stable point, which is due to the application of multi-layered feedback structure and the conversion components on the feedback and feed forward paths. On the other hand, the occurrence of disturbance inevitably brings changes to the armature currents of the rollers, which rises to compensate the speed loss brought by the increased load. When the value of disturbance reaches the same as the normal load, the current even rises as high as two times of its normal value, namely, from below 500 A to over 800 A. These results can only be reliable in the computer simulations and such high current must be carefully monitored and restricted in practice.

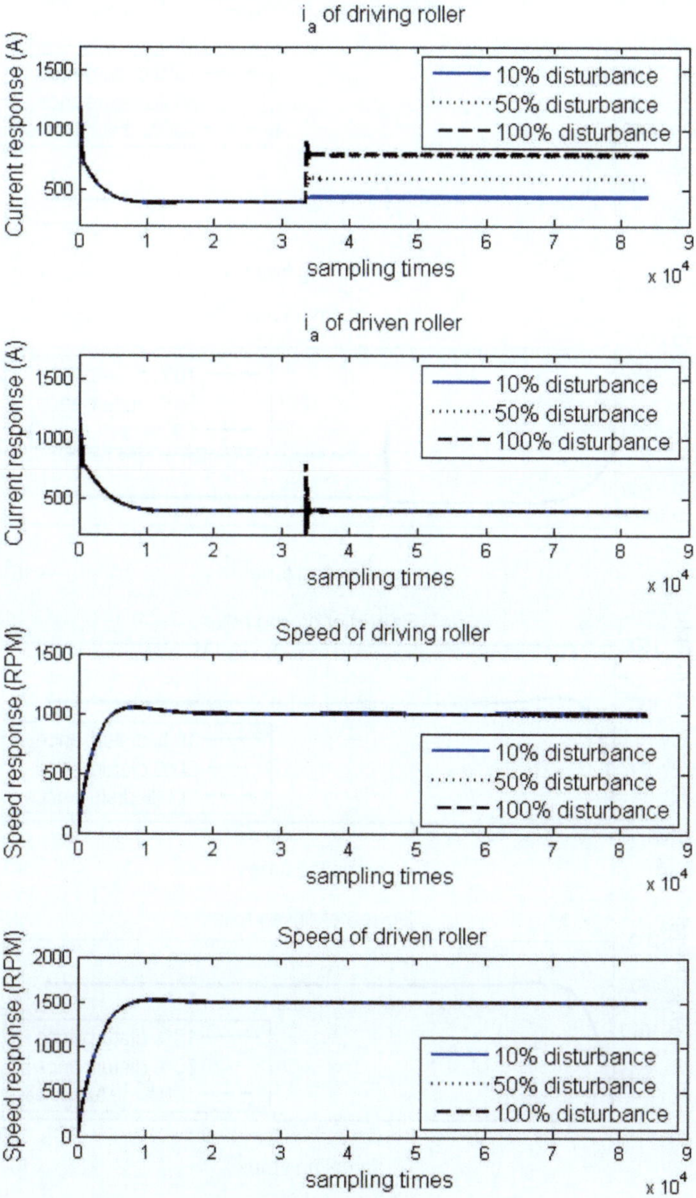

Fig. 3.8 System response with disturbance on the driving motor

3.2.2.2 Device-Based Experimental Results

A further verification for the proposed controller is conducted on a multi-axis stretching experimental platform, as shown in Fig. 3.11, whose basic structure is

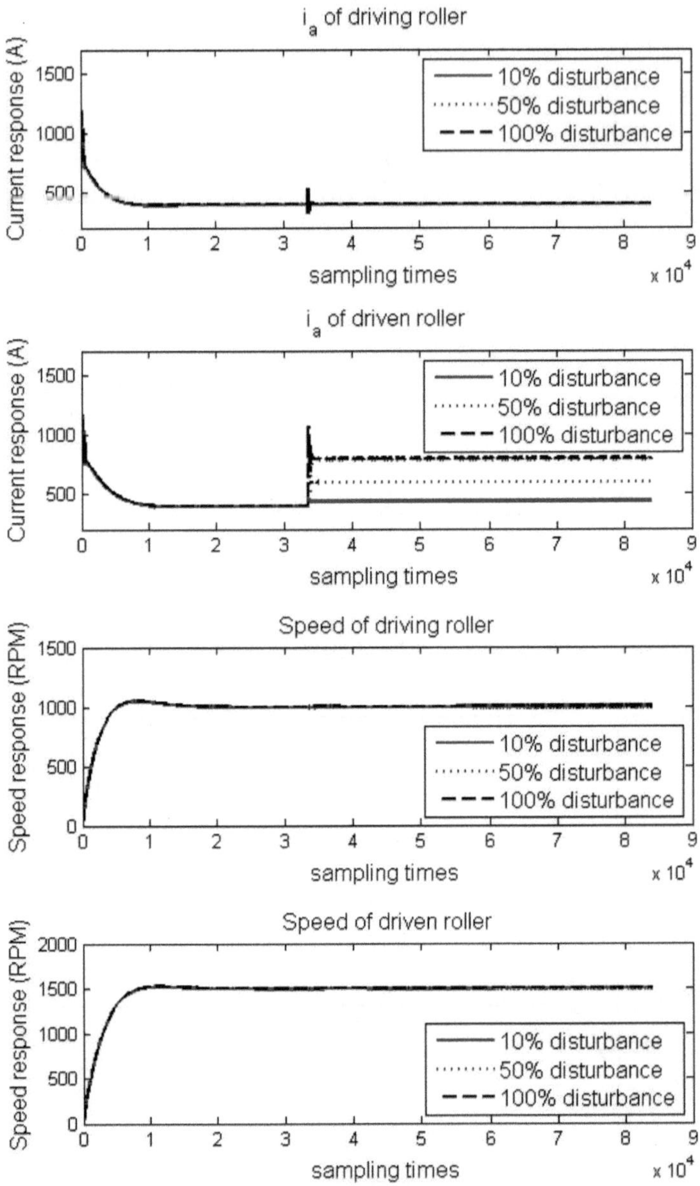

Fig. 3.9 System response with disturbance on the driven motor

illustrated in Fig. 3.5. It consists of a central motion control card (plugged in as an extension card), several rollers and the necessary input and output connections. Each roller is connected to a motor driven by a programmable logical controller (PLC). All the roller controllers are then connected to the motion control card by

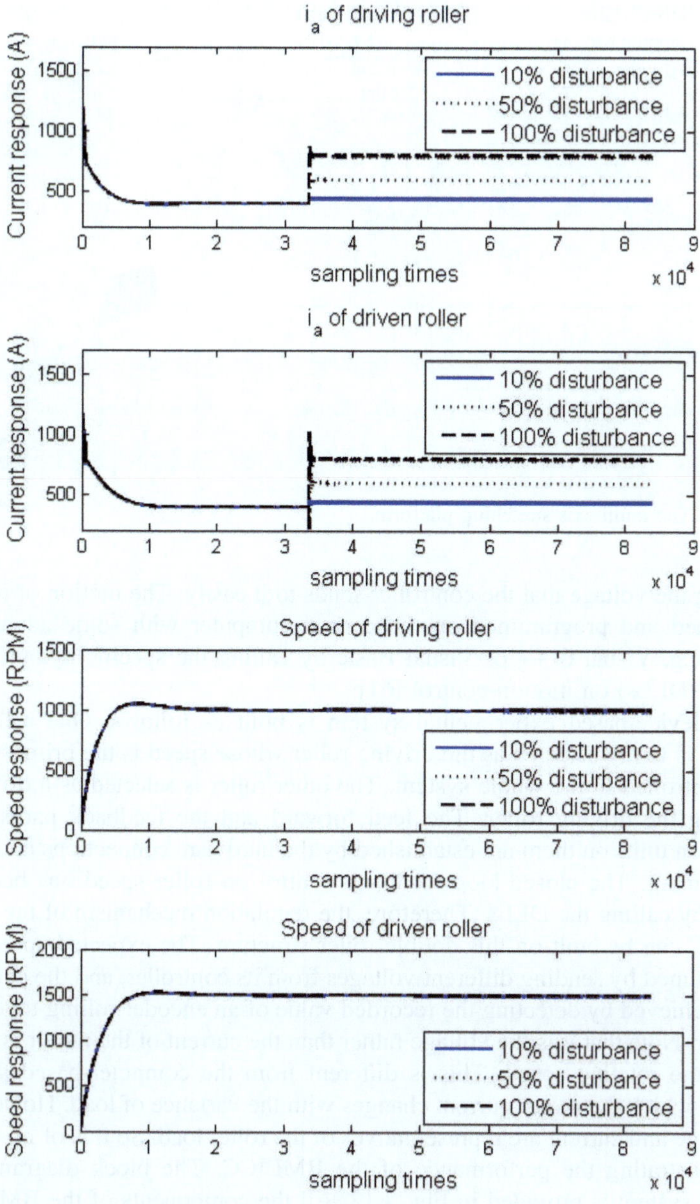

Fig. 3.10 System response with disturbance on both motors

wires so that the feed forward and feedback paths of a control system can be established. The motors connected to the rollers can be of any types. In this platform, two servo motors are applied due to their precise control performance. Another reason for choosing this type of motor is that it can be regulated by

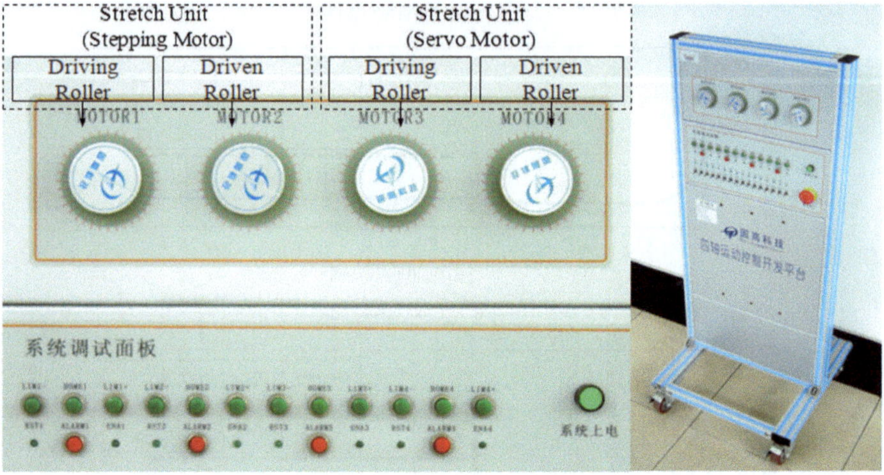

Fig. 3.11 The multi-axis stretching platform

changing the voltage that the controller sends to it easily. The motion of rollers can be planned and programmed on the central computer with some advanced languages, e.g. Visual C ++ or Visual Basic by calling the specific dynamic linking libraries (DLLs) on motion control [61].

The device-based experimental system is built as follows. One roller of the platform is firstly selected as the driving roller whose speed is the primary variable to be controlled in the whole system. The other roller is selected as a driven roller following the driving roller. The feed forward and the feedback paths with the conversion units on them are established by the hardware connections and software programming. The closed-loop feedback control on roller speed has been implemented by calling the DLLs. Therefore, the regulation mechanism of the proposed BMLICC can be built on this double roller structure. The expected speed of each roller is tuned by sending different voltages from its controller, and the actual speed can be achieved by detecting the recorded value of an encoder rolling together with the roller. Note that it is the voltage rather than the current of the motor is regulated to tune the rotating speed. This is different from the computer-based simulation above in which the motor current changes with the variance of load. However, both the voltage and current are representatives of the roller load. So it is of no difference in demonstrating the performance of the BMLICC. The block diagram of such practical system is provided in Fig. 3.12. All the components of the BMLICC are implemented in software, and the related programs are composed using Visual Basic on a motion card-equipped industrial controlling computer with a 2.8 GHz processor and 256 MB memory.

Based on this platform, a two-phase speed regulation process is made to verify the effectiveness of the proposed BMLICC. The stretching ratio of the roller system is fixed to 1.2, namely, the driven roller is 1.2 times faster than the driving one. The

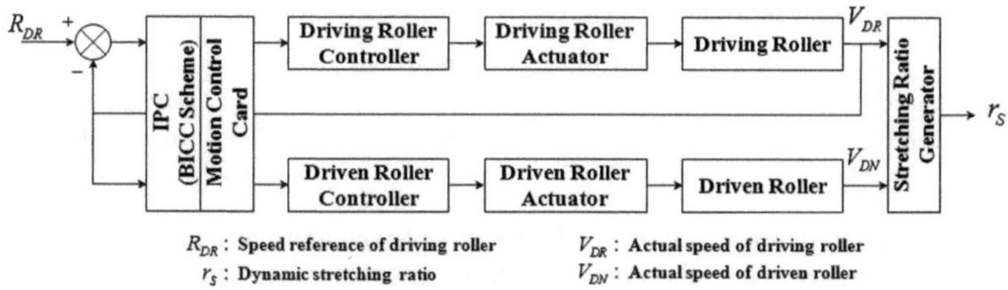

Fig. 3.12 Block diagram of the practical test system

simulation time is measured by control period (CP). A single CP is equivalent to about 10 ms. Each phase of the simulation covers 1,800 CP. The speed of rollers is measured by round per minute (RPM). Detailed simulation plan is provided as below:

1. Stimulation of normal status. The stretching unit starts with an initial speed reference and reaches its stable state. Then the speed of the driving roller is elevated to a faster level, keeping for a given period and decreased to a lower level, which is also kept for a period. (The low speed level is above the initial level.) Both rollers go through three phases with different speeds, and the detailed speed distribution and the corresponding switching time of both rollers are listed in Table 3.3. This experiment aims to verify the ability of the BMLICC in stabilizing the roller system and speed tracking between the driving roller and the driven one. The speeds of both rollers are recorded to show the basic function of the roller system. Meanwhile, the stretching ratio is also calculated and plotted because it is the direct proof for the practical performance of the proposed BMLICC.
2. Stimulation of interfered status. The stretching unit starts with an initial speed reference and reaches its stable state. Then an external disturbance is attached to the driving roller and maintained for a given period. The disturbance takes effect by increasing the load of the roller. After the disturbance on the driving roller is removed, the system waits until both rollers return to their normal running states. Then another disturbance (also in the form of additional load) is attached to the driven roller for a given period. The system takes time to recover from the disturbed status when the disturbance on the driven roller is removed. Detailed interfered time with its objective (driving roller or driven roller) is provided in Table 3.4. This experiment is to verify the robustness of the BMLICC towards

Table 3.3 Speed distribution and switching time

Item	Initial phase	Elevated phase	Decreased phase
Speed (Driving Roller) (RPM)	60	75	50
Speed (Driven Roller) (RPM)	72	90	60
Switching time (CP)	0	600	1,000

Table 3.4 Interfered time with its target

Item	Starting time (CP)	Stopping time (CP)	Target
Interference 1	500	800	Driving roller
Interference 2	1100	1400	Driven roller

external interference and the swiftness of self-regulating to its former working status. In this phase, the control voltages on both rollers are recorded as an indicator of the amount of external interference. The speeds of rollers are also recorded to show the performance of the proposed BMLICC.

Figure 3.13 shows the speed changes of the roller system and the synthesized stretching ratio according to the procedures of the first simulation phase. It can be observed that both rollers follow a standard speed curve to reach their stable working status at the beginning, which mainly results from the application of PID-like controllers as the RC in the BMLICC. The stretching ratio reaches the predefined level, that is, 1.2 times, after both rollers are stable. When the speed of the driving roller is elevated, both rollers reach their new balance points in about 200 CP, namely, about 2 s. The decreased phase of speed costs a little more time because the speed lost (from 75 RPM to 50 RPM, 25 RPM in total) is much larger than the amount that elevated before (from 60 RPM to 75 RPM, 15 RPM in total).

Figure 3.14 illustrates the speed response of the roller system and the corresponding synthesized stretching ratio under interfered circumstance. The changes of the control voltages sent to both rollers to overcome the external interference are

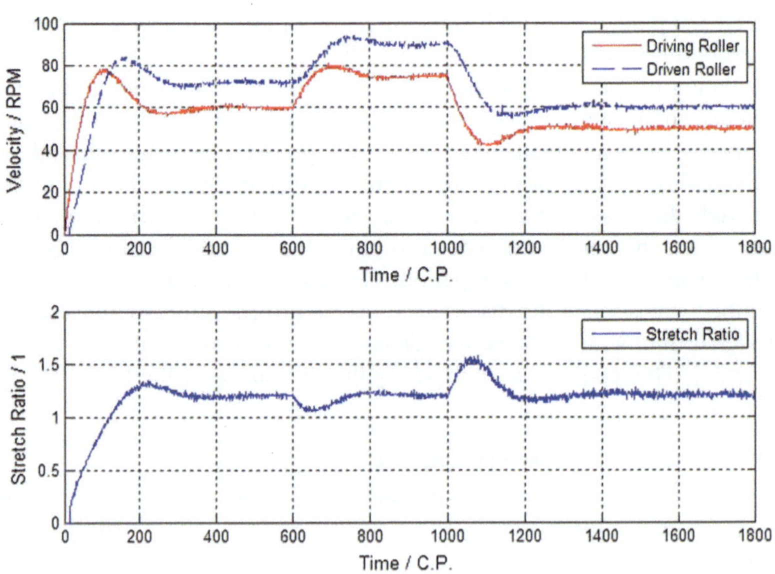

Fig. 3.13 Speed response and stretching ratio in normal working status

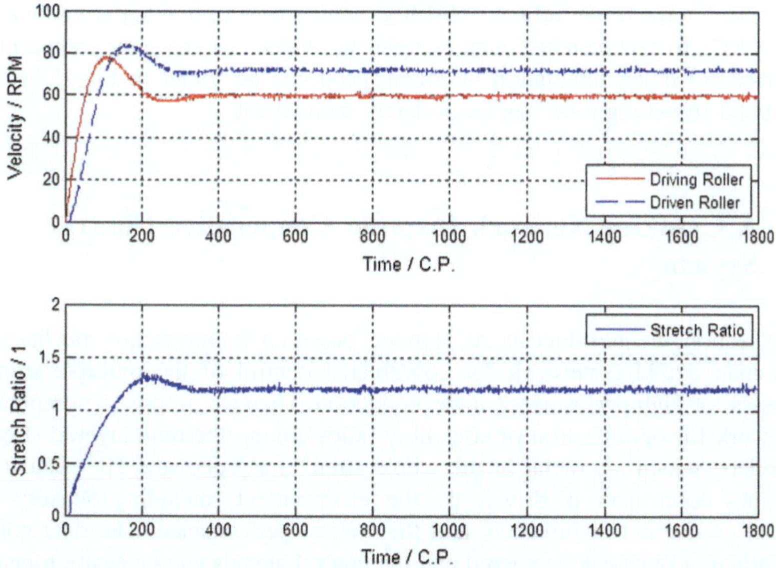

Fig. 3.14 Speed response and stretching ratio in interfered status

also shown in Fig. 3.15. As discussed above, the roller system with the proposed BMLICC encounters two phases of external interference, one to the driving roller and the other to the driven one. Such abnormal phenomenon drives the roller system under the guidance of the BMLICC to react in order to eliminate the negative effects they may bring to the stretching quality, e.g. the speed fluctuations. The changed working condition can be detected and transferred to the RCs of both rollers through the FBPs. Then the RCs generate new control instructions to overcome the external interference, and the instructions are changed control voltages. It can be observed from Fig. 3.15 that the control voltage changes frequently

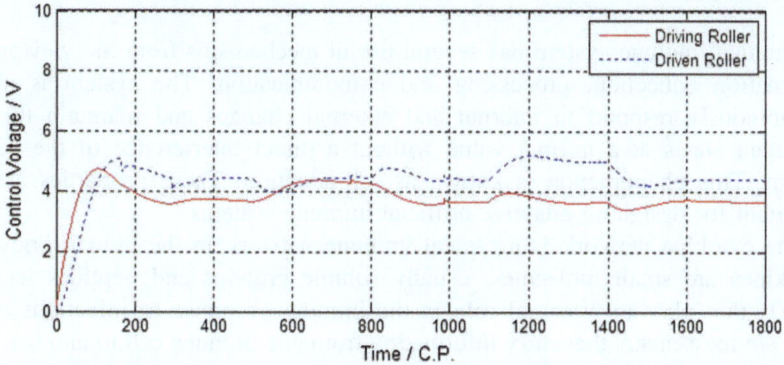

Fig. 3.15 Control voltages to rollers in interfered status

during the interfered period (the BMLICC always tries to find better voltage levels for the RCs to completely counterweight the effect that the interference brings). However, no much fluctuation in roller speed can be found in Fig. 3.14 So the predefined stretching ratio can be perfectly maintained.

3.3 A Cytokine Network Inspired Cooperative Control System

In this section, we introduce an ACN model based on the interaction mechanisms in the human cytokine network for coordinated control of the multiple stretching processes for high performance fiber production. Then we propose a computational framework for optimization of stretching ratios among the multilayered stretching controllers, which is crucial in the whole stretching processes. The regulation of stretching controllers is driven by the environment, including sensory input, external or internal disturbance, and the control performance. The data collected from different units can be shared and the control signals can be easily transmitted within the network, all of which facilitates an adaptive and optimal allocation of the stretching ratios. The experimental results confirm that the system is able to accurately track the set fiber tension and is robust to internal and external disturbances.

3.3.1 Cytokine-Network-Based Stretching Network

In the following, we present the details of the CNSN that aims to achieve the ability for accurate tension tracking and disturbance rejection in multiple stretching zones.

3.3.1.1 Cytokine Network Theory and Its Computational Model

The human immune system has several useful mechanisms from the viewpoint of information collection, processing and communication. The system is able to autonomously respond to internal and external changes and maintain the most important states at a normal value without a direct interference of the nervous system. This phenomenon is known as self-adaption. These properties are very important for designing adaptive artificial immune systems.

The cytokine network is a typical immune network in the human body [62]. Cytokines are small molecules, usually soluble proteins and peptides less than 30 kDa that play an essential role in the immune response to infectious agents. They are messengers that carry information from one immune cell to another, along with these immune cells, constitute the cytokine network, which is one of the

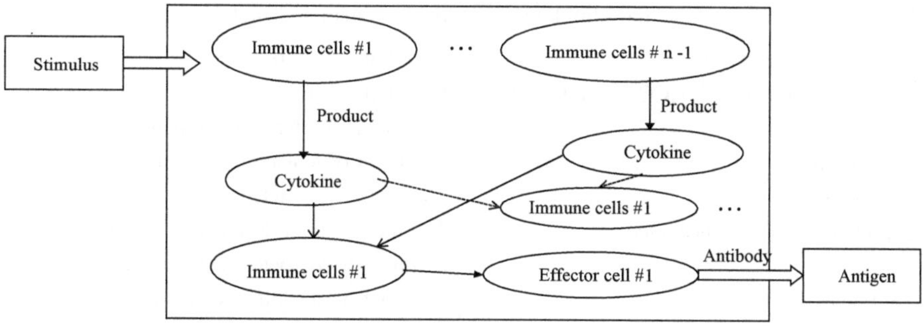

Fig. 3.16 Schematic diagram of immune cells interaction

complex network existing in micro-environment. In this network, varieties of biological entities—immune cells communicate with each other via information exchange and control signal transmitting, resulting in the immunological network, in which cytokine works as signals to convey antibody proliferation information. This immune network can autonomously and accurately detect and identify antigens, and then the immune cells interact with other agents through the messengers, i.e., cytokines, and stimulate corresponding antibodies together against specified antigen.

According to biological findings, the relationship between internal or external antigen and immune cells is very complicated, as shown in Fig. 3.16. The secretion of specified antibodies is an effective defense against antigen, and cytokines as a kind of messenger synthesized for the communication of immune B-cells for finally stimulation of the proliferation of specified antibodies. To simplify this process in engineering applications, we can define a one-to-many relationship among them, namely, several classes antibodies that are secreted by related kinds of locally relevant B-cells including itself can combine with the corresponding antigen. B-cell receptor is an important part in B-cell ontogeny that enables it to specifically recognize antigen type and eventually help to eliminate this kind antigen through interaction of corresponding B-cells with each other.

3.3.1.2 The Immune System Model for Stretching Processes

An immune agent consists of three parts: competent cells which provide environment perception, B-cells which produce cytokines as signal messengers communicated with other agents, as well as effector cells within B-cell which respond to external antigens against invasion by secreting antibodies [63]. These properties are very desirable for control of the stretching system containing multiple coupled units.

Therefore, the proposed CNSN consists of several components: sensors, communication modules, a controller as well as actuators. The mapping relationship between immune system and stretching network is shown in Table 3.5. The effectiveness of the proposed CNSN will be discussed as follows.

Table 3.5 The mapping relationship between immune system and stretching network

Human immune system	Stretching network
Immune competent cell (ICC,T cell,B cell)	Stretching unit
Antigen	External and internal disturbances
Lymphocyte	Detection of tension disturbance
Specific activation	Modulation of stretching process response
Cell clones and disappearance	The increase and decrease of antibody
Cytokines	Medium of competent cells' communication
Antibodies	The control of multilayered stretching units

3.3.1.3 The Construction of the Entire Stretching Network

The control allocation of stretching ratios is crucial to the production of high performance carbon fiber. The CNSN is proposed to achieve coordinated control of the coupled units, which aims to adaptively allocate the optimal ratios combination to different stretching units for continuous and steady stretching processes.

Human immune system regulates information exchange by means of an interaction network consisting of various immune cells which mutually influence each other via hormone-like intercellular molecule messengers called cytokines. CNSN adopts a similar structure of an ACN. On top of that, CNSN has the ability of online optimization that can adaptively reallocate an optimal stretching ratio to each unit in response to environmental changes. In the following, we present this network model and describe the details of the optimization algorithm that determines the set-point of each stretching ratio controller in each unit.

ACN is a new computational framework for simulating the interactions of signaling molecules (cytokines) with immune cells. Ordinary differential equations are used for describing the interactions between immune cells, cytokines, and external stimulus (antigen). Considering each stretching controller input as the information in cytokine network waiting to be optimized between the units by this optimization algorithm, in order to obtain an accurate allocation of these ratios, a specified mathematical function for describing the interactions between these units to compute the re-distributing deviation is presented.

The whole network system is composed of several layers to simulate the multilayered feedback stretching processes. As illustrated in Fig. 3.17, the upper layer is the network layer, which is the most important and capable of receiving all the states, carrying out stretching ratio re-allocation and passing the real-time optimized ratios to the stretching ratio controllers. The next layer is the distributed control layer, which is composed of sensors, inference modules and controllers. The lower layer are the actuators, each composed of one driving roller and one driven roller for regulating the fiber tension in each of the stretching zones and receiving control commands from the network layer. Such a network structure makes it possible to realize a cooperative control of the different stretching units, which is critical for achieving high-performance dynamic of the whole system.

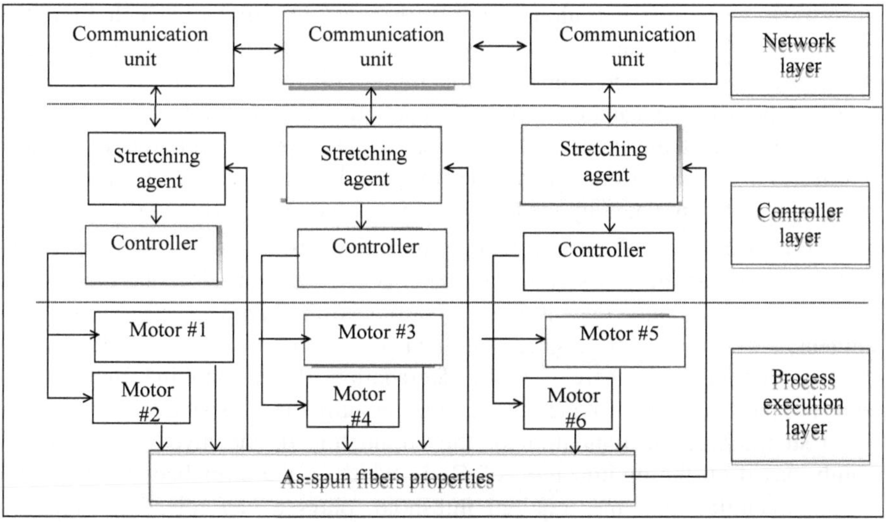

Fig. 3.17 Regulation network of the stretching agents

In the following, from the relationship between the general ACN and the stretching network, the stretching controller inputs can be defined as the different antibody concentrations A_{g1}, A_{g2}, ..., A_{gn}, which are the result of interactions of multiple type cytokines and feedback antigen concentrations. In other words, the controller input deviation can be manipulated from minimizing an objective function that is related to the cytokine concentrations and error between the antigen real-time concentration and the required normal value in human body. In this way, the speed of each driving motor can be regulated at each interval for accurately tracking the final tensile force setting points, which is inspired by this self-adaptive ability of immune system.

$$\dot{C}_j = \varphi_j(A_g, A_{b1}, A_{b2}, \tilde{\theta}_j) - v_j C_j \tag{3.3}$$

$$\varphi_j = \overline{\varphi}_j \times S(w_{j1}A_{b1} + w_{j2}A_{b2} + w_j(A_g - \tilde{\theta}_j)) \tag{3.4}$$

$$\dot{A}_{b1} = f(A_g, A_{b1}, A_{b2}, C_j) \tag{3.5}$$

$$\dot{A}_{b2} = f(A_g, A_{b1}, A_{b2}, C_j) \tag{3.6}$$

$$\dot{A}_g = \psi_j(A_g, A_{b2}, A_{b1}, C_j) \tag{3.7}$$

$$\psi_j(A_g, A_{b2}, A_{b1}) = \overline{\psi}_j \times (A_{g0} - \lambda_1 A_{b1} - \lambda_2 A_{b2}) \tag{3.8}$$

In this modified ACN model, two types antibodies are considered, and A_{b1}, A_{b2} represent the concentration changes of antibodies for eliminating the same invading antigen, A_g is the concentration of this type invading antigens, which should be eliminated by related antibodies. C_j represents the independent cytokine concentration secreted by related immune antigen-presenting cells regarding to invading antigen type, which means the objective function. The final output of this immune system is related to the states of A_{b1} and A_{b2}, which represents different elimination abilities against this same antigen. The function Φ_j expresses the effect to the j-th cytokine that is stimulated by the antigen type and concentrations, and proliferation rate and concentration of corresponding type antibodies. $\tilde{\theta}_j$ is the required threshold of the j-th type antigen. A sigmoid function is used in Φ_j to describe the trend of change of the cytokine concentration according to the elimination requirement of antigens, which is slightly higher than 0 in normal condition for considering the cytokine degradation, where v_j is the degradation of j-th cytokine, and w_{j1}, w_{j2} represent the effect to cytokine stimulation by antibodies, respectively.

To summarize, the regulation of the elimination of the invading antigens is mainly based on the optimization of C_j, in which e, the error between measured antigen quantities and the required threshold, plays a key role. Two types of antibodies concentrations are determined and optimized by C_j respectively, and different representations of C_j leads to the different information allocation results, which also means the different effects on the corresponding antibodies proliferations. These functions facilitate to obtain related cellular immunity elimination against specified antigen A_g. A_{b1}, A_{b2} shows the final effect of re-allocation antibody cell concentrations, respectively, in order to the minimize the antigen concentration, which is inspired from biology.

$$e = A_g - \tilde{\theta}_j \tag{3.9}$$

$$J = \frac{1}{2}\int_{t_0}^{t_1} e^2(\tau)d\tau \tag{3.10}$$

In short, in the immune network, the antigen-presenting cell produces one type of cytokines that can stimulate specific antibodies (usually more than two types and different proliferation rates) proliferation by different type immune cells. In the application of Stretching Network, two different types of immune cells can eliminate one type antigen by means of the related cytokines proliferation, in which two immune cells mean two neighboring driving motors on both sides of a stretching zone, the error between concentration of antigens and the setting value is considered as the tension feedback error, and the distributed control input of two neighboring actuators is modified by this real-time cytokine proliferation rate C_j. The antibody concentration by related immune cells represents the speed of driving motors. The objective function is defined by the feedback error of setting tension value and actual response. To minimize this function, the gradient-based method is used to regulate the two actuators.

In the actual system, the cytokine proliferation rate C_j represents the modified allocation law on these two levels of motor controllers, A_g represents the tension response in each sub-process, and the square of the feedback error is considered as the objective function to regulate the speed of driving motors, which represents the information allocation in these immune cells, specifically, the different dynamics of driving system in the stretching process. For the performance index in (3.10), considering C_j, A_{b1}, A_{b2}, e, the controller inputs are also specified for achieving the minimum function value.

The dynamics of network model can be described by a set of differential Eqs. (3.2–3.9), which can also be used to describe the dynamics of stretching system. We consider this intercellular medium distinct chemical material—cytokines, as the real-time modified states to each sub-process controller at each interval. Generally, we can define $A_{g1}(t)$, $A_{g2}(t)$, …, $A_{gn}(t)$ as the modified inputs of each driving actuator, where $A_{gr}(t)$ represents the input voltage of driving motor. The control vector u of the sub-processes is independently determined by the control allocation strategy, which is different from the conventional PI controller. Therefore, all the control inputs of driving motors are optimized and determined by the control allocation layer for the sake of minimizing of tension feedback error in all sub-processes at the same time, which also bring the final system output T_{en} to a desired target value T_r.

Here we will introduce the operational mechanism of the ACN to be used for the stretching system. Cytokine production is stimulated by the concentration of real-time invading antigens. The density and proliferation rate of this communication medium will also affect the antibody production, which indicates the ability of real-time modification of the corresponding antibody concentrations, and the combined action of antibodies leads to the current antigens elimination. Similarly, considering the interconnections of two driving motors in each process, with the traditional PI control strategy, we can define $A_{gr1}(t)$ and $A_{gr2}(t)$ as the modified control inputs of current r-th stretching zone capable of achieving the real-time optimization for this nonlinear stretching dynamics. This regulation network expresses the iterative effects between multiple types of antibodies and finally antigen elimination based on the proliferation rate of k-th cytokine concentration by means of the production of the corresponding antibodies. So the network control allocation for each unit can be described and implemented in the following form (3.11):

$$u(t) = h(K(t), F(t)) \tag{3.11}$$

Where $K(t)$, $F(t)$ are the real-time control allocation and the PI control law, respectively, both of which comprise of the control input of each stretching sub-process.

To summarize, this network control strategy performs on-line multi-objective optimization to eliminate the tracking errors in all sub-processes at the same time by coordinating the control inputs in each stretching actuator. That is, the modification of each controller input considers the stability and tracking accuracy of the tension on each stretching zone by taking into account of the coupling between the neighboring driving motors,

$$f(e_1, r_1 \ldots e_m, r_m) \rightarrow u_1(t), u_2(t), \ldots, u_m(t)$$

The network behaves essentially like a look-up table which will evolve itself over time, depending on the feedback error and allocation control. In short, Eqs. (3.2–3.7) are used as a platform on which optimization of the inner-loops is implemented and more sub-units can be introduced for collaborated control in the whole system. The optimization method for further re-allocating the set-point of the ratio in each unit will be introduced in the following section.

3.3.2 Allocation Control and Its Optimization Based on the Cytokine Network

In this section, we discuss in detail the online optimization for control allocation, including the problem formulation in the context of stretching processes and control law for each driving actuator. We will show that ratio allocation can be seen as a constrained quadratic optimization and two driving motors on both sides of the stretching zone as directly actuators to realize these stretching ratios.

3.3.2.1 Allocation Control to Stretching Processes

There are many important factors in the stretching process that determine the final fiber quality, such as temperature, pressure, stretching ratios applied in different stretching zones. Traditionally, the stretching ratios are controlled separately, which is not able to be tuned according to the tension requirement for high performance filaments on each level, since various stretching environments require different optimal stretching ratios combination. Without a coordinated regulation of the ratios in these different units only limited performance improvement can be achieved. To resolve this problem, instead of allocating a set of predefined stretching ratios beforehand, online re-allocation of the stretching ratios is necessary to maximize the performance enhancement. In this way, the interaction between the two stretching units can be considered together, and two layers are also defined as the regulation mechanisms in the allocation algorithm within the hardware framework, including network layer in a slower time-scale for modifying the set-points of ratio controllers and the controller layer in a fast time-scale in order to ensure desired overall performance. The primary aim of developing a cooperative ratio control is to allocate the set-points of the control units by optimizing an objective function.

The objectives of the online optimization and control of the stretching ratios include [64, 65]:

1. Ensuring the speed of the inner-loop driving motor close to the desired dynamical states;
2. Restricting the change in the control input compared to the previous sampling instant;
3. After the cooperation of multiple level processes, without affecting the system dynamics, all of which make the system behave the same way as manually regulating the PI coefficients and off-line adjustment of tension tracking.

In this work, we adopt the quadratic programming to describe the optimization problem defined in Eq. (3.12a, 3.12b, 3.12c). The above objectives can be mathematically formulated using the following equations [65, 66]:

$$J_r = \arg\min\left(w_1(T_i(t) - T_{eni}(t)) + w_2\sqrt{K_i(t) - K_i(t - T)}\right) \tag{3.12a}$$

subject to

$$|T_{en}(t - T) - T_s| = B \cdot [u_1(t)\ u_2(t)]^T + \sigma \tag{3.12b}$$

$$K_{\min} \leq K_1(t),\ K_2(t) \leq K_{\max} \tag{3.12c}$$

where the weighting matrix $B \in R^{m \times n}$ represents the transform matrix, in which $m = 2$ is the number of related stretching zones, $n = 2$ is the number of driving actuators in each process which directly regulated by control allocation and can be used to approximately evaluate all desired control inputs with the final tension value. T is the sampling interval. In this case, T_{en1} is the desired tension value of the first stretching zone that should be pre-determined, which can be varied once the tension set-point changes, and T_{en} is the measured tension, σ is a constant used to evaluate the matching degree of control effect and feedback error.

$$K_r(t) = [K_1\ K_2] \tag{3.13}$$

Where $K_r(t)$ is the real control vector for each driving motor, which is feasible with respect to the constraints (3.12c), and will be computed at each sampling instant to minimize the cost function in (3.12a).

The trade-off between the two terms in (3.12a) is determined by the weights w_1 and w_2, in which w_1 aims a faster convergence of the corresponding actuator to its desired position, while w_2 ensures a reasonable moving velocity of actuators.

Hence, it is highly important to determine u_{r} constrained by (3.12c). Using the gradient-based method, and finding a group of online compensation results $K_1,\ K_2 \cdots K_m$ satisfying (3.12a) as the dynamic regulated gains of each driving unit of multiple levels. Tuning of these inputs of the driving motor controllers in the inner loops restricting between -10% and 10% are introduced to prevent the vibration response of system dynamics.

Note that the tension can be approximately considered to be proportional to the combination of control variable of driving motor speeds, which also are regulated by these inner control loops of each driving unit. Based on this assumption, for solving the optimization problem, the resulting control gains in each level stretching can be expressed in the following form:

$$u_r(t) = h_r(K_r(t), F_r(t), e_0) \tag{3.14}$$

For this discrete time dynamic optimization system, h can be found analytically according to the stretching ratio requirement and temperature or pressure. It is also important to point out that although the set-points for the ratios are determined online, no extra time lag will be introduced in solving the optimization problem defined in (3.12a, 3.12b, 3.12c). Consequently, the resulting control input has a linear form as follows [65, 66]:

$$\dot{K}_r = T_{enr} \cdot B \cdot Q \cdot e_0 \cdot [\mathrm{e}_{v1}\ \mathrm{e}_{v2}]^T + \int_0^t T_{enr} \cdot B \cdot P \cdot e_0 \cdot [\mathrm{e}_{v1}\ \mathrm{e}_{v2}]^T dt + K_{r0} \tag{3.15}$$

$$h_r = K_p \cdot e_0 + K_d \frac{de_0}{dt} + [\eta_1 \mathrm{e}_{v1} K_{r1}\ \eta_2 \mathrm{e}_{v2} K_{r2}]^T \tag{3.16}$$

Where $P = \begin{bmatrix} 2 & 1 \\ 1 & 1 \end{bmatrix}$, $Q = 1.5$, e_0 is the feedback tension error of r-th process, e_{v1} and e_{v2} are the inner-loop control error of driving motor velocities. η_1 and η_2 are the related coefficients of two driving motors respectively according to the stretching ratio of current chemical requirements. K_r is 2-D vector representing two driving motor velocity revision factors of r-th stretching zone from the minimization of objective function J. K_p, K_d are also the 2-D vectors. U_{ri} is the inner-loop control law of i-th driving actuator respectively. K_r including K_{r1} and K_{r2} can be updated if needed, namely, especially when the current tension tracking performance is not acceptable. This control method with both self-tuning and conventional PID features is applied in this two-actuator system. In this way, given the desired tension set-point, the continuous or appropriate calculation of valid control inputs for distributed actuators can be determined and simultaneously the collaborative compensation of next level processes are also considered, all of which will be provided with appropriate control inputs and will lead to an approach for overall fiber mechanical performance. This has been demonstrated by the simulation results.

The stretching system frame diagram for realizing this allocation method is shown in Fig. 3.18, including the control allocation block.

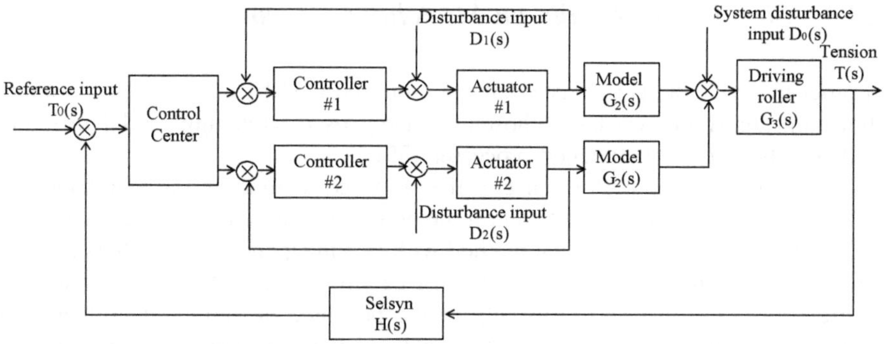

Fig. 3.18 The block diagram of the stretching system

3.3.2.2 Optimization of Control Allocation in the Cytokine Network

As mentioned above, thanks to the online adaptation of the desired stretching ratios, an optimal control of the stretching process can be accomplished, which will result in enhanced fiber quality. This control strategy is easily implemented within the ACN, which has a very similar structure to that of the stretching system. Moreover, the optimized single stage response can be completely fit into the Cytokine Network framework, which is described by (3.2–3.7), where the interaction function φ_j includes a control allocation optimization of the input vector for achieving the optimization results in (3.5) and (3.6). The objective is reformulated in a discrete form as in (3.12a, 3.12b, 3.12c), which aims to ensure smooth operation and accurate tracking of desired tension.

Therefore, the corresponding descriptive functions of each driving motor in the framework of cytokine network can be described as follows [64]:

$$\dot{K}_i(t) = \dot{A}_b(t) \tag{3.17}$$

$$\Psi_{ir}(e, T_{en}, K_j(t), U) = T_{en}U_c + [\eta_1 \nu_1 K_{i1} \; \eta_2 \nu_2 K_{i2}]^T \tag{3.18}$$

Where i represents the number of driving actuator controller, or in the Cytokine Network, the type of antibody in the human body. The function $\Psi_{ir} > 0$ describes the specific and important control law of each driving motors and considers the coupling between the neighboring stretching zones both spatially and temporally, all of which mainly to maintain the tensions of each stretching zone within a certain range.

In this work, the online allocation of the stretching ratios is realized by using the quadratic programming. The obtained ratios are then applied to driving actuators of stretching network model. In the following, the effectiveness of this optimization method embedded in the Cytokine Network will be empirically examined in computer simulations.

3.3.3 Simulation Results on Stretching Network

In order to verify the performance of the proposed CNSN, in particular the online allocation algorithm, we have chosen a widely used stretching model including cold-stretched and hot-stretched processes [35].

To demonstrate the performance of the proposed method, it has been compared with a conventional centralized control method given the same initial conditions. It is expected that the conventional method with two independent controllers can also achieve the desired tension response to a certain degree. However, for a more accurate tension value and especially in the presence of various disturbances in stretching production line, the conventional independent control method will no longer be able to satisfy the requirements. A detailed description of the performance comparison of two control strategies will be given below.

3.3.3.1 The Stretching Process Model

The SIMULINK platform for the stretching process is composed of two main parts: an allocation control algorithm and the network model, as shown in Fig. 3.19.

The state-space representation of the DC motor used for driving the stretching roller can be described as follows:

$$\frac{d}{dt}\begin{bmatrix} i \\ \omega \end{bmatrix} = \begin{bmatrix} \frac{-R}{L} & -\frac{K_b}{L} \\ \frac{K_m}{J} & -\frac{K_f}{J} \end{bmatrix} \times \begin{bmatrix} i \\ \omega \end{bmatrix} + \begin{bmatrix} \frac{1}{L} \\ 0 \end{bmatrix} \times \upsilon_{app}(t) \tag{3.19}$$

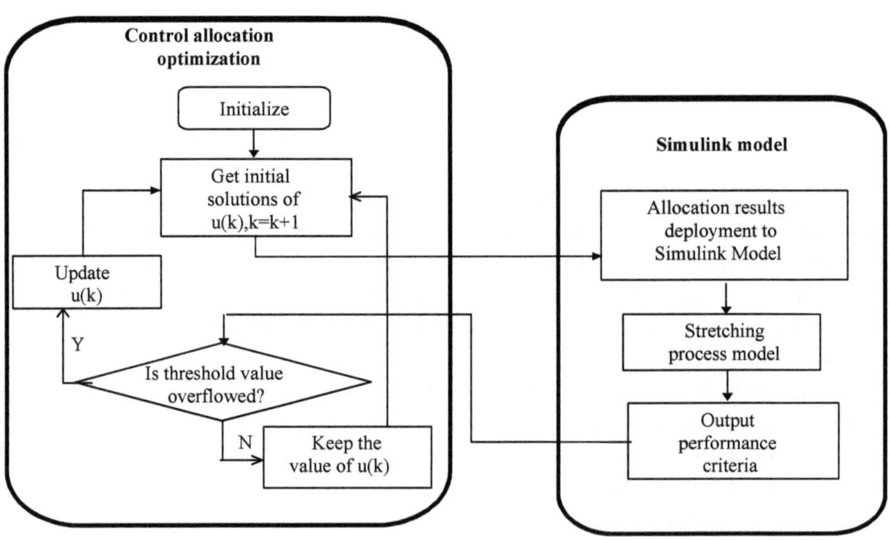

Fig. 3.19 The diagram of simulation platform

Table 3.6 Parameter list of DC motor

Parameter	Value	Unit
R	2.0	Ohms
L	0.5	Henrys
Km	0.015	torque constant
Kb	0.015	emf constant
Kf	0.2	Nms
J	0.02	kg.m^2

Where, i is the current, ω is the angular rate, v_{app} is the applied voltage, R is the resistance, and L is the self-induction.

The parameter settings of the DC motor used in this work are shown in Table 3.6. Given these parameters, we can get the transfer function representation of the DC motor as follows:

$$G_{motor} = \frac{1.5}{s^2 + 14s + 40.02} \tag{3.20}$$

The motion control model has four motors in the stretching system. In this experiment, the two stretch ratios are originally fixed to 1.50 and 3.50, respectively. In this constant stretching ratio control system, we consider that the whole stretching system consists of two spatially separately distributed units with cooperated controllers, and these controllers share the corresponding input instructions from control allocation block. For example, we take the structure of the cold-water stretching process in Fig. 3.20, which is almost the same as the hot-water unit except the environment parameters representing the temperature. The driving roller R1 handles the filaments waiting for stretching. The roller R2 controls the stretching ratio of the filaments by taking a motor as its driver. The velocity ratio of the two motors viewed as the stretching ratio the process needs. The dynamic ratio of the filaments can be maintained by such a feedback loop. The diagram of feedback

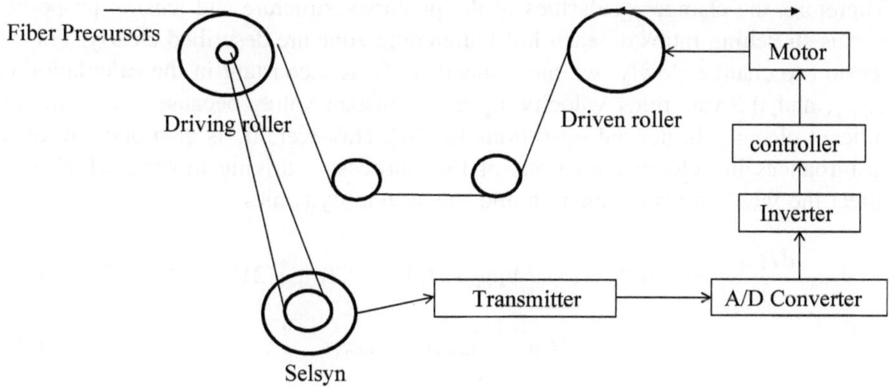

Fig. 3.20 Speed control system of stretching unit for fiber production

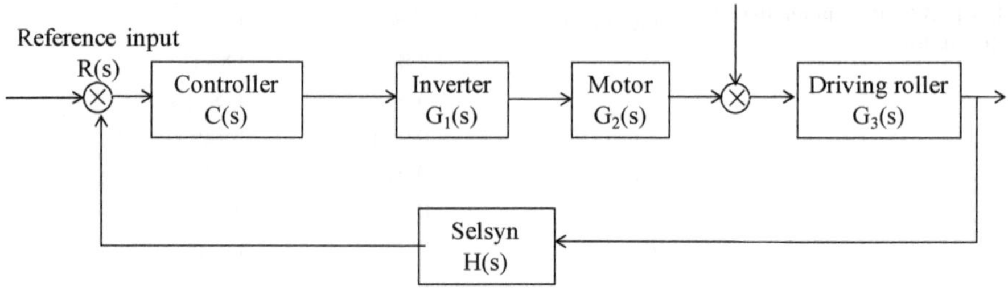

Fig. 3.21 Control diagram of the inner-loop speed regulation system

inner-loops of the separate units with gains allocation is shown in Fig. 3.21 and the closed-loop transfer function of the wholly tension control system can be written as

$$Y(s) = R(s)G_0\left(K_1\frac{(D_1(s)+C(s)G_{21})}{1+C(s)G_{21}(s)}G_{31} + K_2\frac{(D_2(s)+C(s)G_{22})}{1+C(s)G_{22}(s)}G_{32} + D_0\right) \tag{3.21}$$

where $R(s) = 6$ in Eq. (3.21) represents the reference input of stretching unit #1, which further denotes the speed of two driving rollers, $N = 50$ is the interference input, and $H(s) = 0.5$, $G_{21} = G_{22} = \frac{1.5}{s^2 + 14s + 40.02}$, $G_{31}(s) = G_{32}(s) = 7.6/(0.8s + 1)$, and $G_0(s) = \frac{0.6515}{s(s+2)}$ represent the transfer function of the driving motor R2 and driving motor R1, stretching roller, and fiber tension dynamics, respectively.

In the fiber stretching system, for attaining the fiber tension dynamics, according to the Hooke's law and Coulomb's law and considering the special condition of PAN precursors, the transfer function of G_0 is derived from Eq. (3.22) for different fiber spans, temperatures inside the water bath and stretching ratios. The transfer functions for describing precursor dynamics are derived based on hydromechanics of fiber behavior and the dynamics of driving motors used for stretching process. Therefore, the change regularities of the precursor structure and tension properties versus stretching ratios on each level stretching zone are described clearly. For the tension to change slowly, we have specified T_k as a constant in the calculation of T_{k+1}, and the variations velocity η_v as a constant value, because it has no significant changes in normal operations [8, 26]. However, η_v is also one source of disturbances in velocity variations of two successive driving motors, which may affect the whole transfer function and the system dynamics:

$$L_{\text{span}}\frac{dT_{k+1}}{dt} = EA(V_{k+1} - V_k) + T_kV_k - T_{k+1}(2V_k - V_{k+1}) \tag{3.22}$$

$$G_0 = G_{sys1} - G_{sys2} \tag{3.23}$$

where

$$G_{\mathrm{sys1}} = \frac{EA}{L_{span}s + \eta_v}, \quad G_{sys2} = -\frac{EA - T_k}{L_{span}s + \eta_v}, \quad \eta_v = 2V_{k+1} - V_k$$

For this general expression of web stretching system, the modulus of elasticity E and A may change along with the same certain probabilities, the product of both is regarded as parameter disturbance. L_{span} is the length of span between two successive driving rollers in one single stretching zone. To eliminate the variations of E and A of precursors in the plant during the dynamic response, the traditional control strategy that cannot change the PI gains of two driving motor control loops is not enough. Therefore, this novel method has significantly improved the tension tracking performance by gain-tuning in these driving actuators. The complete stretching model including two feedback loops of above system is shown in Fig. 3.22. First, when the cold-water and hot-water stretching units are considered separately, the inner-loop of each unit is independent, and each has its own influence on the output tension. In this case, the controller for each unit can be designed separately. In classical approaches, no coordination between the two units is taken into account, and thus, less accurate tension tracking can be obtained. However, this proposed online optimization method considers the model of velocity set-points reconfiguration of two-level driving motor versus the final fiber tension response for regulation mechanism of this distributed inner-loops allocation. Based on this descriptive optimization model and distributed control of multi-span precursor stretching zones, the inter-connected stretching ratios can be determined for the final tension requirement.

In the simulations, we assume that the roll inertia J_k and roll radius R_k remain constant, and a sticking zone between fibers and the rollers always exists.

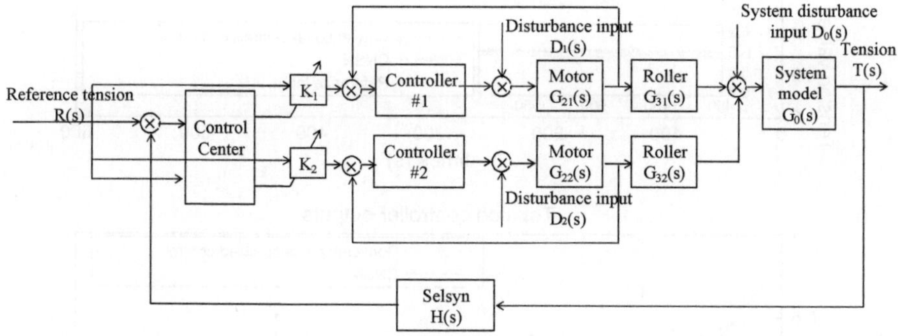

Fig. 3.22 The simulated diagram of overall stretching system

3.3.3.2 Optimization Experiments on the Stretching Network

In this section, experiments of using the proposed CNSN with the optimized stretching ratios allocated to each unit are performed to compare its performance with that of a conventional centralized PID method. In the traditional method, the stretching ratios are often predefined and remain unchanged, and tracking control in the different units is excised separately. Because of the fixed stretching ratios, the traditional method cannot reject disturbances often seen in the system completely, which will degrade the performance of tension tracking. In the proposed method, the stretching ratios are optimized real-time, which enables accurate tracking even in the presence of disturbances.

To demonstrate the effectiveness of the proposed controller, several simulations are conducted. The output responses of the system controlled by the proposed CNSN and the conventional control method are shown in Figs. 3.23, 3.24, 3.25a, b, 3.26a, b, 3.27a, b. Simulations for examining the ability of disturbance rejection of the CNSN are carried out by applying perturbations to the parameters of tension dynamic. Such disturbances are often seen in practical systems, where the states of the stretching system may experience small fluctuations even under normal conditions [5, 30].

Phase I- Tracking Performance

In each simulation, 600 sampling periods are given as the total time in one run and the sampling time $T = 0.05$ s, which is the same as in Phase II presented below. The optimization process can be observed through the experiments, which demonstrates the reason for CNSN to be able to achieve acceptable control

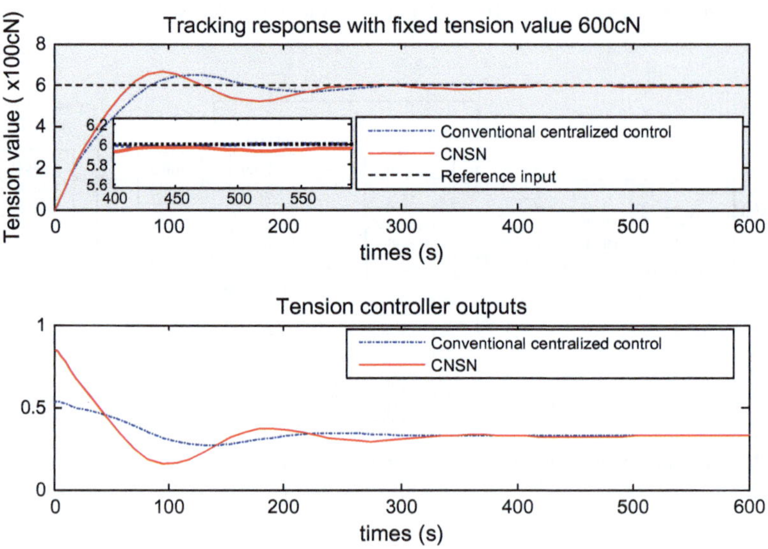

Fig. 3.23 The system response of two control strategy in normal condition

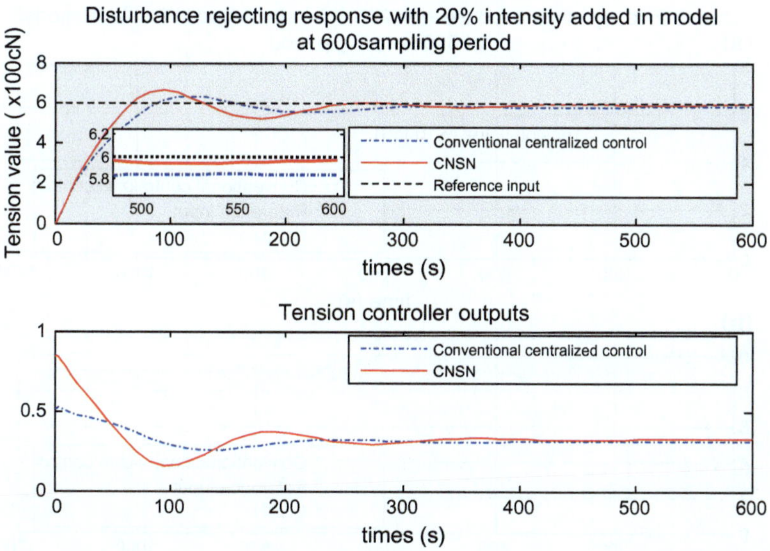

Fig. 3.24 The system response of two control strategies in the presence of disturbances

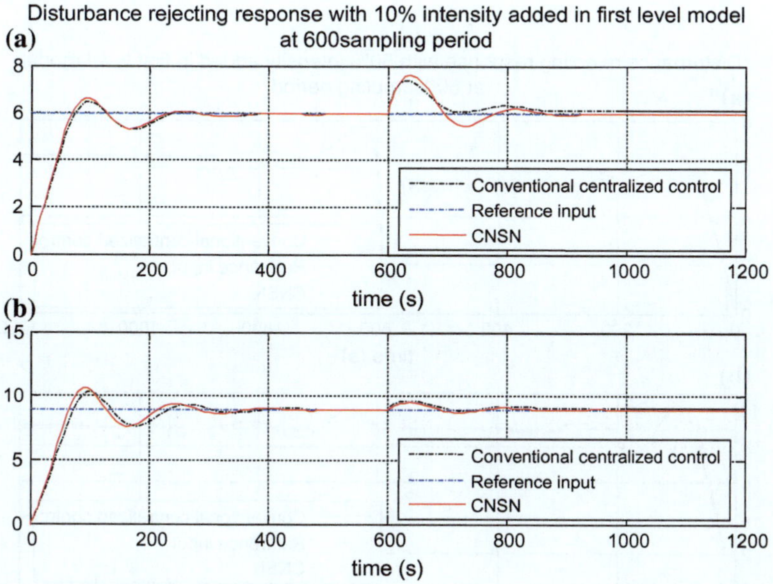

Fig. 3.25 The response curve of 10% disturbance rejection. **a** the first level response; **b** the second level response

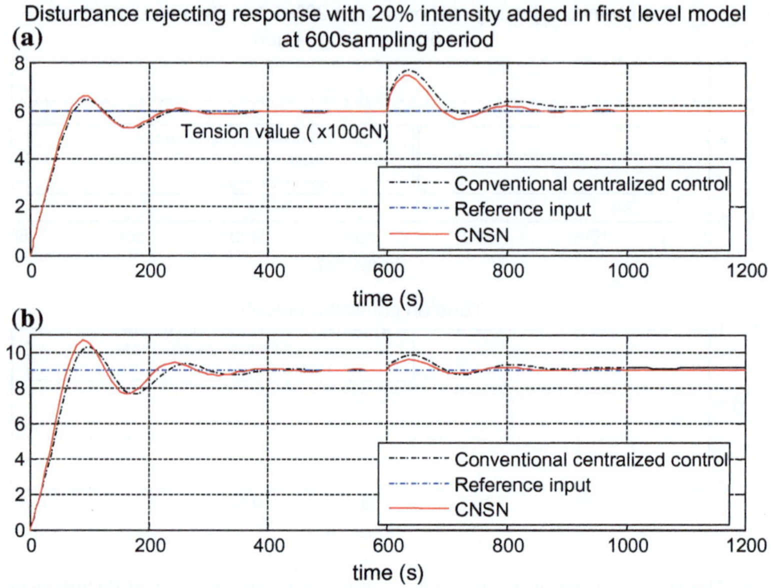

Fig. 3.26 The response curve of 20% disturbance attached. **a** the first level response; **b** the second level response

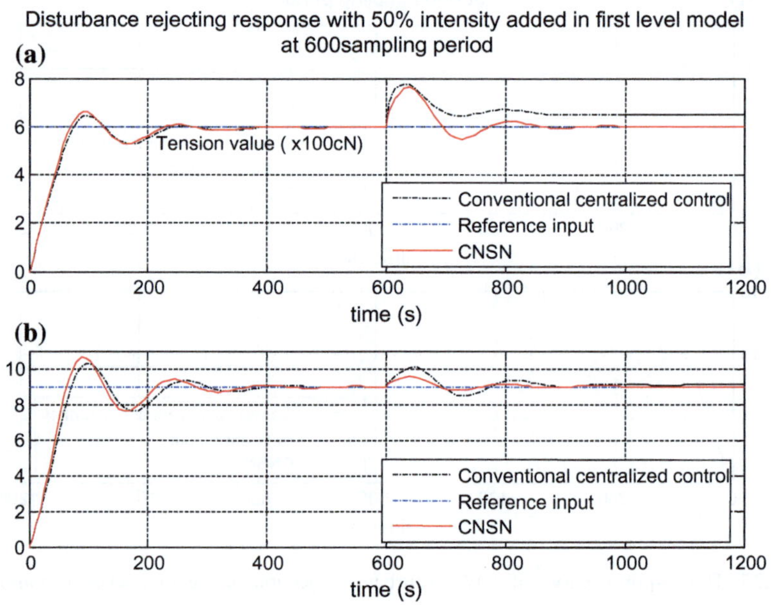

Fig. 3.27 The response curve of 50% disturbance. **a** the first level response; **b** the second level response

performance in normal condition. This also indicates that the introduced online ratio allocation algorithm works very well. Both the proposed CNSN and the conventional controller adopt PID control in each inner-loop unit. The system will first regulate the tension to a stable status with an initial preset final tension. That is to say, the simulation of this phase is regarded as verifying the tracking performance of the two control schemes.

Figure 3.23 shows the response curves of the two controllers, which are almost the same. The actual optimized set-points of stretching ratios are almost the same as the predefined ones. Both outputs satisfy the constraints and produce acceptable tracking of the required tension. So both control methods can achieve acceptable performance in the absence of disturbances.

Then we perturb the parameter η_v with 80% intensity in system model, which is a combination of the velocities of two successive driving motors. As we can see, it seriously affects the tracking performance. In this case, although the tracking response in each separate control loop can be maintained at a desired value, the final performance may not be achieved, which is the most straightforward way to verify the performance of two methods in the presence of disturbances. As seen from Fig. 3.24, compared with the conventional PID controller, the final tension response of the CNSN with coordinated stretching ratio control has exhibited much better tracking accuracy, in which the set-points for the ratio controllers of inner-loops are recalculated according to the objective function and are proved to be an appropriate allocation.

The purpose of the experiments in Phase I is to demonstrate the advantage of the online reallocation of the stretching ratios and show the capability of distribution of cooperative control instructions on these direct driving motors in two-level stretching processes offered by the proposed CNSN in comparison with the conventional PID controller.

Phase II - Disturbance Rejection

Each experimental period for a control loop was chosen to be 1200 simulation time. Three experiments in the presence of different levels of disturbances, namely, 10, 20 and 50%, to the fiber tension dynamics of the cold-water stretching process were conducted using the centralized PID controllers and proposed method, respectively. In these experiments, the driving motor speeds are correspondingly changed for achieving the current level tensile force set-point in the proposed method. In this set of simulations, the initial system state is in a steady state before the regulation begins. Disturbances are then added into the stretching system at the 600-th sampling period to test the robustness properties of the proposed system. In practical fiber production systems, the precursor line is subject to various disturbances. Two types of disturbances can often be found during the normal operation. The first one exists in the inner-loops caused by changing motor loads or unpredictable parameter variations since the radii, fiber material properties and inertia are time-varying. The inner-loop controllers are just used to keep the corresponding stretching ratios stable. The second one is that it leads to tension fluctuations or excess of the allowed maximal tension, the focus of this control allocation is to maintain precursor tension within the pre-set range. Three simulations will be carried out for checking the system response against the second type of disturbances. The aim is to verify the ability of disturbance rejection on two-level tension responses when each inner-loop set-point is online reallocated according to optimization algorithm. Note that in Figs. 3.25a, b, 3.26a, b and 3.27a, b, the steady-state of fiber tension is closely related to the posed control set-points, and their respective responses against disturbance appeared at the 600-th sampling period all have a sharp and quick decrease. These figures show that the control strategy both in the centralized control strategy and in CNSN result in stable and accurate tension tracking, but the former cannot satisfy the tension requirements. This means that the PID controller is able to reject the first type disturbances in the system, but lacks the ability to reject the second type. We will show, however, that the situation will change in the second method, whose results are also given. From these results, we can see that the tracking performance deteriorates due to the fixed ratios in the classical control method after disturbances are introduced, while the CNSN can maintain the steady-state tracking accuracy even in the presence of the disturbances due to the online reallocation of the desired stretching ratios. From the second level stretching process dynamic response in (b) of these figures, the final tensile forces of the two methods also show the different tracking performances, which can also be attributed to the fact that the inner-loop feedback in the classical method just regulates the velocity of their owns, but the influence of driving motor velocity fluctuations in the upper level cannot be eliminated. Although both methods can be regulate the tension after disturbances, but the steady-state tension tracking error of the centralized PI control strategy is much larger. Refer to Fig. 3.25a, b, 3.26a, b, and 3.27a, b for detailed tracking performances of the PID and the proposed controller.

Table 3.7 and 3.8 show additional results that compare the tracking and disturbance rejection performance of the two control schemes, including the overshoot percentage, recovering time and the settling time. The results indicate that both the conventional PID control and the proposed CNSN show very good tracking of the initial setup value and have obtained almost the same overshoot percentage and settling time. But the conventional PID control is more sensitive to the disturbances, especially in the tension tracking. Due to the lack of ability to coordinate the two sub-processes, the conventional PID controller exhibits much weaker disturbance rejection performance, especially when the second type disturbances are present. By

Table 3.7 Numerical performance comparison between two control schemes in normal condition and 80% parameter η_v disturbance attached

	Tracking performance		System parameters interference	
	CNSN	CC-PID	CNSN	CC-PID
Initial overshoot/%	12.7	10.1	12.53	12.58
Settling time/s	300.00	290.00	300.00	290.5
Recover time/s	304	288	300	290
Steady state error/$\lvert\cdot\rvert$	0.30	0.33	0.31	0.61

Table 3.8 Numerical performance comparison at first level between two control schemes in abnormal condition of 10, 20 and 50% disturbance attached

	10% Disturbance rejection		20% Disturbance rejection		50% Disturbance rejection	
	CNSN	CC-PID	CNSN	CC-PID	CNSN	CC-PID
Initial overshoot/%	12.9	11.1	12.8	12.7	12.9	12.9
Settling time/s	213	210	232.0	231.0	215	216
Disturbance overshoot/%	21.6	21.3	23.9	24.5	22.4	24.3
Recover time/s	220	210	220	240	220	230
Steady state error/$\lvert\cdot\rvert$	0.33	0.42	0.32	2.92	0.33	5.12

contrast, the regulation time and the accuracy of set-point tracking of the CNSN are much better than that of the PID controller. Note, however, that this performance gain is at the price of a slightly higher overshoot, more tension fluctuations and a slightly longer recovering time during the redistribution period. To summarize, the proposed CNSN aims to online reallocate the set-points of the stretching ratios in the different units in the presence of disturbances, which leads to more accurate tracking performance and a stronger ability for disturbance rejection.

3.4 Immune-Inspired Collaborative Control Allocation

In this section, we propose a novel immune-based self-adaptive cooperative control allocation (ICCA) method for the multi-level stretching processes [40, 67] , and the ICCA method optimizes the coupled stretching ratios allocation at different levels. In order to demonstrate the effectiveness of the proposed method, the ICCA is applied to the multi-level stretching systems in PANCF production line, and compared with the conventional centralized PI control method and the cytokine network-inspired cooperative control (CNCC) method proposed in [2, 68]. The centralized PI control method performs very well for each single-level stretching process but not the overall performance. The CNCC method has considered more in the coupling effects of tensions between neighboring stretching zones existing in an actual system. Simulation results show that ICCA has better performances in tracking the desired tension, reducing fluctuations, and rejecting disturbances of coupling variables between tensions and driving motor speeds [5, 50, 69].

3.4.1 The Multi-Level Inter-Connected Stretching Model

In this section, the structure of stretching system as part of complete operation of several consecutive driving motors is provided first. Then, taking the multi-level stretching process in fiber production line as an example, an inter-connected mathematical model is built. The dynamics based on the driving actuators are also presented for the whole system. This model has been used to design the controller and determine the selected reference model under optimal conditions with the same order of operated plant.

In what follows, we take the multi-level stretching process in fiber production line as an example and build an inter-connected stretching dynamic model. The dynamic behavior of fiber tension between two successive rollers is considered as the most critical factor in controlling the whole system, which is mainly based on the Hooke's law and the Coulomb's law [8, 43].

According to the Hooke's law, the tension variation is considered as the speed difference between two successive rollers, and the precursor strain ε is related with the tension T:

$$T = EA \cdot \varepsilon = EA\frac{L - L_0}{L_0} \tag{3.24}$$

$$L - L_0 = \eta_v(v_2 - v_1) \tag{3.25}$$

where E is the Young's modulus which is sensitive to the changes of bath and as-spun fiber temperature. A is the fiber precursor section and assumed to be constant. $L - L_0$ is the precise elongation under the stretching stress which is proportional to the velocity difference between two driving motors, and η_v is the conversion coefficient.

Note that the tension variation is related to the velocity difference between the successive rollers. The angular velocity of the driving motor can be directly represented to the speed of the as-spun fibers from the input voltage. The transfer behaviors of the relationship between transporting velocity of precursors on rollers #1, #2 and the tension t_1 are given as follows:

$$L_a\frac{di_a}{dt} + R_a i_a + E = u_a \tag{3.26}$$

$$u_i = K_t i_a - K_t i_l \tag{3.27}$$

$$J_a\frac{d\omega_0}{dt} + f_0\omega_0 = (t_1 - t_0)R + n_0 \cdot u_0 \tag{3.28}$$

where n is the gearing ratio between the driving motor shaft and the roller shaft, $n \cdot u_0$ is the friction torque. L_a is the mutual inductance of field-winding. $E = k_d \cdot \omega$ is the counter-electromotive force. K_t is the torque coefficient. u_i is the input torque from the i-th motor, and $u_i = \frac{k_v}{v}$. $\upsilon = \omega_f \cdot R$ is the fiber velocity, where ω_f is the angular velocity of the motor, R is the radius of the driving roller. It is confirmed that the fiber velocity attached with the roller is related to ω_f. $J_a = n_0^2 J_m + J_f$ is the moment of total inertia, where J_m is the rotating element on the motor side which includes motor armature, driving pulley, driving shaft, etc., and J_f

is the inertia due to viscous material of precursors, f_0 is the viscous friction coefficient.

Based on above description, the tension variation model of each stretching zone can be built as follows.

1. The first process section: The tension variation of the first stretching zone is approximately described as:

$$L_1 \frac{dt_1}{dt} = EA(\upsilon_1 - \upsilon_0) + t_0\upsilon_0 - t_1\upsilon_1 \tag{3.29}$$

where t_1 is the precursor tension state of the first roller, t_0 is a constant as the initial tension before the precursor fiber attaches the first roller, L_1 is the length between two driving rollers in the first section, which is different according to the processing condition of specified stretching.

2. The second process section: The precursor tension and velocities of neighboring rollers are given by

$$L_2 \frac{dt_2}{dt} = EA(\upsilon_2 - \upsilon_1) + t_1\upsilon_1 - t_2\upsilon_2 \tag{3.30}$$

$$J_a \frac{d\omega_1}{dt} + f_0\omega_1 = (t_2 - t_1)R + n_0u_1 \tag{3.31}$$

where t_2 is the precursor tension state of the second roller, L_2 is the length between two driving rollers in the second section

3. The third process section: The precursor tension and velocities of neighboring rollers are given by

$$L_3 \frac{dt_3}{dt} = EA(\upsilon_3 - \upsilon_2) + t_2\upsilon_2 - t_3\upsilon_3 \tag{3.32}$$

$$J_a \frac{d\omega_2}{dt} + f_0\omega_2 = (t_3 - t_2)R + n_0u_2 \tag{3.33}$$

where t_3 is the precursor tension state of the third roller, L_3 is the length between two driving rollers in the third section.

4. The master speed roller: The dynamic of the master speed roller is given by

$$J_a \frac{d\omega_3}{dt} + f_0 = -t_3R + n_0u_3 - \frac{t_w}{2\pi}R^3\omega_3 \tag{3.34}$$

where t_w is the final thickness of fiber precursors.

From above representations, by choosing the dynamic of these sections as $x_1 = [t_1 \upsilon_0]^T$, $x_2 = [t_2 \upsilon_1]^T$, $x_3 = [t_3 \upsilon_2]^T$, $x_4 = \upsilon_3$, $y(t) = [0\,0\,0\,0\,1\,0\,0]x(t)$, $x(t) = [x_1\, x_2\, x_3\, x_4]^T$ is represented to be the state of the nonlinear dynamic system [4, 30].

3.4.2 Immune Self-Adaptive Collaborative Control Allocation

Based on the structure of single stretching zone by controlling input voltage of driving actuators as shown in Fig. 3.28, the ICCA method is proposed for the adaptive tracking as shown in Fig. 3.29. The ICCA consists of two parts, namely a controller level and an optimization level.

In the controller design point of view, the ICCA is a combination of two control methods, including feedback control and the self-adaptive control. The feedback control is mainly used in inner-loops to realize the sub-system stability and eliminate the transient errors. The self-adaptive control is used to minimize the large amplitude fluctuation during the dynamic response. This improved architecture has been promoted by the immune-inspired controller that specifically enables the implementing of optimization results among the actuators of multi-level stretching systems.

On the other hand, the dynamical optimization for searching the optimal stretching ratios is another innovation aspect. It is used for searching the current optimal states under the determined upper optimal tension value. This is also referred to as the Markov property, i.e. the stretching ratio at spatially earlier stage is going to be calculated after the tension value of upper layer has been determined in the optimization process. In the sequel, the self-adaptive control, gain tuning method and control allocation are introduced.

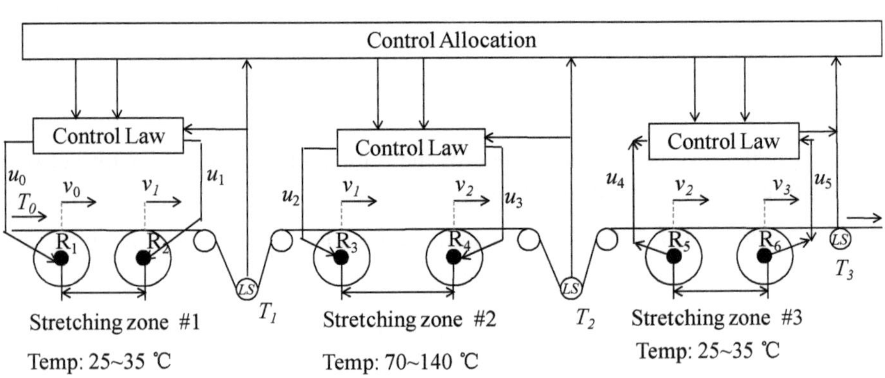

Fig. 3.28 The structure of multi-level stretching system

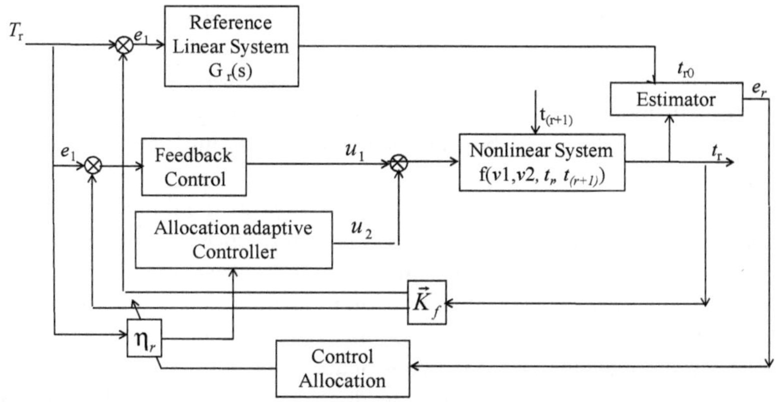

Fig. 3.29 The control structure for the stretching system

3.4.2.1 The Structure of Immune Collaborative Control Allocation

The immune system with the capabilities of antibody acceleration and restrain is to operate the multiple type antigens elimination at the same time. Based on this mechanism, the optimal control allocation of the decentralized driving motors is firstly inspired and achieved mainly by the combination of both the feedback control u_1 and the gain-tuning control u_2.

The output of the ICCA to the nonlinear system is

$$u_r = u_{r1} + u_{r2} \tag{3.35}$$

where $u_1(t)$ is the output of the feedback control, and $u_2(t)$ is the output of the self-adaptive immune allocation control, respectively [58].

$$u_{r2}(t) = k_g c \cdot e_r(t) + \eta_r \cdot e_r(t) \tag{3.36}$$

where r is the number of the current sub-process, $\eta_r(t)$ is the update law from the immune allocation controller, c is a first-order inertia low-pass filter in the u_2 feedback loop to enable the desired transient performance by increasing the adaptive gain. $k_g = \begin{bmatrix} k_{ga} & k_{gb} \end{bmatrix}^T$ are the open-loop tuning gains to two driving motors. The reproducing of reference response is met when the adaptive gain vector $\eta_r(t)$ is chosen suitably, so the estimated error between the system and model output convergences, and the output tension in the current level is within a stable state.

As mentioned earlier, the dynamic of the stretching sub-processes are endowed with nonlinear characteristic, not the variation of system parameters or load disturbance in two neighbor driving motors coupled with each other. For this type of system, there is a manipulated range of each level stretching ratio, which should be optimized online to lead to less fluctuation of tension output. This flexible allocation of stretching ratios refers to an infinite dimensional problem, which is supposed to be an optimization problem using a nonlinear dynamical programming technique. The maintenance of steady and accurate tensions in each stretching zone by re-allocated stretching ratio of relative level is necessary for the whole operation [2]. The above problem will be improved by a reversing optimization and distributed control.

For the stretching inner-loops, the driving motors as the direct actuators of this nonlinear system are used to regulate the stretching ratios as well as the transport gains in each controller. They are important compensations for achieving an acceptable steady state tension responses and ensuring continuous operation with no fold and break of precursors. In the optimal control point of view, this procedure is divided into two levels, inner-loop control and outer-loop optimization. Actually, the speed control in inner-loop includes feedback controller to regulate the speed of these actuators and the immune-inspired controller with adjustable gain tuning. The outer-loop is mainly for recursive convergence by reversing optimization method. Together, both of them constitute the basis of control allocation of ICCA to cope with the parameters variation and nonlinear characteristics in this complex system. Figure 3.30 shows a block diagram of the immune adaptive gain-tuning controller on the speed regulations of two motor-driving rollers in a single level process, where a standard response error e_r of the actual system and the reference model is used to generate the control signals and drive the gain-tuning.

Based on this novel control scheme, the ICCA method is designed as a general method to cope with a type of multiple-level, linearizing or plant-model system. In practice, the stretching ratios may vary within some known bounds, and an acceptable plant operation is not easy to attain because of the changeable conditions. In this situation, the ICCA method generates the most appropriate stretching

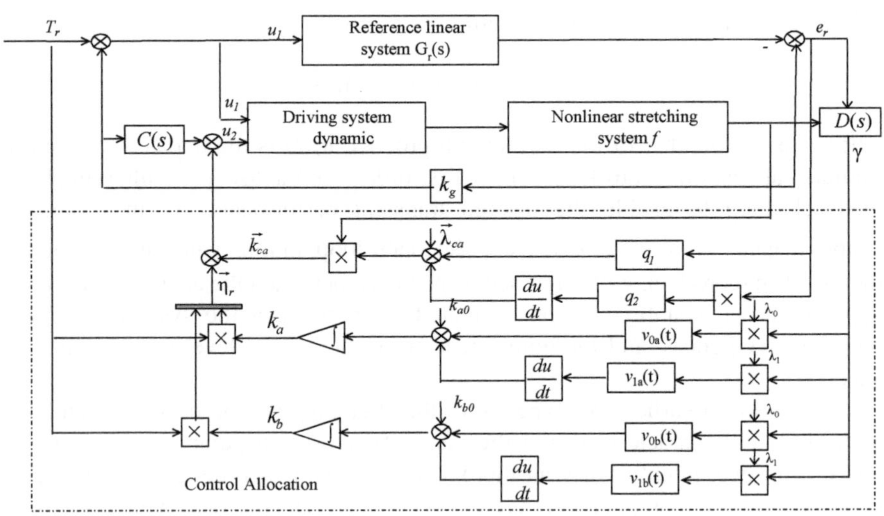

Fig. 3.30 The self-adaptive control structure of single-level stretching process

ratios by dynamical optimization firstly. It is directly driven by the errors between the idealized trajectories and the plant outputs. The calculated solutions are the critical factors for eliminating fluctuations of tension output. Both the calculated gain driven regulation and the conventional inner-loop control are integrated into the self-adaptive collaborative control. Then the updated stretching ratios in a sampling interval around lead to the approaching to optimal tension responses.

In addition, for the self-adaptive controller, a reference model is introduced by assuming that the tensions of neighbor stretching zones are determined by the desired target values. This assumption helps us attain the desired and linear response trajectory to follow, and with the help of this design, the determined tensions can be tracked effectively by the reversing optimization of dynamic programming.

3.4.2.2 The Immune-Inspired Gain-Tuning Controller

The conventional PID control law is widely used for the distributed actuators, which has a compact structure, good dynamic or static characteristics. However, its parameters are unalterable to be difficult in adapting to the changing conditions, especially in dealing with stable tensions tracking under a nonlinear nature. The approach is to design a compensator with the desired self-adaptive states of the sub-systems, and an inspired controller with gain-tuning function is proposed [39].

In this phase, with the control allocation and the output tension evaluation, the immune gain-tuning controller is determined to drive the actuators to handle the dynamic parameters varying with disturbances. As such, the fast dynamic properties and the accurate static properties can be achieved by different parts of the immune-inspired control law described as follows [4, 39].

In order to design a self-adaptive control structure, we imitate the mechanism of immune response against antigens as shown in Fig. 3.31. In Fig. 3.31, the inter-connected biological mechanisms of B-cell, T-cell (including T_s-cell and T_h-cell),

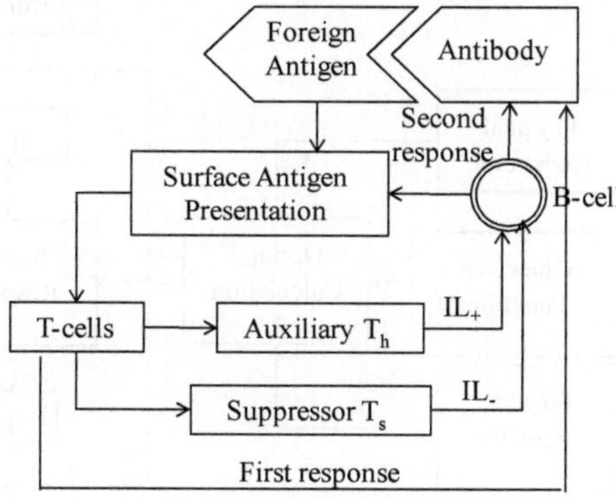

Fig. 3.31 The mechanism of immune response against antigens

antigen and antibody are critical to diagnosing and restoring the balance of complicated internal environments. The immune circulations, identifications and reactions by T-cells in the bloodstream are considered as the primary and fast immune response to the antigens. Along with the accurate adjustment by proliferations of expanded auxiliary and suppressor T-cells, both of them comprise the completely ultimate antigens clearance.

In design of the immune-inspired self-adaptive controller, T_s and T_h represent the online regulations with different functions to the final control input. T_s eliminates the fluctuations of outputs, while T_h mainly helps to tracking the objective response rapidly. The IL_+ and IL_- are used to represent two control effects [59, 70]. The structure of immune-inspired self-adaptive controller is designed as shown in Fig. 3.32.

The immune-inspired self-adaptive controller for online gain regulating in each closed loop of stretching zone can be described as follows [30, 70].

$$e_r(s) = t_r - t_{r0} \tag{3.37}$$

$$u_{r2}(\mathrm{s}) = k_g \cdot C(s) \cdot t_{ir}(s) + \eta_r e_r(s) \tag{3.38}$$

$$\eta_r(t) = g(v_0, v_1, \gamma) \tag{3.39}$$

$$\eta_r(t) = [k_a\, k_b]^T \in R^2 \tag{3.40}$$

where r is the number of the stretching units, $\eta(t) \in R^2$ including the correction factors k_a and k_b is the adaptive law to regulate the secretion speeds of antibodies Th and Ts, and finally converting to the control signals of the secretion of B-cells.

The strictly positive error $e_r(t)$ is obtained based on a linear compensation. $v(t)$ is the solution of the control allocation algorithm k_b from minimizing the objective function (3.41), which is the key factor directly leading to the regulation of k_a. $g(\cdot)$ is the empirical adjustable function according to the reverse revision of dynamics.

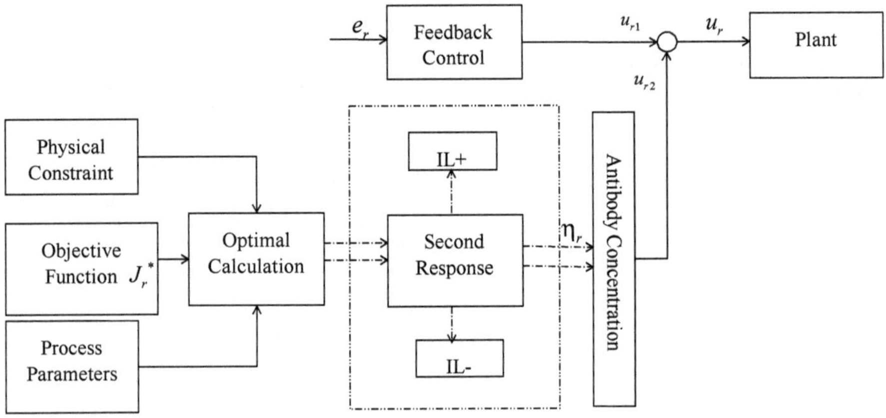

Fig. 3.32 The structure of the immune-inspired self-adaptive controller

For practical application of this strategy, the control structure is designed for implementing the online optimization driven by the real-time dynamical tracking error. Also, it is driven to directly track the desired dynamics of reference by regulating the driving motor speeds.

3.4.2.3 The Collaborative Control Allocation

1. The general optimization of control allocation

Now we consider on how to predict especially online the best combination of the stretching ratios within the acceptable ranges. At first, a special control allocation formulation is constructed in single level for the regulation effect on manipulated variables using standard numerical quadratic programming online. Secondly, the relationship of stretching sub-process is considered. In this section, a reversing optimization from the last stage level #3 to level #2 is also introduced, which is a special method to progressively optimize the final combination of the stretching ratios. For the single stretching zone, with hypothetical mapping from the desired commands of the reference model, a nonlinear estimation between dynamics of these single stretching rollers and the required response is provided. Theoretically, it facilitates to apply these extended actuator constraints. An idealized description of this interconnection system is used here for the generation of the cost function [5, 37].

The control allocation optimization for the single level is a reversing derivation process. It is used to determine the driving motor speeds for tracking the optimal tension value of upper layer by reference model. The optimized tensions lead to the mathematical calculation of idealized dynamical model of the lower layers. It is definitely that the reversing search for the optimal stretching ratios leads to minimize the decoupling effects of neighboring tensions. Therefore, the control allocation optimization needs to develop first the mathematical representation of speeds in multi-level DC motors with tensions. According to the actual working condition, the input voltages of these driving motors are chosen as the manipulated variable η_r, in which the dynamical control allocation directly leads to the desired tension response in all the stretching zones.

For this optimization process, the cost function of single level stretching process for the relations is established, which is designed for reversing optimization of the dynamic and static performance of the follow-up processes as shown in (3.41). The first term is used to describe the effectiveness of the control inputs of the driving motor on the tension errors and speed constraints. The second term is to confine the control energy.

$$\underset{u_r(t)}{J_r(u)} = \min\left\{\int_{t_0}^{t_f} L(e_r,\ u,\ t) + \frac{1}{2}[u_r(t)]^2 dt\right\} \tag{3.41}$$

$$e_r(t) = t_r(t) - t_{r0}(t) \tag{3.42}$$

where t_f is a finite time chosen for the integral approaching to the steady state error, and $L(k)$ describes the loss for alternative combinations of manipulated variables (loss imposed by the deviation with the idealized sub-system outputs $\vec{t}_{ri0}(t)$ at the k-th sampling interval).

For the control law, $u_2(t)$ in (3.38) is the compensation part of control signal to realize the stretching ratios directly, which is obtained by minimizing the objective function (3.41). From (3.41), with all specified degrees of freedom of the driving motors optimized, the multiple stretching ratios are re-allocated to yield an extended vector u_2. The objective is also met based on the control variation $\frac{1}{2}[u_r(t)]^2$ at each control interval and the integral absolute error $e_r(t)$, which are appropriately for smoothness of control inputs to the driving inner-loops and the stable response equal to desired value from t_{r0} to t_r.

Secondly, considering the coupling effect, the physical and special process constraints together, the single objective function (3.41) mapping a manipulated variable vector into precursor tensions is not enough. For example, u_2 in the stretching zone #2 also leads to a new state for the neighboring unit, so the multilevel optimization of these manipulated variables will be considered. In this situation, the tension states of each level need to reach the stable and accurate tracking by the inner-loop control. $t_{ri0}(t)$ is the response of the reference model, $t_{ri}(t)$ is the i-th actual system response in (3.42).

To achieve this, the mathematical representation of the quadratic programming to describe one of the optimization problems is defined in (3.43), where the appropriate allocation of each control signal and convergence speeds are considered together.

$$\min_{u(t)} J_r(\eta_r) = \varphi_r[t_r(N), N] + \sum_{k=0}^{N-1} L[e, x, \eta_r, t] \tag{3.43}$$

$$L[e, x, \eta_r, t] = \frac{1}{2}\alpha^T[t(k) - t_0(k)]^2 + \frac{1}{2}\beta^T[\eta_r(k) - \eta_r(k-1)]^2 + \frac{1}{2}t^2(k) \tag{3.44}$$

where $\eta_r(t) = [k_a\, k_b]^T$ is the direct control inputs of each level, N is the optimization step.

For the solution of this optimization problem, $\eta(k)$ can be attained using the Eulerian-Lagrangian method by choosing the Hamiltonian function $H.\varphi[t(N), N]$ represents the final performance requirements in each level at the N interval. The first term is used to describe the squared sum of total weighted tension loss, α^T is the weighted vector for each step error.

$$H_r = L_r(k) + \overrightarrow{\lambda}(k+1)\left[\vec{f}_r(x(k),\ \eta_r(k)) - \vec{f}_{r0}(x(k),\ \eta_r(k))\right] \tag{3.45}$$

where, $L_r(k)$ is the increment of performance index in the k-th sampling interval. $\overrightarrow{\lambda}$ is the Lagrange multiplier for searching the optimal $\eta_r(k)$ as the co-states estimation. Based on the above definitions, the reversing dynamical optimization can be operated at three stages as follows, which is a reversing process from level #3 to level #1. The final state N of t_{end} is firstly given for solving the t_3^* [36, 54, 71],

$$J_1^*(t_3,\ \eta_3) = J_0^*(t_{end}) + \min_{u(t)}\{J_2^*(t_3,\ \eta_2)\} \tag{3.46}$$

where $J_0^*(t_{end}) = h_3(t_{end}^*)$ is the terminated state of tension. From this representation, the optimal value t_3^* and its minimized value of the objective function $J_1^*(t_3,\ \eta_1)$ can also be calculated.

$$J_2^*(t_2,\ \eta_2) = J_1^*(t_3,\ \eta_3) + \min_{u(t)}\{J_3^*(t_1,\ \eta_1)\} \tag{3.47}$$

where $J_1^*(t_3,\ \eta_3) = h_2(t_3^*,\ t_2,\ \eta_2)$

$$J_3^*(t_1,\ \eta_1) = J_2^*(t_2,\ \eta_2) + \min_{u(t)}\{J_4^*(t_1,\ \eta_1)\} \tag{3.48}$$

where $J_2^*(t_2,\ \eta_2) = h_2(t_2^*,\ t_1,\ \eta_1)$.

All of those are subject to $\eta_{r\min} \le \eta_r(t) \le \eta_{r\max}$. (3.46), (3.47) and (3.48) represent the optimization for tracking the optimized response of corresponding level, respectively, in which the adaptive control with gain-tuning guarantees the smoothness manipulation of the controlled variables $\eta_r(t)$. $h(\cdot)$ is the dynamic of each level. The subscripts of $J_k^*(t_k,\ \eta_k)$ are the stage number of the reversing optimization. t_3^* is also the first solved optimal tension value in a whole optimization interval. In this way, the recursive optimized allocation for the spatial earlier levels t_2^* and t_1^* can be achieved. t_{end}^*, t_3^*, t_2^* and t_1^* are the optimal tension values from the successive calculations, respectively. The overall optimization procedure of the ICCA is proposed in Fig. 3.33.

For this control flow, given the desired reference response, the performance index J_r at each level should be firstly designed to consider the dynamics of the current level. Then the adaptive control law $\vec{u}_2$ is determined from the reversed optimizing for three level objective functions J_1, J_2 and J_3, which are capable of fully utilizing the driving motors. The optimization of t_r is operated in negative gradient direction based on the optimal tension $t_{(r+1)}^*$. The actual control law is derived from the tracking error $\overrightarrow{e}_r$ as the compensation for the feedback control $\vec{u}_1$. By determining the minimum of J_r, the modification of the gains u_2 is implemented for the immune-based self-adaptive controller.

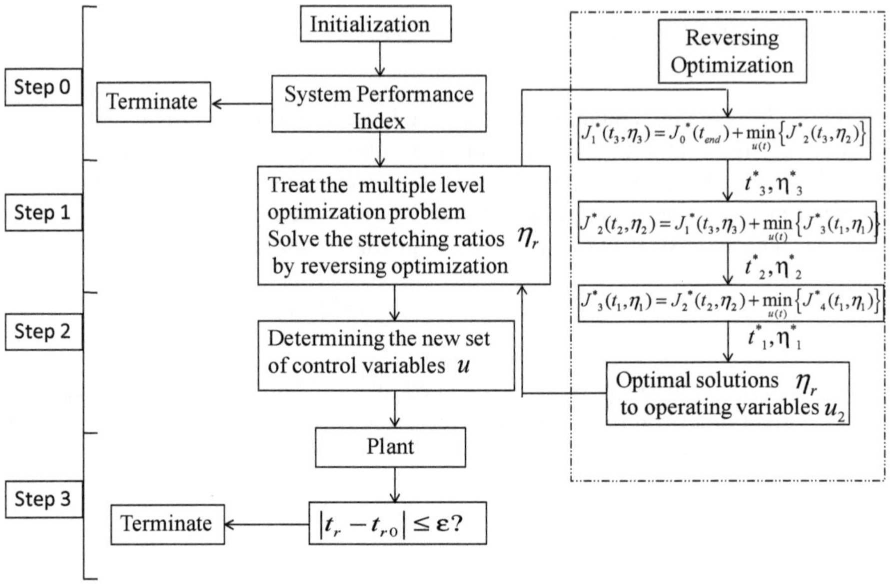

Fig. 3.33 The optimization procedure of the ICCA

(2) The optimization for the inner-loops by the immune-based control allocation

For the purpose of cooperative control of multiple sub-systems, control allocation method is proposed to deal with the uncertainty in the operation and nonlinear characteristics, in which two control laws for the driving motors are merged together. Consider N ($N \geq 2$) sub-processes and motivated from the error elimination strategy aforementioned in self-adaptive immune system, the control allocation algorithm is specifically designed for appropriate compensation for the feedback control. Both the re-allocation control law and the corresponding compatible immune controller are the basis of implementing the collaborative control. This re-allocation and decentralized design for complicated mechanical and material problem will be more appropriate to general problems.

This double-layer structure of the re-allocation and decentralized control strategy is shown in Fig. 3.34, where the tension responses in the specified stretching zones are manipulated by decentralized unidirectional decoupled controller in the inner loop layers, and the set-point of these inner-loops are determined by the optimization layer. The dynamical reversing optimization is based on the objective functions (3.41) and the production restrain conditions [36, 71].

At the optimization design layer, the objective of re-allocation stretching ratios is to select the best combination with the smallest error $\vec{t}_{ri} - \vec{t}_{ri0}$, which is the main factor to evaluate the final performance. Since the effects of roller speeds are capable of being fully used, the adaption to various disturbance conditions is

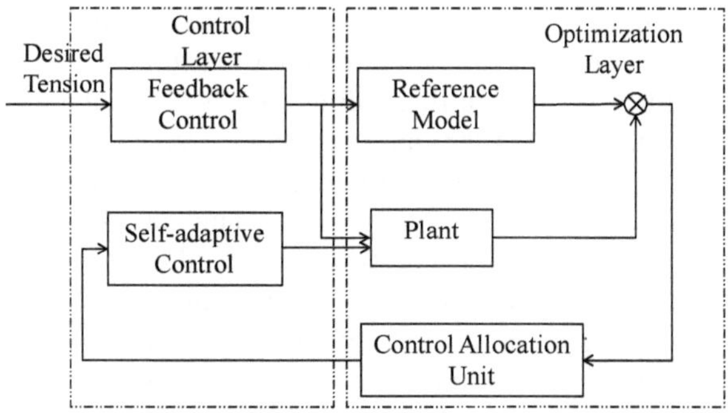

Fig. 3.34 The multiple layers of the stretching system

realized. Therefore, ignoring these initial configurations in each inner-loop first, the phased tensions can be regulated by reallocating all of these manipulated variables.

At the controller design layer, considering the actual smoothness control, a special adaptive control law is gradually utilized for the solution of $J^*(x, \eta_1)$, namely, the combination of the stretching ratios may lead to a satisfied convergence of e_r at the r-th level. The control law of the immune controller with gain-tuning η_r as the compensation of the feedback controller is designed as follows. This is the best way for the multiple-level system to fully use the degrees of freedom of all manipulation variables and meet the different phased dynamical requirements. For the specific control law including two control signals to the neighboring driving motors of one stretching zone, $u_{r1}(t)$ is the feedback control signal, $u_{r2}(t)$ is adaptive control, and there is a weighting factor between them. $\eta_r(t)$ is the regulating law as a part of $u_{r2}(t)$ as shown in (3.28) [30, 70].

$$\eta_r(t) = [k_a \, k_b]^T \tag{3.49}$$

$$k_a(t) = \dot{v}_1 + v_0 + k_{ca} \tag{3.50}$$

$$\dot{v}_1(t) = \lambda_1(\gamma(\mathrm{t}) \cdot p_1 + \dot{\gamma}(\mathrm{t}) \cdot p_2)\vec{t}_r \tag{3.51}$$

$$\dot{v}_0(t) = \lambda_0(\gamma(\mathrm{t}) \cdot p_1 + \dot{\gamma}(\mathrm{t}) \cdot p_2)\dot{\vec{t}}_r \tag{3.52}$$

$$\dot{k}_{ca}(t) = \lambda_{ca}(e_{ri}(t) \cdot q_1 + \dot{e}_{ri}(t) \cdot q_2)T_r \tag{3.53}$$

where r is the number of stretching sub-process, a, b represent two driving motor speeds of the r-th stretching zone, λ_1 and λ_2 are the rates of the tracking approach, whose large value represents the fast regulating speed but leading to excessive accommodation. The difference of them shows the different regulating rates in two driving motors. t_{ri0} (r = 1, 2, …) denotes the desired tension value of the r-th stretching zone, v_i is the correction of the self-adaptive factors of the immune adaptive controller. p_1, q_1 and p_2, q_2 are the coefficients given to $e_{ri}(t)$ and $\dot{e}_{ri}(t)$, respectively. $\vec{k}_{ca}(t)$ is the adaptive gain for eliminating the tracking error $e_{ri}(t)$, which will be activated in the performance deterioration. T_r is a constant value for the set-point of the r-th stretching tension.

Finally, the significant difference between the traditional control and the control allocation is that the latter searching a series of solutions for the sake of searching the best combination of stretching ratios online. This reorganized combination is to attain a more stable final tension value, and the tension is definitely in the required range. This tension state is more helpful to the continuous moving of the precursor, and reduces the chance for fiber folding and breaking. This compensation to the control layer of the current driving actuator directly depends on the results of calculation of optimization layer. According to the represented description of this control laws, the adaptive tracking problem for these actuators is improved to a certain extent.

3.4.3 Simulation Results

In order to demonstrate the effectiveness of the proposed ICCA, a simulation system is proposed for the polyacrylonitrile carbon fiber (PANCF) precursors stretching, implemented with the ratios control allocation algorithm. The PANCF reinforced composite material is a type of structural materials with high strength, high rigidity and low weight [4]. In order to improve these properties, in particular the degree of orientation, multi-level stretching processes are introduced sequentially and verified to be an effective way. Therefore, the stretching ratios combination and the continuous and stable tension responses are the important criteria to reflect the control performance even under different load torque conditions. For the overall systems, the ICCA is proposed by implementing the control allocation to deal with the coupling-loop effects. By means of reversing optimization of control allocation, the optimized results are achieved from searching the best combination of objective functions. The actual non-linear operation is driven by distributed self-adaptive immune controller with gains tuned by optimized results. As follows, we will show the flexibility and performance of the novel ICCA strategy in tension tracking and the fluctuation elimination under load torque disturbances.

3.4.3.1 The Tension Tracking in Stretching Processes with the ICCA

To achieve the tension requirement by the ICCA method, both the stable dynamics and the fast converge of the tension allowable range in each stretching zone need to be considered [59]. To demonstrate the effectiveness of the ICCA, some comparative experiments are conducted including the conventional centralized PI control and the CNCC methods [2].

The simulation experiments are designed with different phases depending on three ingredients: (1) The Simulink control structure; (2) The independent stable tracking for desired reference tension value on the single sub-processes;

(3) Changes of external load torques on the driving motors (acting as an unmeasured disturbance). The tension required values for three methods is the same, which are 9.8 cN at the first level, and the other responses in the second and third levels have almost the same effect with the current level. Note for the initial and the next tension neighbors of the current level are $t_0 = 7.8$ cN and $t_3 = 13.5$ cN, respectively, which are set as constant values in the idealized model. Therefore, the response of reference model is regarded as the desired steady-state response waiting to be tracking. For real-time adaptive tracking of the idealized reference model, the model parameters of the ICCA are revised having a bit of resemblance with those of CNCC. Two methods have the same target to modify the stretching ratios of multi-levels, which are originally fixed to 1.50, 3.50 and 2.50, respectively. The regulating range of stretching ratios is determined from 90 to 110% for single level. Instead, the conventional centralized PI control is also included for comparative purposes by maintaining the constant stretching ratios in the entire process. In the ICCA method, the adaptive gain vector $\overrightarrow{\eta}_i$ is used to operate this online regulation. The output feedback gains $\overrightarrow{k}_{if} = [k_{af}\, k_{bf}]^T$ directly facilitate the system with rapid convergence. The linear compensator $D(s)$ drives the whole control law by means of strictly positive treatment.

In these experiments, the initial tension value t_0 and v_1 in (3.29) are determined by 7.8 cN and 20 $mm \cdot \text{min}^{-1}$, the length of the stretching zone $L = 50$ mm, and the area of cross section of precursor $A = 5 \sim 8\,\mu$m. The simulation results of the ICCAthe CNCC and the centralized PI methods are presented in Fig. 3.35a, b, c. The tracking progresses of three methods are shown in Fig. 3.35a. From Fig. 3.35a, the final stable tension response of the PI control method has a large bias against the initial value. On the contrary, the convergence of the other two methods are still around the tension value within the error of 5.1%. The ICCA method keeps tracking the reference trajectory and induces the three-level stretching ratios compensation. The CNCC and the centralized PI control methods are almost the same at this phase, but the CNCC method starts working when the initial driving motor load υ_3 is not proper. In this phase, the control allocation in the CNCC method plays a critical role in the manipulation with the redundant stretching ratios. Even so, with limited range of speed-adjusting in the CNCC method, both of the methods are of great difficulties to deal with this nonlinear process appropriately. And the vibration appears obviously in the early-time response from the instability of driving motors because the $(\upsilon_{i+1} - \upsilon_i)$ term is not a constant anymore in actual system. Especially in the later-time response, the centralized PI control leads to a larger deviation in tension response due to totally lacking of ability to regulate the stretching ratios.

It can be observed that the ICCA method treats this system more smoothly with the special designed distributed control law, in which the adaptive control and the control allocation are considered together. The comparison of the three methods in convergence speed and steady state error are shown in Fig. 3.35b. The control input signals of the three methods to the stretching dynamics are given to demonstrate the control processes as shown in Fig. 3.35c.

Fig. 3.35 Comparison among the centralized PI control, the CNCC and the ICCA methods. **a** The responses of different control methods. **b** The error between desired value and actual response. **c** The control signals to stretching dynamics

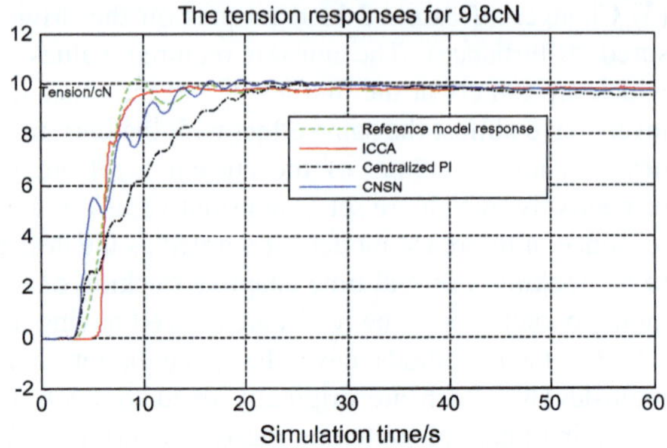

(a) The responses of different control methods.

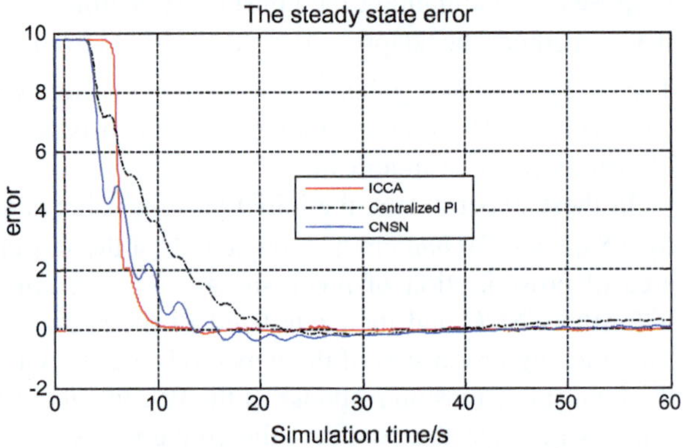

(b) The error between desired value and actual response

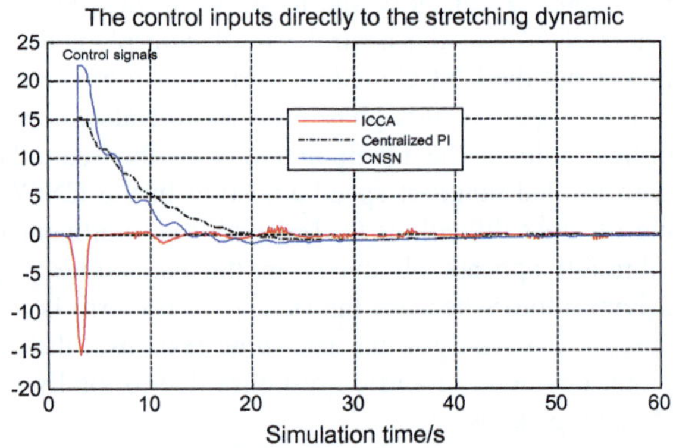

(c) The control signals to stretching dynamics

The centralized PI controller is mainly planted on each driving motor of the inner-loop to carry out functional stretching. Without collaborative regulating on the multi-level stretching processes, it leads to unstable approaching of designate tensions. Furthermore, the presence of various disturbances will lead unpredictable situations, involving high-frequency control signals, large transient errors or tension amplitudes.

In the tracking dynamic responses of reference model, the stable value of tension fluctuations by the ICCA and the CNCC methods in Fig. 3.35a, b, c are apparently smaller than the centralized PI method, which means the algorithms is of better robustness against coupling effects. The CNCC method is the specialized design on a simplified plant situation; the optimization method has restriction to deal with three-level system or above. It has the complementary regulation but without enough space to deal with the extremely load torque disturbance. One of the simplified conditions is that the tension error of two neighboring stretching zones $(t_{i+1} - t_i)$ in system model (3.31), (3.33) and (3.34) is not considered. Therefore, the tension output fluctuation and disturbance rejection in the transient phase apparent in the actual response.

With additional compensation to each control loop, the ICCA method provides a relatively acceptable state tension transient response. In this phase, the combination of control allocation and adaptive control lead to a more flexible performance, which keeps the steady-state response of the tension within a prescribed range.

3.4.3.2 The Self-Adaptive Regulation of the ICCA Method

In this section, we further initialize a particular experiment on the reserving regulation capability of the ICCA method by inserting system uncertainties. In the actual PANCF production line, the appearance of broken filaments on the roller is actually a common phenomenon. This type of load varying will affect the output torque of the driving motors. Such an inevitable situation definitely causes the system uncertainty which is simulated by load torque disturbances of driving motor.

The load torque disturbance in the third speed drive υ_3 is introduced as a simulation to verify the disturbance function. From the stretching dynamic (3.28–3.33), we can easily find out the impact of the input torque variations and the roller driving speeds to the tension response. After inserting 20% abrupt load torque disturbance in driving motor υ_3 in the second stretching zone, the closed-loop trajectories of three methods are illustrated in Fig. 3.36a, b, c, respectively. Apparently, the convergence of the centralized PI control method becomes less flexible to this disturbance. On the contrary, the ICCA and the CNCC methods are still around the tension value within the error of 5.7 and 7.9%. However, there is obvious difference in dynamic response at the presence of abrupt load disturbance compared to normal conditions. Figure 3.36c clearly shows the effect on this condition by the ICCA and the CNCC methods. The tension response of the CNCC method has several major spikes at the speed change, and the stable error is a little larger than that at the smooth speed. In the closed-loop time responses of the ICCA method, the tensions of the three sequential sub-processes converge to the set-points asymptotically and smoothly. It is shown that the responses by three methods are all in an acceptable range, and the ICCA method is the best one in extreme conditions.

Fig. 3.36 The closed-loop responses on the attached system interference. **a** The tension responses of different control methods. **b** The tension error between desired value and actual responses. **c** The control signals in front of stretching dynamics

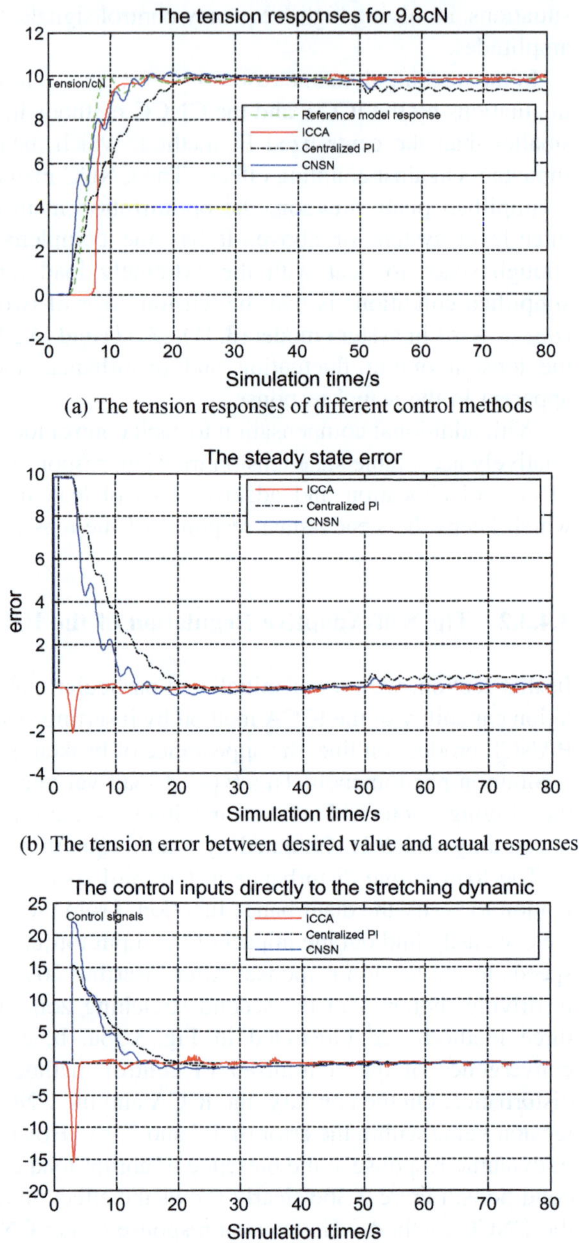

(c) The control signals in front of stretching dynamics

From the control point of view, the ICCA method has a better design process than that of the centralized PI control and the CNCC methods. Firstly, the centralized PI control method with fixed parameters is not capable of changing the gains of the inner-loop driving motors and leads to a restriction of attaining the optimal performance of final fibers. Inherited the merits of traditional methods, the CNCC method with simplified stretching model has a rapid response than that of the centralized PI control method. Therefore, the response with the ICCA method implementing specifically design for non-linear dynamics can meet the tension requirement more acceptable in extreme conditions. Figure 3.36b shows the error between the desired tension value and the actual response of the stretching inner-loops of three methods, respectively, for demonstrating their tracking convergences. Among them, the output spike with the ICCA method is apparently decreased than that with the CNCC method which leads to a more stable and smoothly dynamic process. The control input signals of the three methods to the stretching dynamic are given to demonstrate the control processes as shown in Fig. 3.36c. From the comparison experiments of three methods, we can apparently determine the performance of the ICCA method is better than that of the PI and the CNNC methods with or without the presence of load torque disturbance. What makes the ICCA method being different from other two methods is the adaptive allocation of sub-motors online, in which the dynamical programming can reversely optimize the stretching ratios. Consequently, the tension has a most adjustable range for the self-adaptive tracking errors are directly used to lead to drive the redundant stretching ratios. With self-adaptive regulation to the roller speeds in relevant executive levels at the same time, the acceptable final precursor performance can be touched more easily.

3.5 Conclusions

In this chapter, first, a bio-inspired multi-layered intelligent cooperative controller (BMLICC) is proposed for the stretching process in the industrial fiber production line. Based on the biological neuroendocrine regulation network structure and its working mechanism, an intelligent cooperative controller with the related control structure and cooperative algorithm are proposed to the stretching unit consisting of two rollers. Meanwhile, larger stretching process with many such units can be controlled by the proposed controller with necessary modifications. Then, a novel stretching network based on the cytokine network in human body for cooperative online re-allocation of the desired ratios in multi-stage stretching systems is prestented, resulting in a much smaller steady-state tracking error and better ability to reject disturbances. An optimization algorithm is adopted to tune the set-up points of the stretching ratios in different stretching units. At last, a novel collaborative method with immune mechanism inspired self-adaptive controller combined with allocation optimization on the multi-level stretching at inner-loops is proposed. Based on the real-time dynamical response error between the reference model and the actual plant, the ICCA method is driven by self-adaptive method, in order to cope with the time-varying system plant from eliminating the tracking error. For the industrial application, this kind of online tuning of the inner-loop controllers is also especially recommended for a rational moving system such as a high-precision spindle driver with independent proper speed excitation.

References

1. Liang, X., Ding, Y., Ren, L., Hao, K., Wang, H., Chen, J.: A bioinspired multilayered intelligent cooperative controller for stretching process of fiber production. IEEE. T. Syst. Man. Cy. C. **42**(3), 367–377 (2012)
2. Zhang, T., Jin, Y., Ding, Y., Hao, K.: A cytokine network-inspired cooperative control system for multi-stage stretching processes in fiber production. Soft. Comput. **19**(6), 1523–1540 (2015)
3. Ding, Y., Zhang, T., Ren, L., Jin, Y., Hao, K., Chen, L.: Immune-inspired self-adaptive collaborative control allocation for multi-level stretching processes. Inform. Sciences. **342**, 81–95 (2016)
4. Chen, J., Harrison, I.R.: Modification of polyacrylonitrile (PAN) carbon fiber precursor via post-spinning plasticization and stretching in dimethyl formamide (DMF). Carbon. **40**(1), 25–45 (2002)
5. Liang, X., Ding, Y., Ren, L., Hao, K., Wang, H., Chen, J.: A bio-inspired multilayered intelligent cooperative controller for stretching process of fiber production. IEEE. T. Syst. Man. Cy. C. **42**(3), 367–3773 (2012)
6. Tursini, M., Parasiliti, F., Zhang, D.: Real-time gain tuning of PI controllers for high-performance PMSM drives. IEEE. T. Ind. Appl. **38**(4), 1018–1026 (2002)
7. Kuo, C., Fang, C.: An entire strategy for control of a calendar roller system: Part III intelligent settling time-optimal control. Text. Res. J. **78**(1), 81–87 (2008)
8. Pagilla, P.R., Siraskar, N.B., Dwivedula, R.V.: Decentralized control of web processing lines. IEEE. T. Contr. Syst. T. **15**(1), 106–117 (2007)
9. Shin, K., Kwon, S.: The effect of tension on the lateral dynamics and control of a moving web. IEEE. T. Ind. Appl. **43**(2), 403–411 (2007)
10. Li, Y.G., Rahn, C.D.: Adaptive vibration isolation for axially moving beam. IEEE-ASME. T. Mech. **5**(4), 419–428 (2000)
11. Abjadi, N.R., Soltain, J., Askari, J., Markadeh, G.R.A.: Nonlinear sliding-mode control of a multi-motor web-winding system without tension control. IET. Control. Theory. A. **3**(4), 419–427 (2008)
12. Yerashunas, J.B., Abreu-Garcia, J.A.D., Hartley, T.T.: Control of lateral motion in moving webs. IEEE. T. Contr. Syst. T. **11**(5), 684–693 (2003)
13. Lin, F.J., Chou, P.H.: Adaptive control of two-axis motion control system using interval type-2 fuzzy neural network. IEEE. T. Ind. Electron. **56**(1), 178–193 (2009)
14. Juang, J.G., Huang, M.T., Liu, W.K.: PID control using presearched genetic algorithms for a MIMO system. IEEE. T. Syst. Man. Cy. C. **38**(5), 716–727 (2008)
15. Mehmet, D., Hakan, C., Cengiz, K.M.: Adaptive control of let-off system in weaving. J. Text. I. **100**(2), 186–194 (2009)
16. Lian, K.Y., Chiu, C.S., Liu, P.: Semi-decentralized adaptive fuzzy control for cooperative multi robot system with H∞ motion/internal force tracking performance. IEEE. T. Syst. Man. Cy. B. **32**(3), 269–280 (2002)
17. Wai, R.J.: Robust control for nonlinear motor-mechanism coupling system using wavelet neural network. IEEE. T. Syst. Man. Cy. B. **33**(3), 489–497 (2003)
18. Liu, B., Han, H., Ding, Y.S.: A decoupling control based on the bi-regulation principle of growth hormone. Paper present at Proceedings of the 2005 ICSC Congress on Computational Intelligence Methods and Applications, pp. 15–17 (2005)

19. Cao, J.D., Chen, G.R., Li, P.: Global synchronization in an array of delayed neural networks with hybrid coupling. IEEE. T. Syst. Man. Cy. B. **38**(2), 488–497 (2008)
20. Dinh, N.Q., Afzulpurkar, N.V.: Neuro-fuzzy MIMO nonlinear control for ceramic roller kiln. Simul. Model. Pract. Th. **15**(10), 1239–1258 (2007)
21. Lin, F.J., Hsieh, H.J., Chou, P.H.: Tracking control of a two-axis motion system via a filtering-type sliding-mode control with radial basis function network. IET. Control. Theory. A. **4**(4), 656–671 (2010)
22. Gueaieb, W., Karray, F., Al-Sharhan, S.: A robust hybrid intelligent position/force control scheme for cooperative manipulators. IEEE-ASME. T. Mech. **12**(2), 109–125 (2007)
23. Wang, Z.D., Liu, Y.R., Wei, G.L., Liu, X.H.: A note on control of a class of discrete time-delay stochastic systems with distributed delays and nonlinear disturbances. Automatica. **46**(3), 543–548 (2010)
24. Wang, Z.D., Wei, G.L., Feng, G.: Reliable H∞ control for discrete-time piecewise linear systems with infinite distributed delays. Automatica. **45**(12), 2991–2994 (2009)
25. Wang, Z.D., Ho, D.W.C., Liu, Y.R., Liu, X.H.: Robust H∞ control for a class of nonlinear discrete time-delay stochastic systems with missing measurements. Automatica. **45**(3), 684–691 (2009)
26. Mehraeen, S., Jagannathan, S.: Decentralized optimal control of a class of interconnected nonlinear discrete-time systems by using online Hamilton-Jacobi-Bellman formulation. IEEE. T. Neural. Networ. **22**(11), 1757–1769 (2011)
27. Li, P., Lam, J.: Decentralized control of compartmental networks with h-infty tracking performance. IEEE. T. Ind. Electron. **60**(2), 546–553 (2013)
28. Chien, T.L., Chen, C.C., Huang, C.J.: Feedback linearization control and its application to MIMO cancer immunotherapy. IEEE. T. Contr. Syst. T. **18**(4), 953–961 (2010)
29. Ding, Y.S., Liu, B.: An intelligent bi-cooperative decoupling control approach based on modulation mechanism of internal environment in body. IEEE. T. Contr. Syst. T. **19**(3), 692–698 (2011)
30. Ding, Y.S., Liang, X., Hao, K.R., Wang, H.P.: An intelligent cooperative decoupling controller for coagulation bath in polyacrylonitrile carbon fiber production. IEEE. T. Contr. Syst. T. **21**(2), 467–479 (2013)
31. Chen, J., Cao, X., Cheng, P., Xiao, Y., Sun, Y.: Distributed collaborative control for industrial automation with wireless sensor and actuator networks. IEEE. T. Ind. Electron. **57**(12), 4219–4230 (2010)
32. Nguyen, T.N., Su, S., Nguyen, H.: Robust neuro-sliding mode multivariable control strategy for powered wheelchairs. IEEE. T. Neur. Sys. Ren. **19**(1), 105–111 (2011)
33. Wu, Z.Z.: LMI-Based Multivariable PID controller design and its application to the control of the surface shape of magnetic fluid deformable mirrors. IEEE. T. Contr. Syst. T. **19**(4), 717–729 (2011)
34. Hu, Y.F., Ding, Y.S., Hao, K.R.: An immune cooperative particle swarm optimization algorithm for fault-tolerant routing optimization in heterogeneous wireless sensor networks. Math. Probl. Eng. 2012(1024-123X), 1–19 (2012)
35. Cao, X.H., Cheng, P., Chen, J.M., Sun, Y.X.: An online optimization approach for control and communication codesign in networked cyber-physical systems. IEEE. T. Ind. Inform. **9** (1), 439–450 (2013)
36. Skogestad, S.: Self-optimizing control: The missing link between steady-state optimization and control. Comput. Chem. Eng. **24**(2), 569–575 (2000)
37. Benlatreche, A., Knittel, D., Ostertag, E.: Robust decentralized control strategies for large-scale web handling systems. Control. Eng. Pract. **16**(6), 736–750 (2008)
38. Koc, H., Knittel, D., De Mathelin, M., Abba, G.: Modeling and robust control of winding systems for elastic webs. IEEE. T. Contr. Syst. T. **10**(2), 197–208 (2002)
39. Yang, K.J., Hong, K.S., Matsuno, F.: Robust adaptive boundary control of an axially moving string under a spatiotemporally varying tension. J. Sound Vib. **273**(4), 1007–1029 (2004)
40. Ji, M., Egerstedt, M.: Distributed coordination control of multiagent systems while preserving connectedness. IEEE. T. Robot. **23**(4), 693–703 (2007)

41. Kao, Y.G., Wang, C.H., Xie, J., Karimi, H.R., Li, W.: H∞ sliding mode control for uncertain neutral-type stochastic systems with Markovian jumping parameters. Inform. Sciences. **314**, 200–211 (2015)
42. Sui, S., Tong, S.C., Li, Y.M.: Observer-based adaptive fuzzy decentralized control for stochastic large-scale nonlinear systems with unknown dead-zones. Inform. Sciences. **259** (29), 71–86 (2014)
43. Pagilla, P.R., Dwivedula, R.V., Siraskar, N.B.: A decentralized model reference adaptive controller for large-scale systems. IEEE-ASME. T. Mech. **12**(2), 154–163 (2007)
44. Huang, Y.S., Wang, Z.Y.: Decentralized adaptive fuzzy control for a class of large-scale MIMO nonlinear systems with strong interconnection and its application to automated highway. Inform. Sciences. **274**(8), 210–224 (2014)
45. Lewis, F.L., Zhang, H., Hengster-Movric, K., Das, A.: Cooperative adaptive control for systems with second-order nonlinear dynamics. In: Cooperative Control of Multi-Agent Systems, pp. 259–278. Springer, London (2014)
46. Sun, D., Shao, X., Feng, G.: A model-free cross-coupled control for position synchronization of multi-axis motions: Theory and experiments. IEEE. T. Contr. Syst. T. **15**(2), 306–314 (2007)
47. Lan, W., Thum, C.K., Chen, B.M.: A hard-disk-drive servo system design using composite nonlinear-feedback control with optimal nonlinear gain tuning methods. IEEE. T. Ind. Electron. **57**(5), 1735–1745 (2010)
48. Yu, W., Ren, W., Zheng, W.X., Chen, G., Lv, J.: Distributed control gains design for consensus in multi-agent systems with second-order nonlinear dynamics. Automatica. **49**(7), 2107–2115 (2013)
49. Wang, F.Y., Jin, N., Liu, D., Wei, Q.: Adaptive dynamic programming for finite-horizon optimal control of discrete-time nonlinear systems with-error bound. IEEE. T. Neural. Networ. **22**(1), 24–36 (2011)
50. Yang, J.M., Kwak, S.W.: Output feedback control of asynchronous sequential machines with disturbance inputs. Inform. Sciences. **259**(20), 87–99 (2014)
51. Liang, X., Ding, Y.S., Ren, L.H., Hao, K.R.: Data-driven cooperative intelligent controller based on the endocrine regulation mechanism. IEEE. T. Contr. Syst. T. **22**(1), 94–101 (2014)
52. Luo, Y., Serrani, A., Yurkovich, S.: Model-predictive dynamic control allocation scheme for reentry vehicles. J. Guid. Control. Dynam. **30**(1), 100–113 (2007)
53. Xia, Y.Q., Xie, W., Liu, B., Wang, X.Y.: Data-driven predictive control for networked control systems. Inform. Sciences. **235**(20), 45–54 (2013)
54. Sibielak, M., Raczka, W., Konieczny, J., Kowal, J.: Optimal control based on a modified quadratic performance index for systems disturbed by sinusoidal signals. Mech. Syst. Signal. Pr. **64**, 498–519 (2015)
55. Amrit, R., Rawlings, J.B., Biegler, L.T.: Optimizing process economics online using model predictive control. Comput. Chem. Eng. **58**(45), 334–343 (2013)
56. Vermillion, C., Sun, J., Butts, K.: Predictive control allocation for a termal management system based on an inner loop reference model-design, analysis, and experimental results. IEEE. T. Contr. Syst. T. **19**(4), 772–781 (2011)
57. Wang, X.F., Li, X., Wang, J.H., Fang, X.K., Zhu, X.F.: Data-driven model-free adaptive sliding mode control for the multi degree-of-freedom robotic exoskeleton. Inform. Sciences. **327**(C), 246–257 (2016)
58. Chen, B., Liu, X.P., Liu, K.F., Lin, C.: Adaptive control for nonlinear MIMO time-delay systems based on fuzzy approximation. Inform. Sciences. **222**(3), 576–592 (2013)
59. Zhang, T.P., Xia, X.N.: Decentralized adaptive fuzzy output feedback control of stochastic nonlinear large-scale systems with dynamic uncertainties. Inform. Sciences. **315**(1), 17–38 (2015)
60. Shin, D.H., Kim, Y.D.: Nonlinear discrete-time reconfigurable flight control law using neural networks. IEEE. T. Contr. Syst. T. **14**(3), 408–422 (2006)
61. Googol Corp. Programming guide for GT-series motion controller

62. Farhy, L.S.: Modeling of oscillations of endocrine networks with feedback. Method. Enzymol. **384**(12), 54–81 (2004)
63. Hone, A.: Modeling a cytokine network. Paper present at 2007 IEEE Symposium on Foundations of Computational Intelligence. 389–393 (2007)
64. Wang, C.L.: Multivariable adaptive backstepping control: A norm estimation approach. IEEE. T. Automat. Contr. **57**(4), 989–995 (2012)
65. Bodson, M.: Evaluation of Optimization Methods for Control Allocation. J. Guid. Control. Dynam. **25**(4), 703–711 (2002)
66. Tjonnas, J., Johansen, T.A.: Stabilization of automotive vehicles using active steering and adaptive brake control allocation. IEEE. T. Contr. Syst. T. **18**(3), 545–558 (2010)
67. Wei, Q.L., Song, R.Z., Sun, Q.Y.: Nonlinear neuro-optimal tracking control via stable iterative Q-learning algorithm. Neurocomputing. **168**(30), 520–528 (2015)
68. Zhang, H.W., Lewis, F.L.: Adaptive cooperative tracking control of higher-order nonlinear systems with unknown dynamics. Automatica. **48**(7), 1432–1439 (2012)
69. Liu, M., Cao, X.B., Zhang, S.J., Yang, W.: Sliding mode control of quantized systems against bounded disturbances. Inform. Sciences. **274**(1), 261–272 (2014)
70. Hu, Y., Ding, Y., Hao, K., Ren, L., Han, H.: An immune orthogonal learning particle swarm optimization algorithm for routing recovery of wireless sensor networks with mobile sink. Int. J. Syst. Sci. **45**(3), 337–350 (2014)
71. Huang, L., Suh, I.H., Abraham, A.: Dynamic multi-objective optimization based on membrane computing for control of time-varying unstable plants. Inform. Sciences. **181**(11), 2370–2391 (2011)

Chapter 4
Immune Inspired Collaborative Learning Controllers

In this chapter, three collaborative and learning-type controllers inspired by immune system are introduced. Firstly, a novel reinforcement learning intelligent controller (RLIC) based on primary–secondary response mechanism of immune system is presented [1]. Secondly, an iterative learning control method based on the recognition, response, and memory mechanism of immune system (IRRM-ILC) is proposed [2]. Finally, based on the biological immune mechianisms, a design approach for the immune reconfigurable controller (IRC) is proposed [3].

4.1 Introduction

Learning is a process for an intelligent system to acquire knowledge or experience on the basis of its perception and cognition of the environment and then to act on the environment referring the knowledge/experience so as to improve its behavior performance the next time [4]. The learning control has appeared since 1960s, and has become an important approach in intelligent control research. The learning control system can learn some unknown-information of a certain process and control the process by using the knowledge learned [5]. Since the early 1970s, several structures of learning control have been presented, which include pattern recognition-based learning control [6], reinforcement learning control [7], connectionist learning control [8, 9], and repetitive or iterative learning control [10, 11].

Reinforcement learning control and iterative learning control have attracted rapidly increasing interest in the machine learning and artificial intelligence communities, and have been widely developed in the self-organizing control field. As a trial-and-error scheme, reinforcement learning control decides itself executing action via the estimated feedback information from the control plant. It can find the optimal control strategy through the way of reward or punishment. The reinforcement learning algorithms mainly include Q-learning, SARSA, SARSA (k), Dyna-Q, Q (k)-learning, and fast Q (k)-learning [12, 13]. With the rapid

Y. Ding et al., *Bio-Inspired Collaborative Intelligent Control and Optimization*, Studies in Systems, Decision and Control 118,
https://doi.org/10.1007/978-981-10-6689-4_4

development of biocybernetics, the reinforcement learning control has been integrated with the technology inspired from biological information processing systems, which involves nervous system, genetic system, and immune system. For example, there exist genetic algorithm-based reinforcement learning [14], neural network-based reinforcement learning [15, 16], chaotic theory-based reinforcement learning [17], and fuzzy-based reinforcement learning [18, 19].

Iterative learning control (ILC) is also a well-known advanced control method which utilizes some basic learning principles to achieve control of the plant. The ILC is a recursive algorithm that utilizes the tracking discrepancy/error of the system output from the desired trajectory and iteratively updates its current control command so as to generate an upgraded control command for the next operation, when the system such as chemical processes, robot motions, servo systems, operates repetitively over a fixed finite time interval [20]. It is based on the notion that the performance of a system that executes the same task multiple times can be improved by learning from previous executions (batches, trials, iterations, passes) [21]. It is a data-driven method that uses only the plant's data instead of any knowledge of the plant's model. The ILC is able to gradually improve its control performance by learning from previous actions [22]. Since introduced by Arimoto et al. in 1984 [20], the ILC has drawn much attention from researchers and has been successfully applied in practices, see [4, 21–24] and references therein. It is highly attractive in that the ILC not only requires little prior knowledge of the control system but also controls the plant effectively. ILC has been successfully applied to industrial robots [25], industrial batch process control [26], injection-molding machines [27, 28], aluminum extruders [29], trajectory tracking [30], induction motors [31], chain conveyor systems [32], complex stochastic systems [33], nonlinear system [34], rapid thermal processing [35, 36], and semibatch chemical reactors [37].

Meanwhile, the complex dynamic processes are challenging to control because they are subject to unpredictable disturbances and operated by spatially distributed sensors and actuators. Latent failures of these components may cause instability and safety problem. To meet the performance specifications under such conditions, the control system should include different strategies for each individual operating condition and have adaptive ability to react under uncertain situations [38]. With the purpose of tolerating component malfunctions, the reconfiguration control theory has been developed [39]. Reconfiguration control mainly refers to the scheme of adjusting both structure and parameters of the controller when there is any fault in the control system. Unlike adaptive control strategies which modify the control law directly to improve performance, the reconfigurable control methods modify the system model and then redesign the control strategy based on the new model [40]. The modification could be done either online or offline. Through the structural optimization of the control logic, the reconfigurable controller greatly improves the system's fault-tolerance and reliability. Model matching control is a traditional reconfiguration control scheme which uses a decision element to determine the appropriate model out of multiple system models and switch to the corresponding control strategy when the correct model is determined [41]. In [42], a model-based approach was proposed for fault diagnosis system using a machine-learning technology. In [43], an adaptive reconfiguration scheme to reduce the effect of shadows on solar panels was constructed on the model-based approach. Another effective method to deal with failures is the control reallocation which is critically important for the reconfigurable flight control systems in the case of actuator failures [44]. Several control allocation algorithms have been proposed, such as direct allocation [45] and mathematical programming [46]. [47] applied a robust reconfiguration method based on Petri nets and integer programming. In [48], a modular control strategy that combines model predictive control allocation with the use of an inner loop reference is proposed. Moreover, linear matrix inequality method, sliding model method and neural networks are also widely used in reconfigurable control systems [49–52]. The hybrid approaches are developed rapidly [53]. With development of the intelligent technology, more and more adaptive intelligent control methods are combined with reconfiguration control. Adaptive neural controllers based on back-stepping techniques are designed for different reconfigurable flight control systems [54–56]. In [57], a bio-inspired multi-agent system is proposed for building reconfigurable systems. Virtual components are also designed to compensate for the malfunctioning ones. However, further investigation into novel design techniques which can fit well in industrial applications remains an important issue [58].

4.2 A Immune Reinforcement Learning Intelligent Controller

In this section, a novel reinforcement learning intelligent controller (RLIC) based on the primary–secondary response mechanism of the immune system is presented. The RLIC has the abilities of learning, memory, and evolution. It can learn and produce the control antibodies (CABs) automatically during the period of eliminating control error. The CABs store the corresponding final changing control output that eliminates the control error. The control system has no CABs when the control error appears for the first time. The RLIC can learn and produce the CAB during the period of eliminating the error by using a conventional control algorithm. When the control error appears again, the RLIC can eliminate it rapidly and stably by using the CABs and combining the conventional control algorithm. As a consequence, the RLIC's learning ability and system response become stronger and stronger.

4.2.1 The Immune Primary–Secondary Response Mechanism

The primary–secondary response of the immune system is a learning-memory process. In fact, it is also an adaptive response and reinforcement learning process

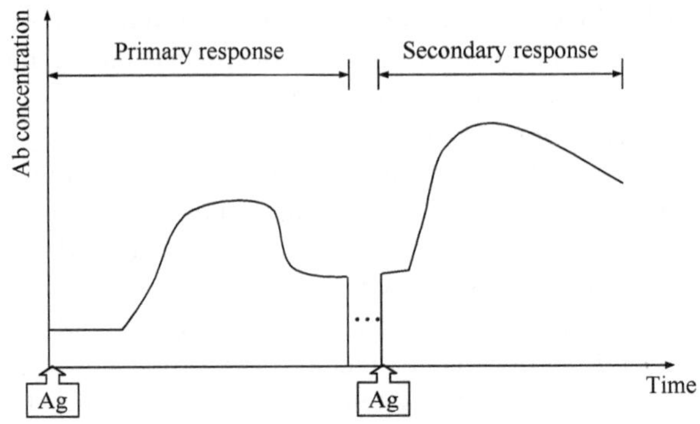

Fig. 4.1 Contrast effectiveness of the primary–secondary response mechanism of the immune system

[59–61]. The contrast effectiveness between the primary response and secondary response of the immune system is as shown in Fig. 4.1 [1]. In Fig. 4.1, Ag and Ab means the antigen invading the body and the antibody to eliminate the antigen, respectively; and the left curve and the right curve denotes the primary and secondary response, respectively.

The primary response has the characteristic of long lag and slow rate of antibody producing, as shown in Fig. 4.1. When the foreign antigen (Ag) invades the body for the first time, the immune system does not have the corresponding antibody to kill it. After a period, the old antibodies in the immune system can produce a new one via evolution, stimulated by the invading antigen. Then the invading antigen can be eliminated quickly. After the antigen is killed, the number of the matched antibodies begins to decline with its death fast. But, a certain number of antibodies, namely memory-cells, are always kept in the body. This is the immune learning process.

When the same foreign antigen invades the body again, the memory-cells in the immune system can make a match with the antigen. If the feature of an antibody can match with that of the antigen, this antibody is called the matched one. Then the matched antibody can propagate rapidly to eliminate the antigen. This is the secondary response and the self-adaptation of the immune system, which is called the immune memory process. Compared with the primary response, the secondary response is characterized by a shorter lag and higher rate of eliminating antigen. And, it has a faster and more efficient response not only to the same antigens but also to the similar ones. As thus, both the speed and accuracy of the secondary response become much stronger after each infection. This response scheme is the intrinsic of a reinforcement learning strategy. The primary–secondary response mechanism of the immune system is shown in Fig. 4.2 [1].

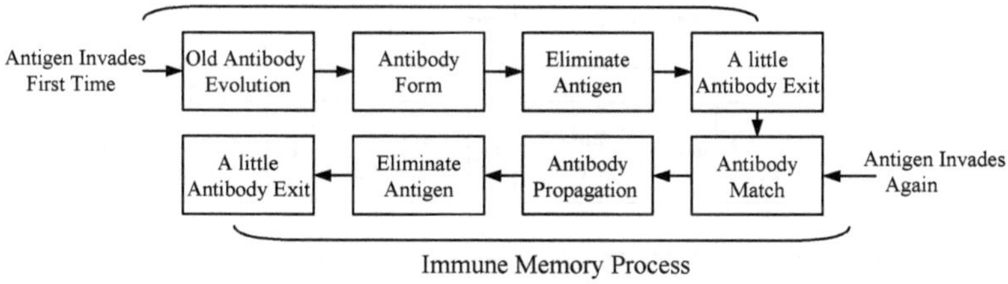

Fig. 4.2 The primary–secondary response mechanism of the immune system

4.2.2 *Design of a Reinforcement Learning Intelligent Controller*

We associate the control error $e(t)$ with the antigen of immune system, and the control strategy to eliminate the $e(t)$ with the antibody. All the CABs in the control system constitute the antibody database, which is denoted as $C_{Ab}(i)$, where i = 1, 2, …, n (n is the total number of all CABs in the antibody database). According to Fig. 4.2, we design a RLIC based on the primary–secondary response mechanism of the immune system. The RLIC is composed of the reinforcement learning control (RLC) unit, CAB recognition (CABR) unit, CAB adaptation (CABA) unit, auxiliary conventional control (ACC) unit, and the CAB database, as shown in Fig. 4.3 [1]. Its learning and evolution process is as shown in Fig. 4.4, where the CCA is the conventional control algorithm.

For the antibody $C_{Ab}(i)$, we define two concepts as follows:

1. **Feature**: It includes the set-point value sp_i (t_0) and the control error e_i (t_0) when the $C_{Ab}(i)$ begins to be produced;
2. **Changing control output memorized:** It is the value of Δu_i:

$$\Delta u_i = u_i(t) - u_i(t_0) \tag{4.1}$$

Fig. 4.3 The structure of the RLIC

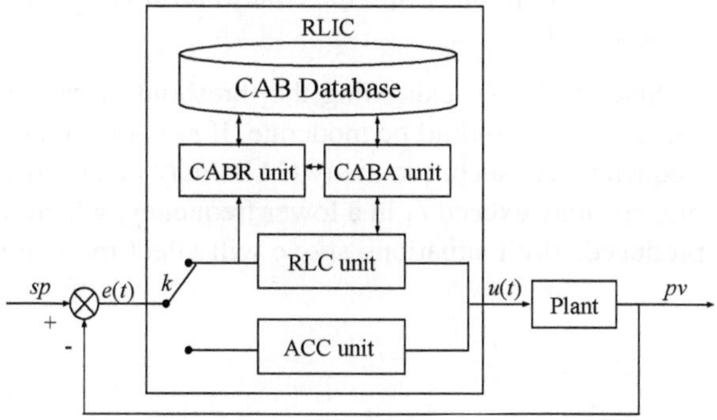

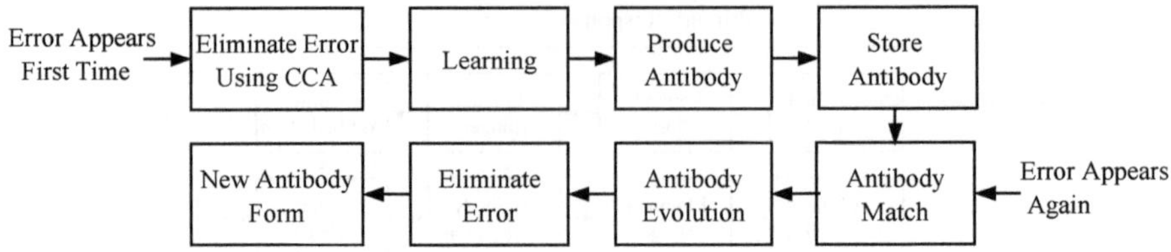

Fig. 4.4 The framework of learning and evolution of the RLIC

where, $u_i(t)$ is the final control output value of the RLIC when the control error is eliminated, and $u_i(t_0)$ is its initial output value when the $C_{Ab}(i)$ begins to be produced.

For the antigen (the control error), its feature includes the set-point value $sp(t)$ and the control error $e(t)$ when the control error appears.

When the antigen appears for the first time, as the primary response of immune system, the RLIC does not have the corresponding antibody. It learns during the process of eliminating the control error by using the conventional control algorithm, and produces the corresponding CAB. When the antigen appears again, the produced CAB begins to take effect. First, the RLIC matches all CABs in the database with the antigen. Then, the most matched CAB is further corrected since sometimes the antigen does not matched with the CAB well. Finally, the RLIC utilizes the corrected CAB to eliminate the antigen. After the antigen is eliminated, the new CAB is produced. The RLIC's learning ability and system response can become stronger and stronger by repeating the above process.

4.2.2.1 The Producing of the First CAB

When the antigen appears for the first time, the RLIC does not have the corresponding antibody. The first CAB is produced by learning from the process of eliminating the control error. The detail process, as shown in Fig. 4.5 [1], is described as follows:

1. **Recording the features**. When the absolute control error exceeds the threshold ε_1, the set-point $sp(t_0)$, the control error $e(t_0)$, and the controller output $u_1(t_0)$ are recorded.

Since only $e(t)$ exceeding the threshold ε_1 can be regarded as the antigen. The value of the ε_1 should be moderate. If ε_1 is too small, $e(t)$ may exceed ε_1 in a higher frequency. As such, too many CABs may be produced. On the contrary, if ε_1 is too big, $e(t)$ may exceed ε_1 in a lower frequency, which may cause too few CABs being produced. Both situations above will affect the control performance.

Fig. 4.5 The producing process of the first antibody

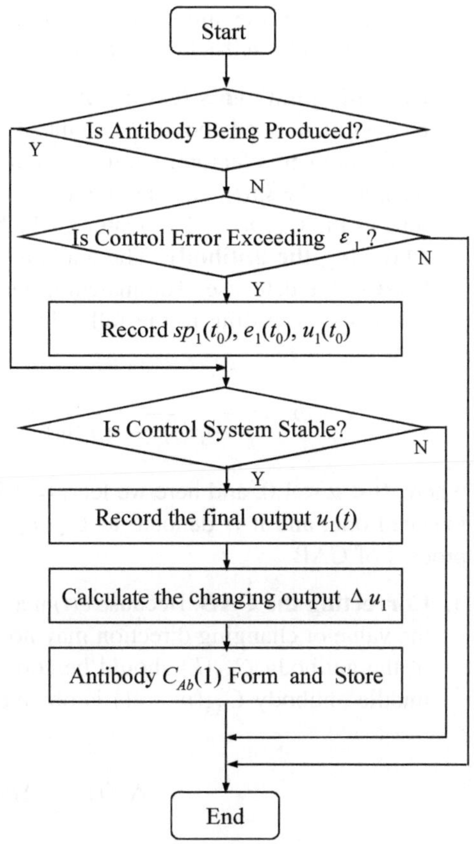

2. **Eliminating the antigen**. The RLIC utilizes the auxiliary conventional control algorithm, such as PID algorithm, to eliminate the control error.
3. **Recording the final control output**. When the control error is eliminated, the final control output value $u_1(t)$ is recorded.
4. **Forming the first CAB**. According to Eq. (4.1), we calculate the changing control output Δu_1, and store the antibody into the database.

4.2.2.2 Producing the Forthcoming CABs

When the antigen appears again, the RLIC will generate the forthcoming CAB via the self-adaptive process. The process is explained in details as follows (assume the total number of antibodies in database to be n).

First, the RLIC matches all CABs in the database with the antigen. Next, it corrects the matched antibody. Finally, it uses the corrected antibody to eliminate the antigen. After the antigen is eliminated, the new antibody is produced. In fact, this is the RLIC's self-adaptation process.

1. **Recording antigen's feature**. It is similar to that of the first CAB's producing process mentioned above. When the absolute control error exceeds the threshold ε_1, the antigen's feature, i.e. the set-point $sp(t)$ and the control error $e(t)$, will be recorded. The $sp(t)$, $e(t)$, and the controller output $u(t)$ are also recorded at same time as the $C_{Ab}(n + 1)$'s feature, which are $sp_{(n+1)}(t_O)$, $e_{(n+1)}(t_O)$, and $u_{(n+1)}(t_O)$.
2. **Matching the antibody**. The antigen's feature is compared with that of every CAB in the database. The matching factor ω_i (i = 1, 2, …, n) of every CAB is calculated according to the following equation:

$$\omega_i = \frac{1}{\alpha|sp(t) - sp_i(t_O)| + (1 - \alpha)|e(t) - e_i(t_O)|}, \tag{4.2}$$

where, $0 < \alpha < 1.0$, and here we let $\alpha = 0.5$. The CAB with the maximum ω_i is the matched one, and it is denoted as $C_{Ab}(k)$, where $k \leq n$. This is also the memory process of CAB.

3. **Correcting the CAB**. Because $e(t)$ may not match with $e_k(t_O)$ well, for example, the value or changing direction may not be uniform; the changing output $\Delta u_k(t)$ of the antibody $C_{Ab}(k)$ should be corrected. The changing output $\Delta u(t)$ of the middle-antibody $C'_{Ab}(n + 1)$ is obtained according to linearization theory,

$$\Delta u(t) = \Delta u_k(t)\frac{e(t)}{e_k(t_O)}, \tag{4.3}$$

where, $e_k(t_O)$ is one feature of the antibody $C_{Ab}(k)$.

4. **Eliminating the antigen**. The RLIC uses the middle-antibody $C'_{Ab}(n + 1)$ to eliminate the antigen. The detail eliminating process will be introduced in Sect. 4.2.2.3.
5. **Forming the new CAB**. After the control error is eliminated, the real changing output $\Delta u_{n+1}(t)$ of the middle-antibody $C'_{Ab}(n + 1)$ is calculated based on Eq. (4.1) and according to the final RLIC's output $u_{n+1}(t)$. Then the new control antibody $C_{Ab}(n + 1)$ is produced. This is also the propagation process of the control body, as shown in Fig. 4.6 [1].

4.2.2.3 The Process of Eliminating the Control Error

Because $\Delta u(t)$ is the value of the final changing control output, in order to eliminate the control error, we use the control antibody algorithm

Fig. 4.6 The producing process of the forthcoming antibodies

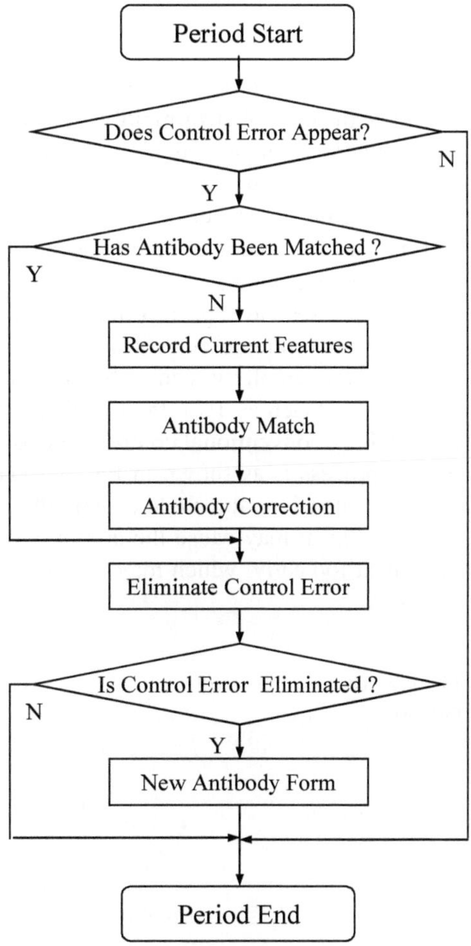

$$u(t) = u(t_0) + \Delta u(t), \tag{4.4}$$

where, $u(t_0)$ is the initial value of the control output to eliminate the control error; and $\Delta u(t)$ is the changing value of the control output of the middle-antibody $C_{Ab}'(n + 1)$. But if only the control antibody algorithm (4.4) is used, the control performance may not be satisfactory. This is because: (1) The low changing rate of the process variable will make the control system longer time to be stable; (2) There is an error between the changing output $\Delta u(t)$ of the middle-antibody and the real final RLIC's changing output.

As to the first aspect, we take a measure of combining the control antibody algorithm and the conventional P (proportional) control algorithm. This measure is named the control antibody and the proportional (CABP) algorithm, its control output is

$$u(t) = u(t_0) + \Delta u(t) + K_{p0}e(t), \quad (4.5)$$

where, K_{p0} is the proportional gain. It can make the output $u(t)$ change fast and eliminate the control error rapidly. When the control error is eliminated, the RLIC's stable output value $u(t)$ equals to $u(t_0)$ plus $\Delta u(t)$ (refer to Eq. (4.4)).

Generally, in Eq. (4.5), the bigger is K_{p0}, the higher is its response speed of eliminating the control error. But if K_{p0} is too big, the control system may oscillate, and the performance may be influenced. The selection of K_{p0} should be careful for different control plants.

As to the second aspect, a measure of combining the CABP algorithm and the auxiliary conventional control algorithm is taken. Here, we choose the PID algorithm as the auxiliary conventional control algorithm. When $e(t)$ exceeds the threshold ε_2 (such as 1%), the CABP algorithm is taken. Otherwise, it switches to the auxiliary conventional control method to eliminate the control error further. The overall process is as shown in Fig. 4.7 [1].

The value of the threshold ε_2 will affect the control effectiveness or precision. If ε_2 is too big, it may cause the auxiliary conventional PID control algorithm come into effect too early, which may weaken the response ability of the control system.

Fig. 4.7 The overall control process

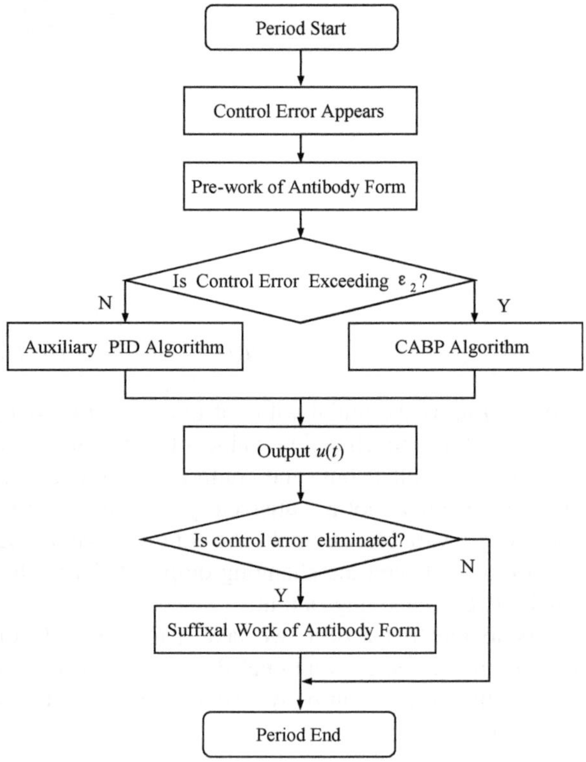

If ε_2 is too small, it may cause the auxiliary PID control algorithm take effect too late or evenly cannot come into effect, which may affect the control precision.

4.2.2.4 Management of the CABs

Along with the running of the RLIC, more and more control errors are eliminated, and the RLIC's speed and accuracy to eliminate the control error are enhanced. However, if all CABs keep alive forever, the number of the CABs in the database may become larger and larger. It will reduce the speed of the program-executing, as such affect the control performance. So we take the following method to manage the CABs.

We divide the overall absolute control error range into 10 intervals, and the set-point range into 10 intervals, as shown in Table 4.1 [1]. In Table 4.1, N_{ij} (i, j = 0, 1, …, 9) is the number of CABs in the corresponding subinterval. By using probability, we may define an approximate value of N_{ij} in every subinterval. If the total CAB number in a certain interval exceeds the maximum number, the oldest CAB will be deleted.

During the CAB matching period, we first locate the corresponding subinterval to let the matching process only execute in this subinterval. If there is no CAB in the current subinterval, the matching process will be done in the nearest subinterval with CABs. As thus, the matching speed keeps very quickly. Moreover, this makes the RLIC more adaptive to the control system. The detail CAB matching process is as shown in Fig. 4.8 [1].

The RLIC has a fine ability of learning and memory, so it has a better self-adaptation. When the control plant is changed, the controller needs to learn at least once to form a new CAB. Along with the increasing of self-learning times and the CAB's quantity, the RLIC can adapt to the new control plant easily.

Table 4.1 The management table of the cabs

Item	Absolute control error (%)										
Set-point value (%)	Interval	0–10	10–20	20–30	30–40	40–50	50–60	60–70	70–80	80–90	90–100
	0–10	N_{00}	N_{01}	N_{02}	N_{03}	N_{04}	N_{05}	N_{06}	N_{07}	N_{08}	N_{09}
	10–20	N_{10}	N_{11}	N_{12}	N_{13}	N_{14}	N_{15}	N_{16}	N_{17}	N_{18}	N_{19}
	20–30	N_{20}	N_{21}	N_{22}	N_{23}	N_{24}	N_{25}	N_{26}	N_{27}	N_{28}	N_{29}
	30–40	N_{30}	N_{31}	N_{32}	N_{33}	N_{34}	N_{35}	N_{36}	N_{37}	N_{38}	N_{39}
	40–50	N_{40}	N_{41}	N_{42}	N_{43}	N_{44}	N_{45}	N_{46}	N_{47}	N_{48}	N_{49}
	50–60	N_{50}	N_{51}	N_{52}	N_{53}	N_{54}	N_{55}	N_{56}	N_{57}	N_{58}	N_{59}
	60–70	N_{60}	N_{61}	N_{62}	N_{63}	N_{64}	N_{65}	N_{66}	N_{67}	N_{68}	N_{69}
	70–80	N_{70}	N_{71}	N_{72}	N_{73}	N_{74}	N_{75}	N_{76}	N_{77}	N_{78}	N_{79}
	80–90	N_{80}	N_{81}	N_{82}	N_{83}	N_{84}	N_{85}	N_{86}	N_{87}	N_{88}	N_{89}
	90–100	N_{90}	N_{91}	N_{92}	N_{93}	N_{94}	N_{95}	N_{96}	N_{97}	N_{98}	N_{99}

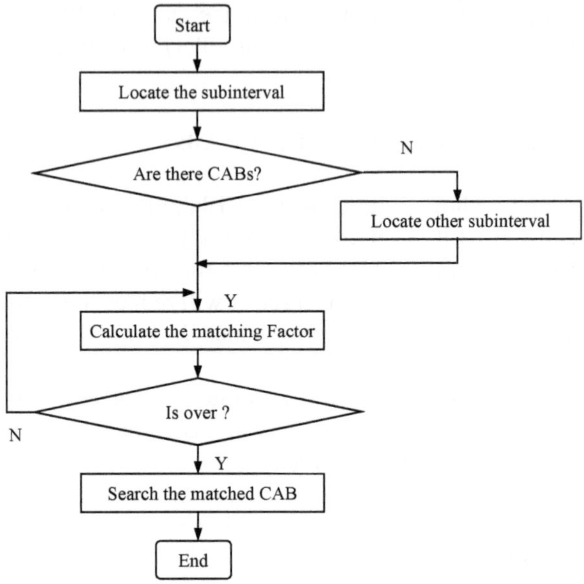

Fig. 4.8 The CAB matching process

4.2.3 *Simulation Experiments*

In order to examine the RLIC's control performance, we carry out our simulation experiments by choosing the following plants

$$G(s) = \frac{e^{-2s}}{(s+2)(s+3)} \tag{4.6}$$

and

$$G(s) = \frac{1+2s}{(1+10)(1+7s)(1+3s)}. \tag{4.7}$$

Plant (4.6) is a second-order one with a time delay in temperature control system, where the liquid temperature is controlled by using the steam through an energy exchanger, and the pressure of the steam is modulated by the condensate level in a tank. Plant (4.7) is a third-order one in level control system, where the water level in the third tank is controlled by the water level in the first tank. As well known, it is difficult to get satisfactory control effectiveness when the two plants are controlled by using conventional control algorithm. Here, we examine the control performance of the RLIC and compare with that of the conventional PID controller. In order to make the contrast effectiveness more clearly, we let the auxiliary conventional control algorithm of the RLIC be the PID controller with the same parameters of the conventional PID controller.

Table 4.2 The control parameters for different plants

Parameters	K_{p0}	K_{p1}	T_{i1}	T_{d1}	ε_1	ε_2
Second-order plant with a time delay	1.5	1.3	50.0	0.0	0.05	0.02
Third-order plant	0.087	5.0	80.0	0.0	0.05	0.02

Through the frequent changing of set-point value, the RLIC can produce CABs. When the control error appears for the first time, the RLIC has no CABs, so it has to use its auxiliary PID algorithm to eliminate the control error. After the first control error is eliminated, the first CAB has been produced. When the control error appears again, the CAB begins to take effect.

The control parameters of the RLIC are as shown in Table 4.2 [1], where K_{p1}, T_{i1}, and T_{d1} are the control parameters of the auxiliary PID algorithm; ε_1 and ε_2 is the RLIC control error threshold and the switching threshold, respectively. The produced CABs are as shown in Table 4.2. From the control performances as shown in Figs. 4.9 and 4.10, it can be found that the RLIC eliminates the control error faster, with no overshoot, and more stable than the conventional PID controller does. That is to say, the RLIC has a better control performance, e.g. faster response and less overshoot, than that of the conventional PID controller.

Why the RLIC eliminates the control error faster than the conventional PID controller does? From the CABP algorithm in Eq. (4.5), we know that the output of the RLIC is equivalent to a step output $\Delta u(t)$ plus P control action. In a certain degree, $\Delta u(t)$ is the final changing output of the controller when the control system becomes stable. Therefore, the RLIC adjusts its output near the stable value, while

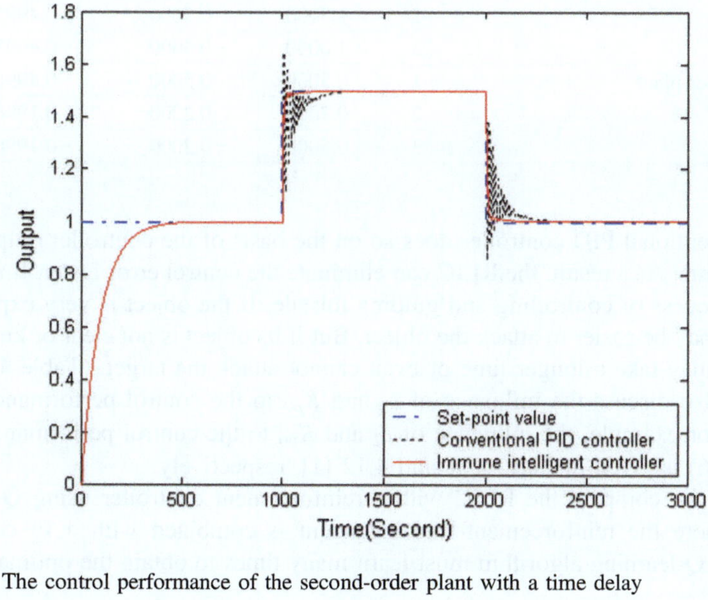

Fig. 4.9 The control performance of the second-order plant with a time delay

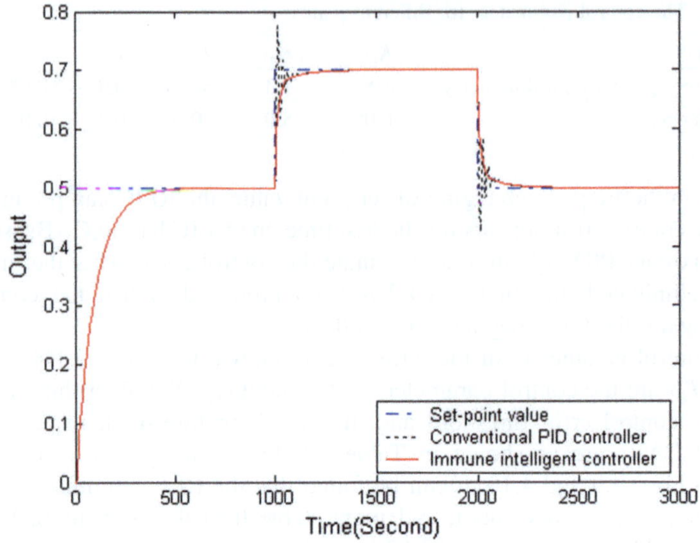

Fig. 4.10 The control performance of the third-order plant

Table 4.3 The resulting cabs for different plants

Name	No.	Set-point	Control error	Changing output
Second-order plant with a time delay	1	1.0000	1.0000	0.9995
	2	1.5000	0.5000	−0.4995
	3	1.0000	−0.5000	−0.4995
Third-order plant	1	0.5000	0.5000	0.4996
	2	0.7000	0.2000	0.1996
	3	0.5000	−0.2000	−0.1996

the conventional PID controller does so on the basis of the controller output value of that time. As a result, the RLIC can eliminate the control error faster. It is similar to the process of controlling and guiding missile. If the object is very explicit, the missile may be easier to attack the object. But if its object is not clear or known, the process may take a longer time or even cannot attack the target. (Table 4.3)

We also discuss the influence of ε_2 and K_{p0} to the control performance of the RLIC. For example, the influence of ε_2 and K_{p0} to the control performance of the plant (4.6) is given in Figs. 4.11 and 4.12 [1], respectively.

Also, we compare the RLIC with a reinforcement controller using Q-learning [16], where the reinforcement-learning agent is combined with a PI controller. Because Q-learning algorithm must learn many times to obtain the optimal control

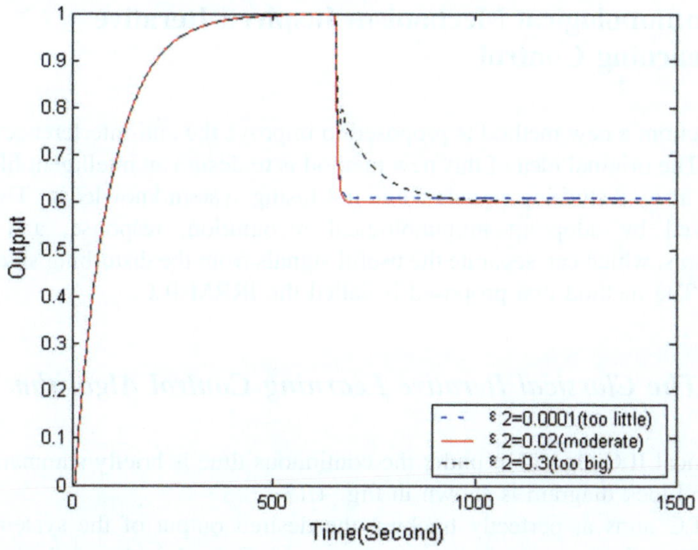

Fig. 4.11 The comparing effectiveness of various ε_2

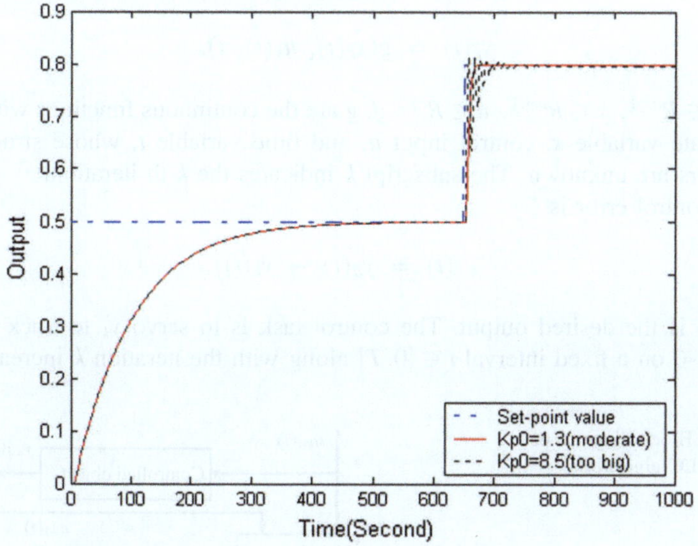

Fig. 4.12 The comparing effectiveness of various K_{p0}

policy. The reinforcement-learning agent needs being trained more than 1000 repetitions of a 500 time-step to improve control performance [16]. While the RLIC needs just only one or several time-steps to learn, and it has better control performance than the reinforcement controller using Q-learning.

4.3 Immunological Mechanism Inspired Iterative Learning Control

In this section, a new method is proposed to improve the anti-interference ability of the ILC. The original idea of this new method is to design an intelligent filter which can filter away disturbing signals while not losing system knowledge. The method is achieved by adopting immunological recognition, response, and memory mechanisms, which can separate the useful signals from the disturbing signals while filtering. The method that proposed is called the IRRM-ILC.

4.3.1 *The Classical Iterative Learning Control Algorithm*

The classical ILC algorithm under the continuous time is briefly summarized. The algorithm block diagram is shown in Fig. 4.13.

The ILC aims at perfectly tracking the desired output of the system without using any information of the state space. Consider the following linear continuous-time system:

$$\begin{aligned} \dot{x}_k(t) &= f(x_k(t),\ u_k(t),\ t), \\ y_k(t) &= g(x_k(t),\ u_k(t),\ t), \end{aligned}$$

where $x \in R^{n\times 1},\ y \in R^{m\times 1},\ u \in R^{r\times 1}$. f, g are the continuous functions with respect to the state variable x, control input u, and time variable t, whose structure and parameters are unknown. The subscript k indicates the k-th iteration.

The control error is

$$e_k(t) = y_d(t) - y_k(t),$$

where y_d is the desired output. The control task is to servo y_k to track y_d, make $e_k(t) \to 0$ on a fixed interval $t \in [0, T]$ along with the iteration k increasing.

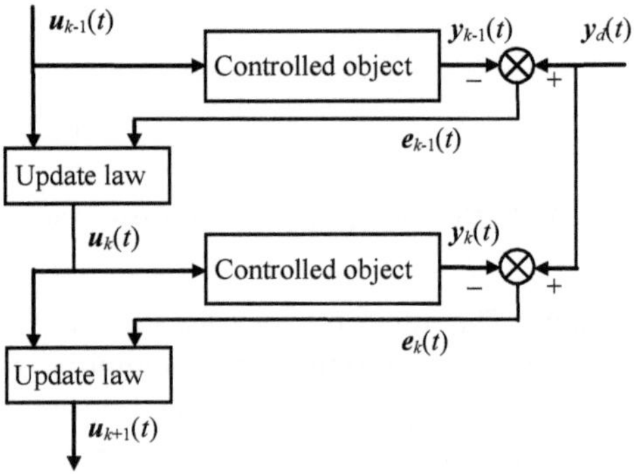

Fig. 4.13 Block diagram of the basic ILC algorithm

For the classical ILC, the following basic postulates are required, although in recent researches some of these postulations have been relaxed [62].

1. Every trial (pass, cycle, batch, iteration, repetition) ends in a fixed time of duration.
2. Repetition of the initial setting is satisfied. That is, the initial state $x_k(0)$ of the objective system can be set to the same point at the beginning of each iteration.
3. Invariance of the system dynamics is ensured throughout the repetition.
4. Output $y_k(t)$ is measured in a deterministic way.
5. The system dynamics are deterministic.

Under these assumptions, the classical ILC with open-loop learning method is given by:

$$u_{k+1}(t) = L(u_k(t), e_k(t)), \tag{4.8}$$

where L is a linear or nonlinear operator. This is the core idea of the ILC. The control input of the $(k+1)$-th iteration is calculated by the control input of the k-th iteration and its control error.

Based on this open-loop learning method, considering the closed-loop control error, a more robust learning method is given by:

$$u_{k+1}(t) = L(u_k(t), e_k(t), e_{k+1}(t)), \tag{4.9}$$

where, L is the update law of the learning process. There are varieties of different research on this update law. The most original and typical one is called "Arimoto-type" update law [20], which is given by:

$$u_{k+1} = u_k + \Gamma \dot{e}_k, \tag{4.10}$$

where Γ is a diagonal learning gain matrix.

Starting from this classical Arimoto-type ILC algorithm, a number of more general expressions have been developed. For instance, in [63], a "PID-like" update law was given as:

$$u_{k+1} = u_k + \Phi e_k + \Psi \int e_k dt + \Gamma \dot{e}_k, \tag{4.11}$$

where Φ,Ψ, and Γ are learning gain matrices.

These algorithms highlighted the perspective that the ILC is a kind of control law which utilized all available past information for the control of a periodic system. The next trial's control input is calculated from the previous trial's control input and transient errors. Traditionally, the applications of the ILC have been focusing on those control systems in where a single, repeated operation is performed. This focus includes many industrial systems in manufacturing, robotic system, and chemical engineering areas, where mass production on assembly lines entails repetition.

However, the capacity of resisting disturbance of the ILC needs further improvement [21]. According to [64], the ILC is an integrator along the iterative axis. All the system responses are integrated by the ILC. This means, all the signals including both systems' useful knowledge and disturbing signals are "learned" by the ILC. Thus, a disturbing signal will be continuously amplified along the iterative axis as a snowball effect. Up to present, there are many research findings on improving the ILC's capacity of disturbance resisting. Most solutions proposed a similar kind of ways which is adding a filter to the ILC, such as second-order Butterworth filter [64], zero-phase filter [65], or more simply, a forgetting factor. However, for these filter-adding solutions, the filters would filter away certain signals no matter they are useful signals or disturbing signals. Therefore, IRRM-ILC, an intelligent filter which can filter away disturbing signals while not losing system knowledge, is proposed.

4.3.2 Design of the IRRM-ILC

4.3.2.1 Definition of Antigen and Antibody

To introduce the IRRM-ILC control algorithm, first, the mathematical expressions of the antigen and antibody are defined as follows.

Antigen

Antigen, the invaded non-self pathogenic microorganisms, is the target of the immune system to be eliminated. In the proposed IRRM-ILC method, a disturbance is regard as an antigen. However, it is hard to mathematically define a disturbance.

An epitope, also known as antigenic determinant, is the part of an antigen that is recognized by the immune system. It is made of several key characteristics of an antigen and defined the antigen. Thus, several state signals of the system are regarded as the key characteristics of a disturbance and are used for defining this disturbance. Hence, we define an antigen as:

$$Ag_i = \{r, u, e, y\},$$

where

$$r = \{r(t),\ r(t-1), \ldots, r(t-n)\},$$

$$u = \{u(t),\ u(t-1), \ldots, u(t-n)\},$$

$$e = \{e(t),\ e(t - 1), \ldots,\ e(t - n)\},$$

$$y = \{y(t),\ y(t - 1), \ldots,\ y(t - n)\},$$

t is the time value when Ag_i disappeared after appeared. From time $(t - n)$ to time t, such definition of an antigen included both historical and current information when this disturbance appeared. This information contained all the key characteristics of the system: system input, controller output, control error, and system output. Thus, the definition of the antigen Ag_i can also be used for recognizing an antigen itself. $i = 1, \ldots, Num_{\max}^{im}$, $Num_{\max}^{im}$ is the maximum number of the defined antigens. It is determined based on the storage capacity and, most importantly, the calculation speed. Since the algorithm will search through the database, the calculation speed will decrease along with $Num_{\max}^{im}$ increasing.

n is the length of the defined antigen. The choice of n will affect the algorithm's performance. The bigger it is, the more historical information of an antigen will contain, and the more kinds of disturbances will be distinguished. It will help provide a more accurate control but lead to a data storage increasing thereby a calculating speed fall. If n is a small number, a disturbance with long duration may be separated into several pieces. The calculating speed will be faster but the control accuracy may decline. A proper n can be achieved base on experience from experiments.

Antibody

Antibody, which evoked by antigen, can identify the antigen for the immune system to eliminate. Since the antibody is used for eliminating antigen, the control input in the IRRM-ILC control system is regarded as the antibody. Therefore, the antibody is defined as:

$$Ab_i = \{u_i^{im}\}$$

where $u_i^{im} = \{u_i^{im}(1), \ldots,\ u_i^{im}(m)\}$.

The antibody under such definition is a set of compensatory control input which generate by the controller to mitigate disturbance. m is the length of the control input set. The choice of m will also affect the algorithm's performance. The bigger the m is, the longer the system will react to the disturbance. Apparently, it is not the longer the better. To choose a proper m can also base on experiments.

Antigen-antibody Relationship

In order to eliminate the antigen accurately, antibody and antigen have a one-to-one corresponding relationship. An antibody contains a paratope which is specific for one particular epitope on an antigen. So, to define this relationship, the variable i of the antigen is also set to be the same in the corresponding antibody. The antigen-antibody relationship is written as $\{Ag_i,\ Ab_i\}$. This antigen-antibody relationship is used as the data structure to store in the database.

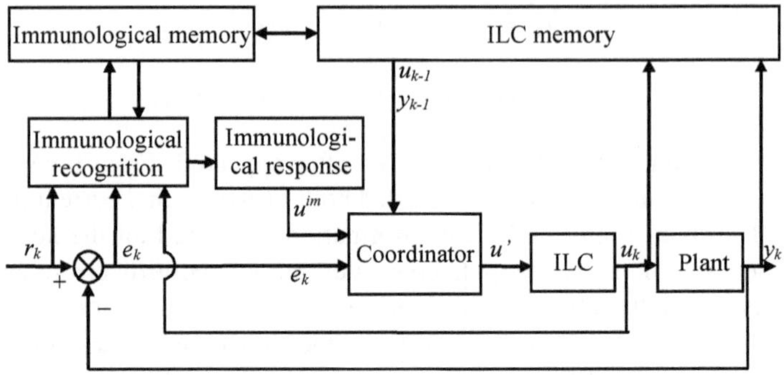

Fig. 4.14 IRRM-ILC control system structure

4.3.2.2 Control System Structure of the IRRM-ILC

The system structure of the proposed IRRM-ILC is shown in Fig. 4.14.

The core controller of the control system is the ILC controller, which contains an ILC updating law. The ILC memory is a database that stores the input and output data of the plant on the last iteration. The controller and the database form the classical ILC control system.

The intelligent algorithm inspired by the immune system is achieved through three components, namely, the immunological recognition unit, the immunological response unit, and the immunological memory unit.

Each unit's function is illustrated as follows:

ILC Controller

The ILC controller unit utilizes the stored data of u_{k-1}, e_{k-1}, and the current closed-loop feedback of control error e_k to calculate the current control input u_k. The update law can be various algorithms with good performance. In this section, we adopted the "PID-like" update law, which given in (4.11), as an experiment method.

ILC Memory

The ILC memory unit stores the data of u_{k-1}, y_{k-1} (the input and output data of the plant) on last trail. Since the memory is only used for stored the data of one trail, the memory's size is small and fixed.

Immunological Recognition Unit

The core of the immunological recognition unit is a recognition algorithm which is made of detecting, searching, and matching processes. The detecting process is for monitoring the real-time system states data to judge whether there is an antigen appeared or not. If an antigen appeared, the searching and matching processes will search through the antigen records in the immunological memory and try to find the match of this antigen. Thus, this recognition unit is able to detect an invaded antigen and then to recognize the antigen.

Immunological Response Unit

The immunological response unit has two kinds of response mechanisms, namely, the primary response and the secondary response. The primary response uses the ILC algorithm to generate a primary antibody. The secondary response is to find a corresponding antibody of the recognized antigen based on the antigen-antibody relationship.

Immunological Memory Unit

The immunological memory unit stores all the ever appeared antigens and their corresponding antibodies. Its stored data grows with the system running along time. This unit also has functions on data operating such as writing, reading, and deleting. Especially, the data deleting function is able to maintain the storage's amount based on a certain forgetting rule.

Coordinator

The function of the coordinate unit is to combine the antibody with the ILC control input to form a united control output during secondary response process. Under the situations of primary response or no antigen appeared, this unit will directly adopt the output data from the ILC memory.

4.3.2.3 IRRM-ILC Control Algorithm

The output of the coordinator is written as:

$$u'_k(t) = \alpha \cdot u_k(t) + (1 - \alpha) \cdot u_i^{im},$$

where $\alpha \in [0, 1]$ is a smooth factor, it helps implanting the antibody into the general control output, avoiding a sharp fluctuate of the controller's output. u_i^{im} is a compensatory control quantity.

The antibody u_i^{im} is also learning and updating along the iterations. The historical antibody will be updated with new antibody and stored into the immunological memory. The updating is defined as:

$$u_i^{im} = u_{k+1}(t).$$

It should be mentioned that, the u_i^{im} is not a function of time t, it can be used whenever the disturbance is appeared.

Finally, the final control output is the combination of the ILC output and the IRRM output, it can be written as:

$$u_{k+1}(t) = u'_k(t) + \Phi e_k(t) + \Psi \int e_k(t)dt + \Gamma \dot{e}_k(t). \tag{4.12}$$

where Φ,Ψ, and Γ are learning gain matrices.

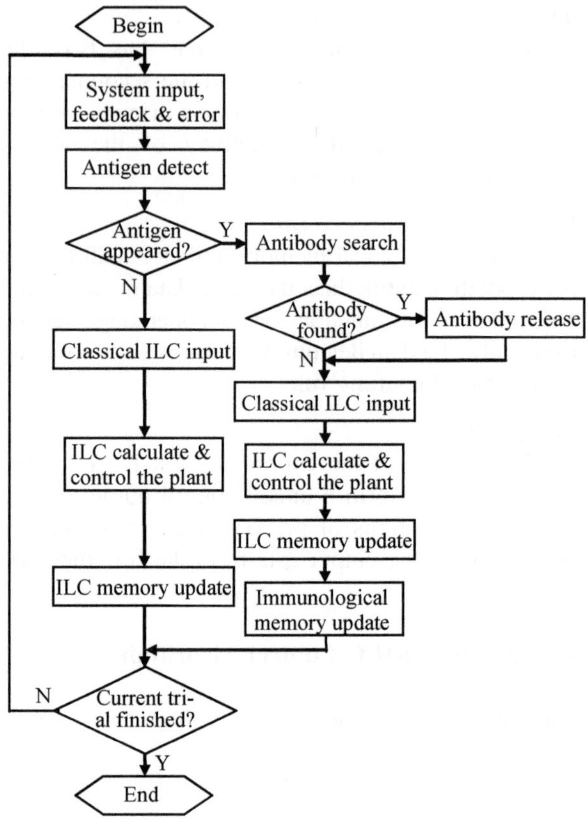

Fig. 4.15 IRRM-ILC control steps of each trial

The flowchart of each trial of the IRRM-ILC algorithm is shown in Fig. 4.15. Theoretically, the same as ILC, the IRRM-ILC is a non-parametric data driven control algorithm. It utilizes the historical input and output data of the control system to calculate the control quantity. However, the difference is, the basic ILC is a simple integrator alone the iteration axis while IRRM-ILC has an intelligent filter before the integrator. The intelligent filter targets the specific disturbances and output a compensatory control quantity to eliminate the disturbances. In general, the antibody inside the immunological memory is the compensatory control quantity. It is the record of the control quantity of a historical best control performance. And then, this antibody will improve itself by learning if this disturbance appears again and again.

Along the time axis of each iteration, first the desired value of the output r_k is the input on the system. It is also the feedback of system's real output y_k. Meanwhile, the real-time r_k, e_k, and u_k are monitored and detected by the immunological recognition unit.

If no antigen is detected, the system will only use the classical ILC controller to control the plant. If there is one or more detected antigen, the other immunological units will be activated.The immunological recognition unit will identify the antigen to see if it is detected the first time. If it is a first detected antigen, primary response of the controller is generated by the ILC. After the control period, the system will store the antigen and the antibody into the immunological memory.

If the antigen has been detected before, the secondary response process will output the historical antibody to modify the controller's output and then eliminate the antigen. The secondary response process is also an iterative learning process. After the immunological memory has been updated, the system would move to the next moment until the current iteration finished. Along the iterative axis, such iterative process circulates with a continual updating database.

4.3.2.4 Recognition and Searching Algorithm

Besides the control algorithm, both the recognition algorithm and the searching algorithm are important parts of the proposed method. The recognition algorithm is used for identifying the invaded antigens. Accurate identification will contribute to the control performance. Otherwise, identification mistakes may lead to unstableness of the control system. The searching algorithm is used for finding a match during identification. Fast and efficient searching algorithm will increase the calculating speed.

The recognition algorithm which adopted in this section is the basic Euclidean distance method. The antigen which defined by r, u, e, and y is a fixed length data set. All the data can be regarded as features and then the data set itself is the feature vector. In order to acquire high accuracy control, the recognition algorithm should maintain the variety of the antigens and distinguish the antigens as detail as possible. Thus, we chose a relatively small value as the threshold of the Euclidean distance of feature vector for separating different antigens. Therefore, the almost exactly same data set will be regarded as the same antigen. Choosing the proper threshold can be achieved by doing a serious of experiment and comparing their results of control performance.

The searching algorithm which adopted in this section is the self-organizing list method. This method rearranges the nodes, keeping the most frequently accessed ones at the top of the list. The data which have been most frequently read would be accessed first. This method helps reducing searching time because, in most application cases, some certain disturbances are reappearing. Meanwhile, if the antigen memory overflowed, the least accessed data would be abandoned.

The self-organizing list searching algorithm is a time consuming algorithm especially when the memory size is large. The control method we proposed here is more time consuming than the classical ILC because of its searching algorithm. It is a method that sacrifices a certain calculation speed to improve the control performance. We believed that such sacrifice is acceptable because of two reasons. On one hand, with the development of computer hardware, the computer's calculating speed is increasing. This development makes the complicated methods such as intelligent algorithm possible in practice. On the other hand, the calculate speed can be increased to a certain extent via improving the algorithms. With limited recourses, the optimization of the searching algorithm of the proposed method is worth researching.

4.3.3 Application to Wet Spinning Coagulation Process

In this section, we use the temperature control system of wet spinning coagulation process as a simulation example to prove the proposed approach is effective. In industrial production, to control the temperature, heating the coagulation bath via vapor heat exchanger is widely used. Vapor heat exchanger is a piece of equipment which is designed for efficient heat transfer from vapor to other medium. In this coagulation process the heat exchanger uses tubes to separate the vapor with the solution in order to avoid material exchanging and keeps the solution still. In Fig. 4.16, components "7"–"9" indicate the vapor heat exchanger in the bath. The high temperature vapor enters the tube entrance, releases its heat during passing through the heat exchange tube and then the cooler vapor leaves the bath via tube exit. Compared to electrical heating, the vapor heating is faster and cheaper, so it has been widely used in industrial production.

The working process of the vapor heat exchanger is a typical batch process. The temperature of the bath will gradually drop along with time because the bath temperature is usually higher than the room temperature. When the temperature reaches a lower limit which is chosen based on the technological requirement, the heat exchanger will start heating the bath to a desired temperature. Such batch process is a suitable plant for applying the IRRM-ILC control. According to [66], the vapor heat exchanger can be simplified and modeled via step response data. We adopt the vapor heat exchanger model (the time delay is set aside for the moment) built in [66] with the transfer function:

$$G(s) = \frac{1}{(s+1)(38s+1)}.$$

It should be mentioned that this transfer function is only used for simulating the plant. After the plant's model is built, it will be regarded as a black box. All the

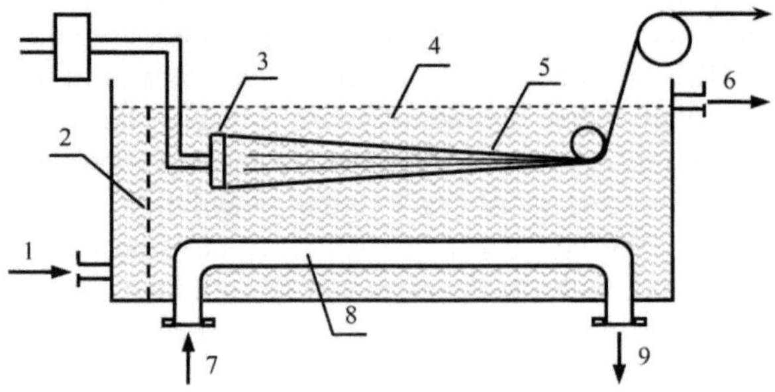

Fig. 4.16 Coagulation bath of wet spinning. (1) water-adding intake; (2) baffle plate; (3) spinneret; (4) coagulation solution; (5) PAN threads; (6) overflow vent; (7) vapor entrance; (8) heat exchange tube; (9) vaporexit

Table 4.4 Parameters of all Experiments

Parameter	Value	Parameter		Value
T_s(s)	0.1	PID-like ILC	Φ	5000
T_l (s)	1000		Ψ	100
K	30		Γ	2000
T_{set} (°C)	60			

Table 4.5 Parameters of IRRM-ILC algorithm

Parameter	Value	Parameter	Value
n	2	$Num_{\max}^{im}$	10,000
m	1	α	0.5

following works including control algorithm design are done without the plant's knowledge because the proposed method in this section is a data-driven method.

A series of simulation experiments based on the temperature control system introduced above has been carried out to verify the effectiveness of the proposed IRRM-ILC. All these experiments shared a group of given parameters, as shown in Table 4.4. Assumed that T_s is the sampling time; T_l is the duration of every iteration; K is the iteration times; T_{set} is the set value of the coagulation bath temperature.

Assumed that the vapor heat exchanger would turn on automatically when the temperature was dropped to 59 °C and then turn off when the bath was heating to 60 °C. With iterative learning process, the control performance would be better and better. On the following experiment, the control performances of the IRRM-ILC with classical ILC have been compared. The related parameters of the IRRM-ILC are listed in Table 4.5.

4.3.3.1 Effectiveness Validation of the ILC

First, an experiment has been conducted to see if the ILC method was feasible for the batch process of temperature control via vapor heat exchanger. The control algorithm is a typical ILC with PID-like update law. The plant is the black box of the vapor heat exchanger.

Because the ILC method is a function based on the assumption that the ideal controller output u_d is existed when $u \rightarrow u_d$ makes $y \rightarrow y_d$. Obviously, in causal control system it is not possible to find such u_d for step input. So the ILC method was commonly proved by using a sinusoidal y_d [20]. Hence, to achieve a step-like signal which the real output can follow, in this experiments a little adjustment on the desired output was made. the reference input r (desired output y_d) was set as:

$$y_d = \begin{cases} T_{set}\sin(6\pi \cdot 10^{-3}t), & t \leq 0.5/6 \cdot 10^3 = 83.\dot{3} \\ T_{set}, & t > 83.\dot{3} \end{cases}.$$

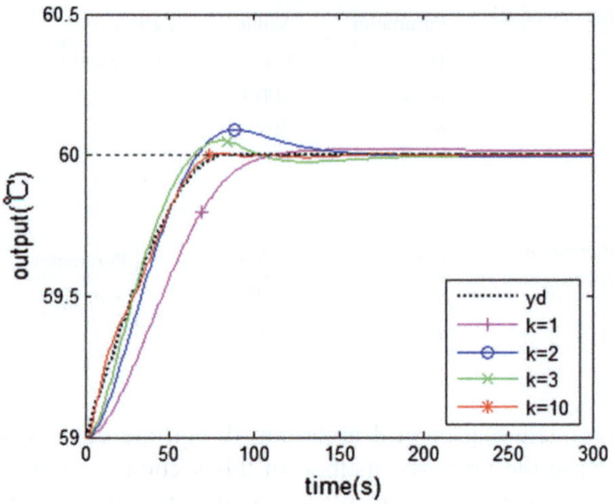

Fig. 4.17 ILC control experiment on vapor heat exchanger

The simulation results within ten iterations are shown in Fig. 4.17. *k* is the iteration times. We can see from the result that the ILC as a data-driven control method is feasible on this temperature control system on vapor heat exchanger.

4.3.3.2 Random Disturbance Experiment

First, the proving experiment of the system's disturbances detecting ability was conducted. In order to present the result clearly, a discrete set of random disturbances was given to the system. The disturbances were within ±10% (±0.1 °C) and with the appearing rate of 1% along the time axis. Figure 4.18a, b gives the detecting result of disturbance on one iteration.

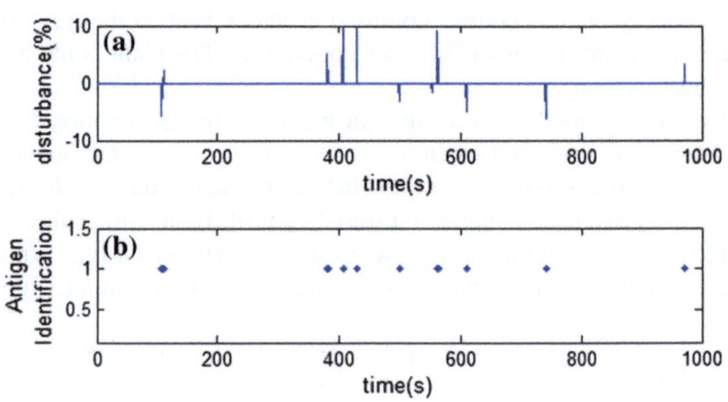

Fig. 4.18 Experiment results of disturbance reorganization

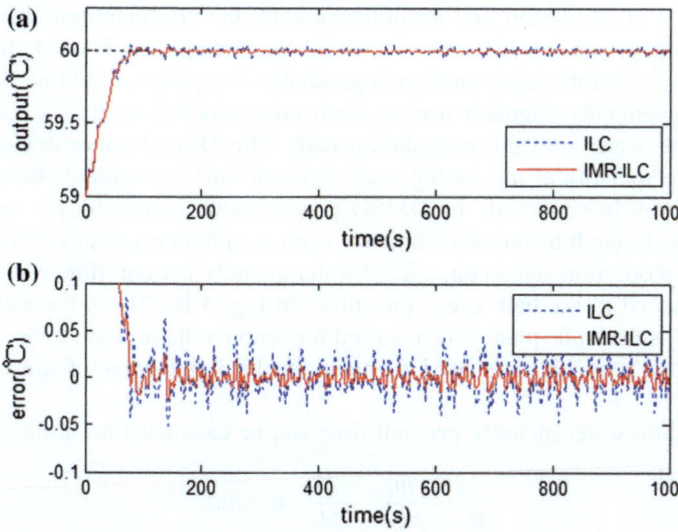

Fig. 4.19 Experiment results with random disturbance

Figure 4.18a plots the added disturbances signals. And then calculation is conducted. The mark bits that set to 1 in Fig. 4.18b means antigen has been detected at that time. Comparing the Fig. 4.18a and Fig. 4.18b we can easily found out that the immunological recognition unit is well functioned. All the disturbances have been correctly detected.

Then, Gaussian white noise has been adopted in the system. The white was generated by the simulation tool. Such disturbance was set to appear along the iterative axis with a probability of 0.5. The simulation results of the 30th iteration of both the classical ILC and the IRRM-ILC under the white noise are shown in Fig. 4.19.

The control performance of the IRRM-ILC against random disturbance is obviously better than what of the conventional ILC. The IRRM improvement of the control system can intelligently identify and target the disturbances and eliminate them efficiently. The control error cause by the random disturbances are maintaining in a small range while the errors under the conventional ILC's control are keep accumulating.

4.3.3.3 Repeated Disturbance Experiment

Besides the temperature, concentration of the coagulation bath is another important factor. With the DMSO continuously diffusing into the coagulation solution, the DMSO concentration of the solution is increasing. The changes of the concentration will affect the following fiber coagulation process. In some simple experimental facilities the concentration was not being controlled because the experimental process was of short-term and small-throughput, the concentration changes were not obvious. However, in actual industrial producing, the DMSO diffusion is continuous so that the concentration is gradually changing and will be far from its set value eventually. General way of such concentration control is achieved by adding pure water into the coagulation bath. The DMSO concentration will be diluted. The quantity of the adding-water per unit time is calculable because when the production line is fixed, the DMSO concentration increasing per unit time is fixed. Thus, through technical parameters such as spinning velocity, PAN solution ingredient proportion and so on, a fixed water quantity per unit time was calculated and then added to the bath every unit time. In Fig. 4.16, “1” is the water-adding intake; “2” is the baffle plate which is used for keeping the solution to be still; “6” is the overflow vent which is used to maintain the liquid-level of the bath to be invariable.

The adding-water quantity per unit time can be calculated according to:

$$\frac{m_d}{M + m_d - M_w} = D_{bath},$$

$$m_d = m_0 - m_1,$$

$$M_w = M_1 - M_0,$$

where

D_{bath} is the mass fraction of DMSO in coagulation bath;
m_d is the mass of DMSO which diffuses into the coagulation bath per unit time;
M is the mass of water that adds into the coagulation bath per unit time;
M_w is the mass of water that permeates into the fiber per unit time;
m_0 and m_1 are the mass of DMSO inside the PAN fiber before and after pass through the coagulation bath, per unit time;
M_0 and M_1 are the mass of water inside the PAN fiber before and after pass through the coagulation bath, per unit time.

Usually, the water that adds into the coagulation bath for concentration control is room temperature water. Thus, such water adding process will cause a disturbance on the temperature however. And since the quantity of water is fixed, the disturbance is a repeated disturbance.

Thus, to build a temperature control system for the coagulation bath should take the repeated disturbance caused by concentration control into consideration. In order to conduct a more targeted research on the repeated disturbance we studied the model of temperature change on coagulation bath due to concentration control water adding first.

This temperature change by adding water is called “convective heat transfer”, which is the transfer of heat from one place to another by the movement of fluids. It is caused by locomotion of material particle and tending to equalize temperature. This process can be modeled as:

$$\frac{dQ}{dt} = h \cdot A(T_{bath} - T_{water}) = h \cdot A\Delta T, \tag{4.13}$$

where,

Q	is the thermal energy in joules;
h	is the heat transfer coefficient (W/m^2 K);
A	is the surface area of the heat being transferred (m^2);
$\Delta T = T_{bath} - T_{water}$	is the time-dependent thermal gradient between the two objects.

The heat transfer coefficient h depends upon physical properties of the fluid and the physical situation in which convection occurs. All the complex factors and mechanisms of the heat transfer process are end up in this heat transfer coefficient. Formulas and correlations are available in many references to calculate heat transfer coefficients for typical configurations and fluids. Empirical value of heat transfer coefficient h is shown in Table 4.6.

Another basic physics law adopted is:

$$Q = Cm\Delta T, \tag{4.14}$$

where, m is the mass of the object and C is the heat capacity.

To calculate the temperature changing process by room temperature water adding into coagulation bath, we combine the thermal energy Q with the heat capacity C.

1. Use (4.13) to calculate the rate of the thermal energy transfer dQ/dt under current difference in temperature between coagulation bath solution (hot water) and added room temperature water (cold water). The surface area of the heat being transferred is equal to the cross-sectional area of the coagulation bath.
2. Calculate the thermal energy Q of unit time.
3. Use (4.14) to calculate the temperature change of hot water and cold water respectively after current unit time. After thermal energy transferred from hot water to cold water, the hot water cools down and the cold water warms up. The thermal energy Q is the same for both hot and cold water but with difference in mass the temperature change is different.
4. Update the current temperature of both hot and cold water and go back to step. (1) until the hot and cold water reach the same temperature.

Table 4.6 Empirical value of heat transfer coefficient

Natural convection (Gas)	Forced convection (Gas)	Natural convection (Water)
5–25	20–100	200–1000
Forced convection (Water)	Vapor condensation	Water boiling
1000–15,000	5000–15,000	2500–25,000

Table 4.7 Experiment setting parameters

Parameter	Value	Parameter	Value
T_{set} (°C)	60	D_{bath}(%)	68
T_{add}(°C)	20	G(kg/h)	25
L(m)	1.5	h(W/m^{2}K)	1000
W(m)	0.5	C_{water}(J/kg)	4.2*10^3
H(m)	0.5	ρ_{water}(kg/m^3)	1*10^3

Table 4.8 Calculation results of water-adding process

Parameter	Value
A(m^2)	0.25
M(kg/h)	9.714
M_{add}(kg/turn)	1.619

We assume that G is the spinning flow rate; L, W, and H are the length, width and height of the coagulation water solution size. All the experiment setting parameters are listed in Table 4.7.

According to [69] the content proportion of the DMSO, PAN, and water of the PAN solution (before coagulation) is 4.01: 1.0: 0.01. And this proportion of the PAN fiber (after coagulation) is 1.65: 1.0: 0.85. And, we suppose M_{add} is the quality of the pure water that is added into the coagulation bath every 10 min. Thus, we can acquire the calculation results in Table 4.8.

With the data in Table 4.7 and Table 4.8, the temperature change in coagulation bath is as shown in Fig. 4.20. Such temperature change pattern is a typical convective heat transfer process. The temperature keep decreasing until stabilize at a certain value and the pattern of decreasing rate is also a decreasing process.

Then, this temperature signal was added into the control system as a repeated disturbance appeared in every iteration. The control results of the 30th iteration are shown in Fig. 4.21a, b. Figure 4.21b is a zoom in Fig. 4.21a at the point when the disturbance appears. When facing repeated disturbance and which is last for a long period, both the IRRM-ILC and conventional ILC appeared to overcome such disturbance with an effort. They took a longer time period to settle down. However, during the adjusting process, the contrastive results indicated that the IRRM-ILC also gave a better perform than the conventional ILC. The temperature control error of conventional ILC is over two times of the IRRM-ILC.

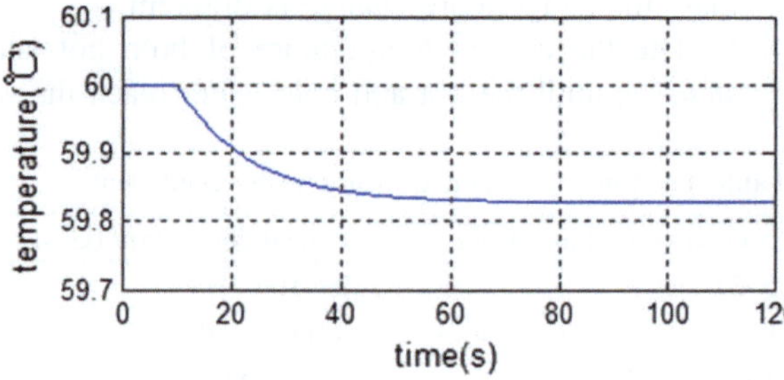

Fig. 4.20 Temperature change caused by water adding

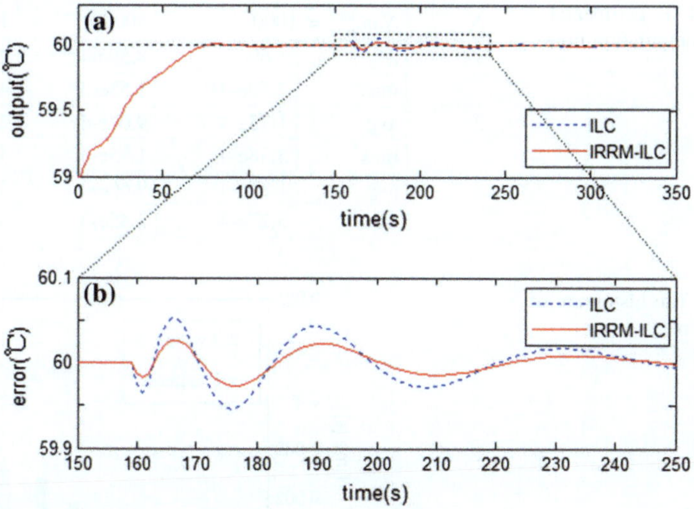

Fig. 4.21 Simulation result with repeated disturbance

4.3.3.4 Experiments of the Computation Time

In order to observe the computation time and prove it is fast enough to meet the demand of control frequency, a group of experiments were carried out.

These experiments have different computation time related parameters of the IRRM-ILC, namely, the single antigen length n and the antigen total number $Num_{\max}^{im}$. In these experiments, the iteration time K was 30 and the control points T_l was 1000, the average computation times were recorded separately. At the same time, due to the randomness of the IRRM-ILC algorithm, the maximal value of the computation time in each experiment has been record. The maximal value reflected the worst situation of computation time when the IRRM-ILC searched the entire database to find the match. To be on the safe side, this maximal value is more crucial than the average when choosing the computation time related parameters. The experiments' computing environment is listed in Table 4.9.

The experiments were separated into three groups with different single antigen length n (n = 1, 2, and 3). In each group, three single experiments were carried out with different antigen total number $Num_{\max}^{im}$ ($Num_{\max}^{im}$ = 1000, 5000, and 10,000). So all together there were 9 experiments. After the experiments, the average and the maximum computation times are shown in Table 4.10. The results are made into a histogram for convenient viewing, as shown in Fig. 4.22.

Table 4.9 Computing environment parameter

CPU	Pentium(R) dual-core	Operating system	Win7 32bit
Clock speed	2.70 GHz	Matlab version	R2009a
RAM	2 GB	Iteration times	30

Table 4.10 Experimental result of computation time

N	$Num_{\max}^{im} = 1000$		5000	10,000
1	avg	1.69e-4	8.59e-4	2.59e-3
	max	2.59e-4	1.55e-3	4.31e-3
2	avg	1.75e-4	9.07e-4	2.97e-3
	max	3.38e-4	1.95e-3	4.74e-3
3	avg	1.87e-4	9.66e-4	3.28e-3
	max	5.27e-4	2.45e-3	5.25e-3

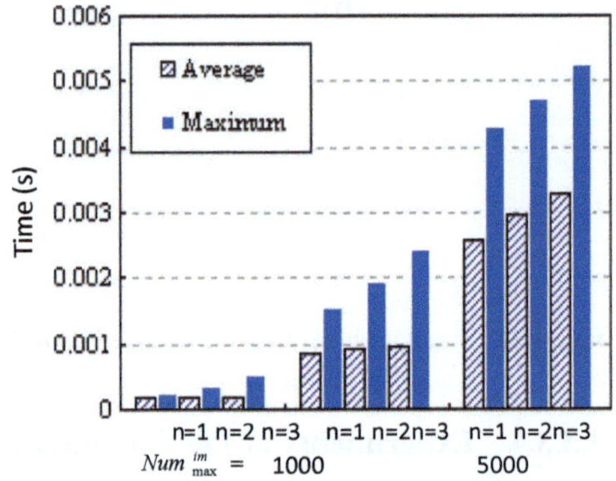

Fig. 4.22 The histogram of computation

To compare the computation time with the basic ILC algorithm, we conducted the tenth experiment: conventional ILC computation time record experiment. Under the same computer environment and iteration parameters, the average computation time of the conventional ILC is 3.23e-7 s (the average value and the maximum value is the same). It is clearly that the proposed intelligent control algorithm sacrifices the control speed for better control performance.

Although the computation time is increased, but from these experiments we can see that when the antigen length n was 2 and the antigen total number $Num_{\max}^{im}$ was 10,000, the control performance was favorable and the maximum computation time was within 0.005 s. If the discrete control system sampling time was longer than 0.005 s, this computation time is acceptable. Normally, temperature control system control frequency is around 0.1 s to 1 s, hence, the IRRM-ILC method will be suitable for use on temperature control system.

4.4 An Immune Reconfigurable Controller

In this section, we propose an immune reconfigurable controller (IRC) by absorbing the most prominent advantages of the biological immune system, because the biological immune system has many features similar to the objectives of the reconfigurable control system, such as self-monitoring, self-defending, self-stabilization and self-learning [67]. The IRC consists of a surveillance unit to imitate the immune surveillance process, a response unit to imitate the immune response process, a memory unit to imitate the immune memory mechanism and a learning unit to imitate the learning ability. By treating the control system as a biological immune system, all these units collaborate with each other and realize the reconfiguration process.

4.4.1 Design of the Immune Reconfigurable Controller

4.4.1.1 Reconfigurable Principle of the Biological Immune System

The biological immune system is able to defend antigen intrusion and maintain the immunologic homeostasis. There are many mechanisms which can be learned from the biological immune system [68]:

1. Immune surveillance mechanism. The immune surveillance mechanism can identify foreign variants by detecting the epitopes on the surface of the antigens.
2. Immune response mechanism. Once the antigens break through the line of defense, the antibodies perform their function by combining with the specific antigens.
3. Immune memory mechanism. Part of the activated antibodies will be permanently stored in the immune system as memory cells. If the same antigen intrudes again, the immune system can give a quick response.
4. Immune self-learning mechanism. The immune self-learning mechanism is incarnated by the clonal selection process. In the response period, the antibodies with high affinity can grow at an index rate through clones.

As we can see, the reconfiguration system is very similar to the immune system which keeps the body stable by removing the intrusive antigens. The antigens are just like component malfunctions which may bring risk to the system stability. Thus, for reconfiguration control systems, the refactoring process can be designed according to the immune response process by defining the actuator failures as antigens and the corresponding control strategies as antibodies.

4.4.1.2 Structure of the Immune Reconfigurable Controller

The structure of the proposed IRC is shown in Fig. 4.23. During initialization, the decoupling controller is set to be the conventional decoupling controller (CDC). In Fig. 4.23, $r(k)$ is the input vector of the control system; $y(k)$ is the output vector of the control system; $e(k)$ is the control error vector; IAE is integrated absolute error index vector; $u_c(k)$ is the control signal vector; $u(k)$ is the actuator output vector (or the input vector of the control plant); $u_{bc}(k)$ and $u_b(k)$ are control signal and output of the faulty actuator (numbered with b); mod_n is the fault model (numbered with n)

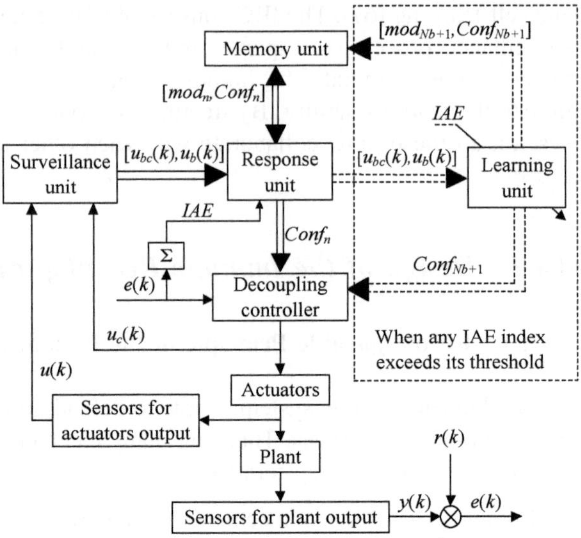

Fig. 4.23 Structure of the IRC

in the faulty actuator's memory cell; $Conf_n$ is the corresponding control strategy to mod_n; mod_{Nb+1} and $Conf_{Nb+1}$ are fault model and its control strategy gained via the learning unit; *Nb* stands for the amount of fault models in the faulty actuator's memory cell before the learning unit starts.

The overall procedure of the proposed IRC scheme is illustrated in Fig. 4.24.

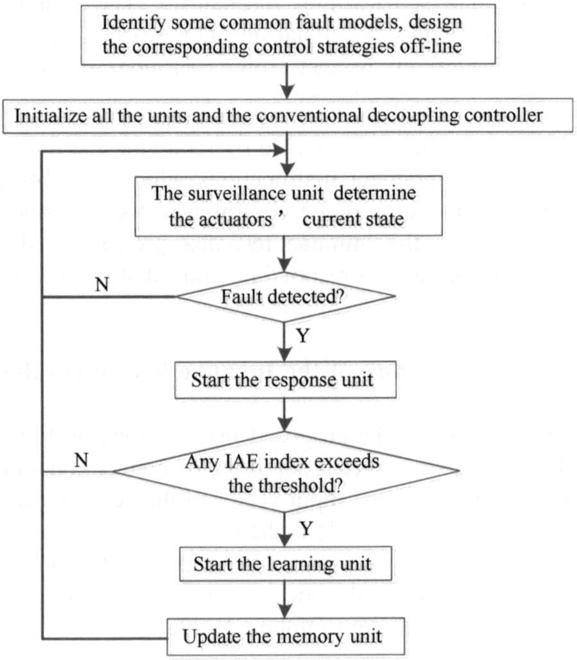

Fig. 4.24 The overall procedures of the IRC

The surveillance unit

The surveillance unit is a model-based diagnosis module which is in charge of fault detection. Define the normal mapping models of the actuators $f_i(\cdot)$ as follows:

$$u_i = f_i(u_{ic})$$

where i stands for the loop numbers (or the actuator numbers), u_{ic} is the control signal of actuator i and u_i is the actual output of actuator i. Determined by the attributes of the practical actuators, the mapping models can be linear or nonlinear.

By comparing the actual output u_i with the theoretical output $f_i(u_{ic})$, the surveillance unit can determine whether the actuator is faulty or not. In order to avoid misjudgment due to disturbances, the diagnosis is based on a continuous time series $T_s = \{k - L + 1, k - L + 2, \ldots, k\}$, where k stands for the present sampling time and L is a given length. Actuator i is considered to be faulty when $u_{ic}(l),\ l \in T_s$ and $u_i(l),\ l \in T_s$ satisfy:

$$\forall l \in T_s,\ |u_i(l) - f_i(u_{ic}(l))| > \delta$$

where δ is a very small positive real constant value. It is a threshold used to evaluate whether the plant's model is changed enough so that a new controller is needed. So the actuator will be treated as faulty only when it deviates from its theoretical output persistently. The random short-term interferences could then be ruled out while the long-term interferences will be regard as equipment failure.

The response unit

When an actuator fault is detected, the response unit will search for a match model in the memory first. Assuming the numbered actuator b (in control loop b) is detected to be faulty at time k, and there are N_b fault models in memory b, an indicator measuring the affinity of the model is designed as:

$$M = \min_{n \in N_s} \left(\sum_{l \in T_s} |u_b(l) - \mathrm{mod}_n(u_{bc}(l))| \right)$$

where $N_s = \{1, 2, \ldots, N_b\}$ is the model number set and $\mathrm{mod}_n(\cdot)$ stands for fault model n. The model corresponding to indicator M is the closest match for the actual fault model and the corresponding control strategy marked by this model is chosen. The current CDC will then be reconfigured according to this control strategy. As decoupling controller, the control strategy configuration consists of two parts: for the decoupling compensators and for the loop controllers.

After the CDC is reconfigured, the control system is stabilized in some degree. Then the response unit will try to further improve this CDC. Since the integrated absolute error (IAE) index is used to evaluate the control performance, it is adopted as optimization objective:

$$J_i = \sum_{s=k-S}^{k-1} |e_i(s)|$$

where S is a given length. A threshold is set for the IAE index:

$$\exists i \in I, J_i > J_{iT}$$

where I is the set of control loop numbers, J_{iT} is the IAE index threshold of loop i. When this condition is satisfied, the learning unit will be enabled.

The learning unit

There are two steps in the learning process. The first step is to adjust the decoupling compensators in accordance with the actual fault model. The faulty actuator's sample data $\{u_{bc}(l),\ u_b(l) \mid l \in T_s\}$ are used to identify the fault model via linear regression analysis. The fault model $u_b = \mathrm{mod}_b(u_{bc})$ is discussed in Section III Part C. Substituting the sample data into the general fault model, we have:

$$\begin{cases} \dfrac{\partial Q}{\partial \hat{b}_b} = -2\sum\limits_{l \in T_s} [u_b(l) - \hat{b}_b - \hat{p}_b \cdot u_{bc}(l)] = 0 \\ \dfrac{\partial Q}{\partial \hat{p}_b} = -2\sum\limits_{l \in T_s} [u_b(l) - \hat{b}_b - \hat{p}_b \cdot u_{bc}(l)] \cdot u_{bc}(l) = 0 \end{cases}$$

where

$$Q = \sum_{l \in T_s} (u_b(l) - \mathrm{mod}_b(u_{bc}(l)))^2,$$

and $\hat{p}_b$ is the approximate value of the damage coefficient and $\hat{b}_b$ is the approximate value of the offset. Then the $\hat{p}_b$ and $\hat{b}_b$ can be achieved by solving:

$$\begin{cases} \hat{p}_b = \dfrac{\sum\limits_{l \in T_s} (u_b(l) - \bar{u}_b)(u_{bc}(l) - \bar{u}_{bc})}{\sum\limits_{l \in T_s} (u_{bc}(l) - \bar{u}_{bc})^2} \\ \hat{b}_b = \bar{u}_b - \hat{p}_b \cdot \bar{u}_{bc} \end{cases}$$

where $\bar{u}_b = \frac{1}{L}\sum\limits_{l \in T_s} u_b(l)$ and $\bar{u}_{bc} = \frac{1}{L}\sum\limits_{l \in T_s} u_{bc}(l)$.

As a result the approximate fault model is obtained. Then the frequency domain transfer function of the faulty actuator can be determined. The compensators should be modified in accordance with the following compensation principle:

$$G_{pij}(s) = \frac{G_{ij}(s)}{G_p(s) \cdot G_{ii}(s)}$$

$$G_{pji}(s) = \frac{G_p(s) \cdot G_{ji}(s)}{G_{jj}(s)}$$

where i and j stand for different loop numbers (as $i \neq j$), $G_{pij}(s)$ is the decoupling compensator in loop i for eliminating the influence of loop j and $G_p(s)$ is the transfer function of the faulty actuator.

The next step is to optimize the parameters of loop controllers so that the IRC can better adapt to the current situation and achieve better performance. Details of the optimization process will be discussed later.

The memory unit

The memory unit is a control strategy repository which consists of different memory cells numbered by the corresponding control loop's number. Each memory cell is in charge of storing fault models of the corresponding control loop's actuator as well as the corresponding control strategies under these fault situations. The specificity of the control strategies is marked by the corresponding fault model along with the memory cell number. To prepare the memory unit, several common fault models and the corresponding control strategies should be pre-stored in each memory cell during the initialization. After the IRC puts into operation, the memory unit can be updated on-line.

4.4.1.3 Multi-objective Optimization Algorithm for the IRC

In order to better meet the requirements of the actual control applications, the parameters of controllers in all loops should be optimized. Since the IAE index is an evaluation of the control performance, it is used as our optimization objective. As a multi-controller system, all the controllers' IAEs are combined so that the problem becomes a multi-objectives problem.

This optimization process is similar to the antibody clone selection process of the immune system. Therefore, for this IRC multi-objective optimization problem, we adopt the multi-objective clonal selection (MOCS) algorithm which described in [69]. The optimization process is conducted on the online identification model of the plant with its fault (identified by learning unit). Its final optimized result will be given to the controllers and controls the real plant.

Due to the adjustment of compensators, the system can be decoupled again. The fault-free loops are almost unaffected by the faulty actuator. Thus for the optimization process, only the controller of the faulty loop needs to be tuned. The antibodies can be coded in the form of [K_P, K_I, K_D] (K_P, K_I and K_D are proportion, differential and integral parameters of the PID controller of faulty loop, respectively).

By applying the MOCS [69] to the IRC, the main steps of the optimization process can be summarized as follows:

Step 1: Code the antibody with binary numbers;

Step 2: By centering on the current PID parameters, randomly generate the initial antibody population Ab whose size is N_{Ab};

Step 3: Calculate each antibody's affinity with the reciprocal of the objective functions and discard the dominated solutions from Ab based on the definition of Pareto dominance, getting the elite population Ab_{non} whose size is N_{non};

Step 4: Clone all individuals in N_{non} by an adaptive clone operator to get the expanded population Ab_{ex} whose size is $N_{non} + \sum_{z=1}^{G_m} (C_z - 1)$ where G_m is the given maximum iteration number. The adaptive clone operator is in accord with crowding distances used in the NSGA-II [70]:
$C_z = round\,[N_C \times D_z]$
where N_C is the maximum number of clones;D_z is the crowding distance of individual z; $z = 1, 2, \ldots, G_m$;

Step 5: Mutate all individuals in Ab_{ex} by a non-uniform mutation operation linked with the iteration:

$$K_{id}^{new} = \begin{cases} K_{id} + \Delta(g, K_{\max} - K_{id}), & r_d > P_m \\ K_{id} - \Delta(g, K_{\max} - K_{id}), & r_d \leq P_m \end{cases}$$

where K_{id} is the parameter randomly chosen from the antibody; $K_{\max}$ is the maximum value of this parameter; r_d is a random variable uniformly distributed between $[0, 1]$;P_m is a given mutation probability and $\Delta(g, \eta)$ is a function related to the current iteration g:

$$\Delta(g, \eta) = \eta\left(1 - r^{\left(1 - \frac{g}{G_m}\right)^2}\right)$$

where r is another random number uniformly distributed between $[0, 1]$. $\Delta(g, \eta)$ decreases with the increase of iteration, which means at the beginning of iteration the individuals mutate at a higher probability to maintain the diversity of population, and by the end of iteration the mutation scale is close to zero;

Step 6: Reselect the non-dominated individuals from Ab_{ex} according to the result of Pareto sorting and update the elite population Ab_{non} with these individuals. When the size of the updated Ab_{non} is larger than N_{non}, discard these individuals who have smaller crowding distances and limit the size to N_{non};

Step 7: Randomly choose an individual from the updated Ab_{non} as the parameters of the next sampling time;

Step 8: If the termination conditions are satisfied, stop the optimization process and return the final results; otherwise, go to Step 3.

After both the compensators and PID controllers are reconfigured, update the corresponding memory cell with the optimized control strategy. Notice that the control strategy consists of the configuration of decoupling compensators and the parameters of PID controllers.

4.4.2 The IRC System for the PANCF Coagulation Bath

The control system for the PANCF coagulation bath includes the control plant, the actuators, the redundant sensors and the IRC along with some special sensors served for the IRC's surveillance unit. Its structure is shown in Fig. 4.25. Since the quality of the as-spun fiber is mostly determined by the parameters, especially the temperature, concentration and liquid-level of the coagulation bath. The control system mainly targets at the stabilization of these three parameters.

In Fig. 4.25, L stands for liquid-level, T stands for temperature and C stands for concentration of the solution. Number the liquid-level loop as Loop 1, the temperature loop as Loop 2 and the concentration loop as Loop 3, then the control signals of the liquid-level loop, temperature loop and concentration loop can be denoted as u_{1c}, u_{2c} and u_{3c}, respectively. And the actuators output of these loops are u_1, u_2 and u_3, respectively.

4.4.2.1 The Control Plant

In Fig. 4.25, we hope to keep the solution at a certain temperature, concentration and liquid-level. There should be three kinds of solutions to consider, which are cold water, hot water and high concentration dimenthylsulfoxide (DMSO) solution

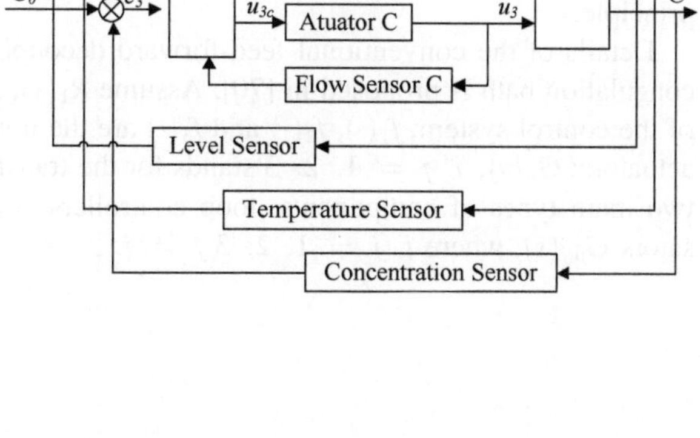

Fig. 4.25 Structure of the control system for the PANCF coagulation bath

to enter into the coagulation bath from three independent ports. The coagulation bath should discharge the spare solution at an appropriate speed from a discharge port so that the liquid-level is maintained. The three main controlled variables have strong interactions with each other thus they cannot be tuned independently and satisfactorily. If the other trace components are ignored, the main ingredients in the coagulation solution are the DMSO and the water.

According to [70], the regulation model of the coagulation bath is shown in Eq. (4.15) based on the principle of the volume conservation, the heat conservation and the solute conservation:

$$\begin{bmatrix} L(s) \\ T(s) \\ C(s) \end{bmatrix} = \begin{bmatrix} G_{11}(s) & G_{12}(s) & G_{13}(s) \\ G_{21}(s) & G_{22}(s) & G_{23}(s) \\ G_{31}(s) & G_{32}(s) & G_{33}(s) \end{bmatrix} \begin{bmatrix} U_1(s) \\ U_2(s) \\ U_3(s) \end{bmatrix} = \frac{1}{Ss + \frac{1}{R}} \begin{bmatrix} e^{-\tau_1 s} & e^{-\tau_2 s} & e^{-\tau_3 s} \\ \frac{T_{DMSO} - T_0}{h_0} e^{-\tau_1 s} & \frac{T_{H_2O,H} - T_0}{h_0} e^{-\tau_2 s} & \frac{T_{H_2O,L} - T_0}{h_0} e^{-\tau_3 s} \\ \frac{D_{DMSO} - D_0}{h_0} e^{-\tau_1 s} & \frac{D_{H_2O,H} - D_0}{h_0} e^{-\tau_2 s} & \frac{D_{H_2O,L} - D_0}{h_0} e^{-\tau_3 s} \end{bmatrix} \begin{bmatrix} U_1(s) \\ U_2(s) \\ U_3(s) \end{bmatrix} \tag{4.15}$$

where S is the area of the slot bottom, and R is the hydraulic resistance of the discharge port; h_0 is the desired liquid-level; T_{DMSO} is the temperature of the high concentration DMSO solution, $T_{H_2O,H}$ is the temperature of the hot water, $T_{H_2O,L}$ is the temperature of the cool water and T_0 is the desired temperature of the solution; D_{DMSO} is the concentration of the DMSO solution, $D_{H_2O,H}$ is the concentration of the hot water, $D_{H_2O,L}$ is the concentration of the cool water and D_0 is the desired concentration of the solution; U_1, U_2 and U_3 are three inputs of the controlled plant (as outputs of three actuators) in the frequency domain; τ_1, τ_2 and τ_3 are the lag coefficients in the regulation of the liquid-level, the temperature and the concentration, respectively.

4.4.2.2 The Conventional Decoupling Controller

Decoupling controller is to reduce or eliminate the coupling influences among various loops via compensations based on the mathematical model or the process mechanism. As a typical realization of this tenet, the feed-forward decoupling method is commonly used for decoupling control. It considers the coupling effect as external disturbances and designs the compensators based on the compensation principle.

Details of the conventional feed-forward decoupling controller for the PANCF coagulation bath is proposed in [70]. Assume $R_1(s)$, $R_2(s)$ and $R_3(s)$ are the inputs of the control system; $f_1(\cdot)$, $f_2(\cdot)$ and $f_3(\cdot)$ are the normal mapping models of three actuators; $G_{ij}(s),\ i,\ j = 1,\ 2,\ 3$ stands for the transfer functions. CDC consists of two main types of components: loop controllers $G_{cii}(s)$ and decoupling compensators $G_{pij}(s)$, where $i,\ j = 1,\ 2,\ 3\ j \neq i$.

For the loop controllers, PID controllers are good choice because of their broad applicability, high reliability, remarkable effectiveness and simple implementation. Each loop controller has two compensators as:

$$G_{pij}(s) = \frac{G_{ij}(s)}{G_{ii}(s)}$$

With the decoupling compensators, all loops can be independent and uncoupled. Parameters of $G_{cii}(s)$ can be tuned separately.

4.4.2.3 Latent Failures

The uncertainty of the MIMO processes results in ineluctable failures. According to the general control experience, frequent cases of actuator faults mainly include freezing, saturation (as a special case of freezing), bias and damage. Thus the general fault model $\mathrm{mod}_b(\cdot)$ can be summed up as follows [71]:

$$u_b(k) = p_b \cdot f_b(u_{bc}(k)) + b_b$$

where $u_b()$ is the output and $u_{bc}()$ is the control signal (as the input) of the faulty actuator,$b \in \{1, 2, 3\}$; $p_b \in [0, 1]$ is the damage coefficient; $b_b \in [0, u_{b\max}]$ is the offset of bias.

Then the frequency domain transfer function of the faulty actuator can be expressed as the following form:

$$G_b(s) = p_b + \frac{b_b}{s}$$

When $p_b = 0$, the actuator is stuck at the position of b_b, and when $p_b = 0$ and $b_b = u_{b\max}$, the actuator is saturated. When $b_b = 0$, the actuator is damaged with the coefficient of p_b. When $p_b = 1$, the actuator has an offset value of b_b, and when $p_b = 1$ and $b_b = 0$, the actuator is normal.

For actuator failures like bias or damage, the faulty actuator can still work to some extent. What happens is that only the output deviates from the expected value with a fixed pattern. However, under the freezing fault cases, the faulty actuator is no longer controllable. As the reconfiguration theory treats the actuator failures as bounded disturbances of system parameters and aims at ensuring the stability of the closed-loop system when parameters float within the setting boundaries, the freezing fault cases are not discussed here since such fault cases may easily cause the parameters float over the boundaries.

Table 4.11 Parameters of the coagulation bath

Name	Value	Name	Value
S/m^2	0.15	$T_{DMSO}/^\circ C$	30
R	1	$T_{H_2O,H}/^\circ C$	65
h_0	0.1	$T_{H_2O,L}/^\circ C$	10
τ_1	1	$T_0/^\circ C$	15
τ_2	1	$D_{DMSO}/\%$	80
τ_3	1	$D_{H_2O,H}/\%$	0.001
		$D_{H_2O,H}/\%$	0.001
		$D_0/\%$	0.65

4.4.3 Simulation Experiments and Results

With the purpose of verifying the effectiveness of the proposed IRC, we specify the parameters of the coagulation bath for the production conditions and the inputs of the control system. Some simulation experiments are conducted under different faulty conditions. All the simulation experiments are carried out with MATLAB R2010b on a workstation computer with a dual-core processor and 2 GB memory.

The specific values are listed in Table 4.11. Note that the concentrations of the hot water and the cold water are both set to 0.01 instead of zero due to the circumstance that the DMSO exists as one kind of impurities in the water [70]. The controlled plant is written as Eq. (4.16) with the parameters substituted by their corresponding values.

$$\begin{bmatrix} L(s) \\ T(s) \\ C(s) \end{bmatrix} = \frac{1}{0.15s + 1}\begin{bmatrix} e^{-s} & e^{-s} & e^{-s} \\ 150e^{-s} & 500e^{-s} & -50e^{-s} \\ 1.5e^{-s} & -6.49e^{-s} & -6.49e^{-s} \end{bmatrix}\begin{bmatrix} U_1(s) \\ U_2(s) \\ U_3(s) \end{bmatrix} \quad (4.16)$$

where the lag coefficients of the temperature, the liquid-level, and the concentration are all 1.0s. The set points for the liquid-level, the temperature and the concentration are 0.1 m, 15 °C and 65%, respectively.

A conventional control scheme which uses single-loop PID controller for each variable respectively is applied for comparison. The initial PID parameters for both the CDC and the IRC are listed in Table 4.12. Both schemes include three sets of PID parameters. The other parameters are given below: $f_1(\cdot) = f_2(\cdot) = f_3(\cdot) = 1$, $L = 10$, $S = 10$, $J_{1T} = 0.04$, $J_{2T} = 5.0$, $J_{3T} = 0.25$, $N_{Ab} = 50$, $G_m = 50$, $P_m = 0.5$. Assume the fault occurs at 20.0 s under the simulation environments.

4.4.3.1 Verification of the IRC with Offline Stored Faults

The IRC is verified to be practical under normal conditions. We set a faulty case for actuator and perform the simulation offline. The fault model chosen to simulate here

Table 4.12 Initial PID parameters

	IRC	Single-loop PID
Liquid-level		
P	0.35	0.5
I	0.8	0.9
D	0	0
Temperature		
P	0.006	0.001
I	0.0015	0.00065
D	0	0
Concentration		
P	0.035	0.1
I	0.1	0.08
D	0	0

is cold water valve bias with $p_3 = 1$, $b_3 = 0.15$. This fault model and its corresponding settings are stored in the memory cells during the initialization. For CDC, we manually adjust the compensators and tune the PID parameters to appropriate values. We also perform a reconfigurable MPC on the same system.

The result of the fault case when the cold water valve has bias of $p_3 = 1$, $b_3 = 0.15$ is shown in Fig. 4.26. Due to the decoupling compensators, the normal loops are not affected by the faulty actuator and the responses of the normal loops are the same as that of the CDC, and thus they are not shown here.

Once the actuator fault is detected, the CDC remains the same (that means it is not reconfigured) while the IRC and the MPC will reconstruct the decoupling controller. The IRC causes less overshoot and can return to the stable state much faster than the CDC. However, when comparing to reconfigurable MPC, the IRC does not have more advantages. So that's why we conduct a further improvement for IRC through online optimization.

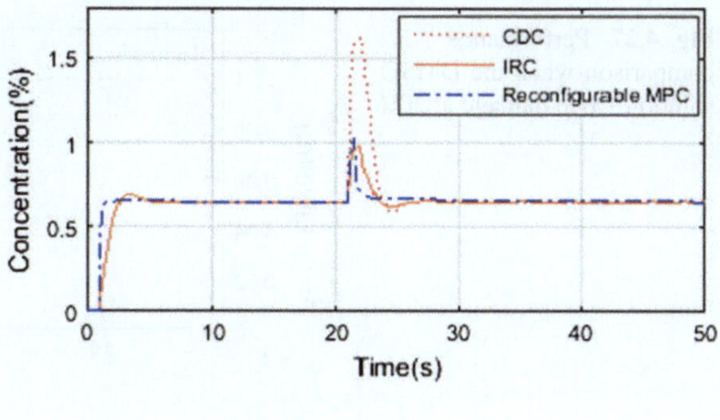

Fig. 4.26 Performance of the IRC when the cold water vavle has a 0.15 bias

4.4.3.2 Simulation with Online Optimization Process

We examine the flexibility of the IRC with online optimization process. Take an example when the DMSO solution valve damages with $p_1 = 0.4$, $b_1 = 0$. But in the memory unit of the IRC, a close match that the DMSO solution damage with $p_1 = 0.6$, $b_1 = 0$ is pre-added, so the learning unit is triggered when $J_1 > J_{1T}$ by this pre-added model.

The performances of the CDC, reconfigurable MPC, the non-optimized IRC and the optimized IRC are shown in Fig. 4.27. And a detailed enlarge figure of Fig. 4.27 is given in Fig. 4.28. The non-optimized IRC is the IRC that uses the closest match in the memory unit to reconstruct itself without further modification while the optimized IRC is the proposed IRC which can reconstruct itself as well as further optimize it with the MOCS algorithm in Section II Part C when needed. The optimized IRC changes the PID parameters of the level loop from $P = 1.3$, $I = 1.3$, $D = 0$ (as stored in the memory cell for the $p_1 = 0.6$, $b_1 = 0$ case) to $P = 1.7035$, $I = 1.9$, $D = 8.9443 \times 10^{-3}$. Compared with the CDC, the non-optimized IRC, and the reconfigurable MPC, the optimized IRC can eliminate the overshoot more quickly and lead to better control performance. The optimized IRC has the best performance among them. Obviously, the non-optimized IRC has greater volatility than the optimized one. This is because the MOCS algorithm can optimize the IRC which uses the closest match in the memory unit to reconstruct itself with further improvement, so that it can better meet the requirements of the actual control systems. This also demonstrates that the IRC can be more effective for unknown fault compared with other reconfiguration schemes.

For the concern of the adopted MOCS may be time-consuming, the computation time is monitored. Under the computer environment mentioned above, the maximum computation time of one control action is 0.03429 seconds, and the average computation time of every control step is 0.00026 seconds. So as we can see, compared to the 0.2s sampling time, the computational complexity of the optimized IRC is acceptable. It is able to finish calculation within every sampling period.

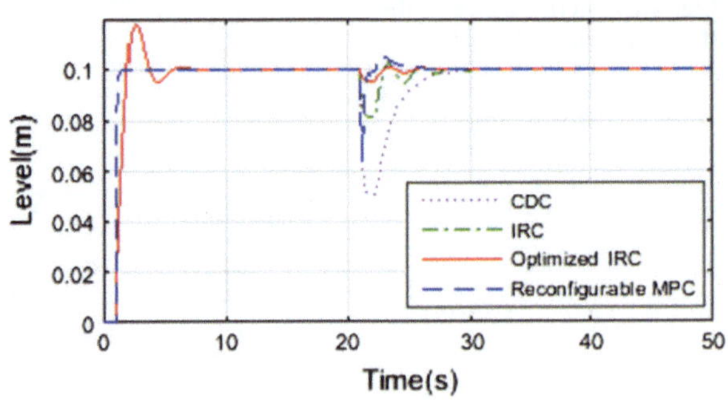

Fig. 4.27 Performance comparison when the DMSO solution vavle damage at 50%

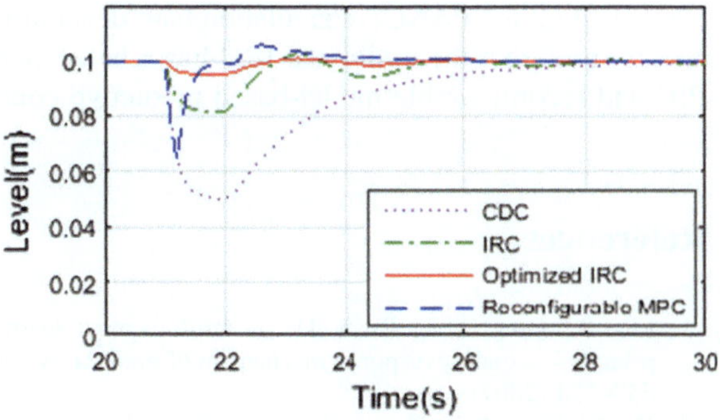

Fig. 4.28 Enlarged detail of Fig. 4.27 from 20 to 30 seconds

4.5 Conclusions

In this chapter, firstly, a novel RLIC based on the primary–secondary response mechanism of the immune system is proposed. The RLIC can self-learn during the process of eliminating the control error, and produce the corresponding CABs automatically. These CABs can be used to eliminate a new control error. After a new control error is eliminated, the new CAB is produced again. In this way, the RLIC can obtain the better control performance. Simulation results demonstrate that the RLIC's control performance is superior to that of the conventional PID controller. Moreover, compared with other reinforcement-learning algorithm, the RLIC can learn faster and has better control performance.

Then, an intelligent iterative learning control method is proposed based on the recognition, response and memory mechanism of immune system. The method is applied to the temperature control system of wet spinning coagulation process. From the simulation results we can conclude that the proposed IRRM-ILC is suitable for the coagulation temperature control. Both random and repeated disturbance can be resisted in some degree. Compared with the traditional control method used on the same plant, the proposed algorithm adopted the data-driven ILC as core so that it has the data-driven advantages. Meanwhile, compared with the basic ILC, the proposed IRRM-ILC has a better ability of disturbance resistibility. The IRRM-ILC has such an improved performance is because it can recognize the disturbance and response to it specifically.

At last, a reconfigurable control method improved by immune system mechanism is proposed. The IRC simulates the process of immune response to antigen to handle the equipment fault. On one hand, the IRC can detect, identify and overcome newly appeared fault. So it is better than regular offline reconfigurable controller, which needs abundant stored pre-designed control models and cannot handle unknown fault. On the other hand, the IRC is able to overcome re-appeared fault much faster due to immune memory mechanism. So it is faster than regular online reconfigurable control method which will carry out the reconfigurable calculation every time fault appears. And then, an online optimization process is added into the IRC to further improve the IRC's performance. Simulation experiments are conducted on the PANCF coagulation bath decoupling control system. Comparison results show that the optimized IRC has a better performance than the decoupling PID and reconfigurable model-based predictive controller (MPC).

References

1. Liu, B., Ding, Y.S., Ruan, D.: A reinforcement learning intelligent controller based on primary–secondary response mechanism of immune system. Dynam. Cont. Dis. Ser. B. **14**(4), 557–574 (2007)
2. Xu, N., Ding, Y.S., Hao, K.R.: Immunological mechanism inspired iterative learning control. Neurocomputing **145**(18), 392–401 (2014)
3. Ding, Y.S., Xu, N., Dai, S.F., Ren, L.H., Hao, K.R., Huang, B.: An immune system-inspired reconfigurable controller. IEEE. T. Contr. Syst. T. (2016)
4. Ruan, X.E., Paka, K.H., Bien, Z.N.: Retrospective review of some iterative learning control techniques with a comment on prospective long-term learning. Contr. Theor. Appl. **29**(8), 966–973 (2012)
5. Ding, Y.S.: Computational Intelligence: Theory, Technique, and Application. Scientific Publishing House, Beijing, China (2004)
6. Kato, T., Ninomiya, Y., Masaki, I.: Preceding vehicle recognition based on learning from sample images. IEEE. T. Intell. Transp. **3**(4), 252–260 (2002)
7. Druet, C., Ernst, D., Wehenkel, L.: Application of reinforcement learning to electrical power system closed-loop emergency control. The 4th European Conference on Principles of Data Mining and Knowledge Discovery: PKDD 2000, Lyon, France. 86–95 (2000)
8. Hougen, D.F., John, F., Maria, G., James, S.: Fast connectionist learning for trailer backing using a real robot. Paper present at IEEE. Int. Conf. Robot. 1917–1922 (1996)
9. Alexander, J.S., Baird, L.C., Baker, W.L., Farrell, J.A.: A design and simulation tool for connectionist learning control systems: Application to autonomous underwater vehicles. Paper present at the Society for Computer Simulation Conference, Baltimore, Maryland.771–776 (1991)
10. Zhou, K.L., Wang, D.W.: Digital repetitive learning controller for three-phase CVCF PWM inverter. IEEE. T. Ind. Electron. **48**(4), 820–830 (2001)
11. Oriolo, G., Panzieri, S., Ulivi, G.: An iterative learning controller for nonholonomic mobile robots. Int. J. Robot. Res. **17**(9), 954–970 (1998)
12. Braga, P.S.: A topological reinforcement learning agent for navigation. Neural. Comput. App. **12**, 220–236 (2003)
13. Leslie, P.K., Michael, L.L., Andrew, W.M.: Reinforcement learning: A survey. J. Artif. Intell. Res. **4**, 237–285 (1996)
14. Gu, D.B., Hu, H.S., Spacek, L.: Learning fuzzy logic controller for reactive robot behaviours. Paper present at IEEE/ASME International Conference on Advanced Intelligent Mechatronics (AIM), Kobe, Japan, 20–24 (2003)
15. Remi, C.: Feedforward neural networks in reinforcement learning applied to high-dimensional motor control. Paper present at Algorithmic Learning Theory: 13th International Conference, ALT 2002, Lubeck, Germany, 24–26 Nov (2002)
16. Anderson, C., Hittle, D., Katz, A., Kretchmar, R.: Synthesis of reinforcement learning, neural networks, and PI control applied to a simulated heating coil. Artif. Intell. Eng. **11**(4), 423–431 (1997)
17. Norihisa, S., Masaharu, A., Makoto, K.: Control of associative chaotic neural networks using reinforcement learning. Lect. Notes. Comput. Sc. pp. 395–400, Springer, Berlin Heidelberg (2004)

18. Cang, Y., Y. Nelson, H.C., Wang, D.W.: A fuzzy controller with supervised learning assisted reinforcement learning algorithm for obstacle avoidance, IEEE. T. Syst. Man. Cyb. 33(1), 17–27 (2003)
19. Esogbue, A.O., Hearnes, W.E., Song, Q.: A reinforcement learning fuzzy controller for set-point regulator problems. Paper present at the 5th IEEE. Int. Conf. Fuzzy, New Orleans, LA, 8–11 Sept (1996)
20. Arimoto, S., Kawamura, S., Miyazaki, F.: Bettering operation of robots by learning. J. Robot. Syst. **1**(2), 123–140 (1984)
21. Bristow, D.A., Tharayil, M., Alleyne, A.G.: A survey of iterative learning control: a learning-based method for high-performance tracking control. IEEE. Contr. Syst. Mag. **26**(3), 96–114 (2006)
22. Owens, D.H., Hätönen, J.: Iterative learning control—an optimization paradigm. Annu. Rev. Control **29**, 57–70 (2005)
23. Ahn, H.S., Chen, Y.Q., Moore, K.L.: Iterative learning control: survey and categorization. IEEE. T. Syst. Man. Cy. C. **37**(6), 1099–1121 (2007)
24. Wang, Y., Gao, F., Doyle III, F.J.: Survey on iterative learning control, repetitive control and run-to-run control. J. Process. Contr. **19**(10), 1589–1600 (2009)
25. Norrlöf, M.: An adaptive iterative learning control algorithm with experiments on an industrial robot. IEEE. T. Robotic. Autom. **18**(2), 245–251 (2002)
26. Jia, L., Shi, J., Chiu, M.S.: Integrated neuro-fuzzy model and dynamic R-parameter based quadratic criterion-iterative learning control for batch process. Neurocomputing. **98**(3), 24–33 (2012)
27. Havlicsek, H., Alleyne, A.: Nonlinear control of an electrohydraulic injection molding machine via iterative adaptive learning. IEEE/ASME. T. Mechatron. **4**(3), 312–323 (1999)
28. Gao, F., Yang, Y., Shao, C.: Robust iterative learning control with applications to injection molding process. Chem. Eng. Sci. **56**(24), 7025–7034 (2001)
29. Pandit, M., Buchheit, K.H.: Optimizing iterative learning control of cyclic production processes with application to extruders. IEEE. T. Contr. Syst. T. **7**(3), 382–390 (1999)
30. Jiang, P., Unbehauen, R.: Iterative learning neural network control for nonlinear system trajectory tracking. Neurocomputing. **48**(1–4), 141–153 (2002)
31. Saab, S.A.: A stochastic iterative learning control algorithm with application to an induction motor. Int. J. Control **77**(2), 144–163 (2004)
32. Barton, A.D., Lewin, P.L., Brown, D.J.: Practical implementation of a real-time iterative learning position controller. Int. J. Control **73**(10), 992–999 (2000)
33. Wang, A., Afshar, P., Wang, H.: Complex stochastic systems modelling and control via iterative machine learning. Neurocomputing. **71**(13–15), 2685–2692 (2008)
34. Li, X., Zhang, W.: Multiple model iterative learning control. Neurocomputing. **73**(13–15), 2439–2445 (2010)
35. Yang, D.R., Lee, K.S., Ahn, H.J., J. H. Lee, J.H.: Experimental application of a quadratic optimal iterative learning control method for control of wafer temperature uniformity in rapid thermal processing. IEEE. T. Semiconduct. M. **16**(1), 36–44 (2003)
36. Gorinevsky, D.: Loop shaping for iterative control of batch processes. IEEE. Contr. Syst. Mag. **22**(6), 55–65 (2002)
37. Mezghani, M., Roux, G., Cabassud, M., Lann, M.V.L., Dahhou, B., Casamatta, G.: Application of iterative learning control to an exothermic semibatch chemical reactor. IEEE. T. Contr. Syst. T. **10**(6), 822–834 (2002)
38. Tousi, M.M., Khorasani, K.: Optimal hybrid fault recovery in a team of unmanned aerial vehicles. Automatica. **48**(2), 410–418 (2012)
39. Zhang, Y., Jiang, J.: Bibliographical review on reconfigurablefault-tolerant control systems. Annu. Rev. Control **32**(2), 229–252 (2008)
40. Rothenhagen, K., Fuchs, F.W.: Doubly fed induction generator model-based sensor fault detection and control loop reconfiguration. IEEE. T. Ind. Electron. **56**(10), 4229–4238 (2014)
41. Giovanini, L.: Robust adaptive control using multiple models, switching and tuning. IET. Control. Theory. A. **5**(18), 2168–2178 (2011)

42. Murphey, Y.L., Abul Masrur, M.: Model-based fault diagnosis inelectric drives using machine learning. IEEE-ASME T. Mech. **11**(3), 290–303 (2006)
43. Nguyen, D., Lehman, B.: An adaptive solar photovoltaic array using model-based reconfiguration algorithm. IEEE. T. Ind. Electron. **55**(7), 2644–2654 (2008)
44. Ma, J., Zheng, Z., Hu, D.: Modular robust reconfigurable flight control systemdesign for an overactuated aircraft. IET. Control. Theory. A. **6**(11), 1620–1632 (2012)
45. Durham, W.C., Cristofaro, A., Johansen, T.A.: Fault tolerant control allocation using unknown input observers. Automatica. **50**(7), 1891–1897 (2014)
46. John, A., Petersen, M., Bodson, M.: Constrained quadraticprogramming techniques for control allocation. IEEE. T. Contr. Syst. T. **14**(1), 91–98 (2006)
47. Li, J., Zhou, M.C., Guo, T., Gan, Y.H., Dai, X.Z.: Robust control reconfiguration of resource allocation systems with Petri nets and integer programming. Automatica. **50**(3), 915–923 (2014)
48. Vermillion, C., Sun, J., Butts, K.: Predictive control allocation for a termal management system based on an inner loop reference model-design, analysis, and experimental results. IEEE. T. Contr. Syst. T. **19**(4), 772–781 (2011)
49. Chen, W., Saif, M.: Observer-based strategies for actuator fault detection, isolation and estimation for certain class of uncertain nonlinear systems. IET. Control. Theory. A. **1**(6), 1672–1680 (2007)
50. Trujillo Rodriguez, C., Velasco de la Fuente, D., Garcera, G., Figueres, E., Guacaneme Moreno, J.A.: Reconfigurable control scheme for a PV microinverter working in both grid-connected and island modes. IEEE. T. Ind. Electron. **60**(4), 1582–1595 (2013)
51. Khomfoi, S., Tolbert, L.M., Leon, M.: Fault diagnosis and reconfiguration for multilevel inverter drive using AI-based techniques. IEEE. T. Ind. Electron. **54**(6), 2954–2968 (2007)
52. Hamayun, M.T., Edwards, C., Alwi, H.: Design and analysis of an fault-tolerant control scheme. IEEE. T. Automat. Contr. **57**(7), 1783–1789 (2012)
53. Staroswiecki, M., Zhang, K., Berdjag, D., Abbas-Turki, M.: Reducing the reliability over-cost in reconfiguration-based fault tolerant control under actuator faults. IEEE. T. Automat. Contr. **57**(12), 3181–3186 (2012)
54. Pashilkar, A.A., Sundararajan, N., Saratchandran, P.: Adaptive back-stepping neural controllerfor reconfigurable flight control systems. IEEE. T. Contr. Syst. T. **14**(3), 553–561 (2006)
55. Shin, D.H., Kim, Y.D.: Reconfigurable flight control system design using adaptive neural networks. IEEE. T. Contr. Syst. T. **12**(1), 87–100 (2004)
56. Shin, D.H., Kim, Y.D.: Nonlinear discrete-time reconfigurable flight control law using neural networks. IEEE. T. Contr. Syst. T. **14**(3), 408–422 (2006)
57. Leitão, P., Barbosa, J., Trentesaux, D.: Bio-inspired multi-agent systems for reconfigurable manufacturing systems. Eng. Appl. Artif. Intel. **25**(5), 934–944 (2012)
58. Zhang, J.F., Khalgui, M., Li, Z.W., Mosbahi, O., AI-Ahmari, A.M.: R-TNCES: A novel formalism for reconfigurable discrete event control systems, IEEE. T. Syst. Man. Cy. A. **43** (4), 757–772 (2013)
59. Dasgupta, D.: Artificial Immune Systems and Their Applications, Springer-Verlag. Inc. (1993)
60. N. D. C. Leandro and T. Jonathan, Artificial Immune System: A New Computational Intelligence Approach, Springer-Verlag. Inc. (2002)
61. Cai, M.-Y.: Medical Immunology. Scientific Publishing House, Beijing, China (2002)
62. Ahn, H.S., Chen, Y.Q., Moore, K.L.: Iterative learning control: brief survey and categorization. IEEE. T. Syst. Man. Cy. C. **37**, 1099–1121 (2007)
63. Moore, K.L.: Iterative learning control for deterministic systems, Advances in Industrial Control. Springer-Verlag, New York (1993)
64. Xu, J., Hou, Z.: On learning control: The state of the art and perspective. Acta Automatica Sinica. **31**(6), 943–955 (2005)
65. Elci, H., Longman, R.W., Phan, M.Q., Juang, J.N.: Simple learning control made practical by zero-phase filtering: Applications to robotics. IEEE. T. Circuit. S-I. **49**(6), 753–767 (2002)

66. Cong, S.B., Yuan, P., Sheng, F.: Dynamic Model Simplification of Heat Exchanger. Petrol. Proces. Petrochem. **27**(10), 5–9 (1996)
67. Stewart, J., Coutinho, A.: The affirmation of self: A new perspective on the immune system. Artif. Life. **10**(3), 261–276 (2004)
68. Dasgupta, D.: Advances in artificial immune systems. IEEE. Comput. Intell. M. **1**(4), 40–43 (2006)
69. Hong, Y.Y., Lin, J.K., Wu, C.P., Chuang, C.C.: Multi-objective air-conditioning control considering fuzzy parameters using immune clonal selection programming. IEEE. Trans. Smart. Grid. **3**(4), 1603–1610 (2012)
70. Deb, K., Pratap, A., Agarwal, S., Meyarivan, T.: A fast and elitist multiobjective genetic algorithm:NSGA-II. IEEE. T. Evolut. Comput. **6**(2), 182–197 (2002)
71. Shen, Q.K., Jiang, B., Cocquempot, V.: Fault-tolerant control for T-S fuzzy systems with application to near-space hypersonic vehicle with actuator faults. IEEE. T. Fuzzy. Syst. **20**(4), 652–664 (2012)

Chapter 5
Data-Driven Neuroendocrine Cooperative Control Systems

The data-driven mechanism can realize the effective control by utilizing the online/offline data of the target system. The neuroendocrine and immune regulating mechanism in human body establishes a rapid responding system to maintain the balance of the body, which can be mathematically derived and therefore provide an inspiration for optimizing the industrial controller. In this chapter, firstly, a novel data-driven cooperative intelligent controller (DDCIC) inspired by the regulating principle of endocrine system in human body is proposed [1]. Secondly, a cooperative control system with data-driven bio-intelligent controller is proposed [2], where the neuroendocrine ultra-short feedback controller (NUC) is adopted to replace the conventional controller for better control performance. Finally, an immune-enhanced unfalsified controller (IEUC) is proposed [3], which is to act as an efficient process to deal with data.

5.1 Introduction

Conventional control schemes, e.g., the proportional-integral-differential (PID) algorithm, are widely used in controlling the industrial systems based on their model-free characteristics. But they cannot solve the problems with complex non-linear systems satisfactorily due to their linear nature [4]. Some advanced control schemes, e.g. the internal model control [5], robust control [6, 7], chaotic control [8, 9] and H-infinity control [10] are therefore proposed but the requirement for accurate plant models is still one of the major barriers to keep these methods from being implemented practically. On one hand, the noises, unstable working conditions and time-varying structures have always been the primary problems that the control system designers should deal with [11]. On the other hand, the lack of accurate model asks for a full use of the running data from which the behaviors of the plant can be extracted to conduct effective control. This is the main motivation for developing the data-driven (or data-based) controllers with algorithms.

Y. Ding et al., *Bio-Inspired Collaborative Intelligent Control and Optimization*, Studies in Systems, Decision and Control 118,
https://doi.org/10.1007/978-981-10-6689-4_5

The natural creatures, especially the human body are good examples of control systems with high efficiency and accuracy [12, 13]. Some bio-inspired controllers have been applied to industrial processes and experienced better performances than traditional schemes [14, 15]. Among all kinds of biological systems, the regulation principles of the subsystems in human body are often taken as an enhancement for improving the conventional controllers [16, 17]. Compared to the independent bio-inspired controller, this is another way to acquire stable performance while not changing the existing controller structure, which would be more convenient to the reform of large and fixed processes [15]. [16] proposed a neuroendocrine ultra-short feedback controller (NUC). It is a nonlinear algorithm that outputs a compensation for a conventional controller using both the system's error feedback and the conventional controller's output. [17] proposed an intelligent bi-cooperative decoupling controller inspired from the modulation mechanism of internal environment in body. However, these types of controllers, although having shown some certain advantages on handling problems of complex systems, still lack of acquiring knowledge from the actual manufacturing process due to their intrinsic characteristics. The research for such kind of controllers can be considered as a prominent aspect if the giant number of running data created during the manufacturing process can be made full use of with certain level of reasonability and efficiency.

Data-driven is a methodology which first derives from computer science. Although there are some similar methodologies in the control field, such as data-based control, model-free control, model-independent control, modeless control etc. Strictly speaking, these concepts are only about employing data in some parts of control process. But data-driven, whose the very beginning and the end of control process are totally based on data, is a more precise illustration of the relationship between data and control system. As a result, data-driven methodology is substantially a series of complex and rigorous mathematical derivation relying on input and output data. In other words, data-driven control is a method which entitles data with the power of describing a control system and requires no mathematic model about control process [18]. The concept of data-driven control takes the data, especially the dynamic and renewable data as the core of the controller designing process [19]. After proposed, data-driven has been given a wide attention. Researches of data-driven control can be divided into three major directions: data-driven controller and its algorithm, data-driven decision-making control system, and data-driven fault diagnosis. Among them, the most extensively-studied one is the data-driven controller and its algorithm, which aims to design an effective controller by using only the online or offline input/output data of the plant. Since the plant model is not always available in industrial control design, and sometimes the plant's mechanism is too complicated to be modeled, one would like to come up with a controller which is able to be tuned directly from measurements of the plant without going through the modeling process [20].

Literatures in recent years have proposed several types of the data-driven controllers and related approaches [18, 21]. The Virtual Reference Feedback Tuning (VRFT) is one of such data-driven control methods. It is a data-driven design methodology that solves a model-reference control problem without the need of any model of the plant. The VRFT was originally proposed in [22] based on a general idea of controller parameters selection and mainly formed in [23]. It has been applied to two degree of freedom controllers building [24], direct [25] nonlinear control design, and design of braking control systems [26]. Other theories, such as Iterative Feedback Tuning (IFT) [27], Model Free Adaptive Control (MFAC) [28], Unfalsified Control ([29]UC), also brought new ideas to data-driven controllers and has been improved in order to adapt to more complex applications [30–35]. Data-driven decision-making control is an important research aspect as well. It focuses on solving decision analysis, classification and optimization problems through data. Cooperative control strategy is an inheritance and development of data-driven decision-making control strategy. On one hand, the concept of cooperation can be embodied in multiple control objectives' cooperation, such as synchronization of robots [36, 37]. On the other hand, it can be applied to one control objective but with an information interchange with other devices [8]. Recently, such cooperative control idea has been introduced into fiber spinning area [38, 39].

5.2 A Data-Driven Endocrine Cooperative Intelligent Controller

In this section, a data-driven cooperative intelligent controller (DDCIC) enhanced by the ultra-short feedback regulation mechanism of the endocrine system in human body is proposed. The principle purpose for proposing the DDCIC is to achieve a combination of the data-driven mechanism, which takes full use of the running data, and the bio-inspired tuning algorithm, which can be taken as guidance for optimizing the data-driven controller as its foundation. In the proposed DDCIC, different combinations and regulation approaches can also be introduced by configuring different components of the proposed controller. A series of computerized experiments on a dynamic tension control system in the acrylic fiber production is applied to verify the performance of the DDCIC. The results show that the mixed mechanism is a good way to realize effective control on the complex plants whose accurate mathematical model is hard to acquire, and the self-tuning functionality should be carefully configured so that better effects can be expected.

5.2.1 Theoretical Foundations

5.2.1.1 The SPSA Data-Driven Mechanism

The simultaneous perturbation stochastic approximation (SPSA) approach for data-driven controller design was firstly proposed by Spall, 1992 [40]. In the control system driven by the SPSA mechanism, the target plant can be linear or nonlinear, and the dynamic characteristics of the plant can be unknown. The controller to be designed is a function approximator, e.g. neural network or polynomial. The SPSA mechanism is responsible for tuning the parameters of such controller, but does nothing to its structure.

Suppose the parameter set of the data-driven controller as θ and θ_k $(k = 1, 2, \ldots)$ are the parameter set at the k-th transient sampling period in a running system. The aim of the SPSA mechanism is to find an optimal parameter set so that a pre-defined performance function,

$$J_k(\theta_k) = E\{[y(\theta_k, k+1) - y_d(k+1)]^2\}, k = 1, 2, \ldots, \tag{5.1}$$

can reach the minimum. Here $J_k(\cdot)$ is the performance function related to the variable θ_k, $y(\theta_k, k+1)$ is the actual output of the plant at the $(k+1)$-th sampling period with the parameter set of the k-th period and $y_d(k+1)$ is the expected output accordingly. Note that $J_k(\cdot)$ is defined as the expectation of the difference between the actual output and the expected one in a control system, so an optimized set of parameters can be worked out theoretically by applying Eq. (5.1) to a derivative calculation. Due to the lack of accurate plant model, the derivative of $J_k(\cdot)$ cannot be solved by applying the following equation

$$J_k(\theta_k) = \frac{\partial y(\theta_k, k+1)}{\partial \theta_k}. \tag{5.2}$$

With the SPSA mechanism, the gradient of the performance function $J_k(\cdot)$ can be replaced by an iterative equation (because the performance function still cannot be solved with the form as shown in Eq. (5.2)) as

$$\widehat{\theta}_k = \hat{\theta}_{k-1} - a_k \hat{g}_k(\widehat{\theta}_{k-1}), \tag{5.3}$$

where $\widehat{\theta}_k$ is the estimation of the parameter set at the k-th period, a_k is a scalar coefficient and $\hat{g}_k(\cdot)$ is called the simultaneous perturbation estimation vector with its l-th component as

$$\hat{g}_{kl}(\widehat{\theta}_{k-1}) = \frac{\hat{J}_k^{(+)} - \hat{J}_k^{(-)}}{2c_k\Delta_{kl}}, \tag{5.4}$$

where

$$\begin{aligned} l &= 1, 2, \ldots, L, \\ \hat{J}_k^{(\pm)} &= (\hat{y}_k^{(\pm)} - y_d(k+1))^2, \\ \hat{y}_k^{(\pm)} &= G_{act}(u_k^{(\pm)}), \\ u_k^{(\pm)} &= G_{ctrl}(\widehat{\theta}_{k-1} \pm c_k\Delta_k, e_k), \\ e_k &= y_d(k+1) - y(\theta_k, k+1), \end{aligned}$$

and

$$\mathbf{\Delta}_k = [\Delta_{k1}, \Delta_{k2}, \ldots, \Delta_{kL}]^T.$$

L is the number of parameters to be tuned, $G_{act}(\cdot)$ is the "virtual" transfer function of the plant (it is called "virtual" because it is unnecessary for the data-driven controller. In fact, the output of the plant can be achieved by the actual runtime data), $G_{ctrl}(\cdot)$ is the transfer function or expression of the controller, c_k is a scalar coefficient and usually converges to zero with the sampling period elapsed, e_k is the dynamic output error and $\mathbf{\Delta}_k$ is a random generated vector which usually satisfies independent bounded symmetric distribution. In the k-th sampling period, the plant error is firstly calculated and some minor "perturbations" are then added to the estimation of the parameter set. Such perturbation can be positive or negative. So there are two possible outputs of the performance function, namely, $\hat{J}_k^{(\pm)}$ and the new set of parameters available. The plant model or any estimation about it is not required during the whole process because all the information can be gained from the input and output. The convergence and stability of such approach is proved by Spall and Christion [41].

5.2.1.2 Endocrine-Based Control Algorithm

The general secretion rules of the hormone [41] reveals that the secreting intensity of the hormones in human body is monotonous, non-negative which has the form of a Hill function as

$$\begin{cases} F_{up}(G) = \dfrac{G^n}{T^n + G^n} \\ F_{down}(G) = \dfrac{T^n}{T^n + G^n} \end{cases}, \tag{5.5}$$

where G is an independent variable determined by the actual secreting velocity of a certain hormone, T is a non-negative threshold and $n(n \geq 1)$ is the order. Both n and T determine the slope of Eq. (5.5).

The hormone regulation mechanism depicted in Eq. (5.5) can be generalized and modified mathematically and then applied to the control systems by embedding an additional bio-inspired section to an existing controller [7, 10]. Figure 5.1 shows the basic structure of such a controller with all the variables in their discrete forms. It consists of a conventional controller unit (CCU), an ultra-short feedback unit (USFU), a signal synthesis unit (SSU) and a signal output unit (SOU).

At the k-th control period (usually the same as the k-th sampling period), the output signal $u(k)$ is firstly sent to the USFU to acquire the ultra-short feedback gain as below

$$u_f(k) = k_f \times u(k), \tag{5.6}$$

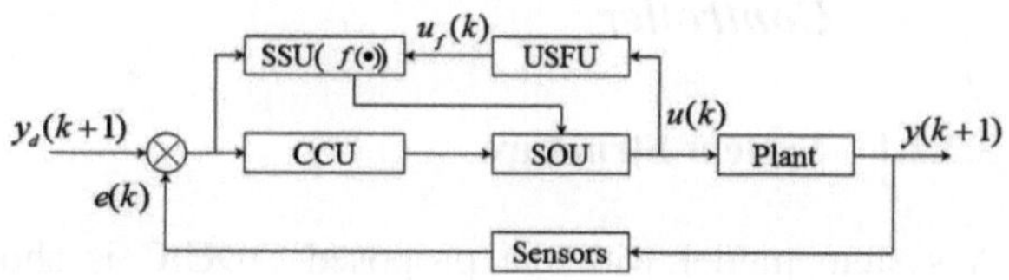

Fig. 5.1 The ultra-feedback intelligent controller

where $u_f(k)$ is the output of the USFU and k_f is the proportional coefficient. Then the $u_f(k)$ together with the plant error $e(k)$ is sent to the SSU to conduct transformation according to the hormone regulation principle, which can be written as follows:

$$\begin{aligned} f(\Delta u_f(k), e(k)) &= \alpha(\frac{|\Delta u_f(k-1)|^n}{\lambda + |\Delta u_f(k-1)|^n} + \beta)L_1 L_2, \\ L_1 &= -\frac{e(k)}{|e(k)|}\frac{\Delta e(k)}{|\Delta e(k)|}, L_2 = \frac{\Delta u_f(k-1)}{|\Delta u_f(k-1)|} \end{aligned} \tag{5.7}$$

where $\Delta u_f(k-1)$ is the variance of $u_f(k)$ and $\Delta u_f(k-1) = u_f(k-1) - u_f(k-2)$. α, β, λ and n are all positive real numbers. Note that $f(\Delta u_f(k), e(k)) = 0$ should be satisfied if $\Delta u_f(k-1) = 0$, so β is always zero. L_1 and L_2 are the controlling direction factors that ensure the output of the controller is always against the changing direction of the plant error so that the error can be effectively eliminated. The values of these factors are either 1 or -1 which satisfy the requirement for eliminating the disturbance. The function of the SOU is to integrate signals of the CCU and the SSU, which can be written as

$$\Delta u(k) = \Delta u_c(k) + f(\Delta u_f(k), e(k)), \tag{5.8}$$

where $\Delta u_c(k)$ is the increment of the CCU, $\Delta u_c(k) = u_c(k-1) - u_c(k-2)$. So the final output of the controller can be written as

$$\begin{aligned} u(k) &= u(k-1) + \Delta u(k) \\ \Delta u(k) &= \Delta u_c(k) + \alpha(\frac{(|\Delta u_f(k-1)|)^n}{\lambda + (|\Delta u_f(k-1)|)^n} + \beta)L_1 L_2 \\ \Delta u(k) &= G_{ctrl}(\theta_k, e(k)) - G_{ctrl}(\theta_{k-1}, e(k-1)) \end{aligned} \tag{5.9}$$

where $G_{ctrl}(\theta_k, e(k))$ is the transfer function of the CCU with a specific parameter set θ_k and plant error $e(k)$ at the k-th controlling period. The control scheme based on the endocrine regulation principle can be regarded as an accelerator for the regular controller. Due to the implementation of such bio-inspired algorithm with predictive feature, the host controller (the data-driven controller in this section) would generate powerful output when necessary and therefore acquire faster speed for responding to the disturbance. The numerical level of the instructions sent to the plant with such mechanism would be raised if the plant error is large or becoming larger, and be eliminated for minor errors.

5.2.2 *Design of the Data-Driven Cooperative Intelligent Controller*

5.2.2.1 System Structure

A system model with the proposed DDCIC is shown in Fig. 5.2 in which the combination of the conventional data-driven controller and a bio-inspired enhancing part, as well as the multiple configurations led by such combination, are the main points that the proposed DDCIC is focusing on. The DDCIC takes charge of collecting and processing runtime data and then generating control signals. It consists of a core controller (CC), an ultra-short feedback controller (UFC), a data-driven optimizer (DDO), an adaptive distributor (AD), and a data collector (DC). The CC can be of any type of the conventional controller. It is essential that the CC should have a series of tunable parameters as the optimizing target of the data-driven mechanism implemented in the DDO. Through the ultra-feedback hormone regulation principle, the UFC is used for enhancing the primitive signal generated by the CC. The final output of the DDCIC is the combination of signals from both the CC and the UFC. Note that the UFC may also have its own tunable parameter set. The DDO is the key component of the DDCIC that is responsible for optimizing the parameters of the CC or/and the UFC (which depends on the specific combination of these two parts) with data-driven mechanism. The AD treats the plant error as criteria to determine when and how much the enhancing signal from the UFC should be attached to the output of the CC to form the final control signal. The AD also provides guidance for the DDO on whether the CC and/or the UFC should be optimized, and the basis for such judgment is the real-time error. The main task of the DC is to build and maintain a database which contains the runtime data in the past several periods. This database is the main data source of the DDO to conduct the data-driven optimization.

5.2.2.2 Control Algorithms

For the proposed DDCIC shown in Fig. 5.2, both the PID controller applied as the CC and the bio-inspired accelerator as the UFC have multiple parameters for tuning, hence several adaptive controller models can be generated by combining the CC and the UFC in different ways. The possibility for generating different combinations out of a single model structure is in fact pointing out another aspect for building the hybrid controllers with more effective optimization. The instances as shown above can be taken as a start point for the comprehensive analysis that would bring more optimized and reasonable controllers.

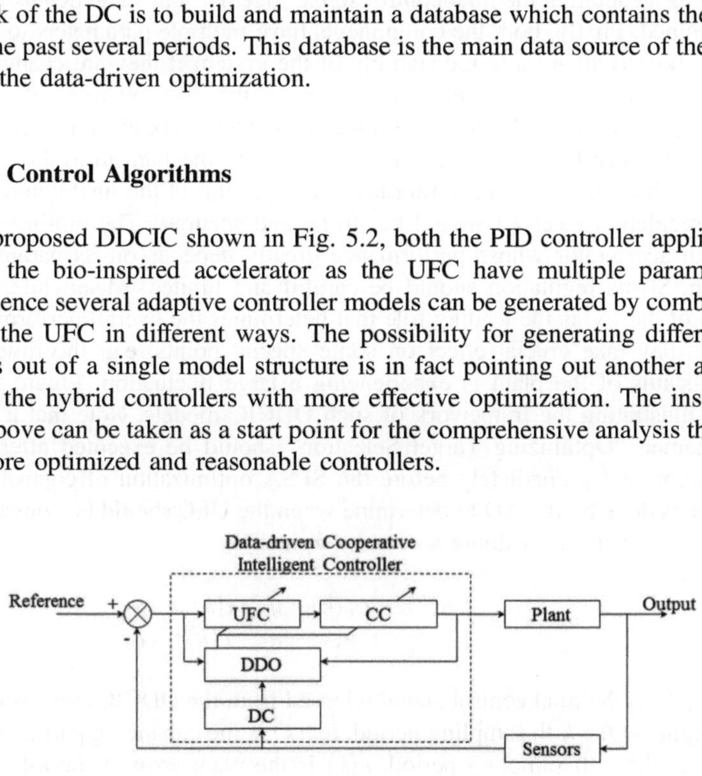

Fig. 5.2 The DDCIC enhanced by the hormone regulation principle

(1) PID-Adaptive DDCIC

The PID-adaptive DDCIC consists of a tunable PID controller as CC and a UFC with fixed parameters. The data-driven tuning approach is done by the DDO and applied to change three PID parameters to realize an optimized control for the plant. The UFC only acts as an accelerator, but its parameters are fixed. The parameters of the PID controller can be originally set to a small value so that the whole system can initiate smoothly. On the contrary, the parameters of the UFC are kept static during the running process and their values are determined by experience of human operators.

(2) UFC-Adaptive DDCIC

Similar to the PID-adaptive DDICC, the UFC-adaptive instance also includes a tunable part and a stable part. But it's the UFC that is selected to be tuned by the data-driven approach rather than the CC. The selection for the original values of parameters is therefore similar to the PID-adaptive instance.

(3) PID-UFC-Adaptive DDCIC

A fully adaptive controller is taken into consideration by tuning both the CC and the UFC for generating the final control signal together, but some details are still to be determined. Firstly, both the components have multiple parameters to be tuned. It brings more challenges to the stability of the system if they are changed simultaneously without any discrimination. Secondly, the different roles of such components in a whole DDCIC determine that they should not receive equal optimizations. The PID controller with a data-driven mechanism as the CC acts as the core of the controller, generating the major portion of the final control signals, and the regulation received should be strong and accurate. The bio-inspired UFC acts as an accelerator whose performance greatly depends on its parameter configuration. So its regulation should be careful and limited. Meanwhile, the optimization of the CC is the leading role that determines the overall performance, but the UFC may take crucial effect on some special points, e.g. the time that the working status of the plant is experiencing a large fluctuation. Figure 5.3 is the diagram illustrating the framework of such DDCIC models. Note that a new procedure named "Optimizing Target Selection" should be executed after the data accumulation and immediately before the SPSA optimization of controllers. This procedure is done by the AD to determine when the UFC should be connected with the CC. The criterion for doing so can be written as

$$u_{fnl}(k) = \begin{cases} u(k) + u_{en}(k) & if \quad e(k) > \varepsilon_{thr} \\ u(k) & if \quad e(k) \le \varepsilon_{thr} \end{cases}, \tag{5.10}$$

where $u_{fnl}(k)$ is the final control signal released from the DDCIC, $u(k)$ is the stored control signal at the k-th sampling period, $u_{en}(k)$ is the enhancing part generated by the UFC at the k-th sampling period, $e(k)$ is the plant error at the k-th sampling period and ε_{thr} is a predefined threshold (set by the operator's experience) to determine whether the output of the UFC should be contributed to the final signal.

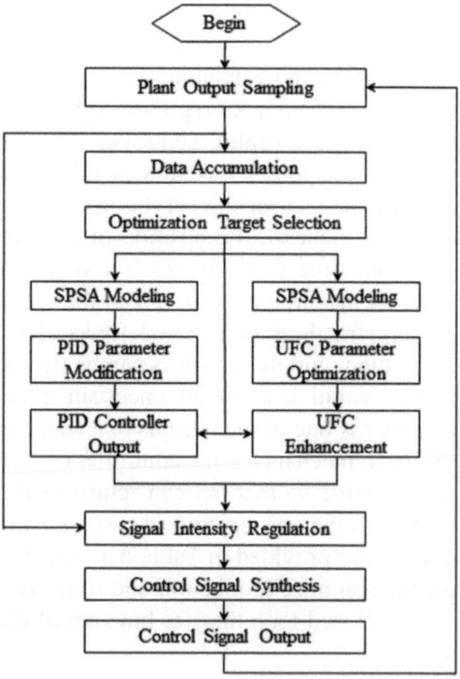

Fig. 5.3 The algorithm flowchart of the DDCIC

The AD here is also responsible for the signal intensity regulation. The detailed distribution plan can be written on the basis of Eq. (5.10) as

$$u_{fnl}(k) = \begin{cases} \alpha u(k) + \beta u_{en}(k) & if \quad e(k) > \varepsilon_{thr} \\ u(k) & if \quad e(k) \leq \varepsilon_{thr} \end{cases}, \qquad (5.11)$$
$$\alpha + \beta = 1, \alpha > 0, \beta \geq 0$$

where α and β are intensity coefficients. In fact, these coefficients can also be taken as part of optimization target of the DDO. The reason for not including them as tuning targets is to prevent the output from vibrating too much.

5.2.3 Simulation Results

5.2.3.1 Constant Tension Control System

The constant tension control system is often applied in the industrial fiber production to acquire stable stretching on the fiber [42]. A classical structure of such system for the acrylic fiber production is depicted as in Fig. 5.4. The free roller R1 handles the fiber to be stretched. The roller R2 controls the stretching speed of the fiber by taking a motor as its driver. The locations of R1 and R2 are both fixed. The roller R3 beneath those two fixed rollers is called the floating roller. It can move vertically so that the tension of the fiber (the tension between R1 and R3, and also R2 and R3) can be dynamically regulated. A selsyn is connected to R3 by a belt so that the position of R3 represented by the angle β can be detected. The angle is then transferred to a central controller. The controller generates corresponding instructions for the motor linked with R2 whose speed can therefore be regulated. The dynamic tension of the acrylic fiber can be maintained by such a feedback mechanism. In general, the position of R3 should be at the same height as that of the selsyn it connects to, and R3 is passively driven by R2 (through the transmission of force on the stretched fiber). But the speed of R3 may encounter disturbances, e.g. random fractions or external forces that result in position changing and tension fluctuation, and these factors can be treated as the uncertainties of the whole system which would lead to an uncertain control model (e.g., the unpredictable fiber diameter is one of the reasons that leads to the load change on the rollers). With the feedback mechanism, the controller can detect the position change and regulates the speed of R2 so that R3 can return to its original location. The diagram for controlling of the tension control system is given by Fig. 5.5, and the parameters of the devices are provided in Table 5.1 (which are collected from an actual set of devices for the spinning production and used for the simulation later).

The closed-loop transfer function of this tension control system can be deducted as

$$Y(s) = \frac{[(R(s) + KN)G_1(s)G_2(s) - N]G_3(s)}{1 + HG_1(s)G_2(s)G_3(s)}, \tag{5.12}$$

and $R(s) = 0.0$, $H(s) = 0.4$, $K = 0.2168$, $N = 200$ $G_1(s) = 7.6/(0.8s + 1,)$ $G_2(s) = 0.06515/[s(0.00521s + 1)]$, and $G_3(s) = 0.45/s$. Note that the $R(s)$ in Eq. (5.12) is the reference input of the whole system. The zero value means that the position of the floating roller (R3 in Fig. 5.4) should be at the same height as that of the selsyn. N is the actual speed of the floating roller where the interference may occur, and equals to 200 for the stable case. Note that the model defined by Eq. (5.12) is actually not necessary for the proposed data-driven-based controllers. These models are listed here only for acquiring the plant response easily in

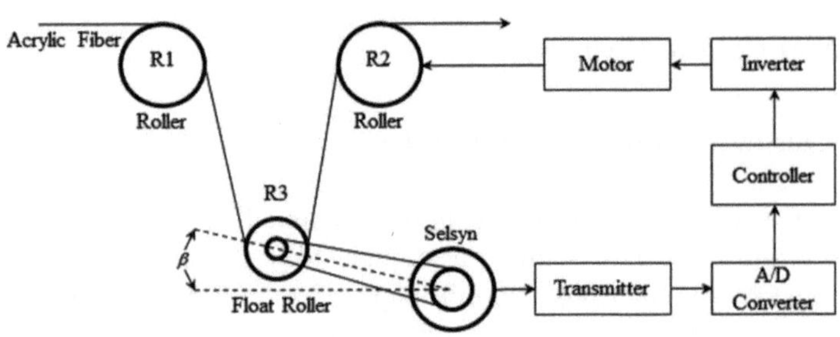

Fig. 5.4 Tension control system for acrylic fiber production

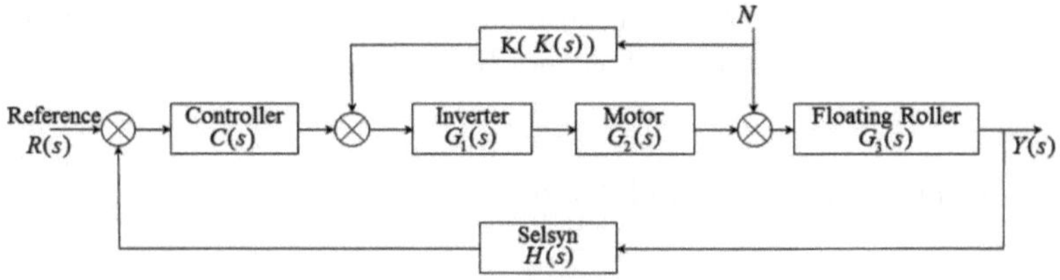

Fig. 5.5 Control diagram of the tension control system

Table 5.1 Device parameters in the tension control system

Item	Value	Item	Value
U_e/v	380	L_M/mH	61.9
I_e/A	2.67	D_{R3}/m	0.31830
$n_p/1$	2	$i_{R3}/1$	0.117647
$J/N \cdot M \cdot s^2$	0.15	D_{R2}/m	0.19304
$C_T/N \cdot M/A$	2.3	$i_{R2}/1$	0.253523
R_M/Ω	11.95	$r/^{\circ}$	0.0

simulation. The controller applied in the system does not include any information about such model, so the plant can still be treated as a model-free one.

5.2.3.2 Experiment Design

A series of simulation experiments based on the tension control system above is carried out to verify the effectiveness of the proposed DDCIC. Four experiments are made by applying different controllers, namely, the PID-adaptive DDCIC (P-type), the UFC-adaptive DDCIC (U-type) and the PID-UFC-adaptive DDCIC (PU-type), plus an SPSA-PID scheme for reference. Each simulation includes two phases as below,

(1) Take the controller to be verified as the core controller of the tension control system, and then start simulation until the status of the system is stable. The speed of the floating roller R3 is kept to 200 rpm. This phase lasts for 25 s.
(2) At the 25-th second, increase the speed of the floating roller R3 from 200 rpm to 220 rpm (a 10% increment as a dynamic noise), then wait until the system returns to its stability. This phase lasts for another 25 s.

The sampling time for all the experiments is set to 0.05 s. The original values of the parameters for each type of controller are listed in Table 5.2. The PID parameters applied in the UFC-adaptive and the PID-UFC-adaptive instances are pre-tuned by a pure SPSA-based PID controller, and the related statistics can be seen at the following experiments. At each phase, some performance indices are recorded to estimate the effectiveness of the controllers, including the amount of overshoot ($A.O.$) and the settling time (T_S). Note that the $A.O.$ is applied instead of the percentage overshoot ($P.O.$) because the reference input of the tension control system is zero, and a general $P.O.$ cannot be calculated.

5.2.3.3 Simulation Results and Analysis

Figure 5.6 presents the system response of the tension control system with four types of controllers, and Fig. 5.7 shows the details of the responses at two transition periods (the starting period and the interference period) Table 5.3 shows the performance indices ($A.O.$ and T_S), and Table 5.4 is the tuned parameters with different control schemes.

All controllers involved can bring the system to stability and the reference input can be acquired. Their respective responses against the noise (the speed changes) appeared at the 25-th second all have a sharp decrease. This is a result of the noise attached to the output immediately and all of them cannot take corresponding actions in the same sampling period.

The response curves of systems with SPSA-PID controller and the P-type DDCIC are almost the same. For the parameter optimization, however, these two controllers have distinctive performances. The proportional coefficient and the differential coefficient of the P-type DDCIC are much smaller than those in the SPSA-PID controller. This phenomenon can be seen as a result of introducing the UFC mechanism to the base controller structure so that the performance is improved. The role of the UFC can also be regarded as a compensation for the control instruction at each sampling period. So the amount of the original PID parameters does not need to rise to a comparatively high level to provide equal controlling intensity. With the UFC mechanism, the amount of overshoot with the P-type DDCIC has a slight increase, and the settling time is increased by 3 s.

Figure 5.8 shows the regulating processes of parameters of different types of controllers. The detailed optimizing effects and intensity by applying different algorithms with parameters can be observed through these figures, which is also an indication on what changes can be brought by introducing the bio-inspired algorithms to the tuning of a controlling scheme (for the process rather than the result). For the controllers that take the PID parameters as the main tuning target (including the SPSA-PID controller and the P-type DDCIC), the values of the parameters are changed at certain sampling periods and keep stable for a period after the changes happen, but the tuning processes in these controllers are different. For the

Table 5.2 The original values of controller parameters

SA-PID	P-type	U-type	PU-type	
K_P	20	20	43.9476	43.9476
K_I	8	8	10.1709	10.1709
K_D	8	8	34.5649	34.5649
α	N/A	1.5	1.5	1.5
γ	N/A	1.0	1.0	1.0
N	N/A	1.2	1.2	1.2

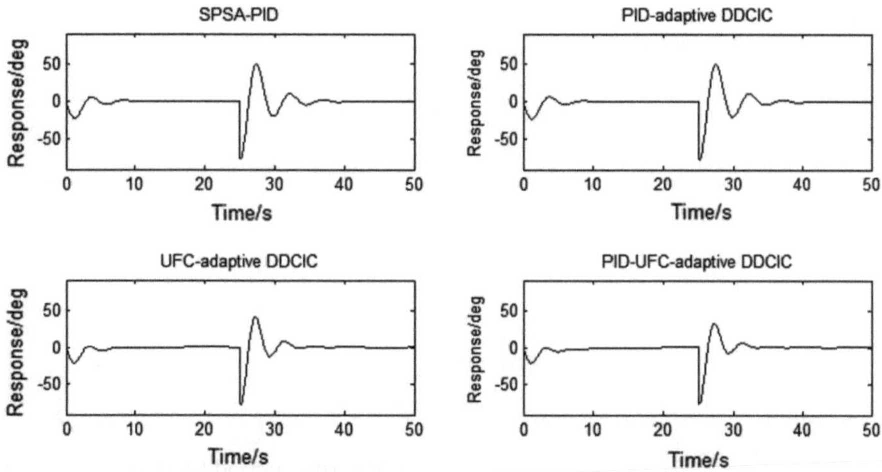

Fig. 5.6 System responses with four different controllers

SPSA-PID controller, they changes several times when the noise is attached, and this phenomenon does not appear with the P-type DDCIC in which only one step is made to deal with the external interference. It is another fact that proves the UFC mechanism can be taken as compensation. For the controllers with tunable UFC parameters, namely, the U-type or the PU-type instance, the regulating process differs with the detailed mechanisms applied. For the U-type one, α decreases from its original value and stays at a lower level. It endures a slight fluctuation against the noise. γ has a large increase, from the original 1.0 to about 7.1. N always stays near zero and also has a fluctuation with the noise, which is similar to α. Such regulations make the overshoot smaller, compared to those PID-tunable controllers, and the settling time also decreases from 39.95 to 36.7 s. For the PU-type one, the rules for how to put the UFC into practice is different from that of the U-type one, that is, a partial but intensive enhancement when the dynamic error exceeds a predefined threshold. So the regulations of its parameters happen in the period at the beginning of the simulation and the period from the 25-th second to the 26-th second, respectively. As a result, the overshoot decrease significantly from over 150 to below 100, compared with the PID-tunable controllers, or from over 120 to below 100, compared with the U-type DDCIC. The settling time also has a slight decrease from 36.7 s with the U-type controller to 36.1 s. It cannot be decreased obviously due to the characteristics of the tension control system as a high-order plant.

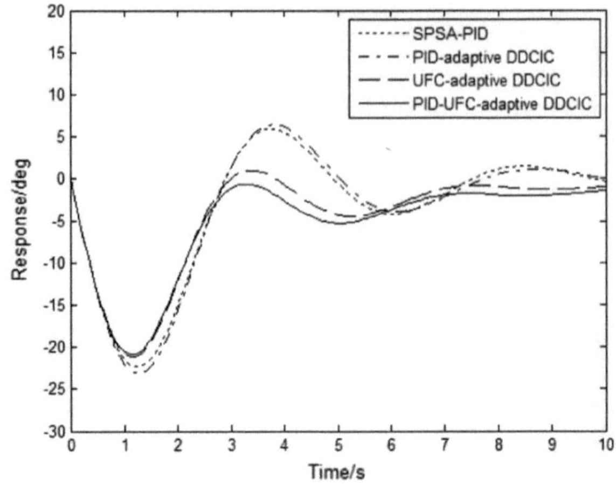

(a) Starting period (0-10s)

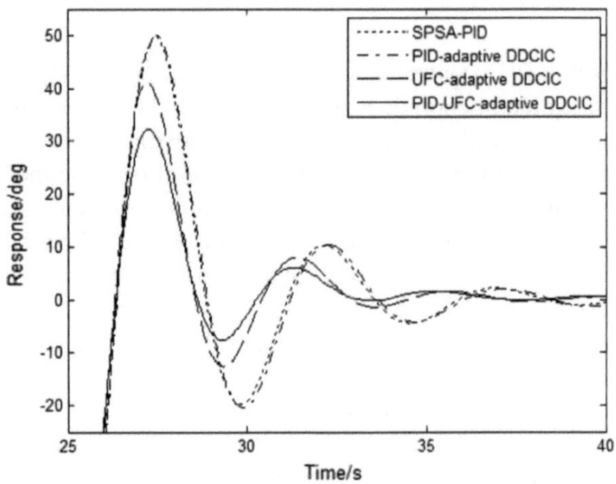

(b) Interference period (25s-40s)

Fig. 5.7 Details of the responses with four different controllers

Table 5.3 Performance indices with different controllers against noise

	SPSA-PID	P-type	U-type	PU-type
A.O.	150.2	151.1	123.6	96.71
T_S/s	39.95	42.95	36.7	36.1

Table 5.4 Optimum values of parameters with different controllers

	SPSA-PID	P-type	U-type	PU-type
K_P	43.9476	28.2220	28.2220	28.2220
K_I	10.1709	16.1392	16.1392	16.1392
K_D	34.5649	4.9110	4.9110	4.9110
α	NA	1.5	0.4642	0.2086
γ	NA	1.0	7.1791	9.9963
N	NA	1.2	0.0164	3.4268e-5

Fig. 5.8 Parameter regulating processes of different controllers

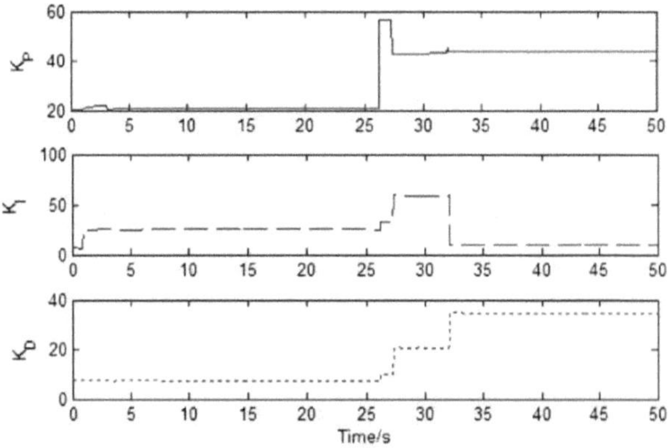

(a) SPSA-PID

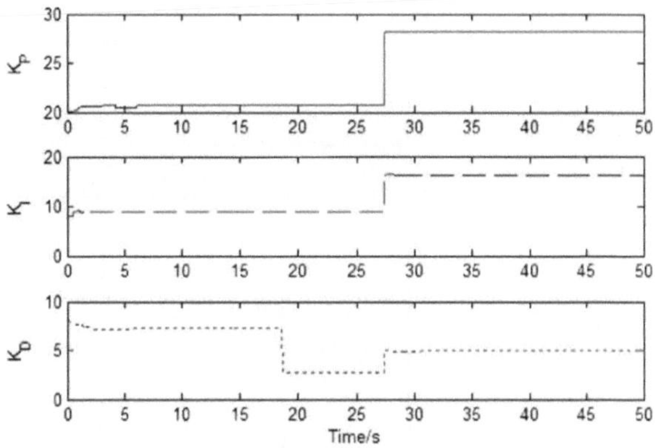

(b) PID-adaptive DDCIC

5.3 A Data-Driven Neuroendocrine Cooperative Control System

In this section, to improve the control performance of a certain industrial production process, a cooperative control system with data-driven bio-intelligent controller is proposed. Firstly, the NUC is adopted to replace the conventional controller to improve the control accuracy. However, the NUC is unpractical in some industrial applications because it is impossible to tune if the model is unknown. And if the model is known but of deviation, the control accuracy will decrease. Thus, a data-driven approach for the NUC (DNUC), which is inspired by the VRFT data-driven control algorithm, is designed. Combining the data-driven decision-making with the data-driven control algorithm, a real-time cooperative control system is proposed. Based on the cooperative optimization strategy, the system utilized real-time data of all related devices in the production process to

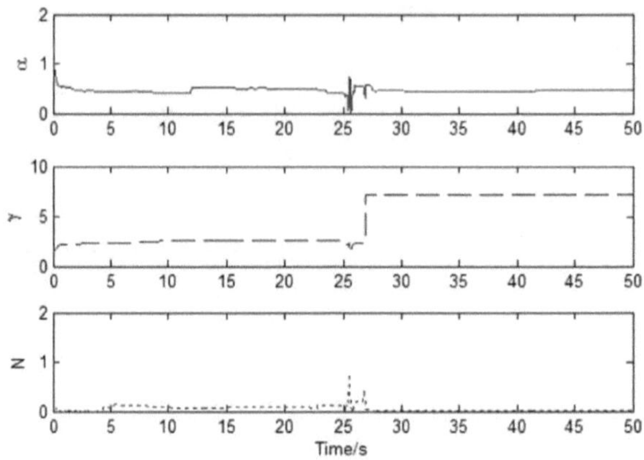

(c) UFC-adaptive DDCIC

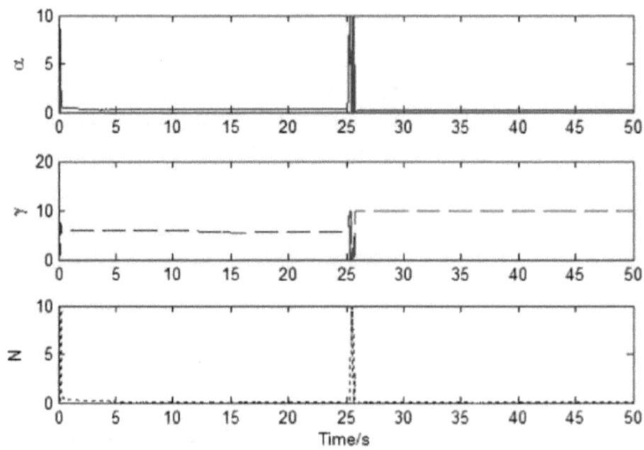

(d) PID-UFC-adaptive DDCIC

Fig. 5.8 (continued)

adjust the control output and maintain the stability. The proposed method different from other similar methods are: (1) The DNUC focuses on single control objective while the controllers in [1, 39] are the decoupling ones for multiple objectives. (2) Different from the control methods which are not data-driven [9, 38], the DNUC doesn't rely on the plant model. (3) Compared with the data-driven bio-intelligent controller designed based on SPSA algorithm [8] which requires a large amount of data to achieve, the DNUC, which is designed by combining the data-driven decision-making with the data-driven control algorithm, is quicker and with less cost to carry out.

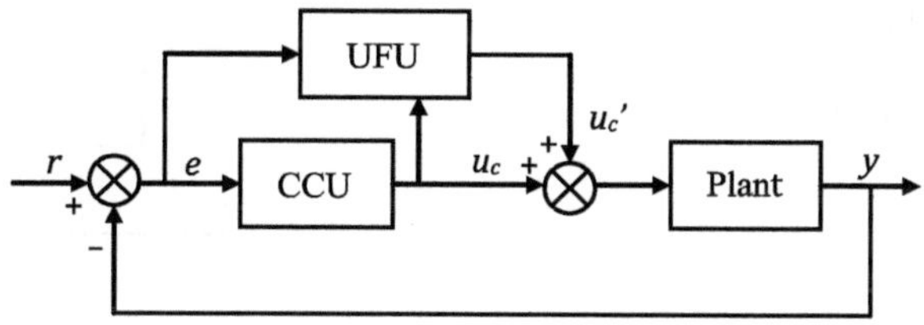

Fig. 5.9 The structure of the NUC control system

5.3.1 *Theoretical Foundations*

5.3.1.1 Neuroendocrine Ultra-Short Feedback Control Method

The NUC combines an additional bio-intelligent control unit with a conventional controller in order to improve the controller's performance. Since the hormone regulation mechanism of neuroendocrine system is a well-preformed self-control process, the NUC method is mathematically generalized and modified the above mechanism and used it to compensate a conventional controller. Figure 5.9 shows the basic structure of such NUC control system. It consists of a conventional controller unit (CCU), an ultra-short feedback unit (UFU), and a control plant [9].

According to the hormone regulation principle [42], the ultra-short feedback unit's algorithm is written as follows [9]:

$$(u_c(k), e(k)) = \alpha\left[\frac{(|\Delta u_c(k)|)^{\xi}}{1+(|\Delta u_c(k)|)^{\xi}} + \beta\right] L_1 \cdot L_2, \tag{5.13}$$

$$L_1 = -\frac{e(k)}{|e(k)|}\cdot\frac{\Delta e(k)}{|\Delta e(k)|}, L_2 = \frac{\Delta u_c(k)}{|\Delta u_c(k)|},$$

where $\Delta u_c(k)$, $k = 1, \cdots, N$ is the variance of $u_c(k)$ and $\Delta u_c(k) = u_c(k) - u_c(k-1)$. $\Delta e(k) = e(k) - e(k-1)$. α, β, and ξ are all positive real numbers. L_1 and L_2 are the controlling direction factors that ensure the output of the controller to be always against the changing direction of the plant error so that the error can be eliminated effectively.

The output of the NUC is an integrated signal of the CCU and the UFU, which is

$$\mathrm{u_{NUC}(k) = u_c(k) + u_c'(k).} \tag{5.14}$$

The control scheme based on the neuroendocrine regulation principle can be regarded as an accelerator for the conventional controller [8].

Note that $u_c'(k) = 0$ should be true if $\Delta u_c(k) = 0$, so β must be set to zero. However, the tuning of the key parameters α and ξ is lack of theoretical guidance. In the existing studies of the NUC, tuning the two parameters was achieved by repeatedly trying on the plant or plant's model, or by using optimization algorithm on plant's model with quantizing the control performance as optimization objective.

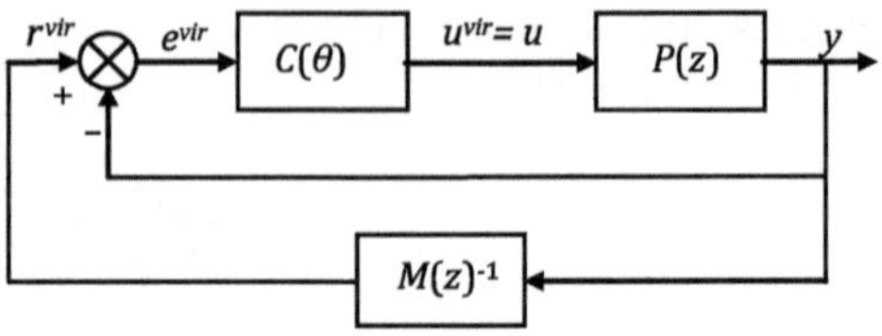

Fig. 5.10 The structure of control system based on VRFT method

5.3.1.2 Virtual Reference Feedback Tuning Method

Among various data-driven algorithms, the VRFT is a very convenient and efficient method. It is suitable for the design of all kinds of parameterized controllers: $C(z;\theta)$, $(C(z;\theta) = \gamma^T(z)\theta)$, including the PID controller. The structure of the VRFT control system is shown in Fig. 5.10. In Fig. 5.10, $\{u(t), y(t)\}$ is a set of measured I/O data of the plant $P(z)$. Which means, even if the $P(z)$ is unknown, we know that it will generate $y(t)$ when feed by $u(t)$. $M(z)$ is the desired reference model for the closed-loop system to design. $r^{vir}(t)$ is a reference signal that makes $y(t)$ be the desired output of the closed-loop system. So we have: $M(z)r^{vir}(t) = y(t)$. The $r^{vir}(t)$ is called "virtual reference" because it is not an actual measurement data. Then, the idea is to search for a controller $C(z;\theta)$ that generates $u(t)$ when feed by $e^{vir}(t)$ (the virtual corresponding tracking error $e^{vir}(t) = r^{vir}(t) - y(t)$). Since both $u(t)$ and $e^{vir}(t)$ signals are known, this task reduces to the identification problem of describing the dynamical relationship between them.

The above idea can be implemented by the following three-step algorithm (where a filtering of data through a user-chosen filter $L(z)$ is also considered). It represents the bulk of the VRFT method [23]. Given a set of discrete measured I/O data of $P(z) : \{u(k), y(k)\}$, $k = 1, \cdots, N$, do the following steps [23]:

(1) Calculate the virtual reference $r^{vir}(k)$ due to $y(k) = M(z)r^{vir}(k)$, and the corresponding tracking error $e^{vir}(k) = r^{vir}(k) - y(k)$.
(2) Filter the signals $e^{vir}(k)$ and $u(k)$ with a suitable filter $L(z).e_L(k) = L(z)e^{vir}(k)$, $u_L(k) = L(z)u(k)$.
(3) Select the controller parameter vector $\hat{\theta}_N$, that minimizes the following criterion:

$$\mathrm{J}_{\mathrm{VR}}^{\mathrm{N}}(\theta) = \frac{1}{\mathrm{N}}\sum\nolimits_{\mathrm{k}=1}^{\mathrm{N}} (\mathrm{u_L(k)} - \mathrm{C}(\mathrm{z};\theta)\mathrm{e_L(k)})^2 \tag{5.15}$$

Note that when $\mathrm{C}(\mathrm{z};\theta) = \gamma^{\mathrm{T}}(\mathrm{z})\theta$, Eq. (5.15) can be given the form

$$\mathrm{J}_{\mathrm{VR}}^{\mathrm{N}}(\theta) = \frac{1}{\mathrm{N}}\sum\nolimits_{\mathrm{k}=1}^{\mathrm{N}} (\mathrm{u_L(k)} - \varphi_{\mathrm{L}}^{\mathrm{T}}(\mathrm{k})\theta)^2 \tag{5.16}$$

$$\varphi_{\mathrm{L}} = \gamma(\mathrm{z})\mathrm{e_L(k)} \tag{5.17}$$

and the parameter vector $\hat{\theta}_N$ is given by

$$\widehat{\theta}_{\mathrm{N}} = \left[\sum_{k=1}^{N} \varphi_{\mathrm{L}}(k)\varphi_{\mathrm{L}}(k)^{\mathrm{T}}\right]^{-1} \sum_{k=1}^{N} \varphi_{\mathrm{L}}(k)u_{\mathrm{L}}(k) \quad (5.18)$$

5.3.2 Design of the DNUC Cooperative Control System

5.3.2.1 Data-Driven Improvement for the NUC

The CCU of the NUC can be all kinds of conventional controllers, and in this section, the classical PID controller is chosen as an example. To build a data-driven PID controller, we adopt the VRFT due to its simplicity, rapidity and accuracy. The specific implementation steps have been introduced in Section II Part B. Since the NUC is a bio-intelligent algorithm, it is crucial yet difficult to build up the non-linear intelligent control unit UFU using only data. Inspired by the VRFT, a possible approach to solve this problem is presented in this section. The idea of VFTR can be concluded in this way: It makes use of the measured data and the reference model to acquire a set of virtual data. The measured data and the virtual data are combined into a full I/O data set of the controller. The controller's structure and the I/O data are used to tune the controller's parameters.

The major tuning method of the UFU can also adopt the above procedures but with one major difference. Because the CCU is working to control the plant while the UFU is used for compensating the CCU. Thus, the data, which is used for tuning the UFU, is collected from a closed-loop system formed by the CCU and the plant after building the CCU via the VRFT, and the open-loop data of the plant.

In the NUC as shown in Fig. 5.9, the UFU unit is for improving the performance of the CCU. Therefore, the tracking error can be used to tune the UFU's parameters. In order to obtain a set of tracking error $e_{cl}(k)$, only one closed-loop experiment of the plant controlled by the CCU was carried out. A set of I/O data of the closed-loop system: $\{r(k), y_{cl}(k)\}$ was acquired.

$$e_{cl}(k) = r(k) - y_{cl}(k) \quad (5.19)$$

Since the output of the UFU is used for compensating $e_{cl}(k)$, we give the UFU a virtual output $u_c^{vir}(k)$ which meets:

$$e_{cl}(k) = P(z)u_c^{vir}(k) \quad (5.20)$$

Using the knowledge of the VRFT illustrated above, there is:

$$\mathrm{M(z)} = \frac{\mathrm{P(z)C(z)}}{1+\mathrm{P(z)C(z)}} \tag{5.21}$$

Based on Eqs. (5.19)–(5.21), $\mathrm{u_c^{vir}(k)}$ can be rewritten as:

$$\mathrm{u_c^{vir}(k)} = \frac{1-\mathrm{M(z)}}{\mathrm{M(z)}}\mathrm{C(z)(r(k) - y_{cl}(k))} \tag{5.22}$$

where both $M(z)$ and $C(z)$ are known and $\{r(k), y_{cl}(k)\}$ is the measured data.

With the input data of the UFU, $(e_{cl}(k), u_c(k))$, the desired output data of the UFU, $(u_c^{vir}(k))$, and the structure of the UFU in Eq. (5.13), we expect the parameters α and ξ of the UFU to minimize the following criterion:

$$\mathrm{J^N}(\alpha, \xi) = \frac{1}{\mathrm{N}}\sum\nolimits_{\mathrm{k}=1}^{\mathrm{N}} \left(\mathrm{u_c^{vir}(k)} - \mathrm{f_{UFU}(u(k), e_{cl}(k))}\right)^2 \tag{5.23}$$

Since the UFU's algorithm is an intelligent one, which is different from linear VRFT tuning method, an optimization algorithm can be used to solve the optimization problem in Eq. (5.23). There are many optimization methods such as genetic algorithm or particle swarm optimization algorithm. Anyone of them would be able to fulfill this tuning task.

At this point, the original bio-intelligent control method NUC has been modified to become the DNUC method.

5.3.2.2 Cooperative Control System

The structure of the cooperative control system is shown in Fig. 5.11. The core controller of the control system is the DNUC. The data-driven model unit is used for tuning the parameters of the DNUC including both the CCU and the UFU.

The feedback unit of performance objective contains the dynamic mechanism model of performance index. It utilizes the real-time data to calculate the real-time performance index of product and then output it into the control system as a feedback signal. The cooperative unit is used for altering the control decision according to the performance's error and other related cooperative devices' data

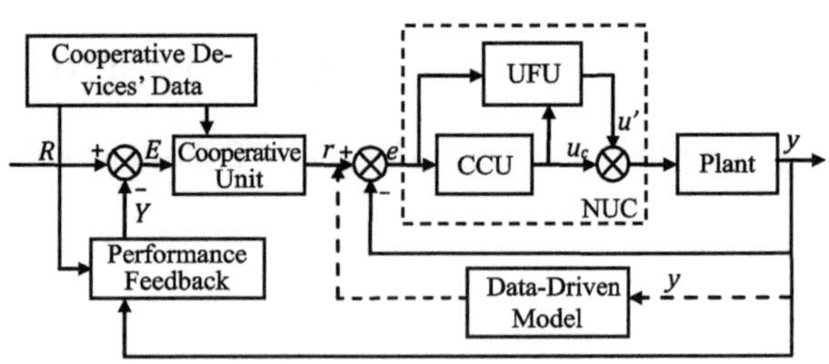

Fig. 5.11 Cooperative control system based on the DNUC

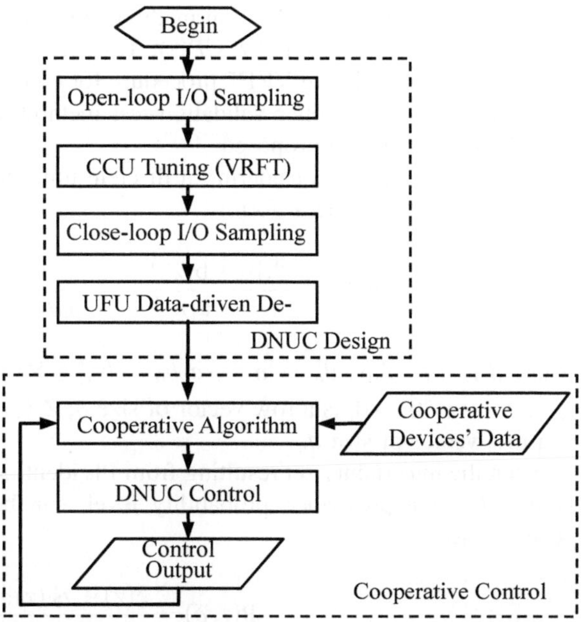

Fig. 5.12 Control steps of the DNUC

based on a certain cooperative mechanism. With the cooperative unit, the system can overcome the disturbance occurred in other related devices, so that a better control performance on the product can be achieved. The cooperative devices' data unit is a database that collects and stores the running data from all the devices which are related to the product's performance index.

The whole system is working coordinately. The cooperative unit provides a control demand and the DNUC unit makes sure that the control performance meets the demand. As shown in Fig. 5.12, achieving this control system contains two major parts, namely, controller tuning and real-time running.

To tune the controller, the CCU is firstly tuned by the standard VRFT. As the standard VRFT steps, a set of plant's open-loop I/O data is collected before the tuning calculation. After the tuning calculation, the designed CCU is then put into closed-loop running. Another set of test data is collected from this closed-loop system. And, with the performance data of the CCU, the parameters of the UFU are tuned. After the UFU controller is tuned, the whole cooperative system is running with the DNUC controller, monitoring database, and a dynamic mechanism algorithm.

5.3.2.3 Controller Validation Test

For the VRFT [43], a controller validation test method is proposed which can ascertain the stability of the closed-loop control system formed by $P(z)$ with the loop closed by $C(z)$. The method is started by identifying $P(z)$ by means of a prediction error (PE) identification method and achieved by the computation of the probability that $C(z)$ destabilizes $P(z)$. This PE identification framework found a specific application in the VRFT procedure for controller validation [43]. Thus, we also adopted this controller validation test method to analysis the stability of the closed loop system we proposed.

According to [44], considering a discrete-time rational transfer function $P_0(z)$ with the following general form:

$$\mathrm{P_0(z) = P(z,\theta_0) = \frac{z^{-d}(b_0 + b_1 z^{-1} + \cdots + b_m z^{-m})}{1 + a_1 z^{-1} + \cdots a_n z^{-n}} = \frac{Z_2(z)\theta_0}{1 + Z_1(z)\theta_0}} \tag{5.24}$$

where d is the delay; $\theta_0^T = [a_1 \cdots a_n b_0 \cdots b_m] \in \mathbf{R}^{q\times 1}$, $(q \triangleq (n+m+1))$; $Z_1(z) = [z^{-1}z^{-2}\cdots z^{-n}0\cdots 0]$ is a row vector of size q; $Z_2(z) = z^{-d}[0\cdots 0\,1z^{-1}z^{-2}\cdots z^{-m}]$ is a row vector of size q.

Then the uncertainty set resulting from PE identification which contains the true system P_0 at a prescribed probability level, can be described in the following generic form:

$$\mathrm{P(z,\delta) = \frac{e(z) + z_N(z)\delta}{1 + Z_D(z)\delta}} \tag{5.25}$$

and

$$\delta \in \mathrm{U} = \{\delta | (\delta - \hat{\delta})^{\mathrm{T}} \mathrm{R} (\delta - \hat{\delta}) < 1\}$$

where $\delta \in \mathbf{R}^{q\times 1}$ is a real parameter vector; $\hat{\delta}$ is the parameter estimation from PE identification; $R \in \mathbf{R}^{q\times q}$ is a symmetric positive definite matrix that is proportional to the inverse of the covariance matrix of $\hat{\delta}$; $Z_N(z)$ and $Z_D(z)$ are the row vectors of known transfer functions; and $e(z)$ is a known transfer function.

Considering a controller $C(z) = X(z)/Y(z)$ that stabilizes the center of the PE uncertainty set, $P\left(z, \hat{\delta}\right)$, then all models in the uncertainty regions are stabilized by $C(z)$ if and only if [44]:

$$\max \mu\left(W\left(e^{j\omega}\right)\right) \le 1$$

where $W(z)$ is defined as

$$W(z) = \frac{-(Z_D + X(Z_N - eZ_D)/(Y + eX))\hat{R}^{-1}}{1 + (Z_D + X(Z_N - eZ_D)/(Y + eX))\hat{\delta}} \tag{5.26}$$

$\hat{R}$ is a square root of the matrix R; $\lambda = \hat{R}(\delta - \hat{\delta})$; $\mu(W(e^{j\omega}))$ is called the stability radius of the loop $[W(z)\lambda]$. For a real vector λ it is computed as follows:

$$\mu\left(W\left(e^{j\omega}\right)\right)=\sqrt{|\mathrm{Re}(W)|_2^2-\frac{\left(\mathrm{Re}(W)\mathrm{Im}(W)^T\right)^2}{|\mathrm{Im}(W)|_2^2}},\ \text{if}\ \mathrm{Im}(W)\neq 0, \tag{5.27}$$

$$\mu\left(W\left(e^{j\omega}\right)\right)=|W|_2,\ \text{if}\ \mathrm{Im(W)}=0. \tag{5.28}$$

With this controller validation method, the stability of the core controller in the system we proposed can be tested. And then, considering the ultra-short feedback intelligent algorithm Eq. (5.13), we assume that by using the optimization algorithm to calculate the parameters α and ξ of the UFU, the results are the best choice for a and ξ. That means, on one hand, the ultra-short feedback intelligent algorithm will be functioned as the hormone regulation process in human body with a convergent performance. On the other hand, with the compensation function of the UFU, the close-loop system formed by the DNUC will be more close to the reference data-driven model $M(z)$ than the close-loop system formed by only the CCU. Based on the definition in Eq. (5.13), if $\Delta u_c(k)\to 0$, the output of the UFU will tend to be 0. Thus, as long as α and ξ are properly chosen and the CCU is validated to be stable through the adopted validation test, we can validate that $\Delta u_c(k)\to 0$ and the whole proposed system is stable.

5.3.3 Simulation Results

5.3.3.1 Cooperative Mechanism of Spinneret Drawing

The control objective is the spinneret draw ratio $\phi_a(k)$. Assuming the solution jets is non-Newtonian elongational flow. Use Δx as a discrete length unit to divide the solution jets along its drawing direction from $x=0$ to $x=L$. For the sake of calculation convenience, let the discretization of bath length L be the same with the discretization of time T. Which means, since the sampling time is T_s, let $T/T_s=L/\Delta x=n$. Then, the value of Δx can be calculated.

Based on the definitions of tensile stress, spinneret draw ratio, elongational viscosity, and the velocity gradient of extension in [45], the dynamic spinneret draw ratio can be written as:

$$\begin{aligned}\phi_a(k)=\sum_{i=1}^{n}\Delta\phi_i(k+i-1)=&\frac{\Delta v_1(k)}{v_s(k)}+\frac{\Delta v_2(k+1)}{v_s(k+1)+\Delta v_1(k+1)}\\&+\cdots+\frac{\Delta v_n(k+n-1)}{v_s(k+n-1)+\cdots+\Delta v_{n-1}(k+n-1)}\end{aligned} \tag{5.29}$$

where $v_r(k)$ is the winding speed of the fiber on coagulation roller. $v_s(k)$ is the solution extrusion speed at the1 spinneret. $\Delta v_i(k)$ is calculable via:

$$\Delta v_1(k) = \dot{\gamma}_i(k)\Delta x = \frac{\sigma(k)}{\eta_i}\Delta x = \frac{v_r(k) - v_s(k)}{\eta_i \sum_{i=1}^{n} \frac{1}{\eta_i}} \tag{5.30}$$

where $\eta_i (i = 1, 2, \cdots, n.)$ is a set of different elongational viscosities along the solution jets.

The cooperative control criterion is the error between the real-time spinneret draw ratio and its desired value. This control target can be written as:

$$\Delta\phi_a(k) = \sum_{k=1}^{n} (\phi_a(k - T) - \phi_{a0}) \tag{5.31}$$

where ϕ_{a0} is the desired value of the spinneret draw ratio and the $\Delta\phi_a(k)$ is the draw ratio error of a point on the solution jets where enters the coagulation bath at time $k - T$ and leaves at time k.

Since the speed of the coagulation roller is mostly responsible for the spinneret draw ratio and it is the last link in the coagulation process, this speed is what we can utilize for adjusting and controlling. Meanwhile, the solution extrusion speed is related to some cooperative facilities. By analyzing the mechanism of the spinning facilities in wet spinning, we find some physical mechanisms that are related to the spinneret draw ratio.

The revolution speed R_r (r/min) of the coagulation roller is $R_r = V_r/(\pi d_r)$, where d_r is diameter of coagulation roller. The metering pump determines the flow rate of the spinning process. It measures the liquid and pumps it into the spinneret continuously. The flow rate of the metering pump is $Q = R_P Q_P$, where R_P is the metering pump revolution speed, Q_P is the metering pump capacity per rev. The spinneret is a circular plate with many orifices. The polymer solution jets are extruded through the orifices and entered the coagulation bath in a certain speed. The total area of the spinneret orifices is

$$\mathrm{S} = \mathrm{n_s}\pi(\mathrm{d_s}/2)^2 \tag{5.32}$$

where n_s is the number of spinneret orifices, d_s is the diameter of spinneret orifices.

Thus, the relationship between the solution extrusion speed and the metering pump is:

$$\mathrm{V_s} = \frac{\mathrm{Q}}{\mathrm{S}} = \frac{\mathrm{R_P Q_P}}{\mathrm{n_s}\pi(\mathrm{d_s}/2)^2} \tag{5.33}$$

The real-time data of the metering pump and the spinneret are monitored and stored into a database. Then the real-time dynamic draw ratio is calculated with these data and the mechanism model Eqs. (5.29)–(5.33). And then, together with the DNUC, the real-time cooperative control system for spinneret drawing process is accomplished.

5.3.3.2 Cooperative Control Simulation

The DNUC is used to control the spinneret draw ratio because it only needs two datasets (one open loop experiment and one closed loop experiment) to tune the controller parameters. The tuning process doesn't need to run the system over and over, thereby reducing the costs.

For the control plant, supposing the roller is a servo motor and its nameplate parameters are: rated voltage = 230V, rated current = 10.4A, rated speed = 377 rad/s, rated power output = 2.0KW, and moment of inertia = 0.027 $J \cdot s^2$. Based on these parameters, the transfer function of this motor is identified as:

$$P_O(s) = \frac{1.786}{0.002218s^2 + 0.1713s + 1} \tag{5.34}$$

It should be mentioned that this transfer function is only used to obtain testing data and give a contrastive result of the control systems with and without using the model. Through the simulation of the proposed method, this transfer function is regarded as being unknown.

Then, the reference model is chosen as:

$$\mathrm{M(z)} = \frac{1-\chi}{\mathrm{z}(1-\chi)} \tag{5.35}$$

where $\chi = 0.8$.

512 pairs of I/O data are collected and the filter that the VRFT presented is: $L = (1 - M(z))M(z)$. After the computation, the VRFT tuning result is $\theta = [0.31510.11450.1163]$, so that the discrete PID controller is built (the sampling time is 0.05 s):

$$\mathrm{C(z)} = \frac{0.5438\mathrm{z}^2 - 0.5448 + 0.1159}{\mathrm{z}^4 - \mathrm{z}^3} \tag{5.36}$$

After tuning the CCU, a closed-loop experiment of the plant control by the CCU is conducted and the closed-loop measurement data is stored. Since the system is settling down within 2.5 *s*, we used the data before 2.5 s and that is 50 pairs of I/O data when the sampling time is 0.05 *s*. With the algorithm proposed in Section II Part C, the final optimal parameters of the UFU are: $\alpha = 0.9013$ and $\xi = 1.1526$.

The control performance of the DNUC is better than that of the classical VRFT controller as shown in Fig. 5.13. The overshoot decreases and both the rise time and settling time are quickened. The DNUC have an especially good control on the overshoot.

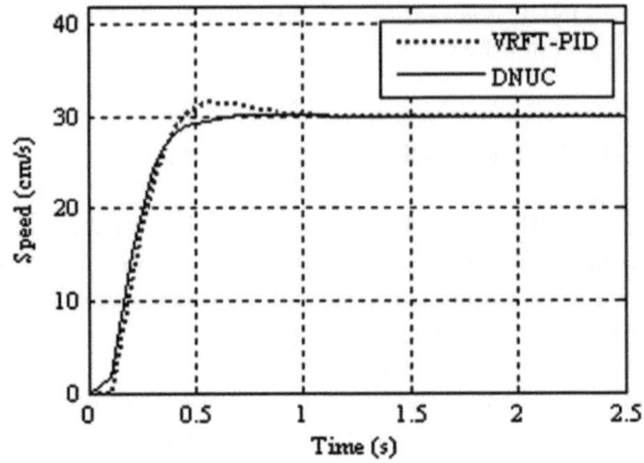

Fig. 5.13 Control performance of the DNUC and the classical VRFT

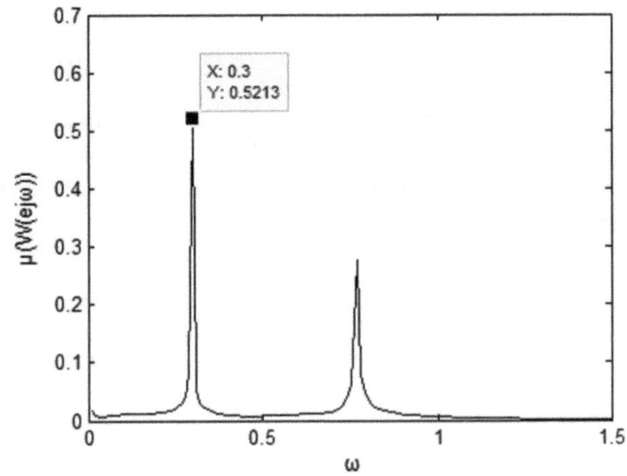

Fig. 5.14 Controller validation test result $\mu(W(e^{j\omega}))$ at each frequency

With the adopted controller validation method, we conducted a validation test on the proposed DNUC method. The result showed that the maximum value of the stability radii $\max\mu(W(e^{j\omega})) = 0.5213 < 1$, so that controller $C(z)$ stabilizes all plants in the uncertainty sets. Consequently, we can also guarantee that the "to-be-validated" controller C(z) stabilizes the true flexible transmission system $P_0(z)$ with probability 95%. The result of $\mu(W(e^{j\omega}))$ is shown in Fig. 5.14.

In order to prove the cooperative control system is effective, two simulation experiments and their results are shown in Fig. 5.15. The experimental parameters are listed in Table 5.5. First, we simulated the situation that with gradual changes occurred in two cooperative devices respectively. As shown in Fig. 5.15(a), after the system was running stably from 10 to 30 s, the metering pump flow was reduced to 50% and then gradually rebounded. Meanwhile, the spinneret orifice clear opening was reduced to 80% and then gradually rebounded. Then, the reaction of the coagulation roller caused by these changes is shown in Fig. 5.15(b). The

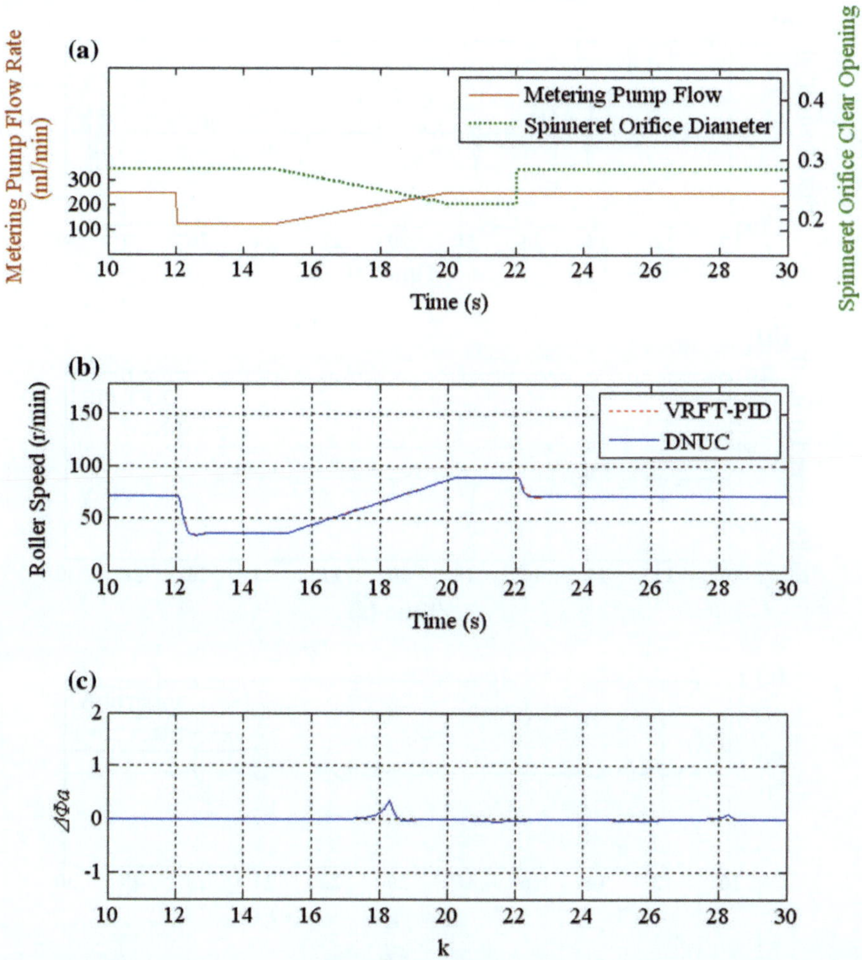

Fig. 5.15 Cooperative control with gradually interference

Table 5.5 Experimental parameters

Item	Symbol	Value
metering pump flow rate / mL	Q	254.43
spinneret orifice diameter / mm	d_s	0.06
spinneret orifice quantity	N	10000
spinneret orifice area / cm^2	S	0.2827
roller diameter / cm	d_r	3.6
spinneret extrusion speed / cm·s^{-1}	V_s	15
winding speed / cm·s^{-1}	V_r	13.5
winding revolution speed / r·min^{-1}	R_r	71.6
spinning drawing ratio	ϕ_a	-10%

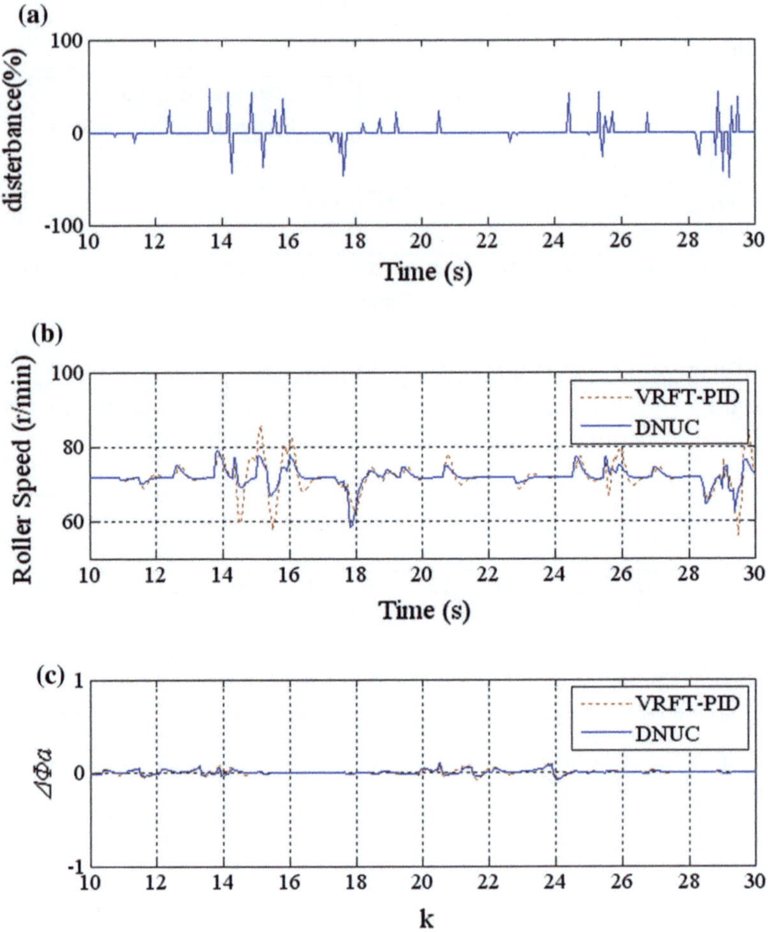

Fig. 5.16 Cooperative control with random disturbance

roller speed is tuning cooperatively according to the changes happened in the metering pump and the spinneret. Such speed tuning has finally maintained the stableness of the spinneret draw ratio. The variation of draw ratio is shown in Fig. 5.15(c). We find that the gradual changes can be completely conquered. The highest fluctuation of $\Delta\phi_a$ is less than 5% and the time period of fluctuation is less than 1 s.

Notice that the independent variable k is the time value when a point where the thread leaves the bath with a draw ratio error $\Delta\phi_a(k)$. And the spinneret drawing is mostly happening at the beginning of the coagulation because liquid condition has a much smaller elongational viscosity. The coagulation time in this simulation was about 6 s. So the sharp speed drop in Fig. 5.15(b) at 12 s caused a draw ratio

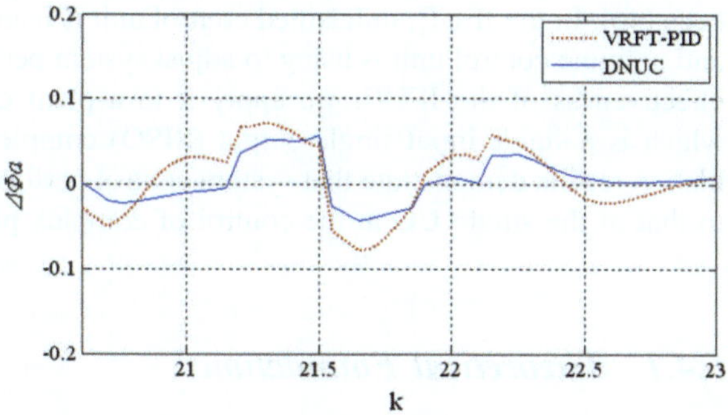

Fig. 5.17 Detail region in Fig. 5.16(c)

fluctuation on the thread where it entered the bath at 12 s and leaved the bath about 6 s later. Thus, the fluctuation in Fig. 5.15(c) was around 18 s.

And then, a simulation with random disturbance occurred in the cooperative devices has been conducted. As shown in Fig. 5.16(a), the disturbance signal occurred 10% along the timeline within a ±50% amplitude. Since all the disturbances in metering pump and spinneret would finally become the disturbances on extrusion speed, we added the disturbance signal directly to the extrusion speed, so that all kinds of possibilities were covered.

Affected by the disturbance, the variation of the roller speed is shown in Fig. 5.16(b). And, the variation of the draw ratio is shown in Fig. 5.16(c). We can see that even with the random disturbance the cooperative control system is able to maintain the spinneret draw ratio to be stable. The roller speed is adjusting around 70r/min in both directions. Since the DNUC has a better control on overshoot, the roller speed controlled by the DNUC can wildly overcome the fluctuations, and make the cooperated control of $\Delta\phi_a$ more stable. The final spinneret draw ration fluctuation is within ±10%.

Figure 5.17 shows an enlarged detail of Fig. 5.16(c) around 21 s to distinguish the performance between the conventional controller and the DNUC. The fluctuation of $\Delta\phi_a$ controlled by the DNUC was much smaller than that by the VRFT on both positive and negative axis. The fluctuation peak value of the DNUC was nearly 50% of the VRFT. It proved to us that the DNUC had a better control performance on maintaining $\Delta\phi_a$ to be steady.

5.4 An Immune-Enhanced Unfalsified Controller

In this section, an immune-enhanced unfalsified controller (IEUC) is presented, which is inspired from biologic principle of immune feedback system and data-driven principle of unfalsified control. The IEUC consists of a unfalfied control unit, an immune control unit and implements through a PID algorithm. The unfalsified control unit relies on evaluation and selection system data [32]. The immune control unit is based on the feedback regulating mechanism of immune cells [46]. In the IEUC, unfalsified control unit is using to select suitable controllers and immune control unit is using to adjust system performance. In order to examine effectiveness of the IEUC, we apply it to a plant of high-speed spinning model which is a single-input-single-output (SISO) complex three-order plant. The simulation results demonstrate that system control performance of the IEUC is superior to that of the single UC in the control of complex plant.

5.4.1 Theoretical Foundations

5.4.1.1 The Unfalsified Control

Unfalsified control is a scheme based on data-driven methodology. For the scope of control system, the unfalsified control scheme (or the unfalsified controller, UC) is a way to build satisfactory controllers with experimental data rather than rely on feigned hypotheses or prejudicial assumptions about the plant [47]. Figure 5.18 reveals the basic procedures of a classical UC. The establishment of a qualified UC is to evaluate each candidate controller from a predefined controller set before put into feedback system which can be visually expressed as "controller sieve". Then, the "controller sieve" makes a strict evaluation of candidate controllers based on input (goal) as well as error (*e*). After the sieve process, candidate controllers are divided into two parts: reject controllers and accept controllers. Then, the reject controllers turn to falsified controllers as well as the accept controllers turn to unfalsified controllers. And the first unfalsified controller is selected to system loop as a current controller. While only the evaluation process relies on input and output data, unfalsified control is a typical data-driven methodology.

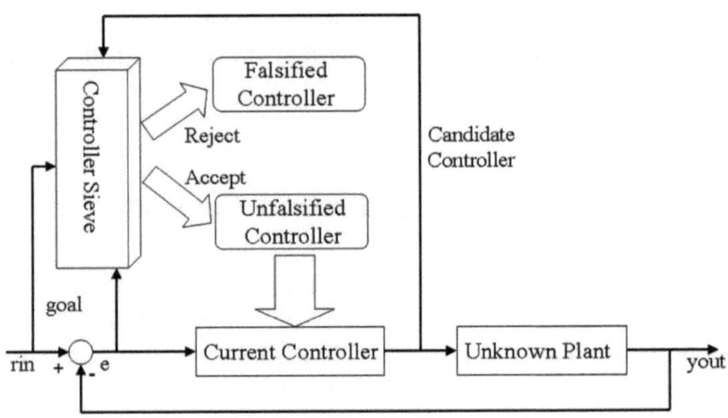

Fig. 5.18 The concept of unfalsified control

5.4.1.2 The Immune Control

Immune system, as a biological concept, is a complexity system which consists of immune executive organs, tissue, cellular and molecular. In general, a important kind of immune executive cellular called "lymphocytes" has the ability to construct a feedback mechanism that enables human being survival in the world full of infection and diseases. These lymphocytes name T cells and B cells. The main cells that are involved in the model are antigen (Ag), antibody (Ab), B cells (B), help T cells (T_H), suppressor T cells (T_s) and antigen presenting cells (APC)[46]. When Ag invades organisms, it is firstly recognized by APC. Then, the APC sends recognition message to T_H, and T_H secretes the interleukin $(IL-2 \,\text{and}\, \text{IL}-14)$ to activate the immune response. Then $IL-2$ stimulates T_s and $IL-14$ stimulates B cells. As the T_s receives stimuli, it secretes another kind of interleukin $(IL-10)$ to inhibit the active of T_H. Meanwhile, as the activated B cells begin to divide, their offsprings (plasma cells) secrete Ab to recognize Ag. After a while, when Ag is prevented by Ab, a dynamic relative equilibrium point can be reached, and the immune response is finished.

5.4.2 Design of the Immune-Enhanced Unfalsified Controller

5.4.2.1 System Structure

Considering the characteristic of unfalsified control and immune control, a novel immune-enhanced unfalsified controller (IEUC) combining both of them is presented as shown in Fig. 5.19. The controller uses unfalsified control to evaluate preset candidate controllers and immune control to regulate control signal and system performance.

The unfalsified control part is based on output $(yout)$, control signal (U), and error (e). And this part transmits selected control's information to immune control part. Then based on P-type immune feedback mechanism [46], the immune controller uses this information as well as error signal of system to modulate the control signal of system. According to the following control algorithm in Sect. 2.4, the unfalsified controller and immune controller can control the plant harmoniously,

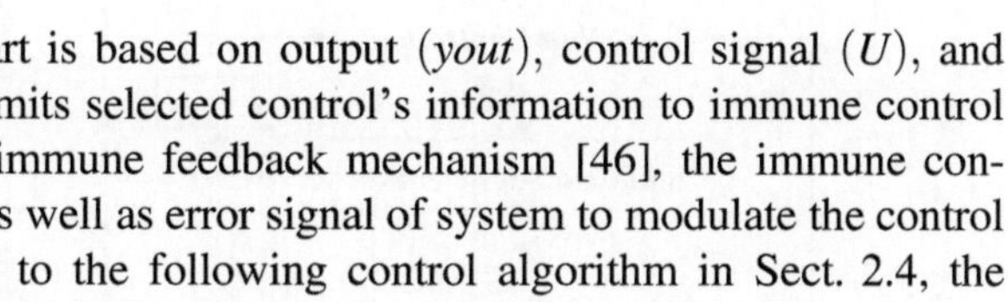

Fig. 5.19 The structure of immune-enhanced unfalsified controller

and the new system could hold both the advantage of data-driven methodology and biologic intelligent methodology.

5.4.2.2 Control Algorithms

As IEUC cannot be easily implemented without a basic control algorithm, a conventional PID type control algorithm mixed with the IEUC is proposed. Global variables of the IEUC are defined as follows: e is error, U is control signal and y is output of the whole system.

The basic algorithm flowchart of unfalsified control with immune controller is shown in Fig. 5.20. The left part of Fig. 5.20 is mainly unfalsified control algorithm and the right part is mainly immune control algorithm.

The initialization process consists of:

A finite set Kr of $Num\,Controller$ candidates $Ki,\ i \in I \triangleq \{1,\ldots,$ $NumController\}$.

the sampling time of control system t_s.
the total time of control system$TotalTime$.
initial time $t = 0$.
initial controller number $i = NumController$.
initial parameters of unfalsified control ω_1, ω_2.
initial parameters of immune control η, l.

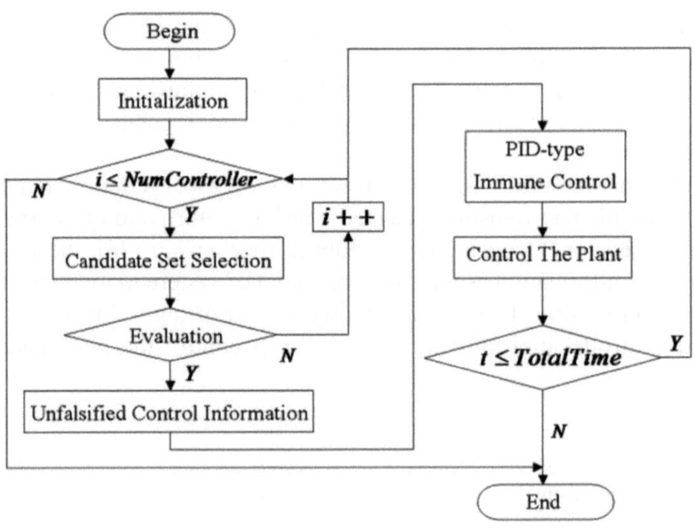

Fig. 5.20 The basic algorithm of unfalsified control with immune controller

The selection and evaluation process for the candidate set is based on the algorithm of unfalsified control [32]: $r_i(t)$ is the fictitious reference signal, combing with a basic PID-type algorithm:

$$r_i(t) = yout_i(t) + \frac{s}{s \cdot K_p + K_i} \cdot (U_i(t) + \frac{s \cdot K_d}{\varepsilon \cdot s + 1} \cdot yout_i(t)) \tag{5.37}$$

where i is current controller number, ε is a small enough value in PID approximate algorithm, K_p, K_i, K_d denotes proportional, integral, and derivative parameter, respectively, and s is the complex variable of Laplace transform. Performance specification set T_{spec} is

$$T_{spec}(r_i(t), yout(t), u(t)) = |\omega_1 * (r_i(t) - y(t))|^2 + |\omega_2 * u(t)|^2 - |r_i(t)|^2 \tag{5.38}$$

where ω_1 and ω_2 are weighting filters depends on user's demand, $*$ denotes the convolution operator.

Moreover, performance evaluation standard E_{value} is

$$\begin{aligned} E_{value}(i, k \cdot t_s) = E_{value}(i, (k-1) \cdot t_s) + \frac{1}{2} \cdot ts \cdot \{T_{spec}(r_i(k \cdot t_s), y(k \cdot t_s), u(k \cdot t_s)) \\ + T_{spec}(r_i((k-1) \cdot t_s), y((k-1) \cdot t_s), u((k-1) \cdot t_s))\} \end{aligned} \tag{5.39}$$

After evaluation process, the falsified controller is discarded and the iteration will continue ($i = i + 1$). The unfalsified controller conveys unfalsified information (in this system are K_p, K_i, K_d) to immune controller. When the process comes to $i > Num\,Controller$, it means no unfalsified controller has been found. The algorithm would terminate, and a new set of candidate controllers are required.

The immune control algorithm is as follows.

The control signal of plant:

$$U(k) = (K_p + K_1 \cdot \frac{1}{z-1} + K_D \cdot \frac{z-1}{z}) \cdot e(k) \tag{5.40}$$

$$\begin{cases} K_p = K_p \cdot \{1 - \eta \cdot f(\Delta U(k))\} \\ K_1 = K_1 \cdot \{1 - \eta \cdot f(\Delta U(k))\} \\ K_D = K_D \cdot \{1 - \eta \cdot f(\Delta U(k))\} \end{cases}$$

where, η is a user design coefficient to adjust the impact of $f(\cdot), f(\cdot)$ is a nonlinear function for considering the effect of the reaction of B cells and the antigens. Also, when $\eta = 0$, the IEUC controller is equal to UC controller. In this section, a fuzzy function has been used to approximate $f(\cdot)$. The fuzzy controller with two inputs and one output is employed here. The immune controller signal $U(k)$ and change of the controller signal $\Delta U(k)$ are two inputs variables; the output of system $yout(k)$ is the output variable. The fuzzy control rules and fuzzy functions are shown in [46].

5.4.3 Simulation Results

In order to examine the control performance of the IEUC, we consider a three-order plant in the high-speed spinning model with delay,

$$G(s) = \frac{1.059e^{-1.5s}}{0.000942s^3 + 0.3316s^2 + 1.1988s} \tag{5.41}$$

where $e^{-1.5s}$ is delay, and gain of this delay is 0.5.

We set the initial parameter as, $Num\,Controller = 45, t_s = 0.1s, TotalTime = 30s$, candidate controllers: $K_p = \{5, 10, 30, 110, 150\}, K_i = \{50, 75, 100\}, K_d = \{20, 10, 4\}$. Performance evaluation: $W_1(s) = \frac{s+20}{2(s+3)}, W_2(s) = \frac{0.01}{1.2(s+1)^3}$. Immune control parameters: $\eta = 0.6, l = 1$. Additionally, based on the IEUC algorithm discussed in Sect. 3.2, when $\eta = 0$, the IEUC controller is equal to UC controller.

Seen from Fig. 5.19, in the unfalsified control element, if the mathematical method is available, the unfalsified set of controllers can be easily obtained. However, based on Eqs. (5.38) and (5.39), as the progress of sieving, many candidate controllers could be evaluated to be falsified and rejected and it is possible that the finite set *Kr* of *Num Controller* candidates would become empty. When the algorithm is being terminated due to this reason, argument the set *Kr* with additional candidate controllers or reset of the performance specification need to be done [32]. In this case, a carefully designed set of PID parameters has been employed to act as candidate controllers and a relatively comprehensive performance specification has been adopted [32]. In the immune fuzzy control element, by adjusting η, l in Eq. (5.40) and fuzzy rules, the fuzzy immune controller can quickly drive the system output to the desired level [46].

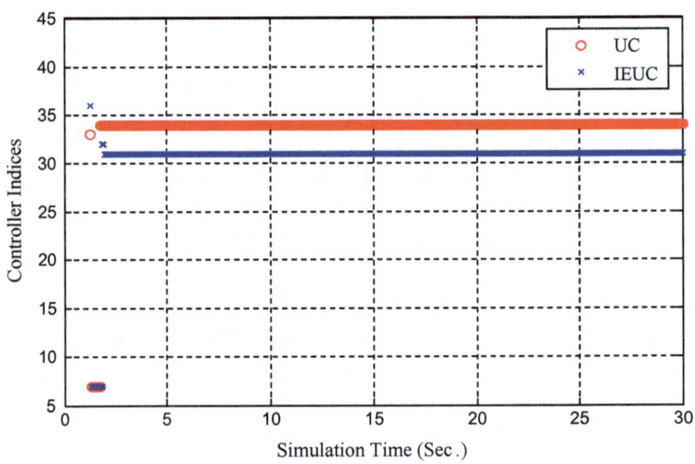

Fig. 5.21 The change of controller

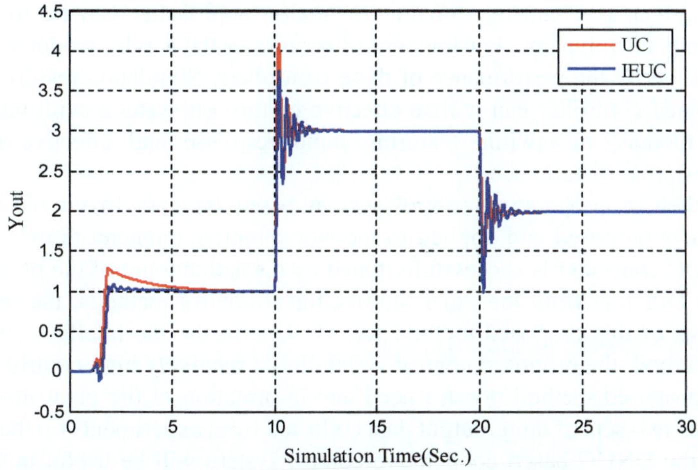

Fig. 5.22 The Performance Comparison between the UC and the IEUC

In practice, a large scale of candidate controllers has been roughly set to judge whether the plant is controllable. Then, elaborative mathematical explanation of desired performance has been used to fix the performance specification. Finally, the impact of immune fuzzy control is regulated by changing η, and l.

Using the IEUC, we can obtain desirable control performance for the system with delay, as shown in Figs. 5.21 and 5.22. The control effectiveness is also illustrated by comparing the performance of the IEUC and the single UC. In order to show the contrast effectiveness of the influence on *yout* with *rin* changing, the set points of *rin* at the 10-*th*, 20-*th*, 30-*th* seconds are changed. Seen from Figs. 5.21 and 5.22, we can see that the overshoot in the IEUC is less than that in the UC, and the influence of *rin* changing is very sharp in the UC, while the IEUC can quickly drive the system output to the desired level with little fluctuation. Seen from Fig. 5.21, we can know that this result can be partly explained as different choices of controllers. The simulation results demonstrate that the IEUC has better control performance and adaptability than that of the UC.

5.5 Conclusions

In this chapter, first, a data-driven cooperative intelligent controller inspired by the ultra-short feedback regulating mechanism of the endocrine system in human body is proposed. The design of the proposed DDCIC can be regarded as an exploratory step of calling the bio-inspired mechanisms for optimizing the controllers of other types, which indicates that different algorithms can also be introduced for solving this problem in a similar way. Through changing the sections that can be tuned by the data-driven mechanism, different regulation approaches can be formed, and experiments on a dynamic tension control system for the acrylic production line are applied to verify the performance of these controllers. Simulation results show that the proposed controller can realize effective control on systems with unknown or varying models, meanwhile featuring rapid response and effective regulation against external disturbance.

And then, a cooperative control system based on data-driven bio-intelligent controller is proposed and applied to the wet spinning spinneret drawing process. The DNUC controller is successfully tuned by a small amount of the plant's offline I/O data. Different from the other bio-intelligent control methods, the most major advantages of this proposed method are: (1) Due to the bio-intelligent and cooperative method, the proposed control system has a relatively high control accuracy; (2) The proposed method doesn't need any information of the plant model; (3) It only needs two sets of input/output data collected from experiments on the plant. As a result, the DNUC based cooperative control system will be useful in those systems which are model-unknown (or inaccurate) and of high costs (time, money, energy etc.) for data collection. To verify the effectiveness of the proposed control system, simulation experiment on wet spinning spinneret draw ratio control process of PAN-base fiber is conducted. Simulation results show that the proposed DNUC is a well-performed controller without requiring any knowledge of the plant. The control performance of spinneret draw ratio is able to cooperatively response to and effectively regulate against disturbances in all related devices.

At last, an immune-enhanced unfalsified controller and its control scheme are presented and applied to control a complex three-order plant in high-speed spinning model with delay. Simulation results demonstrate that the IEUC can rapidly response to the changing of desired level. Moreover, compared with single unfalsified control algorithm, the IEUC can decrease system overshoot as well as reduce rising time easily and successfully.

References

1. Liang, X., Ding, Y., Ren, L., Hao, K., Jin, Y.: Data-driven cooperative intelligent controller based on the endocrine regulation mechanism. IEEE. T. Contr. Syst. T. **22**(1), 94–101 (2014)
2. Ding, Y., Xu, N., Ren, L., Hao, K.: Data-driven neuroendocrine ultrashort feedback-based cooperative control system. IEEE. T. Contr. Syst. T. **23**(3), 1205–1212 (2015)
3. Wang, L., Ding, Y., Hao, K., Liang, X.: An immune-enhanced unfalsified controller for a high-speed spinning process, Knowledge Engineering and Management, pp. 200–202. Springer, Berlin (2011)
4. Rupp, D., Guzzella, L.: Iterative tuning of internal model controllers with application to air/fuel ratio control. IEEE. T. Contr. Syst. T. **18**(1), 177–184 (2010)
5. Montazeri, A., Poshtan, J., Yousefi-Koma, A.: Design and analysis of robust minimax LQG controller for an experimental beam considering spill-over effect. IEEE. T. Contr. Syst. T. **19** (5), 1251–1259 (2011)

6. Hu, J., Wang, Z., Gao, H., Stergioulas, L.K.: Robust sliding mode control for discrete stochastic systems with mixed time delays, randomly occurring uncertainties, and randomly occurring nonlinearities. IEEE. T. Ind. Electron. **59**(7), 3008–3015 (2012)
7. Xu, D., Huang, J.: Robust adaptive control of a class of nonlinear systems and its applications. IEEE. T. Circuits-I. **57**(3), 691–702 (2010)
8. Han, M., Fan, J., Wang, J.: A dynamic feedforward neural network based on Gaussian particle swarm optimization and its application for predictive control. IEEE. T. Neural. Networ. **22**(9), 1457–1468 (2011)
9. Wang, Z., Shen, B., Shu, H., Wei, G.: Quantized control for nonlinear stochastic time-delay systems with missing measurements. IEEE. T. Automat. Contr. **57**(6), 1431–1444 (2012)
10. Hou, Z.S., Xu, J.X.: On data-driven control theory: The state of the art and perspective. Acta. Automat. Sinica. **35**(6), 650–667 (2009)
11. Yang, R., Tarn, T., Zhang, M.: Data-driven feedforward control for electroporation mediated gene delivery in gene therapy. IEEE. T. Contr. Syst. T. **18**(4), 935–943 (2010)
12. Erlhagen, W., Mukovskiy, A., Bicho, E., et al.: Goal-directed imitation for robots: A bio-inspired approach to action understanding and skill learning. Robot. Auton. Syst. **54**(5), 353–360 (2006)
13. Wu, M., Yan, J., She, J.H., Cao, W.H.: Intelligent decoupling control of gas collection process of multiple asymmetric coke ovens. IEEE. T. Ind. Electron. **56**(7), 2782–2792 (2009)
14. Lin, C.L., Hsiao, S.L., Chen, C.K.: Balance of emulated MDOF human postural systems. Int. J. Eng. Sci. **48**(9), 751–770 (2010)
15. Liu, B., Ding, Y.S., Wang, J.H.: An intelligent controller based on ultra-short feedback of neuro-endocrine system. Comput. Sl. **25**(1), 188–191 (2008)
16. Ding, Y.S., Liu, B.: An intelligent bi-cooperative decoupling control approach based on modulation mechanism of internal environment in body. IEEE. T. Contr. Syst. T. **19**(3), 692–698 (2011)
17. Ding, Y.S., Liu, B., Ren, L.H.: Intelligent decoupling control system inspired from modulation of the growth hormone in neuroendocrine system. Dynam. Cont. Dis. Ser. B. **14** (5), 679–693 (2007)
18. Han, J.: From PID to active disturbance rejection control. IEEE. T. Ind. Electron. **56**(3), 900–906 (2009)
19. Darlington, J., Pantelides, C.C., Rustem, B., Tanyi, B.A.: An algorithm for constrained nonlinear optimization under uncertainty. Automatica. **35**(2), 217–228 (1999)
20. Previdi, F., Schauer, T., Savaresi, S.M., Hunt, K.J.: Data-driven control design for neuroprotheses: A virtual reference feedback tuning (VRFT) approach. IEEE. T. Contr. Syst. T. **12**(1), 176–182 (2004)
21. Yamamoto, T., Takao, K., Yamada, T.: Design of a data-driven PID controller. IEEE. T. Contr. Syst. T. **17**(1), 29–39 (2009)
22. Guardabassi, G.O., Savaresi, S.M.: Virtual reference direct design method: An off-line approach to data-based control system design. IEEE. T. Automat. Contr. **45**(5), 954–959 (2000)
23. Campi, M.C., Lecchini, A., Savaresi, S.M.: Virtual reference feedback tuning: A direct method for the design of feedback controllers. Automatica. **38**(8), 1337–1346 (2002)
24. Lecchini, A., Campi, M.C., Savaresi, S.M.: Virtual reference feedback tuning for two degree of freedom controllers. Int. J. Adapt. Control. **16**(5), 355–371 (2002)
25. Campi, M.C., Savaresi, S.M.: Direct nonlinear control design: The virtual reference feedback tuning (VRFT) approach. IEEE. T. Automat. Contr. **51**(1), 14–27 (2006)
26. Formentin, S., De Filippi, P., Corno, M., Tanelli, M.: Data-driven design of braking control systems. IEEE. T. Contr. Syst. T. **21**(1), 186–193 (2013)
27. Hjalmarsson, H., Gunnarsson, S., Gevers, M.: A convergent iterative restricted complexity control design scheme. Paper presented at the 33rd IEEE Conference on Decision and Control, Lake Buena Vista, 14–16 December (1994)
28. Hou, Z., Han, Z.: Parameter estimation algorithm of nonlinear systems and its dual adaptive control. Acta. Automat. Sinica. **21**(1), 122–125 (1995)

29. Safonov, M. G., Tsao, T. C.: The unfalsified control concept: a direct path from experiment to controller, Feedback Control, Nonlinear Systems, and Complexity, vol 202, pp. 196–214. Springer, Berlin (1995)
30. Hildebrand, R., Lecchini, A., Solari, G., Gevers, M.: Asymptotic accuracy of iterative feedback tuning. IEEE. T. Automat. Contr. **50**(8), 1182–1185 (2005)
31. Yan, J., Hou, Z.: Convergence of MFAC based feedback-feedforward ILC systems. Paper presented at the 27th Chinese Control Conference (CCC), Kunming, China, 16–18 July (2008)
32. Jun, M., Safonov, M. G.: Automatic PID tuning: An application of unfalsified control. Paper presented at IEEE CCA/CACSD, Kohala Coast, USA, 22–27 Aug (1999)
33. Van Helvoort, J., de Jager, B., Steinbuch, M.: Direct data-driven recursive controller unfalsification with analytic update. Automatica. **43**(12), 2034–2046 (2007)
34. van Helvoort, J., de Jager, B., Steinbuch, M.: Data-driven controller unfalsification with analytic update applied to a motion system. IEEE. T. Contr. Syst. T. **16**(6), 1207–1217 (2008)
35. Battistelli, G., Mosca, E., Safonov, M.G., Tesi, P.: Stability of unfalsified adaptive switching control in noisy environments. IEEE. T. Automat. Contr. **55**(10), 2424–2429 (2010)
36. Rodriguez-Angeles, A., Nijmeijer, H.: Mutual synchronization of robots via estimated state feedback: a cooperative approach. IEEE. T. Contr. Syst. T. **12**(4), 542–554 (2004)
37. Vergnano, A., Thorstensson, C., Lennartson, B., Falkman, P., Pellicciari, M., Leali, F., Biller, S.: Modeling and optimization of energy consumption in cooperative multi-robot systems. IEEE. T. Autom. Sci. Eng. **9**(2), 423–428 (2012)
38. Liang, X., Ding, Y.S., Ren, L.H., Hao, K.R., Wang, H.P., Chen, J.J: A bioinspired multilayered intelligent cooperative controller for stretching process of fiber production. IEEE. T. Syst. Man. Cy. C. **42**(3), 367–377 (2012)
39. Ding, Y.S., Liang, X., Hao, K.R., Wang, H.P.: An intelligent cooperative decoupling controller for coagulation bath in polyacrylonitrile carbon fiber production. IEEE. T. Contr. Syst. T. **21**(2), 467–479 (2013)
40. Spall, J.C.: Multivariate stochastic approximation using a simultaneous perturbation gradient approximation. IEEE. T. Automat. Contr. **37**(3), 332–341 (1992)
41. Spall, J.C., Cristion, J.A.: Model-free control of nonlinear stochastic systems with discrete-time measurements. IEEE. T. Automat. Contr. **43**(9), 1198–1210 (1998)
42. Farhy, L.S.: Modeling of oscillations in endocrine networks with feedback. Method. Enzymol. **384**, 54–81 (2004)
43. Campi, M.C., Lecchini, A., Savaresi, S.M.: Virtual Reference Feedback Tuning (VRFT): A new direct approach to the design of feedback controllers. Paper presented at the 39th IEEE Conference on Decision and Control. 1, 623–629 (2000)
44. Gevers, M., Bombois, X., Codrons, B., Scorletti, G., Anderson, B.D.: Model validation for control and controller validation in a prediction error identification framework—Part I: theory. Automatica **39**(3), 403–415 (2003)
45. Doi, M., Edwards, S. F.: The theory of polymer dynamics. Oxford University Press (1988)
46. Ding, Y.: A nonlinear PID controller based on fuzzy-tuned immune feedback law. Paper presented at the 3rd World Congress on Intelligent Control and Automation (WCICA), vol. 3, pp. 1576–1580, Hefei, China, 28 June–2 July (2000)
47. Safonov, M.G., Tsao, T.C.: The Unfalsified Control Concept and Learning. IEEE. T. Auto mat. Contr. **42**(6), 843–847 (1997)

Chapter 6
Bio-Network Inspired Cooperative Intelligent Controllers

It is not easy to control most of the complicated nonlinear processes with satisfactory control effectiveness by using conventional control strategies. In this chapter, some bio-network inspired cooperative intelligent controllers and strategies are proposed. Firstly, we propose a novel nonlinear guided intelligent controller (NGIC) inspired by the bi-cooperative regulation mechanism and the regulation characteristics of glucose in human body [1]. Secondly, based on the modulation mechanism of neuroendocrine-immune system, we present a novel nonlinear optimized intelligent controller (NOIC) and provide a method to optimize and adjust its control parameters dynamically [2]. Finally, a new control strategy based on the bi-direction hormone regulating mechanism of neuroendocrine-immune (NEI) network is proposed, and applied to single-phase photovoltaic grid-connected inverter [3].

6.1 Introduction

In order to obtain high product quality and control precision, some nonlinear control techniques, such as neural networks control [4–7], fuzzy control [8–11], expert control [12–15] and other advanced control algorithms [16, 17] have been developed. It also can be found that more and more advanced or intelligent control algorithms will be developed inspired from the bio-information systems [18] in the future. However, some control technologies are developed or designed based on some bio-information models or modulation theories to a degree, and most of these control algorithms are too complicated to apply to the practical processes, which affects their practical application and control effectiveness. Therefore, it is necessary to develop some advanced control algorithms with simple work principle to improve the practical control effectiveness. One of novel ways is to design these controllers based on some special physical regulation mechanism in human body.

In fact, many physical regulation mechanisms or models regulate the physical indices in the body. Among them, the bi-cooperative regulation mechanism (BCRM)

Y. Ding et al., *Bio-Inspired Collaborative Intelligent Control and Optimization*, Studies in Systems, Decision and Control 118,
https://doi.org/10.1007/978-981-10-6689-4_6

is a special and important regulation mechanism with enhanced and inhibited bidirectional modulation actions. The BCRM involves two sub-networks comprising sympathetic nerve network (SNN) and parasympathetic nerves network (PNN). While one physical index in the body changes, the BCRM can make this index rapidly become stable with unusual powerful robustness and stability [19–21] through this enhanced and inhibited cooperative regulating actions. The regulation process of glucose is a typical example modulated by the BCRM. Moreover, the regulation process of glucose has several characteristics such as rapid pulse-regulation output and central regulation output.

The bio-intelligent control algorithms or controllers are inspired form some learning, memory, and modulation mechanism of biological systems [19, 22–25]. The nervous system, endocrine system, and immune system are three important physiological modulation systems in body, and compose the neuroendocrine-immune (NEI) system. The interactions among them are bi-direction and complex [26, 27]. The nervous system can affect the endocrine system and the immune system via neurotransmitters, which is released in the nervous system. Endocrine system can simulate or inhibit the activation of nervous system and immune system through some hormones, e.g. GH, adrenalin. And immune also can modulate the physiological function of nervous system and endocrine via cytokines, which is secreted by immune cells. On the whole, the neuroendocrine-immune (NEI) system becomes a complicated physiological system and has some special modulation mechanism. In this chapter, we just consider the influence of the nervous system on the endocrine and immune system, and the influence of endocrine system on immune system.

The study on the artificial NEI system may inspire some novel methods to solve complicated problems and will also bring a little influence on the conventional control theory. Some researchers have presented several models of NEI [28, 29], such as neuro-immune-cutaneousendocrine network and network of immune-neuroendocrine, but which are only used in the medical field. Several intelligent controllers based on the modulation mechanism of immune system [30, 31], and endocrine or neuroendocrine systems have been reported [32, 33]. However, there are few reports on the intelligent controller based on the NEI system.

6.2 A Bio-System Inspired Nonlinear Intelligent Controller

In this section, we propose a novel nonlinear & guided intelligent controller (NGIC) which is inspired by the special BCRM and the regulation characteristics of glucose in human body. The NGIC is composed of four control units including a cooperative control unit (CCU), an enhanced control unit (ECU), an inhibited control unit (ICU), and an auxiliary control unit (ACU). Wherein, the CCU coordinates the work of the ECU, the ICU and the ACU to improve the control indices such as the rise time, the overshoot, and the settling time of the control system to a degree. The NGIC can independently regulate the dynamic and steady change processes by using different control units. The ECU and the ICU are responsible for adjusting the dynamic change process, and the ACU is for the steady change process. Therefore, we may tune the control parameters of the ECU and the ICU according to the control effect in the dynamic change process, and so do the parameters of the ACU according to the effect in the steady change process.

6.2.1 Bi-Cooperative Regulation Mechanism of Blood Glucose

The bi-cooperative regulation mechanism is of great importance for keeping some body indices in an equilibrium state. The glucose level is one of the most important physiological indices to be modulated via the regulation mechanism.

6.2.1.1 Bi-Cooperative Regulation Mechanism

Under the control of central nervous system, several human physiological regulation systems form a highly complex and stable physiological regulation network. The physiology network maintains its homeostasis through the bi-cooperative regulation mechanism with enhanced and inhibited regulations in human body. The bi-cooperative regulation network comprises two sub-networks, i.e. the sympathetic nerve sub-network (SNN) and the parasympathetic nerves sub-network (PNN). For the same physical index, the SNN and the PNN have enhanced and inhibited regulation actions via secreting a large number of hormones, respectively. The former action can strengthen the immunity and speed up the metabolism activity, while the later one can weaken or slow down the physical activity.

When one index value is lower or higher than its normal level, the regulating action of the SNN first will be much stronger than that of the PNN to make the index rapidly increase or decrease. When the index is close to the normal value, these two opposite regulations may become equal. With the bi-cooperative modulation of the complex bio-network, some physical indices in the body can rapidly become stable with strong robustness and stability [19–22]. The regulation process of glucose is accomplished via the bi-cooperative regulation mechanism as shown in Fig. 6.1, where CNS and BCS are the central nervous system and the blood circulation system, respectively. In Fig. 6.1, the solid line denotes the enhanced modulation signals, and the dashed line denotes the inhibited regulation signals. In addition, the sign of "+" and "−" is denoted as enhanced and inhibited regulation action, respectively. Thus, the bi-cooperative regulation network can maintain the blood glucose level stable and in a normal range. This special regulation mechanism can be inspired for the design and implementation of the NGIC.

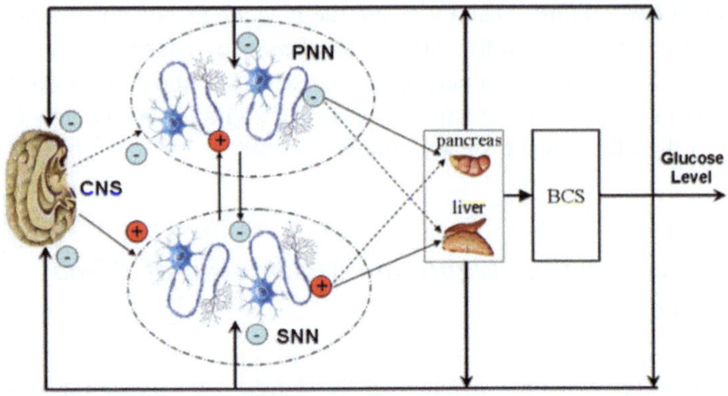

Fig. 6.1 The mechanism of blood glucose regulation

6.2.1.2 Regulation Process of Blood Glucose

The detail regulation process of blood glucose is explained below. When the glucose level is lower than the normal value, the SNN will be more powerful than the PNN. This will cause the secretion of insulin in pancreas to decrease and the secretion of glucagon to increase, which can inhibit the decomposition of glucose and promote the decomposition of hepatic glycogen in liver to make the concentration of glucose rapidly rise. On the contrary, when the glucose level is higher than the normal value, the PNN will be more powerful than the SNN. This will increase the secretion of insulin and reduce the secretion of glucagon, which can promote the decompression of glucose and produce of heparin to make the concentration quickly decrease. When the PNN and the SNN are in equilibrium state, the secretion of glucose and glycogen are nearly equal, which makes the concentration of glucose stable.

During the regulation process of blood glucose, there are several important characteristics to design the NGIC. The first characteristic is the rapid pulse-regulation output, i.e. a rapid increase and then decrease to output of regulation during the regulating process of glucose. In order to regulate the change of blood glucose rapidly, the pancreas releases the insulin as a great pulse with rapid increase and decrease, which makes the glucose quickly approach to the normal value. When a person takes a meal, the concentration of blood glucose may rise quickly. To reduce the glucose level to the normal value, the body may rapidly release great amount of insulin [34, 35]. When the blood glucose decreases to a degree or value, the amount of released insulin may reduce quickly to a relative stable value. The regulation pattern may make blood glucose promptly recover to a normal range.

The second characteristic is that the steady insulin secretion rate can be pre-obtained before the glucose becomes stable, according to the normal glucose level value, and its real time secretion rate is central regulating output based on the steady secretion rate. In fact, the insulin secretion rate $u(G_B)$ of each islet can be given in the following equation [34, 35]:

$$u(G_B) = \frac{N}{\lambda + \exp(-\delta \cdot G_B + \theta)} \quad (6.1)$$

where G_B is the blood glucose level in the body, and N, δ, θ, and λ are positive real number. The steady blood glucose level is the normal glucose value or set-point value. The final secretion rate of insulin released can be easily obtained by Eq. (6.1) based on the steady blood level G_B. The above special regulation characteristics can make blood glucose become stable much quickly.

6.2.2 Design Process of the NGIC

In order to improve the control quality of complex nonlinear control plant, it is necessary to design an advanced nonlinear controller that can eliminate the control error with short rise time, little overshoot and small settling time. The bi-cooperative regulation mechanism and the special regulation characteristics of blood glucose can bring us new ideas to design the NGIC in this section.

6.2.2.1 Overall Structure of the NGIC

In fact, there is close relationship between the bi-cooperative regulation mechanism (BCRM) of blood glucose and some nonlinear process control systems. For example, the blood glucose and the BCRM may correspond to process and controller, respectively. Furthermore, the important regulation characteristics of blood glucose may inspire us with new idea to design the detail control law of NGIC. Therefore, we design the architecture of the NGIC according to Fig. 6.1. The NGIC is comprised of four parts, i.e. a cooperative control unit (CCU), an enhanced control unit (ECU), an inhibited control unit (ICU) and an auxiliary control unit (ACU). The CCU is designed corresponding to the central nervous system, and the ECU and the ICU is corresponding to the PNN and the SSN, respectively. Moreover, both the ECU and the ICU are designed according to the pulse-regulation characteristic of blood glucose. The ACU is considered to improve the control precision of the NGIC. The detail architecture of NGIC is as shown in Fig. 6.2.

According to the architecture in Fig. 6.2, we design the overall control procedure of the NGIC. The NGIC first outputs a great step action when the set-point changes or great error appears. Then the control output rapidly declines to the future steady value of the controller to reduce the overshoot or to avoid oscillation when the control error exceeds a defined threshold value. The general control principle is explained as follows:

(1) The ECU and the ICU are responsible for improving the control indices such as the rise time and overshoot during the dynamic process. When the set-point changes or a great error appears, the ECU will accelerate the rise speed of process variable via a rapid step action. Then, when the control error reaches a

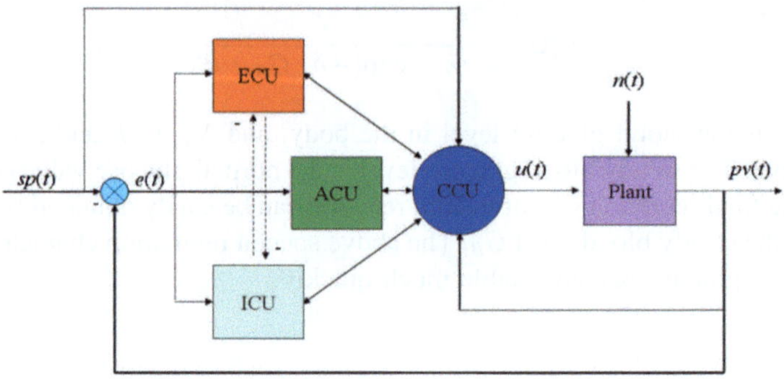

Fig. 6.2 The architecture of the NGIC

prescribed threshold value or the control time exceeds a prescribed limitation, the ICU will decelerate the rise speed of process variable in order to reduce or avoid the control overshoot.

(2) The ACU is responsible for improving the settling time and the control precision in the stable change process. When the control error approaches a relative steady status, the ACU will further improve the control precision of the NGIC through a central proportional control law with a final guided control output [22]. It is also responsible for identifying the model of controlled plant, which is very useful for ensuring the control precision of the NGIC.

(3) The CCU, as the supper manager, will coordinate all the work sequence of the ECU, the ICU and the ACU, which makes the NGIC rapidly eliminate the control error with short rise time, little overshoot, and small settling time.

6.2.2.2 Design Procedure of the NGIC

For a nonlinear model represented by the following equation:

$$\begin{cases} \dot{x}(t) = F(t, x(t), u(t)) \\ y(t) = G(\dot{x}(t), x(t)) \end{cases} \quad t \geq 0 \tag{6.2}$$

where t, $x(t)$, $\dot{x}(t)$, $y(t)$ and $u(t)$ is the control time, process status variable, the first derivation of process status variable, process variable and input variable of controlled plant, respectively. According to the linearization theory, we can obtain the corresponding linear mathematic model at their working point. As we known, most of the linear models can be approximated by a first-order or second-order plant with time-delay. For some high order plants, we can also build this model according to the dominant-pole control theory [36, 37]. And the transfer amplitude factor K, time constant T, and time delay τ can be identified via several identification methods [38, 39].

(1) Design and principle of the ECU

(1) Control algorithm of the ECU

In order to shorten the rise time of control system or to make the process variable quickly reach the set point, we design the ECU to output a rapid step action when the set point changes. This design idea is inspired from the rapid pulse-regulation of blood glucose. In addition, the rapid step action takes effect according to the difference between the initial control output and the ultimate steady control output of the controller when the control system becomes stable. In fact, the ultimate steady control output is tantamount to a guided steady control output. We explain the detail principle of ECU as following.

For the narration convenience, we define t_0 as the initial control time when the set-point of control system changes, $u_{peak}(t)$ as the peak value of the control output of the ECU during the dynamic change process, and t_{peak} as the peak time when the control output of the ECU reaches the peak value $u_{peak}(t)$.

When the control time $t \geq t_0$, the ECU first comes into effect. The ECU outputs a rapid great positive or negative step action according to the real time control error and control action direction. This great step action changes from the initial control output $u_{initial}(t)$ to the control peak value $u_{peak}(t)$. We denote the ultimate steady control output of the NGIC as $u_{steady}(\infty)$ when the control system becomes stable. From the initial output $u_{initial}(t)$, the output peak $u_{peak}(t)$ is designed in the following equation:

$$u_{peak}(t) = u_{initial}(t) + \mu \cdot (u_{steady}(\infty) - u_{initial}(t)) \tag{6.3}$$

where μ is a positive pulse-factor that should ensure the dynamic change output within the reasonable range, i.e. $u_L \leq u_{peak}(t) \leq u_H$, where u_L and u_H is the low and high limitation of control output, respectively. However, when the initial control output $u_{initial}(t)$ is close to the low or high limitation, the $u_{peak}(t)$ often exceeds the output limitation.

As we known, it may be not practical to make the controller output from $u_{initial}(t)$ to $u_{peak}(t)$ in one control period according to Eq. (6.3). Since the actuator cannot follow such a great change of control signal. To solve this problem, we dealt with this step output action by using a filter. The filter is designed according to the regulation law of hormone secretion in endocrine system [19], and is shown in Eq. (6.4):

$$F_{ECU}(t) = \frac{(t - t_0)^m}{T_{ECU} + (t - t_0)^m} \tag{6.4}$$

where T_{ECU} is the time constant of filtering, m is a positive number, time difference $(t - t_0)$ is corresponding to the stimulation factor. When $t = t_0$, $F_{ECU}(t) = 0.0$; when $(t - t_0) \gg 1$, $F_{ECU}(t) \approx 1.0$. Obviously, the bigger the T_{ECU} and the smaller the m, the stronger the filtering effect. According to Eqs. (6.3) and (6.4), we can deduce the control output $u_{ECU}(t)$ of the NGIC,

$$u_{ECU}(t) = u_{initial}(t) + \mu \cdot (u_{steady}(\infty) - u_{initial}(t)) \cdot F_{ECU}(t) \tag{6.5}$$

From Eqs. (6.4) and (6.5), we can see that the initial output of $u_{ECU}(t)$ equals to the initial control output $u_{initial}(t)$ when $t = t_0$. Its final steady output is the peak value $u_{peak}(t)$. During the ECU working period, the control output of the NGIC is equal to $u_{ECU}(t)$.

(2) The method of pre-obtaining the ultimate steady control output

As discussed above, the future steady value of the control output $u_{steady}(\infty)$ is a very important parameter for the NGIC, which just like a remote attacking goal for a guided missile and may determine the control precision. According to the final-value theorem [40], $u_{steady}(\infty)$ can be easily obtained based on the amplitude factor K of the process model. It is well known that the relationship between K and steady value of process variable can be expressed as follows:

$$K = \frac{y_{steady}(\infty) - y_{initial}(t)}{u_{steady}(\infty) - u_{initial}(t)} \tag{6.6}$$

where $y_{initial}(t)$ and $u_{initial}(t)$ is the initial value of process variable and the control output at $t = t_0$, respectively; $y_{steady}(\infty)$ is the steady value of process variable.

Because when the control system becomes stable, the steady value of the process variable will be equal to the set-point value $sp(t)$ of the control system in ideal condition, i.e. $y_{steady}(\infty) = sp(t)$, it is easily deduced from Eq. (6.6) that

$$u_{steady}(\infty) = \frac{sp(t) - y_{initial}(t)}{K} + u_{initial}(t) \tag{6.7}$$

Therefore, as long as we obtain $K, u_{steady}(\infty)$ can be easily derived from Eq. (6.7). For the linear process model, we just need to get the amplitude factor at one set-point according to Eq. (6.6), since the amplitude factor is a constant during the whole range of process variable. However, for the nonlinear control process, it is more difficult to obtain the amplitude factor, because the gain factor often changes with the real-time process variable.

If the set-point of a nonlinear control system does not change, we just identify the amplitude factors at the set-point. However, sometimes the set-point may be modified during a large or even the whole range of process variable. If the accurate process model is available, we can easily calculate the amplitude factors corresponding to the real-time process variable. Practically, the accurate mathematic model of process is not available.

Here, we use the following method to solve this problem. First, we identify an amplitude factor during every divided small range of the process variable. Assuming the obtained data pairs of amplitude factors with their corresponding real-time process variables are (K_1, y_1); (K_2, y_2); $\ldots(K_n, y_n)$, where K_i, y_i is the correlative amplitude factor and process variable in every small process range, respectively. Then, we can obtain the approximate relationship between the amplitude factors and the process variables by using some fitting technologies such as least square method, artificial neural network. In this section, we use the fitting tool in Matlab to fit their relationship. Assuming their relationship fitted is as shown in the following equation:

$$K(t) = G(y(t)) \tag{6.8}$$

where $K(t)$, $y(t)$ is the real-time amplitude factor and process variable, respectively. According to Eq. (6.8), we can easily get the correlative amplitude factor via replacing the process variable by set-point value. Therefore, we can calculate the correlative $u_{steady}(\infty)$ of the control output of the NGIC according to Eq. (6.7).

(2) Design and principle of the ICU

(1) Control algorithm of the ICU

Based on the quick decrease specificity of rapid pulse-regulation output of glucose, we design the control law of the ICU. If we keep the ECU outputting the final peak value $u_{peak}(t)$ for a long time, a great overshoot or dramatic violent oscillation may appear during the dynamic change process. Therefore we design the ICU with an opposite step action with respect to the output of the ECU.

We define t_{down} as the time when the NGIC changes its output from the peak value $u_{peak}(t)$ to the opposite direction. The ICU will output an opposite step action with respect to the output direction of the ECU, when $t \geq (t_0 + t_{down})$. This opposite step action can weaken the control action of the ECU, which can slow down the rise speed of process variable to approach the steady value and reduce or avoid the great overshoot and even oscillation.

The opposite step output action starts from zero and ends at the difference between the prospective $u_{peak}(t)$ and $u_{peak}(t)$ of the NGIC. In order to be practical and ensure the satisfactory control effect, we also design a filter for the ICU according to the regulation law of hormone secretion [19]. The filter for the ICU is designed as shown in the following equation:

$$F_{ICU}(t) = \left\{ \frac{(t-(t_0+t_{down}))^n}{T_{ECU}+(t-(t_0+t_{down}))^n} + \frac{(|E(t)|-1)^n}{T_{ECU}+(|E(t)|-1)^n} \right\} / 2 \tag{6.9}$$

where T_{ICU} and n is similar to T_{ECU} and m in Eq. (6.4), respectively. We also choose the relative control error $E(t)$ as a filter factor to ensure the control effect, where $E(t) = e(t)/e(t_{down})$, and $e(t_{down})$ is the control error at $t = t_0 + t_{down}$. The time difference between $t - (t_0 + t_{down})$ and $|E(t)| - 1$ is denoted as the stimulation factor of hormone secretion. Obviously, when $t = t_0 + t_{down}$, $e(t) = e(t_{down})$, and $E(t) = 1.0$, we can obtain $F_{ICU}(t) = 0.0$; when $t \gg t_0 + t_{down}$ or $E(t) \to 0$, we can get $F_{ICU}(t) \to 1.0$. Moreover, the bigger the T_{ICU} and the smaller the n, the stronger the filtering effect. Compare Eq. (6.9) and Eq. (6.4), we can see that both them are filters designed according to the regulation law of hormone, but both their initial value and final value are different.

According to the afore-analysis together with Eq. (6.9), we can deduce the control law of the ICU as shown in the following equation:

$$u_{ICU}(t) = (u_{steady}(\infty) - u_{peak}(t)) \cdot F_{ICU}(t) \tag{6.10}$$

where $u_{ICU}(t)$ is the control output of the ICU. From Eqs. (6.9) and (6.10), we can see that when $t = t_0 + t_{down}$, or $E(t) \to 1$, the output of the ICU is equal to zero, i.e. $u_{ICU}(t) = 0.0$, which is the weakest inhibited action to the ECU; when $t \gg t_0 + t_{down}$, or $E(t) \to 0$, its output is close to the difference between $u_{peak}(t)$ and $u_{steady}(\infty)$, i.e. $u_{ICU}(t) \approx (u_{steady}(\infty) - u_{peak}(t))$, which is the strongest inhibited action to the ECU.

When $t \geq (t_0 + t_{down})$, both the ECU and the ICU are in work status, and the control output of the NGIC is the summation of $u_{ECU}(t)$ and $u_{ICU}(t)$. From the control principle of the ECU, we can see that the $u_{ECU}(t)$ equals $u_{peak}(t)$ during this period. Therefore, by Eqs. (6.9) and (6.10), the control law of the NGIC during this period can be derived in the following equation:

$$u_{NGIC}(t) = u_{peak}(t) + (u_{steady}(\infty) - u_{peak}(t)) \cdot F_{ICU}(t) \tag{6.11}$$

According to Eq. (6.9), we find that when $t = t_0 + t_{down}$ or $E(t) \to 1$, the output of the NGIC is equal to $u_{peak}(t)$, i.e. $u_{NGIC}(t) = u_{peak}(t)$; when $t > > t_0 + t_{down}$ or $E(t) \to 0$, its output will be close to the steady output $u_{steady}(\infty)$ of the NGIC, i.e. $u_{NGIC}(t) = u_{steady}(\infty)$. Actually, when the positive step action of the ECU is added to the negative action of the ICU, the correlative addition result is the output of NGIC. Thus, the ICU action can weaken the ECU action through a right opposite step output, which can make the NGIC eliminate the control error with short rise time, small or no overshoot to a degree.

As discussed above, $u_{steady}(\infty)$, which is a steady guided output value and can make the control system become stable more quickly with small rise time, little overshoot, is a very important parameter of the NGIC. However, it is a realistic problem to pre-obtain $u_{steady}(\infty)$ of the NGIC. We will discuss this problem in Sect. 6.2.3.1.

(2) The way to deal with the pulse-output exceeded

As discussed in Sect. 6.2.2.2, the ECU needs output a great step action to improve the rise time, and the pulse-factor parameter μ should guarantee the control output $u_{peak}(t)$ between the low and high limitation. However, it is difficult to do so when the initial control output $u_{initial}(t)$ is very close to the low or high limitation. This means that sometime the output of the NGIC may exceed the output limitation. If we do not compensate the limited outputs, the control effectiveness of the NGIC will be seriously influenced. Therefore, we apply the following compensation method to solve this problem.

We assume that $u_{peak}(t)$ is the initial calculated peak value of the control output during the dynamic change process after the set-point changes. If the calculated peak value $u_{peak}(t)$ exceeds the output limitation, it may be greater than u_H or less than u_L. We denote t_{down} as the initial set time when the NGIC changes its output from the peak value $u_{peak}(t)$ to the opposite direction. In fact, the control output should be limited by the high or low limitation. We design a compensation method of equivalent area to compensate for the limited output. The detail compensation algorithm is introduced as follows.

Obviously, the cutoff or limit value from the peak value $u_{peak}(t)$ is either $u_{peak}(t) - u_H$ or $u_L - u_{peak}(t)$. The original limited output area S_0 between the control time t_0 and $t_0 + t_{down}$ is either $S_0 = t_{down} \cdot (u_{peak}(t) - u_H)$ or $S_0 = t_{down} \cdot (u_L - u_{peak}(t))$. When the control output changes from $u_{peak}(t)$ to the opposite direction, the actual compensated time is equal to $t'_{down} = t_{down} + \Delta t_{down}$, where Δt_{down} is the compensated change time. The compensated limited input area S_c is $\Delta t_{down} \cdot (u_H - u_{steady}(\infty))$ or $\Delta t_{down} \cdot (u_{steady}(\infty) - u_L)$. Therefore, according to the equivalent area principle, we can calculate the compensated change time Δt_{down}. Here we take the output limited by u_H as an example in Fig. 6.3 to explain the calculation process.

The compensated input area S_c should be equal to the initial input area S_0, i.e.:

$$t_{down} \cdot (u_{peak}(t) - u_H) = \Delta t_{down} \cdot (u_H - u_{steady}(\infty)) \tag{6.12}$$

We can further obtain the compensated change time through the following equation:

$$\Delta t_{down} = t_{down} \cdot (u_{peak}(t) - u_H)/(u_H - u_{steady}(\infty)) \tag{6.13}$$

Fig. 6.3 The compensated process of the time t_{down}

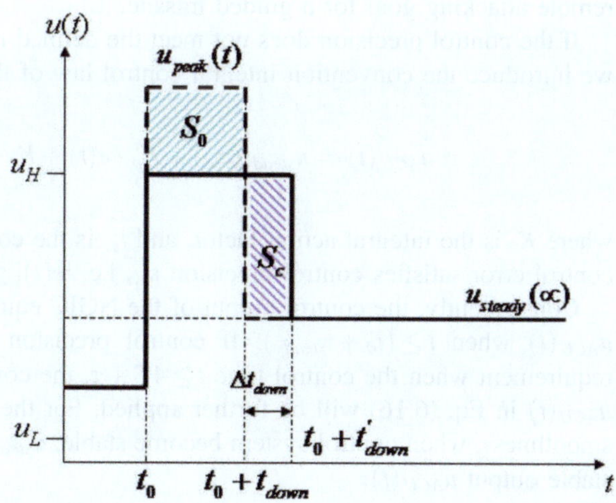

So we can get the actual compensated time t_{down} bellow:

$$t'_{down} = t_{down} + \Delta t_{down} = (1 + (u_{peak}(t) - u_H)/(u_H - u_{steady}(\infty))) \cdot t_{down} \quad (6.14)$$

(3) Design and principle of the ACU

(1) Control algorithm of the ACU

In fact, we cannot pre-obtain the ideal steady output $u_{steady}(\infty)$ of the NGIC. Therefore, if only the control laws of the ECU and the ICU are applied, the control precision of the NGIC will be affected seriously. When the control error reaches a relative steady status, we should further consider other advanced strategies to improve the control precision.

We define t_{steady} as the time when the control system reaches a pre-defined steady status that is the control error and its dynamic change value satisfies the following condition $|e(t)| \leq \varepsilon_1$, and $|\Delta e(t)| \leq \varepsilon_2$, where ε_1, ε_2 are positive factors, and $\varepsilon_2 < \varepsilon_1$. For a general first-order or second-order controlled model with time-delay, the process variable should be very close to the final steady process value when $t \geq 3T + \tau$. Hence, we further design the ACU to improve the control precision of the NGIC when $t \geq t_0 + t_{steady}$ or $t \geq 3T + \tau$. Inspired by the central regulating output of blood glucose, we devise a central proportional control law with an ultimate steady control output for the ACU as follows [22]:

$$u_{ACU}(t) = u_{steady}(\infty) + K_p \cdot e(t) \quad (6.15)$$

where K_p is the proportional action factor. From Eq. (6.15), we find that when the control error changes, the central output of the ACU always is $u_{steady}(\infty)$. That means when the control error is equal to zero, the control output of the ACU is equal to $u_{steady}(\infty)$, i.e. $u_{ACU}(t) = u_{steady}(\infty)$. In fact, $u_{steady}(\infty)$ is just like a remote attacking goal for a guided missile.

If the control precision does not meet the defined requirement when $t \geq 4T + \tau$, we introduce the convention integral control law of the ACU as follows:

$$u_{ACU}(t) = u_{steady}(\infty) + K_p \cdot e(t) + K_i \cdot \int_{4T+\tau}^{t_{ss}} e(t)dt \quad (6.16)$$

where K_i is the integral action factor, and t_{ss} is the control time when the absolute control error satisfies control precision ε_{ss}, i.e. $|e(t)| \leq \varepsilon_{ss}$.

Consequently, the control output of the NGIC equals to the output of the ACU $u_{ACU}(t)$ when $t \geq (t_0 + t_{steady})$. If control precision does not meet the defined requirement when the control time $t \geq 4T + \tau$, the control law with integral action $u_{ACU}(t)$ in Eq. (6.16) will be further applied. For the control patterns switch with smoothness, when control system become stable, $u_{steady}(\infty)$ is updated by the final stable output $u_{ACU}(t)$.

(2) The method of identifying the model parameters

As explained in Sect. 6.2.2.2, the model parameters such as the amplitude factor K, time constant T, and time delay τ are crucial for the NGIC. Among them, the amplitude factor K has close relationship with the control precision, and it can be calculated according to Eq. (6.6). The time constant T and time delay τ are used to determine the switch time for the ACU to work. Actually, approximated time constant T and time delay τ are enough for guaranteeing the control precision of the NGIC.

In this section, we apply the two-point step identification [31, 32] to identify the above model parameters. First, two points at time t_1, t_2 are identified from the sampled data of step response. The t_1, t_2 shall satisfy the following conditions: $y(t_1) = 0.384 \cdot y_{steady}(\infty)$ and $y(t_2) = 0.632 \cdot y_{steady}(\infty)$, where $y_{steady}(\infty)$ is the steady-state value of the process variable. Then, the model parameters T and τ of the controlled plant can be calculated according to the following equation:

$$\begin{cases} T = 2 \cdot (t_2 - t_1) \\ \tau = 2 \cdot t_1 - t_2 \end{cases} \tag{6.17}$$

(4) Cooperative process of the CCU

Corresponding to the management role of central nervous system, we design the cooperative law of the CCU. The CCU should coordinate all the working status of the ECU, the ICU, and the ACU to ensure that the NGIC has satisfactory control performances. With the coordinated instructions of the CCU, these four control units can cooperatively work with each other. The detailed cooperative process is explained as follows.

(1) The ECU first works when $t_0 \le t < (t_0 + t_{down})$. The output of the NGIC is the output of the ECU, i.e. $u_{NGIC}(t) = u_{ECU}(t)$, referred to Eq. (6.5).
(2) Both the ECU and the ICU come into effect when $(t_0 + t_{down}) \le t < (t_0 + t_{steady})$. The output of the NGIC is the summation of the output of the ECU and the ICU, i.e. $u_{NGIC}(t) = u_{ECU}(t) + u_{ICU}(t)$, referred to Eq. (6.11).
(3) When $t \ge (t_0 + t_{steady})$ or $t \ge (3T + \tau)$, only the ACU takes effect. The control output of the NGIC is the output of the ACU referred to Eq. (6.15). If the control precision does not meet the defined requirement when $t \ge 4T + \tau$, the control output of the NGIC shall refer to Eq. (6.16).

Thus, the ECU, the ICU, and the ACU can cooperatively work well under the coordinated instructions of the CCU. The work status and output laws of all the control units are shown in Table 6.1, where "×" means being not in work status, and "√" means being in work status. Their cooperative work process is shown in Fig. 6.4, from which we can find that the ECU works during the period of $t \in [t_0,\ t_0 + t_{steady})$, the ICU works during the period of $t \in [t_0 + t_{down},\ t_0 + t_{steady}]$, and the ACU works during the period of $t \in [t_0 + t_{steady},\ \infty]$. When control

Table 6.1 The work status of the ngic

Period	Status			Control law
	ECU	ICU	ACU	
$[t_0, t_0 + t_{down})$	√	×	×	Equation (6.5)
$[t_0 + t_{down}, t_0 + t_{steady})$	√	√	×	Equation (6.11)
$[t_0 + t_{steady}, \infty)$	×	×	√	Equation (6.15) or (6.16)

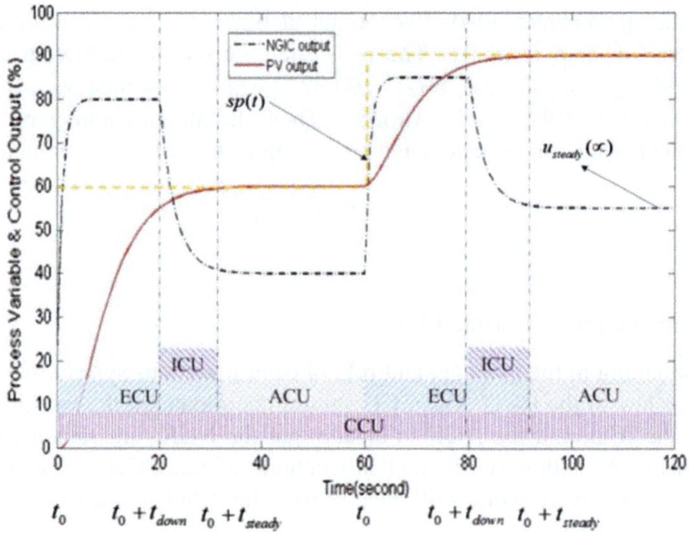

Fig. 6.4 The overall work process of the NGIC

set-point changes again, the control time t is reset to t_0. The NGIC will repeat control steps from step 1 to 3.

6.2.2.3 The Approach to Adjust the Control Parameters

As discussed in Sect. 6.2.2.2, the ECU and the ICU are responsible for regulating the process variable in dynamic change process, and the ACU is for that in steady change process. That is to say, the ECU, the ICU, and the ACU can independently adjust the process variable in the dynamic change process and steady change process. Hence, the parameters of the ECU, the ICU, and the ACU can be adjusted independently according to the different control effect in the dynamic or steady process.

The control parameters of the NGIC include μ, t_{down}, K_p, K_i, T_{ECU}, T_{ICU}. Among them, T_{ECU}, T_{ICU}, m and n only affect the filtering effect and have less influence on the control indices of dynamic change process. Hence, only μ, t_{down}, K_p, K_i are the key parameters of the NGIC. In fact, only μ, t_{down} affect the control indices of dynamic change process, and only K_p, K_i have influence on the control indices of steady change process. Therefore, the parameters μ and t_{down} are independent of the parameters K_p and K_i, which make it easy to tune the control parameters of the NGIC. The tuning methods of these parameters are explained as follows.

(1) The parameters μ and t_{down} are tuned according to the increasing speed and the overshoot amount of the process variable in the dynamic change process. If there is an excessive overshoot or an overtly rapid rise, parameters μ and t_{down} can be reduced. Otherwise, parameters μ and t_{down} shall be increased if there is an inadequate overshoot or an overtly slow rise.
(2) The proportional factor K_p and the integral factor K_i are tuned according to the changes of process variable in the steady change process. When tuning K_p, if the process variable approaches the steady state too slowly, K_p will be increased; if it is too fast or even oscillatory, K_p will be decreased. Similarly, when tuning K_i, if the control precision is poor, K_i will be increased; if there is a oscillation of steady-state, K_i will be decreased.
(3) Other control parameters T_{ECU}, T_{ICU}, m and n are set according to the general control condition within the defined specified range of values, i.e.$T_{ECU} = 0 \sim 5 \cdot T_s$, $T_{ICU} = 0 \sim 0.5 \cdot (T + \tau)$, where T_s is the control period. In addition, m and n are set from 1 to 5.

6.2.2.4 Stability Analysis of the NGIC Control System

According to the above analysis, we can get the integrated control algorithm of the NGIC as shown in Eq. (6.18):

$$u_{NGIC}(t) = \begin{cases} u_{initial}(t) + \mu \cdot (u_{steady}(\infty) - u_{initial}(t)) \cdot F_{ECU}(t), \\ \qquad t \in [t_0, t_0 + t_{down}) \\ u_{peak}(t) + (u_{steady}(\infty) - u_{peak}(t)) \cdot F_{ICU}(t), \\ \qquad t \in [t_0 + t_{down}, t_0 + t_{steady}) \\ u_{steady}(\infty) + K_p \cdot e(t), \\ \qquad t \in [t_0 + t_{steady}, \infty) \\ u_{steady}(\infty) + K_p \cdot e(t) + K_i \cdot \int_{4T+\tau}^{t_{ss}} e(t)dt, \\ \qquad t \in [4T + \tau, \infty) \end{cases} \tag{6.18}$$

According to the work principle of the NGIC, we know that the ECU and the ICU (refer to the two former equations in Eq. (6.18)) are only responsible for adjusting the process variable in the dynamic process, and the ACU (refer to the later two equations in Eq. (6.18)) is only for adjusting the steady process. According to the Lyapunov stability theory [41, 42], the stability of a control system is only determined by the control status or conditions when the control time $t \to \infty$. Hence, the ECU and the ICU have no relationship with the stability of the whole control system. Therefore, only the ACU affects the system stability. When the control time $t \to \infty$, the control output of the NGIC is

$$u_{NGIC}(t) = \begin{cases} u_{steady}(\infty) + K_p \cdot e(t), \ t \in [t_0 + t_{steady}, \infty) \\ u_{steady}(\infty) + K_p \cdot e(t) + \\ \qquad + K_i \cdot \int_{4T+\tau}^{t_{ss}} e(t)dt, \ t \in [4T + \tau, \infty) \end{cases} \tag{6.19}$$

If the control precision satisfies the defined requirements, only the first equation in Eq. (6.19) is applied to the control of dynamic change process. Otherwise, only the later equation in Eq. (6.19) is used. Obviously, only K_p or K_i may affect the system stability. Hence, it is very easy to guarantee the control system in stable state.

6.2.3 Experimental Results

To testify the control performance of the NGIC, we choose a bio-reactor plant as the controlled model. The bio-reactor is a typical nonlinear process with time-varied amplitude factor and time constant [43, 44], as shown in Fig. 6.5.

The bio-reactor plant is a bio-chemical fermentation process with strong nonlinear features. The alcoholic fermentation is one of the popular and important bio-chemical processes, since its product, the ethanol, could represent an alternative energy source used as a partial substitute for gasoline. In Fig. 6.5, C_s is glucose concentration; C_p is ethanol concentration; F_{ag} is the flow of cooling agent; T_C is the temperature controller; T_{in} is the temperature of input flow; T_r is the temperature of bio-reactor; T_e is the temperature of outflow; O_2 is oxygen; F_i is the input flow; and F_e is the exceeding flow.

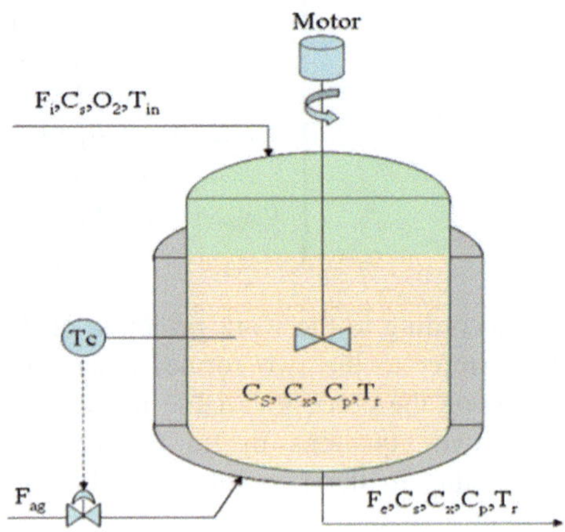

Fig. 6.5 The bio-reactor process

The mathematical model of bio-reactor is given as follows [43, 44].

$$
\begin{cases}
\dfrac{dC_x}{dt} = \mu_x C_x \dfrac{C_s}{K_s + C_s} e^{-K_p C_p} - \dfrac{F_e}{V} C_x \\
\dfrac{dC_p}{dt} = \mu_p C_x \dfrac{C_s}{K_{s1} + C_s} e^{-K_{p1} C_p} - \dfrac{F_e}{V} C_p \\
\dfrac{dC_s}{dt} = -\dfrac{1}{R_{sx}} \mu_x C_x \dfrac{C_s}{K_s + C_s} e^{-K_p C_p} \\
\qquad - \dfrac{1}{R_{sp}} \mu_p C_x \dfrac{C_s}{K_{s1} + C_s} e^{-K_{p1} C_p} + \dfrac{F_i}{V} C_{s,in} - \dfrac{F_e}{V} C_s \\
\dfrac{dC_{o_2}}{dt} = (k_{la})(C_{o_2}^* - C_{o_2}) - r_{o_2} \\
\dfrac{dT_r}{dt} = \dfrac{F_i}{V}(T_{in} + 273) - \dfrac{F_e}{V}(T_r + 273) \\
\qquad + \dfrac{r_{o_2} \Delta H_r}{32 \rho_r C_{heat,r}} + \dfrac{K_T A_T (T_r - T_{ag})}{V \rho_r C_{heat,r}} \\
\dfrac{dT_{ag}}{dt} = \dfrac{F_{ag}}{V_j}(T_{in,ag} - T_{ag}) + \dfrac{K_T A_T (T_r - T_{ag})}{V_j \rho_{ag} C_{heat,ag}}
\end{cases}
\tag{6.20}
$$

where, T_{ag} is the temperature of cooling agent in the jacket; C_{o_2} is the dissolved concentration of oxygen in liquid, $C_{o_2}^*$ is the equilibrium concentration of oxygen in liquid. Other parameters are as shown in Table 6.2. Moreover, all the parameters in Table 6.2 are just for this bio-reactor model, not for the practical control field.

For both the process models and practical control process, the application process of NGIC is similar. The first step is to identify process parameters, such as the amplitude factor K, time constant T, and time-delay τ by using different identification method. The next step is to design the NGIC according to the identified process parameters. Here, we carry out the step response several times in the range of control output. Furthermore, we use the two-point step identification, i.e. Equations (6.6) and (6.17) to identify the corresponding amplitude factor K, time constant T, and time-delay τ. The sampled response data and the identified model parameters are shown in Table 6.3, where T_r is the temperature of reactor. The parameters of practical control field can be identified according to the same method as above. The amplitude factor K is very important for the control precision, and the time constant T and time-delay τ can be used to calculate the time when the ACU takes effect.

By using the fitting tools in Matlab (Rev. 2009b), we get the relationship between $K(t)$ and $T_r(t)$:

Table 6.2 Model parameters of bio-reactor process [43]

No	Parameters	Meaning	Values
1	F_i	Input flow	$51l/h$
2	F_e	Output flow	$51l/h$
3	T_{in}	Input flow temperature	$25°C$
4	T_e	Output flow temperature	$25°C$
5	$T_{in,ag}$	Temperature of input cooling agent	$15°C$
6	$C_{s,in}$	Concentration of glucose of input follow	$60g/l$
7	k_{la}	Mass transfer coefficient for oxygen	$38 \cdot (1.024)^{T_r-20}$
8	r_{o_2}	Rate of oxygen consumption	$0.516 \cdot C_{xo_2} \cdot C_{o_2}/(8.86 + C_{o_2})$
9	V	Total volume of reaction medium	$V = 1000l$
10	μ_x	Maximum specific growth rate	h^{-1}
11	K_s	Constant in the substrate term for growth	$1.030g/l$
12	K_p	Constant of growth inhibition by ethanol	$0.139g/l$
13	K_{p1}	Constant of fermentation inhibition by ethanol	$0.070g/l$
14	R_{sx}	Ratio of cell produced per glucose consumed for growth	0.607
15	R_{sp}	Ratio of ethanol produced per glucose consumed for fermentation	0.435
16	μ_p	Maximum specific fermentation rate	$1.790h^{-1}$
17	K_{s1}	Constant in the substrate term for ethanol production	$1.680g/l$
18	ΔH_r	Reaction heat of fermentation	$518kl/molO_2$
19	ρ_r	Density of the mass of reaction	$1080g/l$
20	$C_{heat,r}$	Heat capacity of mass of reaction	$4.18Jg^{-1}K^{-1}$
21	K_T	Heat transfer coefficient	$3.6 \times 105Jh^{-1}m^{-2}K^{-1}$
22	A_T	Heat transfer area	$1m^2$
23	V_j	Volume of the jacket	$50l$

$$K(t) = -25.68 * T_r(t)^{0.7398} + 243.6 \tag{6.21}$$

Moreover, the relationship between the real-time time constant $T(t)$ and $T_r(t)$ is as follows:

$$T(t) = -748.9 * T_r(t)^{-0.12} + 529.9 \tag{6.22}$$

If this model is controlled by conventional control schedule, the control effectiveness is always not satisfactory or even divergent. In our simulation experiments, we first compare the control effect of the NGIC with that of conventional PID

Table 6.3 The identified model parameters

$u(100\%)$	$T_r(°C)$	K	$T(s)$	$\tau(s)$
0.0	36.0296			
0.01	34.9017	−111.5480	40	−1
0.05	31.6498	−87.3479	36	−2
0.1	29.2065	−68.1064	30	−1
0.2	26.5961	−47.1055	24	0
0.3	25.2288	−35.9612	22	−1
0.4	24.3885	−29.0717	20	−1
0.5	23.8199	−24.3947	18	0
0.6	23.4095	−21.0128	16	1
0.7	23.0994	−18.4540	16	0
0.8	22.8568	−16.4505	16	0
0.9	22.6619	−14.8393	14	1
1.0	22.5018	−13.5154	14	1

controller, internal model controller (IMC) and artificial neural network (ANN) controller. To make the comparison results more clearly, we optimize all the control parameters in the NGIC, the PID, the IMC and the ANN controllers by using the same optimization software. The IMC is designed according to the identified model as shown in Eqs. (6.21) and (6.22). Moreover, the ANN controller includes three layers, i.e. input layer, middle layer and output layer. Among them, the input layer consists of control error, the change rate of control error and the accumulation of the control error of three control periods, and the output layer only comprises the control output. We choose the GATOOL in Matlab as optimization software. Therein, the individual number of initial populations is 50, the maximum search generation is 500, the low limitations of all searching space are zero, the initial searching space is [0,1], and the other parameters are default values. The optimized parameters are as follows,

The optimized parameters of the PID are $K_p = 1.3330$, $K_i = 0.0892$, $K_d = 0.3051$.

(1) The optimized parameters of the ANN controller are as follows. The neuron number of the hidden layer is 13 and the learning efficiency factor of the ANN controller is 0.2549.
(2) The optimized control parameter of the IMC is 0.6028.
(3) The optimized parameters of the NGIC are $K_p = 6.6934$, $K_i = 0.6043$, $u = 15.9293$, $t_{down} = 0.1042$, $m = 3$, $n = 1$.

In the experiments, we first change the set-point at 20 h from 29.82 to 33, then change it from 33 to 30 at time 120 h, and next change it from 30 to 28 at time 220 h, and finally change it from 28 to 29 at time 320 h. To make the comparison of these control methods more clearly, we intercept only part of the results, of which the time is from the beginning to the 300 h.

6.2.3.1 Basic Control Ability

The compared control effectiveness of model (6.20) is as shown in Figs. 6.6 and 6.7. Figure 6.6 is the comparison of the output of process variables between the NGIC and the PID, the IMC and the ANN controller, and Fig. 6.7 is that of control output among them.

(1) The comparisons of control performance index. From Fig. 6.6, we can find that when we first change set-point from 29.8 °C to 33 °C, the overshoot, settling time, and control precision of the NGIC are much better than those of the PID, the IMC and the ANN controller. From Fig. 6.7, we can also find that the output of the NGIC is much smoother than that of others.
(2) The comparisons of control robustness. When we change the set-points within the full range of process variable, first from 33 °C to 30 °C, and then change it

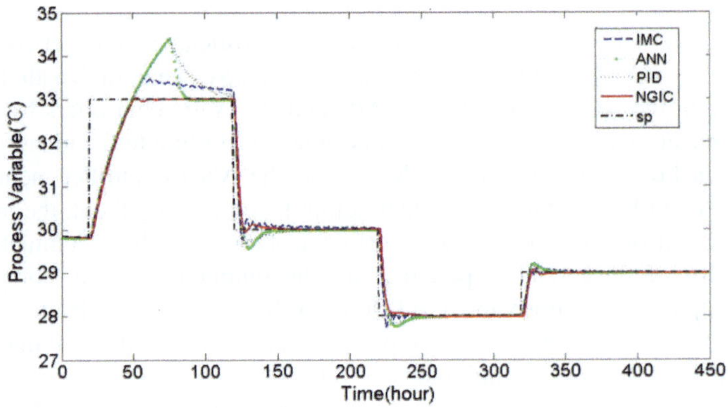

Fig. 6.6 The comparison of process variable

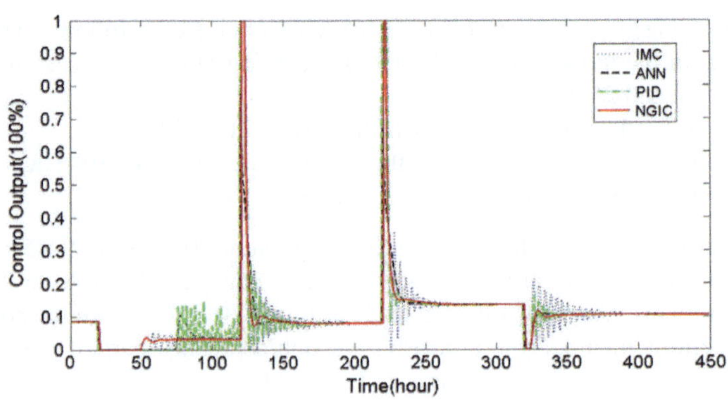

Fig. 6.7 The comparison of control output

Table 6.4 Control indexes of controllers

Control indexes	PID (h)	IMC (h)	ANN (h)	NGIC (h)
t_{r1}	29.8	29.6	29.5	29.5
$\sigma\%_1$	4.3	1.45	4.3	0.11
t_{s1}	100	Non	71	41
t_{r2}	5.9	5.7	5.8	6.2
$\sigma\%_2$	1.24	0.94	1.3	0.12
t_{s2}	34	31.5	24.2	24.1
t_{r3}	25.5	25.3	25.2	25.8
$\sigma\%_3$	0.69	0.48	0.68	0.21
t_{s3}	27.1	28.9	26.2	19.5

again from 30 °C to 28 °C, the amplitude factor $K(t)$ changes from about −110 to −64, and the time constant $T(t)$ changes from about 40 to 24. The detailed control indexes of the controllers are as shown in Table 6.4. Where, t_{rn}, $\sigma\%_n$, and $t_{sn}(n = 1, 2, 3)$ are rise time, overshoot and steady time, respectively. From Fig. 6.6 and Table 6.4, we can find that the control robustness of the NGIC is superior to that of the PID, the IMC and the ANN controller. As the NGIC can keep satisfactory control effectiveness at different set-points with so great changes of the model parameters. However, the other controllers control the process plant with greater overshoot and longer settling time.

Hence, the basic control ability of the NGIC is much better than those of the PID, the IMC and the ANN controller.

6.2.3.2 Noise Control

The NGIC can distinguish the set-point change error from the noise error, according to the change status of the set-point value. If the set-point does not change, the NGIC only output the definition control output even some noise may exit in the control system. Therefore, the NGIC has better control performance to restrain or eliminate the noise error. In order to testify the ability of restraining disturbance of the NGIC, we inject white random noise into the process plant. The amplitude of the white noise is from −0.15 to 0.15, and its mean value is zero. The original optimized parameters of the NGIC, the PID, the IMC and the ANN controllers keep unchanged. Figures 6.8 and 6.9 are the comparison of process variables and that of control outputs with noise, respectively (Fig. 6.9).

From, we can find that when the same noise injects to these different control systems, the control effect of the NGIC is better than that of the others. As the control effectiveness of the others is more oscillated. From Fig. 6.8, we can find that the control output of the NGIC system is more stable than that of other control systems. We can also find that the output of the NGIC can keep stable with no influence from the noise before the central proportional or integral control law in the

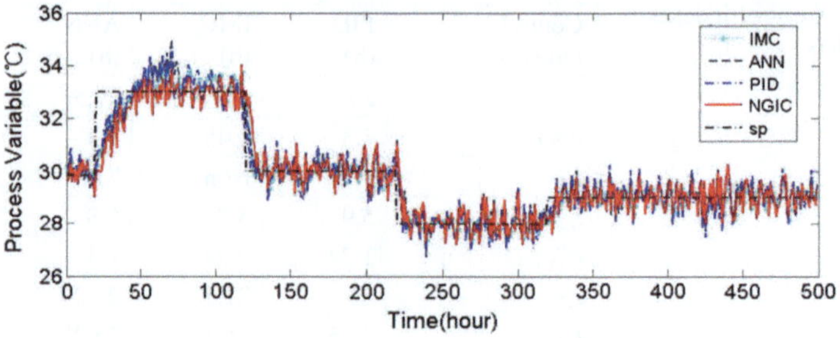

Fig. 6.8 Comparison of control effect with noise

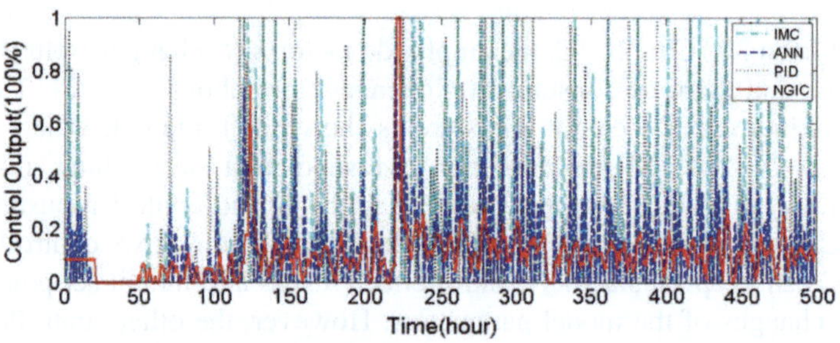

Fig. 6.9 Comparison of control output with noise

NGIC is applied. Hence, the ability of restraining noise of the NGIC is much better than that of the others.

Thus, from the experimental results we can conclude that the control performance of the NGIC with more simple structure is much better than that of the PID, the IMC and the ANN controller.

6.3 An Intelligent Controller Inspired from Neuroendocrine-Immune System

Based on the regulation mechanism of NEI system, we present a novel nonlinear optimized controller (NOIC) in this section. The NOIC is composed of a presentation unit (PU), an antibody control unit (ACU), main control unit (MCU), optimization unit (OU), and identification unit (IU). The ACU is composed of several antibody controller entities (ACEs). According to the real-time control error, the PU first pretreats control error dynamically based on the mechanism of immune presentation, and the ACU can adjust the number of ACE to eliminate control error. MCU is designed to regulate the control action of PU and ACU. And the OU and IU can optimize the parameters of the PU and ACU based on the genetic algorithm

(GA). Thus the control performance can be improved. Virtually, the NOIC is a complicated nonlinear proportional action, which can eliminate control error without integral action.

6.3.1 Modulation Mechanism of Neuroendocrine-Immune System

In this section, we just consider the influence of the nervous system on the endocrine and immune system, and the influence of endocrine system on immune system.

To defend the body against the foreign invaded antigen is the primary function of the immune system. When antigens invade body, antigen-presenting cells (APC) (i.e. Macrophages) ingest and partially digest antigens and then present the antigen on their surfaces of the antibodies (B and T lymphocytes). This process is called antigen presentation. The presented antigen is more immunogenic than unprocessed antigen. In other word, the antibodies would eliminate the presented antigen more quickly. At the beginning, there may be very lots of antibodies to eliminate the antigens. Along with the decrease of the number of the invading antigens, the corroding antibody may also decrease. The statement of immune system can be feedback to nervous system and endocrine system via cytokines. Then both the nervous system and the endocrine system can simulate or inhibit the process of eliminating the antigens via neurotransmitters and hormones, respectively. Furthermore, nervous system also modulates the endocrine system via neurotransmitters. As thus, the antigens are eliminated rapidly.

Corresponding to complex control system, nervous system, endocrine system, and immune system may be regarded as optimization level, coordinator level, and executive level, respectively. Then we can abstract this mechanism to the special modulation architecture as shown in Fig. 6.10 [2].

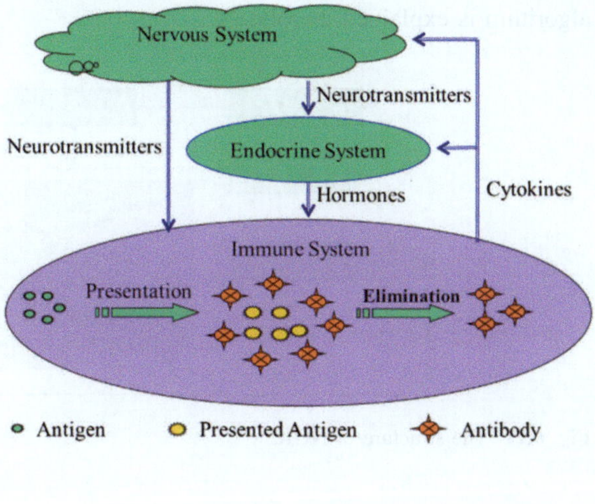

Fig. 6.10 The regulation mechanism of NEI system

6.3.2 *Design of the NOIC*

Inspired the modulation mechanism of NEI, we design a nonlinear optimized controller (NOIC). In this section, we will describe the design and implementation process of the NOIC in detail. First, the structure of the NOIC is introduced. Then the control algorithms of PU, ACU, and MCU are described, respectively.

6.3.2.1 Structure of NOIC

According to the Fig. 6.10 [2], we design a nonlinear optimized intelligent controller (NOIC) as shown in Fig. 6.11, which is composed of a presentation unit (PU), an antibody control unit (ACU), a main control unit (MCU), an optimized unit (OU), and an identification unit (IU). The ACU is composed of several antibody controller entities (ACEs). The OU and IU are mapped to nervous system. The MCU is mapped to endocrine system, and PU and ACU are mapped to immune system.

In this section, control error $e(t)$ is regarded as antigen. Control error $e(t)$ is first pretreated in PU, and the presented error is denoted as $E(t)$. Then $E(t)$ is controlled by ACU. The MCU adjusts the control parameters of PU and ACU. The IU is used to identify the model of the plant. And the OU optimizes all the parameters of MCU.

6.3.2.2 Control Algorithm of NOIC

(1) Presentation Unit

Based on the presentation mechanism of immune system mentioned in Sect. 6.3.1, we design a processed algorithm for the control error. Being processed using this method, the control error may be eliminated stable and rapidly. This algorithm is explained as follows.

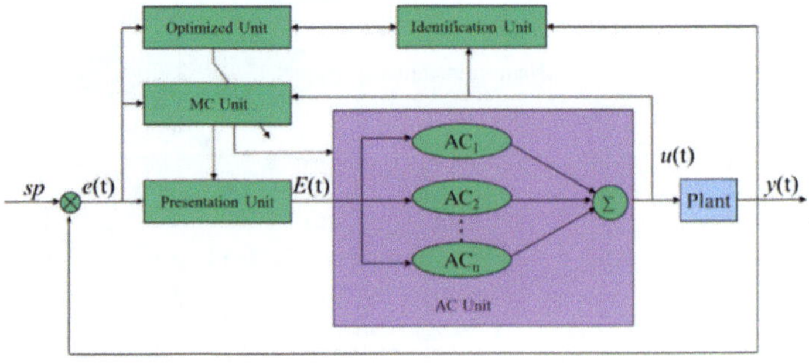

Fig. 6.11 The structure of NOIC

Without integral action, control error is difficult to eliminate when it is very little. From the dynamical changing output equation of proportional action $\Delta u(k) = K_p(e(k) - e(k-1))$, we can find that the reason the proportional action can not eliminate the little control error is the term of $e(k) - e(k-1)$ is nearly equal to zero when control error is very little. Aim to solve this problem, the presentation algorithm is designed. The presentation process is that the control error is magnified while the absolute value of control error is less than a threshold ε_1; in contrast, it may remain the original value. The magnified rule is as follows:

$$\begin{cases} E(t) = e(t) \cdot F(e(t)) & \\ F(e(t)) > 1 & |e(t)| < \varepsilon \\ F(e(t)) = 1 & |e(t)| \geq \varepsilon \end{cases} \tag{6.23}$$

where $F(e(t))$ is a $e(t)$ function, determined by MCU; and $E(t)$ is the presented control error. Obviously, the bigger is threshold ε, the more sensitive is the PU.

(2) Antibody control unit

According to the eliminating antigen mechanism of immune system, the number of ACE should be changed along with the real-time control error. The output of ACU is the summary of all the outputs of ACE,as shown in Eq. (6.24):

$$u(t) = \sum_{i=1}^{n} K_{pi} \cdot E(t), \tag{6.24}$$

where $E(t)$ is the presented control error, K_{pi} is proportional action gain of ACE_i, and n is the total number of the ACE in ACU. In this paper, proportional parameters of the all ACEs in ACU are same. So we have

$$u(t) = n \cdot K_p \cdot E(t). \tag{6.25}$$

Actually, it is more practical that n is positive real number, instead of being positive integer. Hence, we select n as real number, and n is determined by MCU.

(3) Main control unit

Mapped to endocrine system, the MCU is designed to regulate the control action of PU and ACU. The parameters of PU and ACU may be regarded as the hormone of endocrine system.

The MCU can modulate the presentation degree of PU via the presentation function $F(e(t))$, which modulates the sensitivity of PU. The presentation function $F(e(t))$ is designed as follows,

$$F(e(t)) = \begin{cases} \frac{\alpha \cdot}{|\exp(|e(t)| - \beta)|} & |e(t)| < \varepsilon \\ 1 & |e(t)| \geq \varepsilon \end{cases} \tag{6.26}$$

where $\alpha, \beta, \varepsilon$ are all positive factors, the ranges of β and ε are all from 0.0 to 1.0. Furthermore, the selection of the ranges of $\alpha, \beta, \varepsilon$ should ensure $F(e(t))$ larger than 1.0.

MCU can strengthen or weaken the control action of ACU through regulating the number of ACE. According to the control theory, the control action should be enhanced when the control error is big, vice versa. The adjusting algorithm is designed as follows,

$$n = \lambda \cdot |e(t)| \tag{6.27}$$

where λ is positive factors and larger than 1.0, and the selection of λ should ensure n more than 1.0.

Hence, the general control algorithm of NOIC is

$$\begin{cases} u(t) = n \cdot K_p \cdot E(t) \\ E(t) = \begin{cases} \frac{e(t)\cdot\alpha}{|\exp(|e(t)|)-\beta|} & |e(t)| < \varepsilon \\ e(t) & |e(t)| \geq \varepsilon \end{cases} \\ n = \lambda \cdot |e(t)| \end{cases} \tag{6.28}$$

6.3.2.3 Optimization of Control Parameters

The nervous system may regulate the endocrine system and immune system in the general [26, 27]. It can regulate the endocrine system and immune system harmoniously to make the body eliminate the antigen quickly and stable. According to this mechanism, the OU and IU are designed. The IU is used to identify the mathematical model of plant online. The OU may optimize the presentation degree and the control action of ACU, i.e. enhancing or weakening the control action, via detecting the control error, the process value, and the output of the ACU. All parameters in the NOIC are optimized in the limited space by adopting GA [32]. The optimization steps are as follows.

(1) **Individual encoding.** We define the assembly of all the parameters to be optimized in the NOIC as the individual. Every parameter is encoded in a real number.
(2) **Initial individual population.** To accelerate the simulation, we determine the search interval of the initial PID control parameters based on the Z-N method or experiment. Then the individual population is produced in the GA.
(3) **Identification.** The mathematical model can be obtained offline. But, in order to be more practical, the mathematical model must be corrected on line. There are some methods to correct the model. For example, the neural networks, fuzzy, the least square method, and so on [33, 45]. In this section, we just use the least square method, since it is easily implemented.

(4) **Fitness function definition.** We define the fitness function according to the requirement of control performance. First, we select the integral of the absolute value of control error $e(t)$ as one part of the objective function. To prevent the dynamic control output from being too great, the square term of the control output is added in the objective function. Furthermore, the overshoot δ and adjust time t_s are also added to ensure the stability and rapidity of the control system. Thus, we get the objective function as follows,

$$J = \int_0^\infty (g_1|e(t)| + g_2u^2(t))dt + g_3t_s + g_4\delta \tag{6.29}$$

where, g_1, g_2, g_3, and g_4 are weights, which are determined according to various requirement of control performance. Then, we define the fitness function as Eq. (6.30). The optimization process is shown in Fig. 6.12 [2].

$$f = 1/\{\int_0^\infty (g_1|e(t)| + g_2u^2(t))dt + g_3t_s + g_4\delta\} \tag{6.30}$$

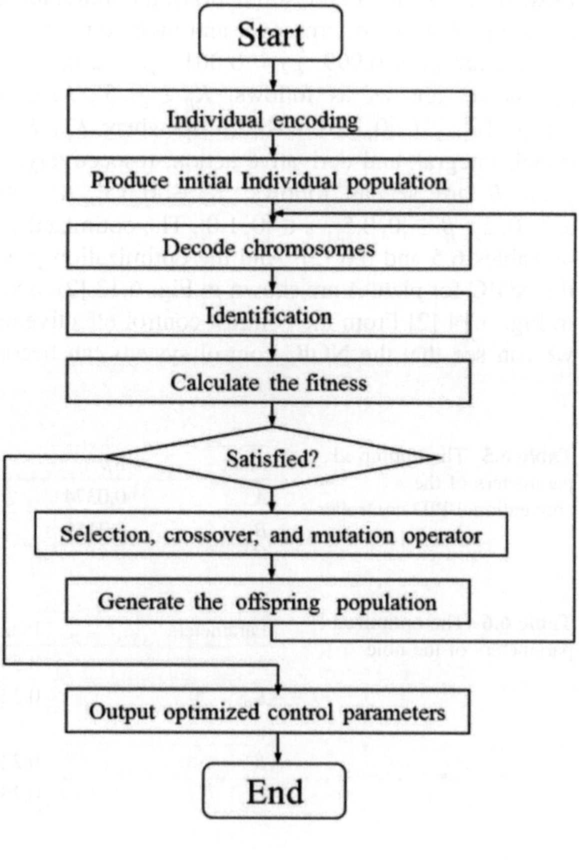

Fig. 6.12 The optimization process

6.3.3 Simulation Results

In order to examine the NOIC's control performance, we choose plant *A* and *B* for the simulations. Plant *A* is a first-order plant with large delay and $\tau/T \gg 1$ in large load control system, whose transfer function is:

$$G(s) = \frac{2}{60s+1}e^{-100} \tag{6.31}$$

As well known, plant *A* is very difficulty to be controlled using conventional control algorithm. B is a nonlinear plant, whose discretization equation. (Ts = 0.5 s) is:

$$y(k) = 0.8y(k-1) - 0.15(y(k-2))^2 + 0.85u(k), \tag{6.32}$$

where *u*(*k*), *y*(*k*) are input and output variable, respectively. We also compare its control performance with that of the conventional PID controller. To make the contrast effectiveness more clearly, the control parameters of the conventional PID controller are also optimized by adopting the GA and with the same conditions as those of the NOIC. In the simulations, the individual number of every population is 50, the probability of crossover and mutation is 0.9 and 0.05, respectively, and the weights are $g_1 = 0.999,\ g_2 = 0.001,\ g_3 = 2.0,\ g_4 = 100$. For plant *A*, the control parameters are set as follows: $K_p \in [0,5],\ K_i \in [0,10],\ K_d \in [0,5],\ \lambda \in [1,10]$, $\alpha \in [0,10],\ \beta \in [0,1.0],\ \varepsilon \in [0,1.0]$, where K_p, K_i, K_d, are the gains of proportional, integral, and derivative action, respectively. And the control parameters of plant B are set as follows: $K_p \in [0,15],\ K_i \in [0,10],\ K_d \in [0,5],\ \lambda \in [1,20]$, $\alpha \in [0,2],\ \beta \in [0,0.5],\ \varepsilon \in [0,1.0]$. The optimized control parameters are as shown in Tables 6.5 and 6.6 [2]. And the optimization processes of all the parameters in the NOIC for plant *A* are shown in Fig. 6.13 [2], and those of plant *B* are as shown in Fig. 6.14 [2] From the contrast control effectiveness in Figs. 6.15, and 6.16 [2], we can see that the NOIC control system can become stable faster and with less

Table 6.5 The optimized parameters of the conventional PID controller

Plant	*Kp*	*Ki*	*Kd*
A	0.0374	0.1043	4.9271
B	0.3183	0.1037	0.2559

Table 6.6 The optimized parameters of the noic

Parameters	Plant *A*	Plant *B*
λ	8.5697	4.5534
K_p	0.3235	0.3615
α	1.0624	3.8855
β	0.2437	0.0591
ε	0.3571	0.4070

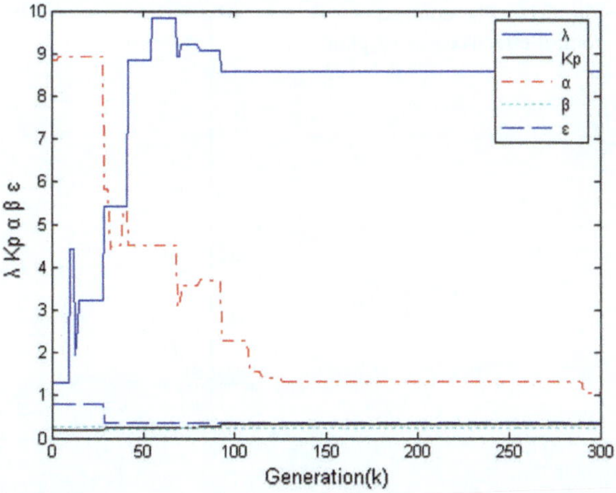

Fig. 6.13 The parameter optimization process of Plant A for NOIC

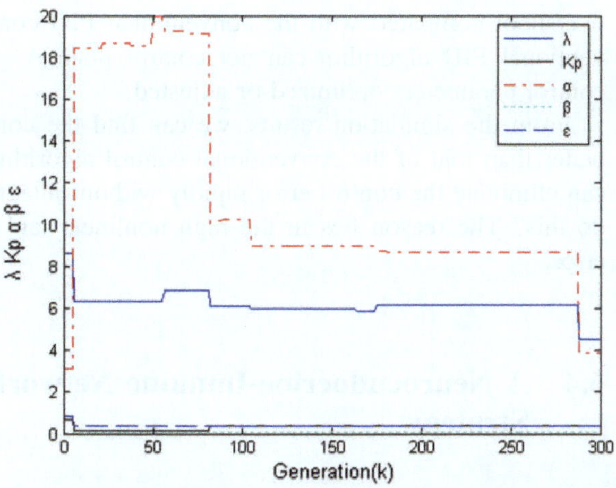

Fig. 6.14 The parameter optimization process of plant B for NOIC

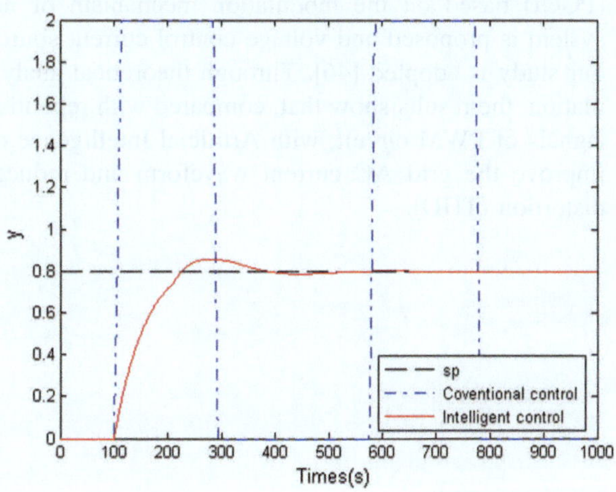

Fig. 6.15 The contrast control effectiveness of Plant A

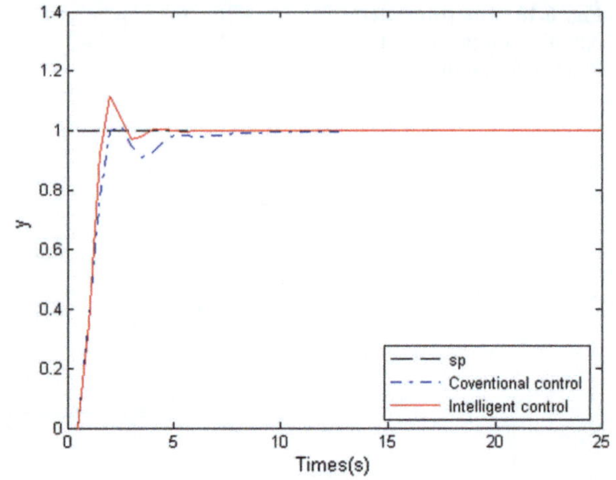

Fig. 6.16 The contrast control effectiveness of plant B

overshoot compared with the conventional PID controller. Furthermore, the conventional PID algorithm can not control plant *A* stably, no matter how are the control parameters optimized or adjusted.

From the simulation results, we can find the control performance of NOIC is better than that of the conventional control algorithm. Also, can found the NOIC can eliminate the control error rapidly without integral action. Why the NOIC can do this? The reason lies in the high nonlinear and right presentation for control error.

6.4 A Neuroendocrine-Immune Network Based Control Strategy

In this section, a new control method for photovoltaic grid-connected inverter (PGCI) based on the modulation mechanism of neuroendocrine-immune (NEI) system is proposed and voltage control current source (VCCS) inverting circuit in our study is adopted [46]. Through theoretical analysis and Matlab/Simulink simulation, the results show that, compared with repetitive PI controller, the controlled signals of PWM circuit, with Artificial Intelligence methods, can more effectively improve the grid AC current waveform and reduce grid current total harmonic distortion (THD).

6.4.1 The Model of Photovoltaic Grid-Connected System

Photovoltaic grid-connected system (PGCS) converts solar energy into DC electrical power by solar photovoltaic array, and then the photovoltaic inverter converts the DC electrical power into AC current with the same frequency and phase. AC electrical power can not only supply to the load in an electrical system, but also to power grid. PGCI is a core component of the system. At present, the infeed circuit of photovoltaic power inverter has two primary ways: the current source and the voltage source.

The direct current mode of inverting circuit is used widely in PGCS, such as, Hysteresis current controller, two loops control of voltage and current, space vector PWM (SVPWM) controller, deadbeat control algorithm with repetitive controller, and predictive control algorithms [47–50]. In recent years, some intelligent control methods, such as fuzzy control, neural networks, and genetic algorithms [51–53] have been applied widely in PGCI. However, how to improve the performance of these intelligent controllers has become a more urgent problem to be solved for PGCI.

A DC/AC stage convert of 3KW single-phase PGCS with an H full-bridge inverted circuit and VCCS current tracking inverter circuit is shown in Fig. 6.17. Where U_{dc} is the DC voltage of DC/DC boost topology circuit, C_{dc} is the stabilized voltage capacitor between the DC/DC stage convert and the DC/AC stage convert, L_{ac} is the AC current filter inductance, I_{ac} is the output current of inverter, R_L is the total resistance of inductance and circuit wires and U_{ab} is the output voltage of inverter. We can control the PWM signal of the H full-bridge inverted circuit and make the grid current and voltage with the phase and frequency.

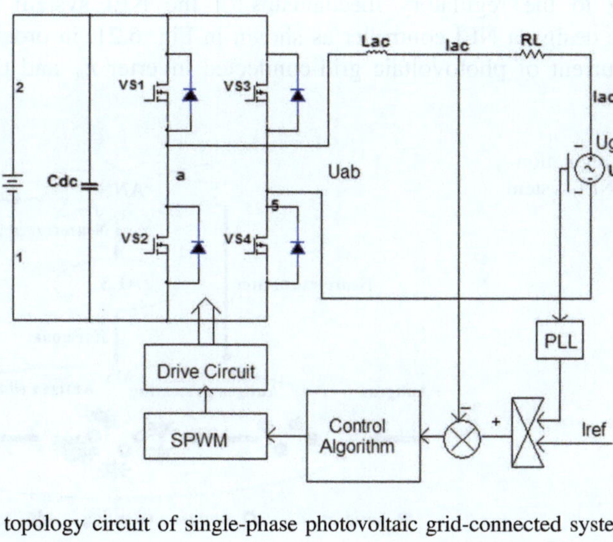

Fig. 6.17 The topology circuit of single-phase photovoltaic grid-connected system

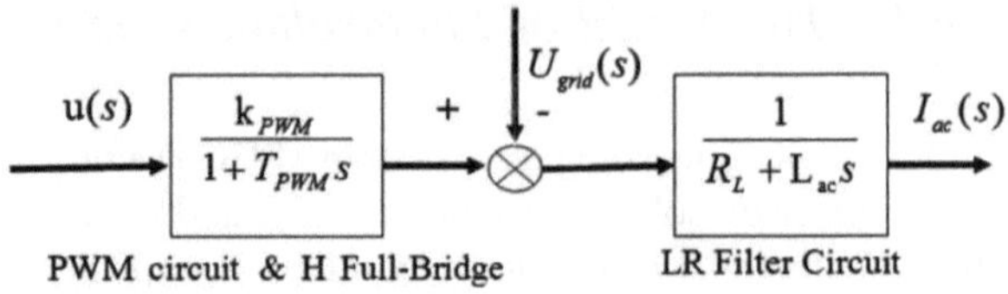

Fig. 6.18 The mathematical model of single-phase photovoltaic grid-connected system

From Fig. 6.17, we can obtain the mathematical model of single-phase PGCS by means of the voltage balance equation theory and the Laplace transform [54, 55], as shown in Fig. 6.18. Where, k_{pwm} is the voltage gain for the inverter circuit, and T_{PWM} is the cycle of triangular wave in PWM circuit.

6.4.2 The Design of Neuroendocrine-Immune Controller

The NEI system includes three major sub-systems, neural network, immune system, and endocrine system. As shown in Fig. 6.19, the two-way regulatory mechanism of these three sub-systems makes various physiological indicators in a highly stable by the carrier of cytokines, hormones and neurotransmitter [56].

Hormone regulating mechanism is a kind of closed-loop negative feedback mechanism [57]. First, the neural tube system promotes hypothalamus to release TH, and then TH promotes pituitary to release RH. As shown in Fig. 6.20, TH and RH correspond to corticotrophin-releasing hormone (CRH) and adrenocorticotrophic hormone (ACTH), respectively.

According to the regulatory mechanisms of the NEI system as shown in Fig. 6.19, we design a NEI controller as shown in Fig. 6.21, in order to make the output AC current of photovoltaic grid-connected inverter I_{ac} and the voltage of

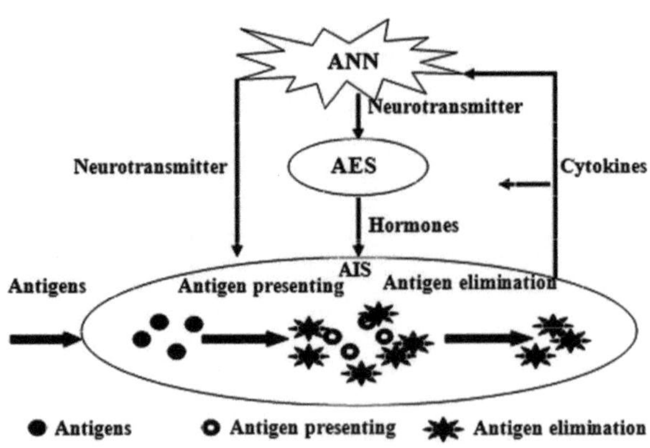

Fig. 6.19 The regulation mechanism of NEI system

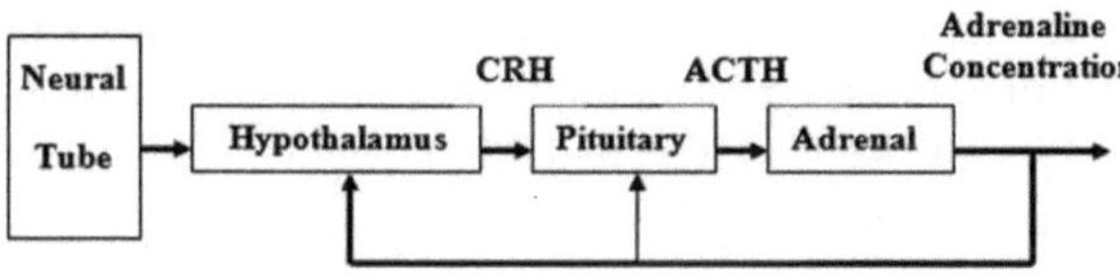

Fig. 6.20 The regulation loop of adrenal system

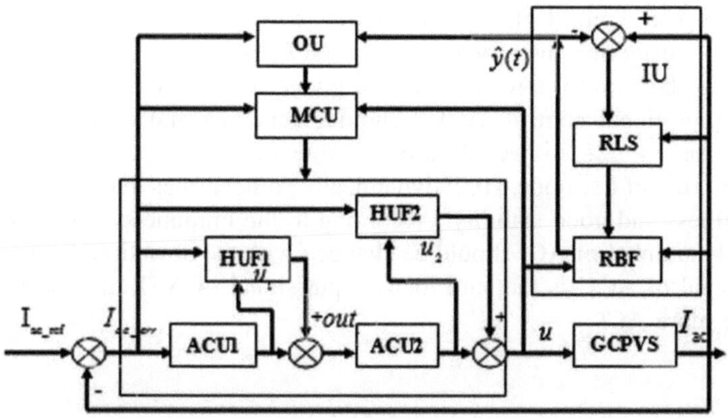

Fig. 6.21 The structure diagram of NEI controller

power grid U_{grid} have the same phase and frequency. In the NEI control system, the deviation current I_{ac_err} between targets AC current I_{ac_ref} and I_{ac} is taken as an antigen.

The controller includes two hormone feedback units (HFU1, HFU2), two antibody control units (ACU1, ACU2), one main controller unit (MCU), one optimization unit (OU), and one identification unit (IU) and other components. The ACU is constituted with several ACE entities. OU and IU, which are corresponding to the neural tube system, achieve the aim of optimization design for the structural parameters of MCU, HFU and ACU.

MCU and HFU, which are corresponding to the endocrine system, achieve the aim of regulating the concentration and the number of antibodies. ACU, which is corresponding to the immune system, achieves the aim of elimination of antigen.

Control error I_{ac_err} is defined as the antigen. Firstly, the num of antibodies is controlled by ACU1, and obtains a hormone concentration TH u_1, which is also affected by the target hormone I_{ac} through the feedback regulation. Therefore, HFU1 affects u_1, so out_1 is the actual released value of hormone concentrations of TH. Similarly, ACU2 and HFU2 are affected both by a hormone TH and by target hormone I_{ac} and get the actual released value of hormone concentrations of RH out_2, that is u too. ACU is used to adjust the number of Antibody Control Individual (ACI) and eliminate antigens, while MCU is mainly used to regulate or coordinate the structural parameters of HFU and ACU, which plays the key role of integrated control strategy. IU achieves the aim of identification the mathematical model of PGCS online. OU mainly based on immune algorithm (IA) realizes optimizing the structural parameters of ACU, HUF and MCU, which can greatly improve the control performance of NEI controller.

Generally, in a photovoltaic grid-connected system, good quality of inverter output waveform, fast dynamic response, and strong anti-interference are required. And in a cascade control system (CCS), the output of main controller is the given input of the deputy controller. Because of the rapidly adjust capacity of vice control loop, CCS can control the main control error variable quickly and steady. Therefore, considering the structure of the NEI controller, ACU1 should dynamically change the control error as a part of given input of ACU2, so we choose the role of proportional control. ACU2 should work fast and steady elimination of control error I_{ac_err}, so we choose the role of PI control. According to the adjustment rule of hormone, HUF dynamically compensates the deviation of output with swiftness and good stability. According to the immune system of eliminating antigen, the number of ACI should be changed with the deviation dynamically. The output signal of ACU is the sum total output signal of ACI and the greater deviation, the more ACI.

6.4.2.1 The Antibody Control Unit ACU1

The output of ACU1 is changed with the control error I_{ac_err} dynamically. According to proportional control algorithm, the output expression of every $ACI1_i$ is

$$u_{1i}(k) = Kp_{1i}(I_{ac\_err}(k)) \tag{6.33}$$

where k the serial number of sampling points is, $u_{1i}(k)$ is the out output signal of $ACE1_i, I_{ac\_err}$ is the control error and Kp_{1i} is the proportional gain of $ACE1_i$. So the expression of ACU1 is

$$u_1(k) = \sum_{i=1}^{n_1} Kp_{1i}(I_{ac\_err}(k))(n_1 \in R^+) \tag{6.34}$$

where, n_1 is the number of ACI in the ACU1.

If all Kp_{1i} equal, then

$$u_1(k) = n_1 Kp_1(I_{ac\_err}(k))(n_1 \in R^+) \tag{6.35}$$

6.4.2.2 The Antibody Control Unit ACU2

According to PI control algorithm, the output expression of every $ACI2_j$ is

$$u_{2j}(k) = Kp_{2j}(out_1(k)) + Ki_j \sum_{j=1}^{k} out_1(j) \tag{6.36}$$

If all proportional gains of Kp_{2j} equal, as well as all integral gain Ki_j, then the output expression of ACU2 is

$$u_2(k) = n_2(Kp_2(out_1(k)) + Ki \sum_{j=1}^{k} out_1(k)(n_2 \in R^+) \tag{6.37}$$

where, n_2 is the number of $ACI2_j$ in the ACU2.

6.4.2.3 The Hormone Feedback Unit HFU1

Farhy has proposed a general theoretic model of glandular secretion of hormones. Hormone change is a monotonic and non-negative regulation function, and the rise or fall follows the laws of Hill [19]. According to the law regulating hormone secretion, the change rate of output signal of ACU1 $\Delta u_1(k)$ is the input signal of HFU1. At the same time, the sign of the control error signal I_{ac_err} determines the direction of hormones rise or fall. The expression of algorithms as follows:

$$\begin{cases} \Delta u_1(k) = n_1 Kp_1(I_{ac\_err}(k)) - I_{ac\_err}(k-1) \\ a = \begin{cases} +1, I_{ac\_err}(k) \geq 0 \\ -1, I_{ac\_err}(k) < 0 \end{cases} \\ F_1(\Delta u_1(k), a) = a \frac{(|\Delta u_1(k)|)^m}{\lambda + (|\Delta u_1(k)|)^m} \end{cases} \tag{6.38}$$

where λ and m determine the rate of hormone feedback regulation and a determines the rise or fall of hormones. When the target hormone secretion I_{ac} is insufficient, HFU1 stimulates the secretion of hormone concentration TH, otherwise, it inhibits the secretion.

Finally, the output expression of HFU1 is

$$out_1(k) = u_1(k) + F_1(\Delta u_1(k), a) \tag{6.39}$$

6.4.2.4 The Hormone Feedback Unit HFU2

Similarly, the change rate of output signal of ACU2, $\Delta u_2(k)$ is the input signal of HFU2. At the same time, the sign of the control error signal I_{ac_err} determines the direction of hormones rise or fall. The expression of algorithms is as follows:

$$\begin{cases} \Delta u_2(k) = n_2(Kp_2(out_1(k) - out_1(k-1)) + Kiout_1(k)) \\ a = \begin{cases} +1, I_{ac\_err}(k) \geq 0 \\ -1, I_{ac\_err}(k) < 0 \end{cases} \\ F_2(\Delta u_2(k), a) = a\frac{(|\Delta u_2(k)|)^m}{\lambda + (|\Delta u_2(k)|)^m} \end{cases} \tag{6.40}$$

At last, the output expression of HFU2 is

$$u(k) = u_2(k) + F_2(\Delta u_2(k), a) \tag{6.41}$$

6.4.2.5 The Main Controller Unit MCU

Corresponding to the endocrine system, the main control unit (MCU) can adjust the structural parameters of HFU and ACU. The controlled parameters includes the number of AEC1 in ACU1 n_1, proportional gain Kp_1, the number of AEC2 in ACU2 n_2, the parameters of PI controller Kp_2 and Ki, both λ and m in HUF. Based on immune algorithm, OU mainly optimizes the above-mentioned structural parameters $\{n_1, n_2, \lambda, m, Kp_1, Kp_2, Ki\}$, thereby improving the performance of the NEI controller.

6.4.2.6 The Identification Unit IU

Before optimize the NEI structural parameters, we have to identify the PGCS online. RBF neural network with forgetting factor and recursive least squares (RLS) can achieve an online self-tuning, adaptive ability, model identification and prediction estimation [58, 59].

From Fig. 6.18, in PGCS, the input signal of PWM control circuit is $u(s)$ and I_{ac} is the output inverted current. The RBF network model of the photovoltaic system is identified, as shown in Fig. 6.22.

$$\widehat{I}_{ac}(k) = f[I_{ac}(k-1), I_{ac}(k-2), \ldots, I_{ac}(k-n); u(k), u(k-1), \ldots, u(k-m+1)] \tag{6.42}$$

6.4.2.7 The Optimization Unit OU

Nervous system can regulate the whole endocrine system and the function of immune system, and can quickly and effectively eliminate the invasion of antigens in vivo. Based on it, we designed the OU and IU. IU is used for online identification of the mathematical model of PGCS. According to the indexes of PGCS, such as AC current deviation I_{ac_err}, bias voltage deviation V_{err}, frequency deviation f_{err}, total harmonic distortion (THD) and power factor (PF), MCU uses immune optimization

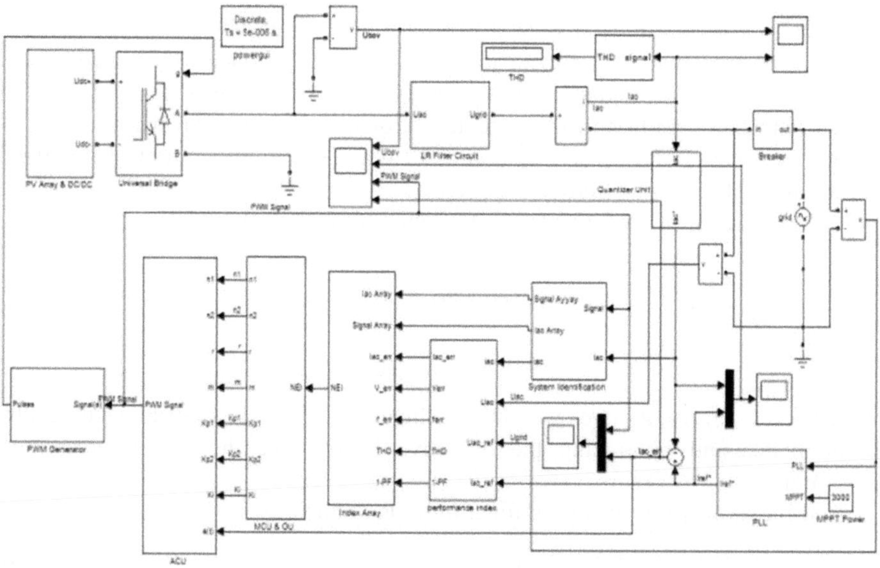

Fig. 6.22 The simulink simulation of photovoltaic grid-connected system

algorithm to optimize the NEI structural parameters $\{n_1, n_2, \lambda, m, Kp_1, Kp_2, Ki\}$ in a limited space.

(1) **Individual Genes Encoding:**

Define all NEI structural parameters $\{n_1, n_2, \lambda, m, Kp_1, Kp_2, Ki\}$ to be optimized as a chromosome. As the problem is continuous space parameters optimization, this paper uses real number for each individual structure domain to simulate clonal expansion, super mutants, receptor editing, and operation coding. As the range of the selected parameters is different, so uniform processing is needed. The individual encoding cell is defined as

$$(f_{n_1}, f_{n_2}, f_{\lambda}, f_m, f_{Kp_1}, f_{Kp_2}, f_{Ki}), f(i = n_1,\ n_2,\ \lambda,\ m, Kp_1,\ Kp_2,\ Ki) \in (0,\ 1) \quad (6.43)$$

Then the corresponding structural parameters of the NEI controller is

$$i = i_{\min} + f_i \times (i_{\max} - i_{\min}) \quad (6.44)$$

where $i_{\max}, i_{\min}(i = n_1, n_2, \lambda, m, Kp_1, Kp_2, Ki)$ corresponding to the minimum and maximum value respectively.

(2) **The identification model of RBF Neural Network:**

In order to evaluate performance index of structural parameters of the NEI controller, we use RBF neural network with forgetting factor and RLS, which can achieve identification and forecast estimation of PGCS online.

(3) **Fitness Index Function:**

PGCS should be synchronized with the grid, and frequency of power grid is 50 Hz in China. The frequency deviation f_{err} should be limited in ± 0.5 Hz. The higher harmonics do harm to the connected device and the total harmonic of AC current of inverter should be less than 5%. When the power output of inverter is greater than 50% of rated power output, the average PF shall be not less than 0.9. So the expression of performance index function is as follows:

$$J_{\min} = \int_0^{\infty} (g_1|e(t)| + g_2|V_{err}(t)| + g_3|f_{err}(t)| + g_4 THD(t) + g_5(1 - PF(t))) \tag{6.45}$$

where g_1, g_2, g_3, g_4, g_5 are weight coefficients, selected according to control requirements and optimizing the objective is to make the performance index $J \to \min$.

And the expression of fitness function F is defined as

$$F = 1/(J_{\min} + x) \tag{6.46}$$

where $x \in R^+$, in order to protect against any spillage and here $x = 0.1$.

(4) **Immune Optimization Algorithm:**

The following model adopts medical immune optimization algorithm [59, 60] to optimize the NEI structural parameters $\{n_1, n_2, \lambda, m, Kp_1, Kp_2, Ki\}$ in the limited space. Where a is the selection factor and $a \in (0, 1)$, r for the short-range jump of ball field radius; b for remote jump factor and $b \in (0, 1)$, R for remote jump radius; c for the update factor and $c \in (0, 1)$.

6.4.3 Simulation Results

In order to verify the control performance of the NEI controller, the simulation model of 3KW PGCS based on Matlab/Simulink simulation model, is shown in Fig. 6.22.

The DC bus voltage of DC/DC Boost circuit is about 360 V, the AC output current of inverter synchronously runs with a 220 V/50 Hz AC power grid. And the simulating effects are described.

In the simulation process, the range of structural parameters of immune optimization algorithm is shown in Table 6.7. The weighting factors of fitness index function are shown in Table 6.8. The key performance indicators of the PGCS includes AC current deviation I_{ac_err}, voltage deviation V_{err}, frequency deviation f_{err}, THD, and PF.The range of structural parameters of the NEI controller for the PGCS $PF\{n_1, n_2, \lambda, m, Kp_1, Kp_2, Ki\}$ is shown in Table 6.9.

Table 6.7 The range of structural parameters of immune optimization algorithm

Pop size	Selection factor	Short-range jump radius	Remote factor	Remote jump radius	Update factor
50	0.2	0.02	0.8	[0.2 0.3]	0.45

Table 6.8 The weighting factor of fitness index function

I_{ac_err} (g_1)	V_{err} (g_2)	f_{err} (g_3)	THD (g_4)	PF (g_5)
0.1	0.01	10	80	2

Table 6.9 The range of structural parameters of nei controller

n_1	n_2	λ	m	Kp_1	Kp_2	Ki
[0 10]	[0 10]	[0 5]	[0 20]	[0 100]	[0 100]	[0 10000]

Table 6.10 The range of structural parameters of nei controller

n_1	n_2	λ	m	Kp_1	Kp_2	Ki
0.9365	1.0593	1.251	5.143	1.9881	1.1066	2515

Based on immune optimization algorithm, the optimization parameters of the NEI controller is shown in Table 6.10.

Comparing with the result of repetitive PI controller based on internal model control theory [61, 62], the NEI integrated controller has better control effect as shown in Figs. 6.23, 6.24, 6.25, 6.26.

As shown in Figs. 6.23 and 6.24, the output voltage of inverter U_{cov}, the actual inverter current I_{ac}, and target AC current I_{ac_ref}, the output of signal the PWM circuit control controlled by the NEI controller, and bias AC current deviation I_{ac_ref} respectively. The results show that the value AC current deviation I_{ac_ref} in the NEI controller is in the range $\pm$ 0.25, while $\pm$ 0.5 in PI controller. It is obvious that the NEI controller has better control effect than PI controller in tracking error.

In theory, the power grid with voltage peak voltage 311 V and frequency 50 Hz for 3KW PGCS, both its effective inverter current I_{ac} and target AC current I_{ac_ref} should be 13.64A. As shown in Fig. 6.25, the fundamental current value of I_{ac} is 13.67A and harmonic THD 1.68%, while in Fig. 6.26 the value is 13.64A and THD 0.53% respectively.

Two control algorithms reach the requirement that the output of inverter voltage the peak voltage of 311 V, frequency 50 Hz. But comparing with the result of the PI controller based on internal model control theory, the NEI integrated controller can more quickly eliminate the control error, and dynamically and effectively restrain total harmonic distortion of inverter current. Both the output voltage u_{cov}

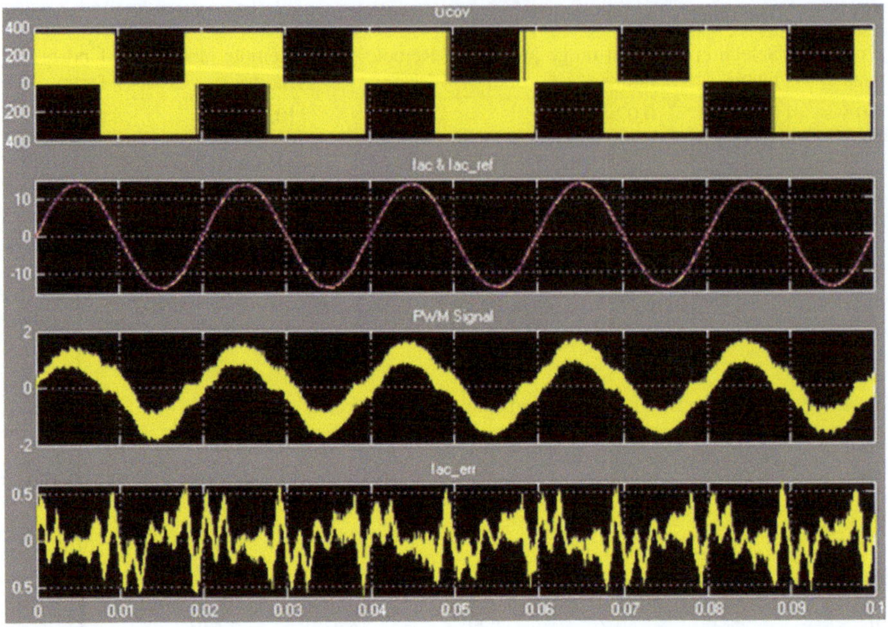

Fig. 6.23 The simulation results of photovoltaic grid-connected system with repetitive PI controller

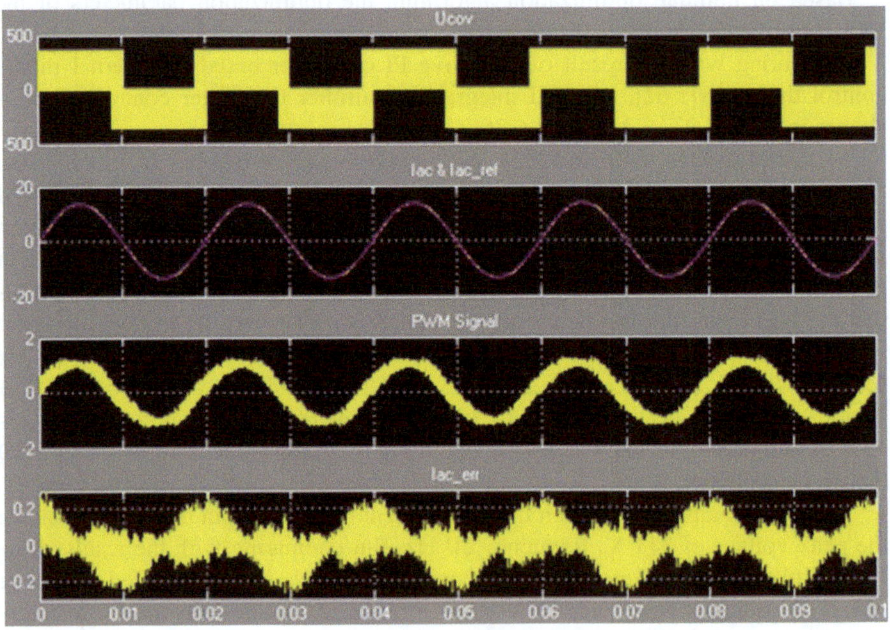

Fig. 6.24 The simulation results of photovoltaic grid-connected system with NEI controller

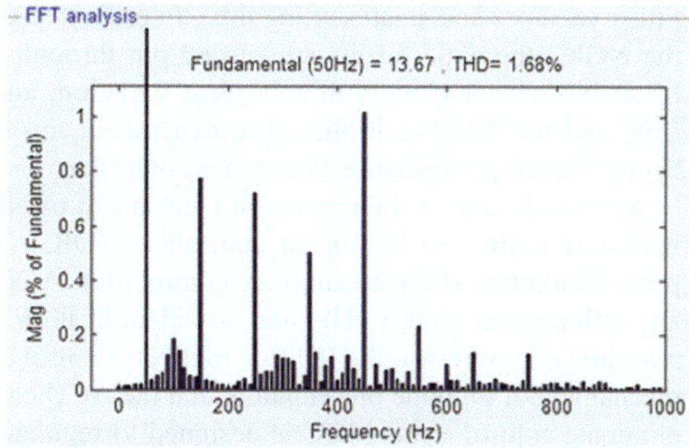

Fig. 6.25 The FFT analysis results of photovoltaic grid-connected system with repetitive PI controller

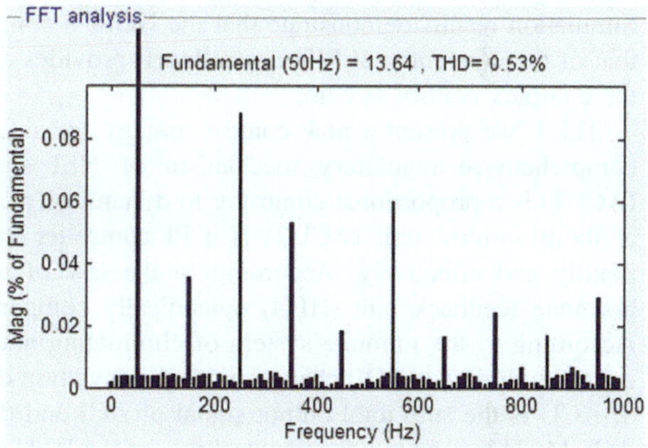

Fig. 6.26 The FFT analysis results of photovoltaic grid-connected system with NEI controller

and current I_{ac} can achieve amplitude and phase tracking with power grid and make the power factor PF close to 1. The output waveform keeps as a pure sine wave with little current harmonic and reaches the requirements of grid-connected system. Therefore, the NEI controller for PGCS gets better controlling effect than the PI controller.

6.5 Conclusions

First, we propose a novel NGIC based on the bi-cooperative regulation mechanism and the regulation characteristics of the blood glucose in the body. The NGIC can independently regulate the dynamic change process and steady change process by using different control unit, which makes it easy to tune the control parameters independently according to the different control effect of dynamic or steady change processes. The NGIC can rapidly eliminate the control error with short rise time, little overshoot and small settling time. In order to testify the control performance of the NGIC, the experiments are carried out through applying the NGIC to a nonlinear bio-reactor plant with great time variation, and compared with the PID, the IMC and the ANN controller. The experiment results indicate that the NGIC has better control performance than other controllers.

Second, based on the regulation mechanism of NEI system, we present a novel nonlinear optimized intelligent controller (NOIC). The NOIC is composed of a presentation unit (PU), an antibody control unit (ACU), a main control unit (MCU), an optimization unit (OU), and an identification unit (IU). According to the real-time control error, the PU first pretreats control error dynamically based on the mechanism of immune presentation, and the ACU can adjust the number of ACE to eliminate control error. MCU is designed to regulate the control action of PU and ACU. And the OU and IU can optimize the parameters of the PU and ACU based on the genetic algorithm (GA). Thus the control performance can be improved. Simulation results demonstrate that the NOIC's control performance is better than that of the conventional PID controller. It provides a new and efficient method for the complex control system.

Third, we present a new control strategy for single-phase PGCS based on the comprehensive regulatory mechanism of NEI system. Antibody control unit (ACU1) is a proportional controller to dynamically change the control error, while antibody control unit (ACU2) is a PI controller to elimination of control error rapidly and effectively. According to the law of regulating hormone secretion, hormone feedback unit (HFU) dynamically compensate for the output of ACU. According to the immune system of eliminating antigen, the number of antibody control individual (ACI) changes with the deviation dynamically. The output signal of ACU is the sum total output signal of ACI and the greater deviation, the more ACI. MCU is used to adjust control role of the HFU and ACU. IU achieves the aim of identification the mathematical model of photovoltaic grid-connected inverter system online. Based on immune algorithm, OU is mainly used to optimize the structural parameters of the NEI controller, thereby improving the NEI controller performance. At last, through the simulation of Matlab/Simulink, the results show that the NEI controller can more quickly eliminate the control error, dynamically and effectively restrain total harmonic distortion of inverter current and has greater anti-interference ability. Therefore, intelligent technology is a new solution in the field of photovoltaic grid-connected system.

References

1. Liu, B., Ding, Y., Gao, N., et al.: A bio-system inspired nonlinear intelligent controller with application to bio-reactor system. Neurocomputing. **168**(C), 1065–1075(2015)
2. Liu, B., Ding, Y.: An intelligent controller inspired from neuroendocrine-immune system. Dynam. Cont. Dis. Ser. B. **13**(b), 31–35(2006)
3. Wang, L., Ding, Y., Liang X., Hao K.R., Sun Y.Z.: A neuroendocrine-immune network based control strategy for single-phase photovoltaic grid-connected inverter. In: The 23rd Chinese

Control and Decision Conference (2011 CCDC), Mianyang, China, pp. 2492–2497, 23–25 May (2011)
4. Vrabie, D., Lewis, F.: Neural network approach to continuous-time direct adaptive optimal control for partially unknown nonlinear systems. Neural Networks. **22**(3), 237–246 (2009)
5. Fu, L.J., Cao, J.G., Liao, C.R., et al.: Study on neural networks control algorithms for automotive adaptive suspension systems. In: International Conference on Neural Networks and Brain, 2005. Icnn&b. 1795–1799(2005)
6. Chen, P., Qin, H., Sun, M., et al.: Global adaptive neural network control for a class of uncertain non-linear systems. Iet. Control. Theory. A. **5**(5), 655–662 (2011)
7. Hu, H., Woo, P.Y.: Fuzzy supervisory sliding-mode and neural-network control for robotic manipulators. IEEE. T. Ind. Electron. **53**(3), 929–940 (2006)
8. Shaocheng, T., Changying, L., Yongming, L.: Fuzzy adaptive observer back stepping control for MIMO nonlinear systems. Fuzzy. Set. Syst. **160**(19), 2755–2775 (2009)
9. Hwang, C.: Decentralized fuzzy control of nonlinear interconnected dynamic delay systems via mixed h. IEEE. T. Fuzzy. Syst. **19**(2), 276–290 (2011)
10. Wang, S.C., Liu, Y.H.: A modified pi-like fuzzy logic controller for switched reluctance motor drives. IEEE. T. Ind. Electron. **58**(5), 1812–1825 (2011)
11. Tao, C.W., Taur, J., Chuang, C.C., et al.: An approximation of interval type-2 fuzzy controllers using fuzzy ratio switching type-1 fuzzy controllers. IEEE. T. Syst. Man. Cy. B. **41**(3), 828–839 (2011)
12. Etik, N., Allahverdi, N., Sert, I.U., et al.: Fuzzy expert system design for operating room air-condition control systems. Expert Syst. Appl. **36**(6), 9753–9758 (2007)
13. Chee, F., Fernando, T.L., Savkin, A.V., et al.: Expert PID control system for blood glucose control in critically ill patients. IEEE. T. Inf. Technol. B. A Publication of the IEEE. Eng. Med. Biol. **7**(4), 419–425 (2003)
14. Valenzuela, L.M.A., Bentley, J.M., Lorenz, R.D.: Expert system for integrated control and supervision of dry-end sections of paper machines. IEEE. Pulp. P. **40**(2), 145–156(2003)
15. Park, Y.M., Lee, K.H.: Application of expert system to power system restoration in sub-control center. IEEE. T. Power. Syst. **17**(6), 407–415 (1995)
16. Liang, X., Ding, Y.S., Ren, L.H., et al.: A bioinspired multilayered intelligent cooperative controller for stretching process of fiber production. IEEE. T. Syst. Man. Cy. C. **42**(3), 367–377 (2012)
17. Liang, X., Ding, Y., Ren, L., et al.: Data-Driven cooperative intelligent con troller based on the endocrine regulation mechanism. IEEE. T. Contr. Syst. T. **22**(1), 94–101 (2014)
18. Lei, G., Atakelty, Hailu.: Comprehensive learning particle swarm optimizer for constrained mixed-variable optimization problems. Int. J. Comput. Int. Sys. **3**(6), 832–842(2012)
19. Farhy, L.S.: Modeling of oscillations in endocrine networks with feedback. Method. Enzymol. **384**(12), 54–81 (2004)
20. Brazzini, B., Ghersetich, I., Hercogova, J., et al.: The neuro-immuno-cutaneous-endocrine network: Relationship between mind and skin. Dermatol. Ther. **16**(2), 123–131 (2003)
21. Payne, J.K.: A neuroendocrine-based regulatory fatigue model. Biol. Res. Nurs. **6**(2), 141–150 (2004)
22. Yao, X., Liu, Y.: Making use of population information in evolutionary artificial neural networks. IEEE. T. Syst. Man. Cy. B. A Publication of the IEEE. Syst. Man. Cyb. **28**(3), 417–25(1998)
23. Tay, J.C., Jhavar, A.: CAFISS: a complex adaptive framework for immune system simulation. ACM Symposium on Applied Computing. pp. 158–164(2005)
24. Stepney, S., Clark, J.A., Johnson, C.G., et al.: Artificial immune systems and the grand challenge for non-classical computation, pp. 204–216. Artificial Immune Systems. Springer, Berlin, Heidelberg (2003)
25. Vargas, P., Moioli, R., Castro, L.N.D., et al.: Artificial homeostatic system: a novel approach, pp. 754–764. Advances in Artificial Life. Springer, Berlin, Heidelberg (2005)

26. Wommack J.C., Salinas, A., Jr, M.R., et al.: Behavioural and neuroendocrine adaptations to repeated stress during puberty in male golden hamsters. J. Neuroendocrinol. **16**(9), 767–775 (2004)
27. Haake, P., Schedlowski, M., Exton, M.S., et al.: Acute neuroendocrine response to sexual stimulation in sexual offenders. Can. J. Psychiat. **48**(4), 265–271 (2003)
28. Liu, B., Ding, Y.S.: An Intelligent controller based on primary–secondary responding mechanism of immune system. In: Proceeding of 7th International Conference on Computational Intelligence and Natural Computing, pp. 467–470. Salt Lake City, USA, June (2005)
29. Dong, H.K.: PID controller tuning of a boiler control system using immune algorithm typed neural network. In: Computational Science - Iccs 2004, International Conference, pp. 695–698. Kraków, Poland, 6–9 June (2004)
30. Liu, B., Ren, L., Ding, Y.: A novel intelligent controller based on modulation of neuroendocrine system. Computer Simulation. **23**(2), 129–132 (2006)
31. Shen W.M., Salemi, B., Will, P.: Hormones for self-reconfigurable robots. In: Proceedings of the International conference on Intelligent Autonomous Systems (IAS-6), pp. 918–925. Venice, Italy(2000)
32. Ajlouni, N., Alhamouz, S.: Genetic design of fuzzy mapped pid controllers for non-linear plants. Inform. Technol. J. **3**(1), 2004
33. Ferrari, S., Stengel, R.F.: Smooth function approximation using neural net works. IEEE. T. Neural. Networ. **16**(1), 24–38 (2005)
34. Fournier, R.L.: Basic transport phenomena in biomedical engineering. CRC Press, Taylor & Francis Group (2012)
35. Sturis, J., Polonsky, K.S., Mosekilde, E., et al.: Computer model for mechanisms underlying ultradian oscillations of insulin and glucose. Am. J. Physio. **260**(1), 801–809 (1991)
36. Duncan, G.A.: Digital control system design for a unique nonlinear MIMO process using QFT technique. Iee. P-Contr. Theor. Ap. **142**(5), 466–474 (1995)
37. Chao, W., Yao, Z.: Approach on nonlinear control theory for designing STATCOM controller. Paper presented at IEEE International Conference on Grey Systems and Intelligent Services. pp. 871–875(2007)
38. Ljung, L.: System Identification. Theory for the user. Epfl. **16**(1), 9–11 (2002)
39. W.C. Shaoyuan Li, Industry process identification and control, in, Chemical industry press, Beijing (2010)
40. Doi, M., Mori, Y.: A Study on robust asymptotic tracking property for generalized minimum variance control. Amer. Contr. Conf. 1472–1477(2002)
41. Shevitz, D., Paden, B.: Lyapunov stability theory of nonsmooth systems. IEEE. T. Automat. Contr. **39**(9), 1910–1914 (1994)
42. Man, Z., Wu, H.R., Liu, S., et al.: A new adaptive backpropagation algorithm based on Lyapunov stability theory for neural networks. IEEE. T. Neural. Networ. **17**(6), 1580–1591 (2006)
43. Nagy, Z.K.: Model based control of a yeast fermentation bioreactor using optimally designed artificial neural networks. Chem. Eng. J. **127**(1), 95–109 (2007)
44. Aiba, S., Shoda, M., Nagatani, M.: Kinetics of product inhibition in alcohol fermentation. Reprinted from Biotech. Bioeng. **X** (6), 845–864 (1968); Wiley, New York (2000)
45. Yu, W., Li, X.O.: Fuzzy Identification Using Fuzzy Neural Networks With Stable Learning Algorithms. IEEE. T. Fuzzy. Sys. **12**(3), 411–420 (2004)
46. Shibata, T., Tashima, T., Tanie, K.: Emergence of emotional behavior through physical interaction between human and robot. **4**, 2868–2873 (1999)
47. Buso, S., Fasolo, S., Mattavelli, P.: Uninterruptible power supply multi-loop control employing digital predictive voltage and current regulators. IEEE. T. Ind. Appl. **37**(6), 1846–1854 (2001)
48. Zhang, L.W., Liu, J.: A novel algorithm of SVPWM inverter in the over-modulation Region based on fundamental voltage amplitude linear output control. In: Proceedings of Chinese Society for Electrical Engineering (CSEE), vol. 25, no. 19, pp. 12–18. IEEE(2005)

49. Premrudeepreechacharn, S., Poapornsawan, T.: Fuzzy logic control of predictive current control for grid-connected single phase inverter. In: Record of the 28th IEEE Photovoltaic Specialists Conference, pp. 1715–1718. IEEE, Anchorage, AK, USA. Piscataway, NJ, USA, 15–22 September(2000)
50. Kjaer, S.B., Pedersen, J.K., Blaabjerg, F.A.: Review of single-phase grid-connected inverters for photovoltaic modules. IEEE. T. Ind. Appl. **41**(5), 1292–1306 (2005)
51. Sakhare, A., Davari, A., Feliachi, A.: Fuzzy logic control of fuel cell for stand-alone and grid connection. J. Power Sources **135**(1–2), 165–176 (2004)
52. He, S.F., Yi, L.Z, Tan, P., et al.: Research of space vector PWM inverter based on the neural network. In: Proceedings of the 11th International Conference on Electrical Machines and Systems, vol. 4, pp. 1802–1805(2008)
53. El-Tamaly, H.H., Mohammed, A.A.E.: Optimal operation of photovoltaic/utility grid interconnected electrical power system using neural network. Power Systems Conference, 2006. MEPCON 2006. Eleventh International Middle East. IEEE. 1–6 (2006)
54. Robert, W., Erickson, Dragan, Maksimovic.: Fundamentals of power electronics (second edition). Kluwer Academic Publishers, New York(2001)
55. Zheng, S.C., Ming, D., Su, J.H., et al.: Simulation and experiment research of photovoltaic generation system and its islanding. Acta Simulata Systematica Sinica(2005)
56. Ding, Y.S., Liu, B.: An intelligent bi-cooperative decoupling control approach based on modulation mechanism of internal environment in body. IEEE. T. Contr. Syst. T. **19**(3), 692–698 (2011)
57. Walsh, G.C., Ye, H., Bushnell, L.: Stability analysis of networked control systems. IEEE. T. Contr. Syst. T. **10**(3), 438–446 (2002)
58. Narendra, K.S., Parthasarathy, K.: Identification and control of dynamical systems using neural networks. IEEE. T. Neural. Networ. **1**(1), 4–27 (1990)
59. Farmer, J.D., Packard, N.H., Perelson, A.S.: The immune system, adaptation, and machine learning. The Organization of information systems for government and public administration. Unesco, 187–204(1979)
60. Castro, L.N.D., Timmis, J.: An artificial immune network for multimodal function optimization. Evolutionary Computation, 2002. CEC '02. Proceedings of the 2002 Congress on. 699–704(2002)
61. Liang, J., Green, T.C., Weiss, G., et al.: Repetitive control of power conversion system from a distributed generator to the utility grid. International Conference on Control Applications. vol.1, pp. 13–18. IEEE(2002)
62. Weiss, G., Zhong, Q.C., Green, T.C., et al.: H ∞, repetitive control of DC-AC converters in microgrids. IEEE. T. Power. Electr. **19**(1), 219–230 (2004)

Chapter 7
Gene Regulatory Network Inspired Cooperative Controllers

Comparing to neural networks, genetic regulatory networks (GRNs) have two variables, mRNA and proteins, interacting in a very complex way. This special structure of GRNs promises better adaptive performance. In this chapter, cooperative controllers inspired by GRNs are proposed. First, the robust H_∞ state estimation problem is investigated for a class of discrete-time stochastic GRNs with probabilistic measurement delays [1]. Then, a bio-inspired cooperative controller via evolution of GRN is presented to make the performance index of each stage stable over the whole study horizon and the overall performance index close to the desired value against vibrations simultaneously [2].

7.1 Introduction

It is well known that genetic regulatory networks (GRNs) play a key role in systems biology as they explain the interactions between genes (mRNA) and proteins. GRNs are collections of DNA segments in a cell which interact with each other indirectly but with their mRNA, protein expression products and other substances. The genetic information in DNA can be transferred to mRNAs (transcription), which is used as a template to synthesize proteins (translation). These proteins will accumulate to give the genes particular structural properties and some proteins can regulate the transcription process in turn. Understanding the nature and functions of various GRNs is very interesting and crucial important for the treatment of many diseases (such as cancer and acquired immune deficiency syndrome) which stem from the malfunctions of GRNs of the corresponding cell lines.

By constructing the GRNs models inspired by biological mechanisms in living organisms, we can gain indirect access to the desired quantitative biological relationships, represented by the structure and parameters of these models. Theoretical studies can also contribute to engineering applications, such as developing circuits and systems with biotechnological design principles of synthetic genetic regulatory

Y. Ding et al., *Bio-Inspired Collaborative Intelligent Control and Optimization*, Studies in Systems, Decision and Control 118,
https://doi.org/10.1007/978-981-10-6689-4_7

networks. Generally speaking, two main types of GRNs models have been proposed and utilized, i.e., Boolean models and differential equation models, to describe GRNs structures [3–7]. Theoretical analysis and experiment investigation on GRNs have attracted increasing attention, and a large amount of important results have been reported in the literature, making significant contributions for understanding both static and dynamic behaviors of biological systems in details [3–27], and references therein.

GRNs models are unavoidably affected by modeling uncertainties, including parametric error, time delays and stochastic noises. One simple reason is that the parameters and the saturation functions of GRNs cannot be measured exactly [13]. Time delays exist inevitably in transcription, translation, and translocation processes because of the slow reaction process. Gene regulation is an intrinsically noisy process due to intracellular and extracellular noise perturbations, which are derived from random births and deaths of individual molecules and environmental fluctuations. When modeling uncertainties appear in continuous GRNs, the stability issue, as the most important properties for any dynamics systems, has been intensively exploited [9–11, 13, 22–26]. In [22], the authors studied sufficient conditions for the stochastic stability of the genetic networks with disturbance attenuation. A simple delay-independent or delay-dependent measure of attenuation for a class of GRNs is developed in [9] developed. Robust stochastic stability of uncertain stochastic GRNs with disturbance attenuation was investigated and a delay-dependent robust stability condition was derived in [26].

By the requirement in implementing and application of GRNs for computer-based simulation, discrete-time GRNs become more important than their continuous-time counterparts in today's digital world while most GRNs studied are assumed to act in continuous-time manner. Therefore, it is of vital importance to formulate the discrete-time GRNs so as to study the dynamical characteristics [8, 12, 16, 28, 29]. Therefore, the stability analysis for stochastic discrete-time GRNs become increasingly significant and has received particular research interests. For example, the robust stability problem [8], the global exponential stability problem [16], and the robustly exponentially stability problem [12] have been addressed.

In the literature mentioned above, the states of the system are fully known. However, it should be pointed out that in practice, for the complexity of large-scale networks, only partial information about the gene states is available in the network outputs. The state variables determine the GRNs dynamics and become important for subsequent analysis, such as signal processing and control engineering. Meanwhile, the biologists are interested in the steady-state values of the actual network components, that is, the concentrations of the mRNA and protein, for the goal of identifying genes of interest and designing drugs. Hence, state estimation techniques have been raised to tackle the mentioned problem, which could make the gene states fully available from the measurement data. For instance, the state estimation problem for GRNs to approximate the true concentrations of the mRNA and protein has been investigated in continuous-time manner [14, 15, 17, 18, 21, 27].

7.2 Robust State Estimation for Genetic Regulatory Networks

In this section, we deal with the robust H_∞ state estimation for a class of stochastic discrete-time GRNs with probabilistic measurement delays.

In order to reflect a more realistic situation, probabilistic measurement delays have been introduced into the state estimation problem, which may occur in a random way for probabilistic causes, for example, network congestion, accidental packet loss, and a failure in measures [30]. It is definitely true that the network measurements are always delayed due primarily to the finite signal transmission speed. Therefore, the network measurements could not contain information about the current gene states, while the delayed network measurement could. The most popular way to describe probabilistic measurement delay or other kind of missing measurement is to take it as a Bernoulli distributed white sequence [20, 28, 31–34].

When both modeling uncertainties and probabilistic missing measurements exist in discrete-time GRNs models, how to design robust estimators from the available measurements and obtain a new sufficient criteria to ensure the existence of the estimator are challenging issues. The measurement delays are described by a binary switching sequence satisfying a conditional probability distribution. The GRNs under considering involves the modeling uncertainties, including parameter error, stochastic disturbance and time-varying delays, and the activation function are characterized by sector-bounded nonlinearities. By utilizing Lyapunov stability theory and stochastic analysis techniques, sufficient conditions are first established to ensure the existence of the desired estimators in the terms of a linear matrix inequality (LMI). These conditions are dependent on the lower and upper bounds of the time-varying delays. Then, the explicit expression of the desired estimator is shown to ensure the estimation error dynamics to be robustly exponentially stable in the mean square for the addressed system.

Throughout this section, N^+ stands for the set of nonnegative integers; R^n denotes the n-dimensioned Euclidean space. The superscript "T" represents the matrix transposition. The notation $X \geq Y$ (respective $X > Y$), where X and Y are symmetric matrices, means that $X - Y$ is positive semi-definitive (respective positive definite). I is the identity matrix with compatible dimension. In symmetric block matrices, the symbol "*" is used to denote term that is induced by symmetry. Moreover, $E\{\cdot\}$ stands the expectation operator. $L_2[0, \infty)$ is the space of square-integrable vector functions over $[0, \infty).|\cdot|$ stands for the Euclidean vector norm. Matrices, if not explicitly, specified, are assumed to have compatible dimensions. Sometimes, the arguments of function will be omitted in the analysis when no confusion can arise.

7.2.1 Problem Formulation

Discrete-time GRN with time-varying delays and random noise can be described by the following difference equations:

$$\begin{cases} x(k+1) = -(A+\Delta A(k))x(k) + (B+\Delta B(k))g(y(k-\tau(k))) + \delta(k,x(k))w(k) + L_x v_x(k) \\ y(k+1) = -(C+\Delta C(k))y(k) + (D+\Delta D(k))x(k-\sigma(k)) + L_y v_y(k) \end{cases} \tag{7.1}$$

where $x(k) = [x_1(k),\ldots,x_n(k)]^T \in \mathbb{R}^n$, $y(k) = [y_1(k),\ldots,y_n(k)]^T \in \mathbb{R}^n$, $x_i(k)$ and $y_i(k)$ $(i = 1,\ldots,n)$ denote the concentrations of mRNA and protein of the ith node at time k, respectively; $A = diag\{a_1,\ldots,a_n\}$, $C = diag\{c_1,\ldots,c_n\}$, $D = diag\{d_1,\ldots,d_n\}$ are constant matrices for a fixed system mode with a_i, c_i, d_i, representing the decay rates of mRNA, protein and the translation rate of the ith gene, respectively; $B = (b_{ij}) \in \mathbb{R}^{n\times n}$ is the coupling matrix of the genetic network; for more details about biological interpretation of B, we refer readers to Refs. [8, 13, 16, 27]; $\Delta A(k)$, $\Delta B(k)$, $\Delta C(k)$, and $\Delta D(k)$ represent the parameter uncertainties; the exogenous disturbance signals $v_x(k)$, $v_y(k) \in \mathbb{R}^n$ satisfy $v_i(\cdot) \in L_2[0,\infty)$. L_x and L_y are the known real constant matrices. $\tau(k)$ and $\sigma(k)$ are time-varying delays denoting, respectively, the feedback regulation delay and translation delay which satisfy

$$0 \le \tau_m \le \tau(k) \le \tau_M,\ 0 \le \sigma_m \le \sigma(k) \le \sigma_M \tag{7.2}$$

where the lower bound τ_m, σ_m and the upper bound τ_M, σ_M are known positive integers.

Furthermore, $g(y(k-\tau(k))) = [g_1(y_1(k-\tau(k))),\ldots,g_n(y_n(k-\tau(k)))]^T \in \mathbb{R}^n$ is a nonlinear function with $g_i(\cdot)$ denoting the feedback regulation of the protein on the transcription, which is used to a monotonic function such as the Hill form, that is, $g_i(s) = s^{h_j} / (1+s^{h_j})$, $(j = 1,\ldots,n)$, where h_j is the Hill coefficient. In this section, $g_i(\cdot)$ is assumed to satisfy the following more general sector-bounded conditions. $\delta(k,x(k)) : \mathbb{R} \times \mathbb{R}^n \to \mathbb{R}^n$ represents a noise intensity function vector, satisfying

$$\delta^T(k,x(k))\delta(k,x(k)) \le x^T(k)Hx(k) \tag{7.3}$$

where $H > 0$ is a known matrix. *w(k)* is a scalar Wiener process (Brownian motion) with $\mathrm{E}\{w(k)\} = 0$, $\mathrm{E}\{w^2(k)\} = 1$, $\mathrm{E}\{w(i)w(j)\} = 0\ (i \ne j)$.

Assumption 7.1 The admissible parameter uncertainties are assumed to be of the form:

$$[\Delta A(k)\ \Delta B(k)\ \Delta C(k)\ \Delta D(k)] = QF(k)[W_1\ W_2\ W_3\ W_4]$$

where Q, W_i $(i = 1, 2, 3, 4)$ are the known real constant matrices and *F(k)* is the unknown time-varying matrix-valued function satisfying

$$F^T(k)F(k) \le I,\ \forall k \in \mathbb{N}^+.$$

Assumption 7.2 The vector-valued activation function $g_i(\cdot)$ is assumed to satisfy the following sector-bounded condition, namely, for $\forall x, y \in \mathbb{R}^n$:

$$[g(x) - g(y) - U_1(x-y)]^T[g(x) - g(y) - U_2(x-y)] \le 0$$

where U_1, U_2 are known real constant matrices, and $U = U_1 - U_2$ is symmetric positive definite matrix.

For the complexity of large-scale networks, only partial information about the gene states is available in the network outputs. Therefore, we need to estimate the gene states from available measurements for the goal of obtaining the true states of the GRNs (concentrations of the mRNA and protein). Similar to Refs. [17, 27], we can assume network measurements to be given as follows:

$$z_x(k) = Mx(k),\ z_y(k) = Ny(k) \tag{7.4}$$

where *M, N* are known constant matrices. $z_x(t),\ z_y(t) \in \mathbb{R}^l$ are the ideal measurements or outputs of the network. The actual network outputs are usually subjected to probabilistic delays that can be described preferably by

$$\tilde{z}_x(k) = \alpha_k z_k(k) + (1-\alpha_k)z_x(k-1),\ \tilde{z}_y(k) = \beta_k z_k(k) + (1-\beta_k)z_y(k-1) \tag{7.5}$$

where stochastic variable α_k, $\beta_k \in \mathbb{R}$ are Bernoulli distributed white sequences governed by

$$\begin{aligned} &\text{prob}\{\alpha_k = 1\} = \text{E}\{\alpha_k\} = \alpha_0,\ \text{prob}\{\alpha_k = 0\} = 1 - \text{E}\{\alpha_k\} = 1 - \alpha_0 \\ &\text{prob}\{\beta_k = 1\} = \text{E}\{\beta_k\} = \beta_0,\ \text{prob}\{\beta_k = 0\} = 1 - \text{E}\{\beta_k\} = 1 - \beta_0 \end{aligned} \tag{7.6}$$

Here α_0, $\beta_0 > 0$ are known constants. Obviously, for α_k, β_k, the variance $\sigma_\alpha = \alpha_0(1-\alpha_0)$, $\sigma_\beta = \beta_0(1-\beta_0)$.

The main objective of this section is to estimate the concentrations of mRNA and protein in (1) from the available network outputs in (7.5). The GRNs state estimator to be considered is given as follows:

$$\begin{cases} \hat{x}(k+1) = -A_x\hat{x}(k) + B_x\tilde{z}_x(k) \\ \hat{y}(k+1) = -A_y\hat{y}(k) + B_y\tilde{z}_y(k) \end{cases} \tag{7.7}$$

where $\hat{x}(k)$, $\hat{y}(k) \in \mathbb{R}^n$ are the estimations of $x(t)$ and $y(t)$, and A_x, A_y, B_x, B_y are the estimator gain matrices to be determined.

Let the estimation error vectors be $\tilde{x}(k) = x(k) - \hat{x}(k)$ and $\tilde{y}(k) = y(k) - \hat{y}(k)$, the estimation error dynamics can be described as following from (7.1), (7.5) and (7.7) by

$$\begin{cases} \tilde{x}(k+1) = -(A + \Delta A(k) + \alpha_k B_x M - A_x)x(k) + (B + \Delta B(k))g(y(k - \tau(k))) \\ \qquad + \delta(k, x(k))w(k) - A_x\tilde{x}(k) - (1 - \alpha_k)B_x Mx(k-1) + L_x v_x(k) \\ \tilde{y}(k+1) = -(C + \Delta C(k) + \beta_k B_y N - A_y)y(k) + (D + \Delta D(k))x(k - \sigma(k)) \\ \qquad -A_y\tilde{y}(k) - (1 - \beta_k)B_y Ny(k-1) + L_y v_y(k) \end{cases} \tag{7.8}$$

For convenience, we denote

$$\bar{x}(k) = \begin{bmatrix} x(k) \\ \tilde{x}(k) \end{bmatrix} \bar{y}(k) = \begin{bmatrix} y(k) \\ \tilde{y}(k) \end{bmatrix},$$
$$\bar{x}(j) = \psi(j), j = -\sigma_M, -\sigma_M + 1, \ldots, -1, 0, \ \bar{y}(j) = \varphi(j), j = -\tau_M, -\tau_M + 1, \ldots, -1, 0,$$

where $\psi(j), j = -\sigma_M, -\sigma_M + 1, \ldots, -1, 0$ and $\varphi(j), j = -\tau_M, -\tau_M + 1, \ldots, -1, 0$ are the initial conditions, which assumed to be independent of the process $w(k)$.

Before formulating the estimator to be investigated, we may first introduce the following stability concepts for the augmented system (7.8).

Definition 7.1 System (7.8) with $v_x(k) = 0$ and $v_y(k) = 0$ is said to be a globally robustly exponential state estimator of the GRNs (7.1) with measurements (7.5) in the mean square sense, if there exist constants $\alpha > 0$ and $0 < \mu < 1$ such that

$$\mathrm{E}\left\{ |\bar{x}(k)|^2 + |\bar{y}(k)|^2 \right\} \leq \alpha\mu^k \left(\max_{-\sigma_M \leq i \leq 0} |\bar{x}(k)|^2 + \max_{-\tau_M \leq i \leq 0} |\bar{y}(k)|^2 \right)$$

Definition 7.2 System (7.8) is said to be a robust H_∞ state estimator of the GRNs (7.1) with measurements (7.5) in the mean square sense with zero initial conditions, if there exists a scalar $\gamma > 0$, such that

$$\mathrm{E}\sum_{k=0}^{\infty} \left\{ |\bar{x}(k)|^2 + |\bar{y}(k)|^2 \right\} \leq \gamma^2 \mathrm{E}\sum_{k=0}^{\infty} \left(|v_x(k)|^2 + |v_y(k)|^2 \right)$$

for all nonzero $v_x(k), v_y(k) \in L_2[0, \infty)$.

Remark 1 The essential difficulty to define the robust H_∞ state estimator of the GRNs (7.1) with measurements (7.5) in the mean square sense is that there are two variables, the concentrations of mRNA and protein, instead of only one variable, the neural state, in neural networks. It took our great efforts to tackle the difficulty. Definition 7.2 is similar to Definition 7.1 (ii) in [14] while in the discrete sense.

The aim of this section is to develop techniques to deal with the robust H_∞ state estimation issue for uncertain discrete stochastic GRNs (7.1) with probabilistic measurement delays. More specifically, with admissible uncertainties and time-delays, we would like to design the parameters $A_x, A_y,\ B_x, B_y$ of the estimator such that the estimation error system (7.8) is stochastic stable with a prescribed H_∞ disturbance rejection attenuation under zero initial condition and (7.8) with $v_x(k) = 0$ and $v_y(k) = 0$ is robustly exponentially stable. In the following, we will utilize the Lyapunov stability theory to design a proper estimator of the form (7.7) for the GRNs (7.1) with measurements (7.5) in the terms of LMI.

7.2.2 Main Results

The following lemmas are essential in establishing our main results:

Lemma 1 *(see* [27, 28]*) Let D and W be real constant matrices, matrix F(k) satisfies* $F^T(k)F(k) \le I$. *Then we have:*

(i) *For any* $\varepsilon > 0,\ DF(k)W + W^T F^T(k)D^T \le \varepsilon^{-1}DD^T + \varepsilon W^T W$

(ii) *For any* $P > 0,\ \pm 2x^T y \le x^T P^{-1} x + y^T P y$.

Lemma 2 *(Schur complement) Given the constant matrices* Ω_1, Ω_2 *and* Ω_3, *where* $\Omega_1^T = \Omega_1$ *and* $\Omega_2^T = \Omega_2 > 0$, *then* $\Omega_1 + \Omega_3^T \Omega_2^{-1} \Omega_3 < 0$, *if and only if*

$$\begin{bmatrix} \Omega_1 & \Omega_3^T \\ \Omega_3 & -\Omega_2 \end{bmatrix} < 0 \text{ or } \begin{bmatrix} -\Omega_2 & \Omega_3 \\ \Omega_3^T & \Omega_1 \end{bmatrix} < 0.$$

We are now ready to give the main results as follows. A sufficient condition is derived in the form of LMI so as to guarantee the robust mean-square exponential stability for the system (7.8) *with* $v_x(k) = 0$ and $v_y(k) = 0$.

Theorem 7.1 *Let the estimator parameters* A_x, B_x, A_y *and* B_y *be given and the admissible conditions hold. Then, the estimation error system* (7.8) *with* $v_x(k) = 0$ *and* $v_y(k) = 0$ *is robustly exponential stable in the mean square if there exist eight positive definite matrices* P_{11}, P_{12}, P_{21}, P_{22}, P_{31}, P_{32}, P_{41}, P_{42}, *and three positive constant scalars* λ, ε_1, *and* ε_2 *such that the following LMI holds:*

$$\Psi_1 = \begin{bmatrix} \Psi'_{11} & * & * \\ R_1 & T_1 & * \\ 0 & \bar{Q}_1^T & -\varepsilon_1 I \end{bmatrix} < 0,\ \Psi_2 = \begin{bmatrix} \Psi'_{22} & * & * \\ R_2 & T_2 & * \\ 0 & \bar{Q}_2^T & -\varepsilon_2 I \end{bmatrix} < 0, \qquad (7.9)$$

where

$$\Psi'_{11} = \begin{bmatrix} \Pi_1 + 2\varepsilon_1 W_1^T W_1 & 0 & 0 & 0 \\ 0 & -P_{21} & 0 & 0 \\ 0 & 0 & -P_{31} & 0 \\ 0 & 0 & 0 & -P_{41} + \varepsilon_1 W_4^T W_4 \end{bmatrix}$$

$$\Psi'_{22} = \begin{bmatrix} \Pi_2 + 2\varepsilon_2 W_3^T W_3 & 0 & 0 & 0 & 0 \\ 0 & -P_{22} & 0 & 0 & 0 \\ 0 & 0 & -P_{32} & 0 & 0 \\ 0 & 0 & 0 & -P_{42} - \lambda I + \varepsilon_2 W_2^T W_2 & -\lambda \breve{U}_2 \\ 0 & 0 & 0 & -\lambda \breve{U}_2^T & -\lambda \breve{U}_1 \end{bmatrix}$$

$$R_1 = \begin{bmatrix} -\sqrt{2} P_{11} A & 0 & 0 & 0 \\ -\sqrt{2} P_{21}(A + \alpha_0 B_x M - A_x) & -\sqrt{2} P_{21} A_x & -\sqrt{2} P_{21}(1 - \alpha_0) B_x M & 0 \\ \sqrt{\sigma_\alpha} P_{21} B_x M & 0 & \sqrt{\sigma_\alpha} P_{21} B_x M & 0 \\ 0 & 0 & 0 & \sqrt{2}(P_{12} + P_{22}) D \end{bmatrix}$$

$$R_2 = \begin{bmatrix} -\sqrt{2} P_{12} C & 0 & 0 & 0 & 0 \\ -\sqrt{2} P_{22}(C + \beta_0 B_y N - A_y) & -\sqrt{2} P_{22} A_y & -\sqrt{2} P_{22}(1 - \beta_0) B_y N & 0 & 0 \\ \sqrt{\sigma_\beta} P_{22} B_y N & 0 & \sqrt{\sigma_\beta} P_{22} B_y N & 0 & 0 \\ 0 & 0 & 0 & \sqrt{2}(P_{11} + P_{21}) B & 0 \end{bmatrix}$$

$$T_1 = diag(-P_{11}, -P_{21}, -P_{21}, -(P_{12} + P_{22}))$$
$$T_2 = diag(-P_{12}, -P_{22}, -P_{22}, -(P_{11} + P_{21}))$$

$$\overline{Q}_1 = \begin{bmatrix} -\sqrt{2} P_{11} Q & 0 & 0 & 0 \\ -\sqrt{2} P_{21} Q & 0 & 0 & 0 \\ 0 & 0 & 0 & 0 \\ 0 & 0 & 0 & \sqrt{2}(P_{12} + P_{22}) Q \end{bmatrix} \quad \overline{Q}_2 = \begin{bmatrix} -\sqrt{2} P_{12} Q & 0 & 0 & 0 & 0 \\ -\sqrt{2} P_{22} Q & 0 & 0 & 0 & 0 \\ 0 & 0 & 0 & 0 & 0 \\ 0 & 0 & 0 & \sqrt{2}(P_{11} + P_{21}) Q & 0 \end{bmatrix}$$

$$\Pi_1 = -P_{11} + P_{31} + (\sigma_M - \sigma_m + 1) P_{41} + H P_{11}$$
$$\Pi_2 = -P_{12} + P_{32} + (\tau_M - \tau_m + 1) P_{42}$$
$$\breve{U}_1 = (U_1^T U_2 + U_2^T U_1)/2, \ \breve{U}_2 = -(U_1^T + U_2^T)/2.$$

Proof For the stability analysis of the system (7.8), we construct the following Lyapunov-Krasovskii function:

$$V(k) = V_1(k) + V_2(k) + V_3(k) + V_4(k) + V_5(k) \tag{7.10}$$

where

$$V_1(k) = x^T(k)P_{11}x(k) + y^T(k)P_{12}y(k);\ V_2(k) = \tilde{x}^T(k)P_{21}\tilde{x}(k) + \tilde{y}^T(k)P_{22}\tilde{y}(k);$$
$$V_3(k) = x^T(k-1)P_{31}x(k-1) + y^T(k-1)P_{32}y(k-1);$$
$$V_4(k) = \sum_{i=k-\sigma(k)}^{k-1} x^T(i)P_{41}x(i) + \sum_{i=k-\tau(k)}^{k-1} y^T(i)P_{42}y(i);$$
$$V_5(k) = \sum_{j=-\sigma_M+1}^{-\sigma_m}\sum_{i=k+j}^{k-1} x^T(i)P_{41}x(i) + \sum_{j=-\tau_M+1}^{-\tau_m}\sum_{i=k+j}^{k-1} y^T(i)P_{42}y(i)$$

Calculating the difference of $V_i(k)$, $i = 1,2,3,4,5$ along the system (8) and taking its mathematical expectation yield

$$\mathbb{E}\{\Delta V(k)\} = \mathbb{E}\{\Delta V_1(k)\} + \mathbb{E}\{\Delta V_2(k)\} + \mathbb{E}\{\Delta V_3(k)\} + \mathbb{E}\{\Delta V_4(k)\} + \mathbb{E}\{\Delta V_5(k)\} \tag{7.11}$$

where, it is not difficult to check that

$$\begin{aligned}\mathbb{E}\{\Delta V_1(k)\} &= \mathbb{E}\{V_1(k+1) - V_1(k)\}\\ &= \mathbb{E}\Big\{[-(A+\Delta A(k))x(k) + (B+\Delta B(k))g(y(k-\tau(k)))]^T\\ &\qquad P_{11}[-(A+\Delta A(k))x(k) + (B+\Delta B(k))g(y(k-\tau(k)))]\\ &\quad + \delta^T(k,x(k))P_{11}\delta(k,x(k)) - x^T(k)P_{11}x(k) - y^T(k)P_{12}y(k)\\ &\quad + [-(C+\Delta C(k))y(k) + (D+\Delta D(k))x(k-\sigma(k))]^T\\ &\qquad P_{12}[-(C+\Delta C(k))y(k) + (D+\Delta D(k))x(k-\sigma(k))]\Big\}\end{aligned} \tag{7.12}$$

$$\begin{aligned}&\mathbb{E}\{\Delta V_2(k)\} = \mathbb{E}\{V_2(k+1) - V_2(k)\}\\ &= \mathbb{E}\{[-(A+\Delta A(k)+\alpha_0 B_xM - A_x)x(k) + (B+\Delta B(k))g(y(k-\tau(k))) - A_x\tilde{x}(k) - (1-\alpha_0)B_xMx(k-1)]^T P_{21}\\ &\cdot[-(A+\Delta A(k)+\alpha_0 B_xM - A_x)x(k) + (B+\Delta B(k))g(y(k-\tau(k))) - A_x\tilde{x}(k) - (1-\alpha_0)B_xMx(k-1)]\\ &+ \sigma_\alpha[B_xMx(k) + B_xMx(k-1)]^T P_{21}[B_xMx(k) + B_xMx(k-1)] - \tilde{x}^T(k)P_{21}\tilde{x}(k)\\ &+ [-(C+\Delta C(k)+\beta_0 B_yN - A_y)y(k) + (D+\Delta D(k))x(k-\sigma(k)) - A_y\tilde{y}(k) - (1-\beta_0)B_yNy(k-1)]^T P_{22}\\ &\cdot[-(C+\Delta C(k)+\beta_0 B_yN - A_y)y(k) + (D+\Delta D(k))x(k-\sigma(k)) - A_y\tilde{y}(k) - (1-\beta_0)B_yNy(k-1)]\\ &+ \sigma_\beta[B_yNy(k) + B_yNy(k-1)]^T P_{22}[B_yNy(k) + B_yNy(k-1)] - \tilde{y}^T(k)P_{22}\tilde{y}(k)\}\end{aligned} \tag{7.13}$$

$$\begin{aligned}&\mathbb{E}\{\Delta V_3(k)\} = \mathbb{E}\{V_3(k+1) - V_3(k)\}\\ &= \mathbb{E}\{x^T(k)P_{31}x(k) - x^T(k-1)P_{31}x(k-1) + y^T(k)P_{32}y(k) - y^T(k-1)P_{32}y(k-1)\}\end{aligned} \tag{7.14}$$

$$\begin{aligned}\mathbb{E}\{\Delta V_4(k)\} &= \mathbb{E}\{V_4(k+1) - V_4(k)\}\\ &\leq \mathbb{E}\left\{x^T(k)P_{41}x(k) - x^T(k-\sigma(k))P_{41}x(k-\sigma(k)) + \sum_{i=k-\sigma_M+1}^{k-\sigma_m} x^T(i)P_{41}x(i)\right.\\ &\quad \left. + y^T(k)P_{42}y(k) - y^T(k-\tau(k))P_{42}y(k-\tau(k)) + \sum_{i=k-\tau_M+1}^{k-\tau_m} y^T(i)P_{42}y(i)\right\}\end{aligned} \tag{7.15}$$

$$\begin{aligned}&\mathbb{E}\{\Delta V_5(k)\} = \mathbb{E}\{V_5(k+1) - V_5(k)\}\\ &= \mathbb{E}\left\{(\sigma_M - \sigma_m)x^T(k)P_{41}x(k) - \sum_{i=k-\sigma_M+1}^{k-\sigma_m} x^T(i)P_{41}x(i) + (\tau_M - \tau_m)y^T(k)P_{42}y(k) - \sum_{i=k-\tau_M+1}^{k-\tau_m} y^T(i)P_{42}y(i)\right\}\end{aligned} \tag{7.16}$$

Substituting (7.12)–(7.16) into (7.11) results in

$$\begin{aligned}\mathbb{E}\{\Delta V(k)\} \leq \mathbb{E}\{&\xi_0^T(k)\Psi_{11}\xi_0(k) + \sigma_\alpha\xi_0^T(k)\widehat{C}_{01}^T P_{21}\widehat{C}_{01}\xi_0(k) + 2\xi_0^T(k)\widehat{F}_{01}^T(k)P_{11}\widehat{F}_{01}(k)\xi_0(k)\\ &+ 2\xi_0^T(k)\widetilde{F}_{01}^T(k)P_{21}\widetilde{F}_{01}(k)\xi_0(k) + \eta_0^T(k)\Psi_{12}\eta_0(k) + \sigma_\beta\eta_0^T(k)\widehat{C}_{02}^T P_{22}\widehat{C}_{02}\eta_0(k)\\ &+ 2\eta_0^T(k)\widehat{F}_{02}^T(k)P_{12}\widehat{F}_{02}(k)\eta_0(k) + 2\eta_0^T(k)\widetilde{F}_{02}^T(k)P_{22}\widetilde{F}_{02}(k)\eta_0(k)\\ &+ 2\xi_0^T(k)\widehat{F}_{11}^T(k)(P_{12}+P_{22})\widehat{F}_{11}(k)\xi_0(k) + 2\eta_0^T(k)\widehat{F}_{12}^T(k)(P_{11}+P_{21})\widehat{F}_{12}(k)\eta_0(k)\}\end{aligned} \tag{7.17}$$

where Π_1 and Π_2 are defined in the theorem, and

$$\begin{aligned}\xi_0(k) &= \left[x^T(k), \tilde{x}^T(k), x^T(k-1), x^T(k-\sigma(k))\right],\\ \eta_0(k) &= \left[y^T(k), \tilde{y}^T(k), y^T(k-1), g^T(y(k-\tau(k))), y^T(k-\tau(k))\right]\end{aligned}$$

$$\Psi_{11} = \begin{bmatrix}\Pi_1 & 0 & 0 & 0\\ 0 & -P_{21} & 0 & 0\\ 0 & 0 & -P_{31} & 0\\ 0 & 0 & 0 & -P_{41}\end{bmatrix} \Psi_{12} = \begin{bmatrix}\Pi_2 & 0 & 0 & 0 & 0\\ 0 & -P_{22} & 0 & 0 & 0\\ 0 & 0 & -P_{32} & 0 & 0\\ 0 & 0 & 0 & -P_{42} & 0\\ 0 & 0 & 0 & 0 & 0\end{bmatrix}$$

$$\begin{aligned}\widehat{C}_{01} &= [B_xM, 0, B_xM, 0],\ \widehat{C}_{02} = \left[B_yN, 0, B_yN, 0, 0\right],\\ \widehat{F}_{01}(k) &= [-(A+\Delta A(k)), 0, 0, 0],\ \widehat{F}_{02}(k) = [-(C+\Delta C(k)), 0, 0, 0, 0],\\ \widetilde{F}_{01}(k) &= [-(A+\Delta A(k)+\alpha_0 B_xM - A_x), -A_x, -(1-\alpha_0)B_xM, 0],\\ \widetilde{F}_{12}(k) &= [0, 0, 0, B+\Delta B(k), 0],\\ \widetilde{F}_{02}(k) &= \left[-(C+\Delta C(k)+\beta_0 B_yN - A_y), -A_y, -(1-\beta_0)B_yN, 0, 0\right],\\ \widehat{F}_{11}(k) &= [0, 0, 0, D+\Delta D(k)]\end{aligned}$$

From Assumption 7.2, we have

$$\begin{bmatrix} y(k-\tau(k)) \\ g(y(k-\tau(k))) \end{bmatrix}^T \begin{bmatrix} \breve{U}_1 & \breve{U}_2 \\ \breve{U}_2^T & I \end{bmatrix} \begin{bmatrix} y(k-\tau(k)) \\ g(y(k-\tau(k))) \end{bmatrix} \leq 0 \tag{7.18}$$

where $\breve{U}_1$, $\breve{U}_2$ are defined in the theorem.

Then, from (7.17) and (7.18), it follows that

$$\begin{aligned} \mathbb{E}\{\Delta V(k)\} &\leq \mathbb{E}\{\Delta V(k)\} - \mathbb{E}\left\{\lambda \begin{bmatrix} y(k-\tau(k)) \\ g(y(k-\tau(k))) \end{bmatrix}^T \begin{bmatrix} \breve{U}_1 & \breve{U}_2 \\ \breve{U}_2^T & I \end{bmatrix} \begin{bmatrix} y(k-\tau(k)) \\ g(y(k-\tau(k))) \end{bmatrix}\right\} \\ &= \mathbb{E}\Big\{\xi_0^T(k)\Big[\Psi_{11} + \sigma_\alpha \widehat{C}_{01}^T P_{21} \widehat{C}_{01} + 2\widehat{F}_{01}^T(k) P_{11} \widehat{F}_{01}(k) + 2\widetilde{F}_{01}^T(k) P_{21} \widetilde{F}_{01}(k) + 2\widehat{F}_{11}^T(k)(P_{12}+P_{22})\widehat{F}_{11}(k)\Big]\xi_0(k) \\ &\quad + \eta_0^T(k)\Big[\Psi_{22} + \sigma_\beta \widehat{C}_{02}^T P_{22} \widehat{C}_{02} + 2\widehat{F}_{02}^T(k) P_{12} \widehat{F}_{02}(k) + 2\widetilde{F}_{02}^T(k) P_{22} \widetilde{F}_{02}(k) + 2\widehat{F}_{12}^T(k)(P_{11}+P_{21})\widehat{F}_{12}(k)\Big]\eta_0(k)\Big\} \end{aligned} \tag{7.19}$$

where

$$\Psi_{22} = \begin{bmatrix} \Pi_2 & 0 & 0 & 0 & 0 \\ 0 & -P_{22} & 0 & 0 & 0 \\ 0 & 0 & -P_{32} & 0 & 0 \\ 0 & 0 & 0 & -P_{42} - \lambda I & -\lambda \breve{U}_2 \\ 0 & 0 & 0 & -\lambda \breve{U}_2^T & -\lambda \breve{U}_1 \end{bmatrix}$$

Notice that, since $\Psi_1 < 0$, $\Psi_2 < 0$, there exist two sufficiently small scalars $\varepsilon_{01} > 0$, $\varepsilon_{02} > 0$ such that

$$\widehat{\Psi}_1 := \Psi_1 + \varepsilon_{01} \begin{bmatrix} I_{2n\times 2n} & 0 \\ 0 & 0 \end{bmatrix} < 0,\ \widehat{\Psi}_2 := \Psi_2 + \varepsilon_{02} \begin{bmatrix} I_{2n\times 2n} & 0 \\ 0 & 0 \end{bmatrix} < 0 \tag{7.20}$$

In order to deal with the exponential stability, we should first prove that (7.20) implies

$$\begin{aligned} &\Psi_{11} + \varepsilon_{01} \begin{bmatrix} I_{2n\times 2n} & 0 \\ 0 & 0 \end{bmatrix} + \sigma_\alpha \widehat{C}_{01}^T P_{21} \widehat{C}_{01} + 2\widehat{F}_{01}^T(k) P_{11} \widehat{F}_{01}(k) + 2\widetilde{F}_{01}^T(k) P_{21} \widetilde{F}_{01}(k) + 2\widehat{F}_{11}^T(k)(P_{12}+P_{22})\widehat{F}_{11}(k) < 0 \\ &\Psi_{22} + \varepsilon_{02} \begin{bmatrix} I_{2n\times 2n} & 0 \\ 0 & 0 \end{bmatrix} + \sigma_\beta \widehat{C}_{02}^T P_{22} \widehat{C}_{02} + 2\widehat{F}_{02}^T(k) P_{12} \widehat{F}_{02}(k) + 2\widetilde{F}_{02}^T(k) P_{22} \widetilde{F}_{02}(k) + 2\widehat{F}_{12}^T(k)(P_{11}+P_{21})\widehat{F}_{12}(k) < 0 \end{aligned} \tag{7.21}$$

In fact, by Lemma 2, the above equalities are equivalent to

$$\Psi_3(k) = \begin{bmatrix} \widehat{\Psi}_{11} & * \\ R_1(k) & T_1 \end{bmatrix} < 0, \ \Psi_4(k) = \begin{bmatrix} \widehat{\Psi}_{22} & * \\ R_2(k) & T_2 \end{bmatrix} < 0 \tag{7.22}$$

where

$$\widehat{\Psi}_{11} = \Psi_{11} + \varepsilon_{01} \begin{bmatrix} I_{2n\times 2n} & 0 \\ 0 & 0 \end{bmatrix},$$

$$\widehat{\Psi}_{22} = \Psi_{22} + \varepsilon_{02} \begin{bmatrix} I_{2n\times 2n} & 0 \\ 0 & 0 \end{bmatrix},$$

$$R_1(k) = \begin{bmatrix} \sqrt{2}P_{11}\widehat{F}_{01}(k) \\ \sqrt{2}P_{21}\widetilde{F}_{01}(k) \\ \sqrt{\sigma_\alpha}P_{21}\widehat{C}_{01} \\ \sqrt{2}(P_{12}+P_{22})\widehat{F}_{11}(k) \end{bmatrix}$$

$$= \begin{bmatrix} -\sqrt{2}P_{11}(A+\Delta A(k)) & 0 & 0 & 0 \\ -\sqrt{2}P_{21}(A+\Delta A(k)+\alpha_0 B_x M - A_x) & -\sqrt{2}P_{21}A_x & -\sqrt{2}P_{21}(1-\alpha_0)B_x M & 0 \\ \sqrt{\sigma_\alpha}P_{21}B_x M & 0 & \sqrt{\sigma_\alpha}P_{21}B_x M & 0 \\ 0 & 0 & 0 & \sqrt{2}(P_{12}+P_{21})(D+\Delta D(k)) \end{bmatrix}$$

$$R_2(k) = \begin{bmatrix} \sqrt{2}P_{12}\widehat{F}_{02}(k) \\ \sqrt{2}P_{22}\widetilde{F}_{02}(k) \\ \sqrt{\sigma_\beta}P_{22}\widehat{C}_{02} \\ \sqrt{2}(P_{11}+P_{21})\widehat{F}_{12}(k) \end{bmatrix}$$

$$= \begin{bmatrix} -\sqrt{2}P_{12}(C+\Delta C(k)) & 0 & 0 & 0 & 0 \\ -\sqrt{2}P_{22}(C+\Delta C(k)+\beta_0 B_y N - A_y) & -\sqrt{2}P_{22}A_y & -\sqrt{2}P_{22}(1-\beta_0)B_y N & 0 & 0 \\ \sqrt{\sigma_\beta}P_{22}B_y N & 0 & \sqrt{\sigma_\beta}P_{22}B_y N & 0 & 0 \\ 0 & 0 & 0 & \sqrt{2}(P_{11}+P_{21})(B+\Delta B(k)) & 0 \end{bmatrix}$$

$$T_1 = diag(-P_{11}, -P_{21}, -P_{21}, -(P_{12}+P_{22})),$$
$$T_2 = diag(-P_{12}, -P_{22}, -P_{22}, -(P_{11}+P_{21})).$$

Note that $R_1(k)$ and $R_2(k)$ can be decomposed as

$$R_1(k) = R_1 + \Delta R_1(k), \ R_2(k) = R_2 + \Delta R_2(k) \tag{7.23}$$

where

$$R_1 = \begin{bmatrix} -\sqrt{2}P_{11}A & 0 & 0 & 0 \\ -\sqrt{2}P_{21}(A+\alpha_0 B_x M - A_x) & -\sqrt{2}P_{21}A_x & -\sqrt{2}P_{21}(1-\alpha_0)B_x M & 0 \\ \sqrt{\sigma_\alpha}P_{21}B_x M & 0 & \sqrt{\sigma_\alpha}P_{21}B_x M & 0 \\ 0 & 0 & 0 & \sqrt{2}(P_{12}+P_{22})D \end{bmatrix}$$

$$\Delta R_1(k) = \begin{bmatrix} -\sqrt{2}P_{11}\Delta A(k) & 0 & 0 & 0 & 0 \\ -\sqrt{2}P_{21}\Delta A(k) & 0 & 0 & 0 & 0 \\ 0 & 0 & 0 & 0 & 0 \\ 0 & 0 & 0 & \sqrt{2}(P_{12}+P_{22})\Delta D(k) & 0 \end{bmatrix}$$

$$\Delta R_2(k) = \begin{bmatrix} -\sqrt{2}P_{12}\Delta C(k) & 0 & 0 & 0 & 0 \\ -\sqrt{2}P_{12}\Delta C(k) & 0 & 0 & 0 & 0 \\ 0 & 0 & 0 & 0 & 0 \\ 0 & 0 & 0 & \sqrt{2}(P_{11}+P_{21})\Delta B(k) & 0 \end{bmatrix}$$

$$R_2 = \begin{bmatrix} -\sqrt{2}P_{12}C & 0 & 0 & 0 & 0 \\ -\sqrt{2}P_{22}(C+\beta_0 B_y N - A_y) & -\sqrt{2}P_{22}A_y & -\sqrt{2}P_{22}(1-\beta_0)B_y N & 0 & 0 \\ \sqrt{\sigma_\beta}P_{22}B_y N & 0 & \sqrt{\sigma_\beta}P_{22}B_y N & 0 & 0 \\ 0 & 0 & 0 & \sqrt{2}(P_{11}+P_{21})B & 0 \end{bmatrix}$$

From Assumption 7.1, it follows readily that

$$\Delta R_1(k) = \overline{Q}_1 F(k)\overline{W}_1, \ \Delta R_2(k) = \overline{Q}_2 F(k)\overline{W}_2 \tag{7.24}$$

where

$$\overline{Q}_1 = \begin{bmatrix} -\sqrt{2}P_{11}Q & 0 & 0 & 0 \\ -\sqrt{2}P_{21}Q & 0 & 0 & 0 \\ 0 & 0 & 0 & 0 \\ 0 & 0 & 0 & \sqrt{2}(P_{12}+P_{22})Q \end{bmatrix} \quad \overline{W}_1 = \begin{bmatrix} W_1 & 0 & 0 & 0 \\ W_1 & 0 & 0 & 0 \\ 0 & 0 & 0 & 0 \\ 0 & 0 & 0 & W_4 \end{bmatrix}$$

$$\overline{Q}_2 = \begin{bmatrix} -\sqrt{2}P_{12}Q & 0 & 0 & 0 & 0 \\ -\sqrt{2}P_{22}Q & 0 & 0 & 0 & 0 \\ 0 & 0 & 0 & 0 & 0 \\ 0 & 0 & 0 & \sqrt{2}(P_{11}+P_{21})Q & 0 \end{bmatrix} \quad \overline{W}_2 = \begin{bmatrix} W_3 & 0 & 0 & 0 \\ W_3 & 0 & 0 & 0 \\ 0 & 0 & 0 & 0 \\ 0 & 0 & W_2 & 0 \end{bmatrix}$$

Note that $\Psi_3(k)$ and $\Psi_4(k)$ can be decomposed as follows:

$$\Psi_3(k) = \Psi_3 + \Delta\Psi_3(k), \ \Psi_4(k) = \Psi_4 + \Delta\Psi_4(k) \tag{7.25}$$

where $\Psi_3 = \begin{bmatrix} \widehat{\Psi}_{11} & * \\ R_1 & T_1 \end{bmatrix} < 0, \ \Delta\Psi_3(k) = \begin{bmatrix} 0 & * \\ \Delta R_1(k) & 0 \end{bmatrix}$,
$\Psi_4 = \begin{bmatrix} \widehat{\Psi}_{22} & * \\ R_2 & T_2 \end{bmatrix} < 0, \ \Delta\Psi_4(k) = \begin{bmatrix} 0 & * \\ \Delta R_2(k) & 0 \end{bmatrix}$.

Let $\widetilde{Q}_1^T = \left[0, \overline{Q}_1^T\right]$, $\widetilde{W}_1 = \left[\overline{W}_1, 0\right]$, $\widetilde{Q}_2^T = \left[0, \overline{Q}_2^T\right]$, $\widetilde{W}_2 = \left[\overline{W}_2, 0\right]$. Using Lemma 1 (i), $\Delta\Psi_3(k)$ and $\Delta\Psi_4(k)$ can be rewritten as

$$\begin{aligned}\Delta\Psi_3(k) &= \widetilde{Q}_1 F(k)\widetilde{W}_1 + \widetilde{W}_1^T F^T(k)\widetilde{Q}_1^T \le \varepsilon_1^{-1}\widetilde{Q}_1\widetilde{Q}_1^T + \varepsilon_1 \widetilde{W}_1^T\widetilde{W}_1 \\ \Delta\Psi_4(k) &= \widetilde{Q}_2 F(k)\widetilde{W}_2 + \widetilde{W}_2^T F^T(k)\widetilde{Q}_2^T \le \varepsilon_2^{-1}\widetilde{Q}_2\widetilde{Q}_2^T + \varepsilon_2 \widetilde{W}_2^T\widetilde{W}_2\end{aligned} \tag{7.26}$$

It is implied from (7.25) and (7.26) that

$$\Psi_3(k) \le \Psi_3' + \varepsilon_1^{-1}\widetilde{Q}_1\widetilde{Q}_1^T,\ \Psi_4(k) \le \Psi_4' + \varepsilon_2^{-1}\widetilde{Q}_2\widetilde{Q}_2^T \tag{7.27}$$

where

$$\Psi_3' = \begin{bmatrix} \Psi_{11}' + \varepsilon_{01}\begin{bmatrix} I_{2n\times 2n} & 0 \\ 0 & 0\end{bmatrix} & * \\ R_1 & T_1 \end{bmatrix} \quad \Psi_4' = \begin{bmatrix} \Psi_{22}' + \varepsilon_{02}\begin{bmatrix} I_{2n\times 2n} & 0 \\ 0 & 0\end{bmatrix} & * \\ R_2 & T_2 \end{bmatrix}$$

It follows from Lemma 2 (Schur complement) that (7.21) is equivalent to the fact that the right sides of (7.27) is negative definite. Thus, we arrive at the conclusion $\Psi_3(k) < 0$, $\Psi_4(k) < 0$ and there for (7.21) holds. Furthermore, the combination of (7.19) and (7.21) leads to

$$\mathbb{E}\{\Delta V(k)\} \le -\varepsilon_{01}\mathbb{E}\left\{|\bar{x}(k)|^2\right\} - \varepsilon_{02}\mathbb{E}\left\{|\bar{y}(k)|^2\right\} \tag{7.28}$$

Now, let us proceed with the exponential stability of the system (7.8). First, from (7.10), it is readily verified that

$$\mathbb{E}\{\Delta V(k)\} \le \rho_{11}\mathbb{E}\left\{|\bar{x}(k)|^2\right\} + \rho_{21}\sum_{i=k-\sigma_M}^{k-1}\mathbb{E}\left\{|\bar{x}(i)|^2\right\} + \rho_{12}\mathbb{E}\left\{|\bar{y}(k)|^2\right\} + \rho_{22}\sum_{i=k-\tau_M}^{k-1}\mathbb{E}\left\{|\bar{y}(i)|^2\right\} \tag{7.29}$$

where

$$\begin{aligned}\rho_{11} &= \max\{\lambda_{\max}(P_{11}), \lambda_{\max}(P_{21})\} \\ \rho_{11} &= (\sigma_M - \sigma_m + 1)(\lambda_{\max}(P_{31}) + \lambda_{\max}(P_{41})) \\ \rho_{12} &= \max\{\lambda_{\max}(P_{21}), \lambda_{\max}(P_{22})\} \\ \rho_{22} &= (\tau_M - \tau_m + 1)(\lambda_{\max}(P_{32}) + \lambda_{\max}(P_{42}))\end{aligned}$$

For any scalar $\mu > 1$, the above inequality, together with (7.28), implies that

$$\mu^{k+1}\mathbb{E}\{V(k+1)\} - \mu^{k}\mathbb{E}\{V(k)\} = \mu^{k+1}\mathbb{E}\{\Delta V(k)\} + \mu^{k}(\mu-1)\mathbb{E}\{V(k)\}$$
$$\leq -\mu^{k+1}\left(\varepsilon_{01}\mathbb{E}\left\{|\bar{x}(k)|^2\right\} + \varepsilon_{02}\mathbb{E}\left\{|\bar{y}(k)|^2\right\}\right) + \mu^{k}(\mu-1).$$
$$\left(\rho_{11}\mathbb{E}\left\{|\bar{x}(k)|^2\right\} + \rho_{21}\sum_{i=k-\sigma_M}^{k-1}\mathbb{E}\left\{|\bar{x}(i)|^2\right\} + \rho_{12}\mathbb{E}\left\{|\bar{y}(k)|^2\right\} + \rho_{22}\sum_{i=k-\tau_M}^{k-1}\mathbb{E}\left\{|\bar{y}(i)|^2\right\}\right)$$
$$= \mu^{k}\omega_{11}(\mu)\mathbb{E}\left\{|\bar{x}(k)|^2\right\} + \mu^{k}\omega_{21}(\mu)\sum_{i=k-\sigma_M}^{k-1}\mathbb{E}\left\{|\bar{x}(i)|^2\right\} + \mu^{k}\omega_{21}(\mu)\mathbb{E}\left\{|\bar{y}(k)|^2\right\} + \mu^{k}\omega_{21}(\mu)\sum_{i=k-\tau_M}^{k-1}\mathbb{E}\left\{|\bar{y}(i)|^2\right\} \tag{7.30}$$

where

$$\omega_{11}(\mu) = -\mu\varepsilon_{01} + (\mu-1)\rho_{11},$$
$$\omega_{21}(\mu) = (\mu-1)\rho_{21},$$
$$\omega_{12}(\mu) = -\mu\varepsilon_{02} + (\mu-1)\rho_{12},$$
$$\omega_{22}(\mu) = (\mu-1)\rho_{22}.$$

Furthermore, for any integer $N \geq \max\{\tau_M, \sigma_M\} + 1$, summing both sides of (7.30) from 0 to $N-1$ with respect to k, we have

$$\begin{aligned}\mu^{N}\mathrm{E}\{V(N)\} - \mathrm{E}\{V(0)\} \leq\ & \omega_{11}(\mu)\sum_{k=0}^{N-1}\mu^{k}\mathrm{E}\left\{|\bar{x}(k)|^2\right\} + \omega_{21}(\mu)\sum_{k=0}^{N-1}\sum_{i=k-\sigma_M}^{k-1}\mu^{k}\mathrm{E}\left\{|\bar{x}(i)|^2\right\}\\ & + \omega_{12}(\mu)\sum_{k=0}^{N-1}\mu^{k}\mathrm{E}\left\{|\bar{y}(k)|^2\right\} + \omega_{22}(\mu)\sum_{k=0}^{N-1}\sum_{i=k-\tau_M}^{k-1}\mu^{k}\mathrm{E}\left\{|\bar{y}(i)|^2\right\}\end{aligned} \tag{7.31}$$

Note that for $\sigma_M, \tau_M \geq 1$,

$$\begin{aligned}\sum_{k=0}^{N-1}\sum_{i=k-\sigma_M}^{k-1}\mu^{k}\mathrm{E}\left\{|\bar{x}(i)|^2\right\} &\leq \sigma_M\mu^{\sigma_M}\max_{-\sigma_M \leq i \leq 0}\mathrm{E}\left\{|\psi(i)|^2\right\} + \sigma_M\mu^{\sigma_M}\sum_{i=0}^{N-1}\mu^{i}\mathrm{E}\left\{|\bar{x}(k)|^2\right\}\\ \sum_{k=0}^{N-1}\sum_{i=k-\tau_M}^{k-1}\mu^{k}\mathrm{E}\left\{|\bar{y}(i)|^2\right\} &\leq \tau_M\mu^{\tau_M}\max_{-\tau_M \leq i \leq 0}\mathrm{E}\left\{|\varphi(i)|^2\right\} + \tau_M\mu^{\tau_M}\sum_{i=0}^{N-1}\mu^{i}\mathrm{E}\left\{|\bar{y}(k)|^2\right\}\end{aligned} \tag{7.32}$$

Then from (7.31) and (7.32), one has

$$\begin{aligned}\mu^{N}\mathrm{E}\{V(N)\} \leq\ & \mathrm{E}\{V(0)\} + [\omega_{11}(\mu) + \sigma_M\mu^{\sigma_M}\omega_{21}(\mu)]\sum_{k=0}^{N-1}\mu^{k}\mathrm{E}\left\{|\bar{x}(k)|^2\right\}\\ & + \sigma_M\mu^{\delta_M}\omega_{21}(\mu)\max_{-\delta_M \leq i \leq 0}\mathrm{E}\left\{|\psi(i)|^2\right\}\\ & + [\omega_{12}(\mu) + \tau_M\mu^{\tau_M}\omega_{22}(\mu)]\sum_{k=0}^{N-1}\mu^{k}\mathrm{E}\left\{|\bar{y}(k)|^2\right\}\\ & + \tau_M\mu^{\tau_M}\omega_{22}(\mu)\max_{-\tau_M \leq i \leq 0}\mathrm{E}\left\{|\varphi(i)|^2\right\}\end{aligned} \tag{7.33}$$

Let $\rho_{01} = \min\{\lambda_{\min}(P_{11}), \lambda_{\min}(P_{21})\}$, $\rho_1 = \max\{\rho_{12}, \rho_{22}\}$, $\rho_2 = \max\{\rho_{12}, \rho_{22}\}$, $\rho_{02} = \min\{\lambda_{\min}(P_{12}), \lambda_{\min}(P_{22})\}$.

It is obvious that

$$\mathrm{E}\{V(N)\} \geq \rho_{01}\mathrm{E}\left\{|\bar{x}(N)|^2\right\} + \rho_{02}\mathrm{E}\left\{|\bar{y}(N)|^2\right\} \tag{7.34}$$

It also follows readily from (7.29) that

$$\mathrm{E}\{V(0)\} \leq \rho_1 \max_{-\delta_M \leq i \leq 0} \mathrm{E}\left\{|\psi(i)|^2\right\} + \rho_2 \max_{-\tau_M \leq i \leq 0} \mathrm{E}\left\{|\varphi(i)|^2\right\} \tag{7.35}$$

In addition, it can be verified that there exist a scalar $\mu_0 > 1$ such that

$$\omega_{11}(\mu_0) + \sigma_M \mu_0^{\delta_M} \omega_{21}(\mu_0) = 0, \omega_{12}(\mu_0) + \tau_M \mu_0^{\tau_M} \omega_{22}(\mu_0) = 0 \tag{7.36}$$

Substituting (7.34)–(7.36) into (7.33), we can get

$$\begin{aligned}\rho_{01}\mathrm{E}\left\{|\bar{x}(N)|^2\right\} + \rho_{02}\mathrm{E}\left\{|\bar{y}(N)|^2\right\} &\leq \left(\rho_1 + \sigma_M \mu_0^{\delta_M} \omega_{21}(\mu_0)\right) \max_{-\delta_M \leq i \leq 0} \mathrm{E}\left\{|\psi(i)|^2\right\} \\ &+ \left(\rho_2 + \tau_M \mu_0^{\delta_M} \omega_{22}(\mu_0)\right) \max_{-\tau_M \leq i \leq 0} \mathrm{E}\left\{|\varphi(i)|^2\right\}\end{aligned} \tag{7.37}$$

This indicates that the estimation error system (7.8) with $v_x(k) = 0$ and $v_y(k) = 0$ is robustly exponentially stable in the mean square. And the proof of Theorem 7.1 is completed.

Remark 2 It should be pointed out that the filtering problem for the discrete-time stochastic GRNs with probabilistic measurement delays is of the form as in (7.7), which is similar to [27] while in discrete sense.

Next, we are proceeding to consider the H_∞ performance of the estimation error system (7.8) under zero initial condition for all nonzero$v_x(k), v_y(k) \in L_2[0, \infty)$.

Theorem 7.2 *Let the estimator parameters A_x, B_x, A_y, B_y and $\gamma > 0$ be given. Then, the estimation error system* (7.8) *is robustly stochastically stable with disturbance attenuation γ, if there exist eight positive definite matrices P_{11}, P_{12}, P_{21},P_{22}, P_{31}, P_{32}, P_{41}, P_{42}, and three positive constant scalars λ, ε_1, and ε_2 such that the following LMI holds:*

$$\psi_1 = \begin{bmatrix} \psi'_{11} & * & * & * \\ 0 & -\gamma^2 I & * & * \\ R'_1 & 0 & T_1 & * \\ 0 & 0 & Q_1^{-T} & -\varepsilon_1 I \end{bmatrix} < 0, \; \psi_2 = \begin{bmatrix} \psi'_{22} & * & * & * \\ 0 & -\gamma^2 I & * & * \\ R'_2 & 0 & T_2 & * \\ 0 & 0 & Q_2^{-T} & -\varepsilon_2 I \end{bmatrix} < 0 \tag{7.38}$$

and the other variables are the same as defined in Theorem 7.1.

Proof Consider the same Lyapunov-Krasovskii function (7.10) as in Theorem 7.1. For a given $\gamma > 0$, we define

$$J(n) = \mathrm{E}\sum_{k=0}^{n}\left[\bar{x}^T(k)\bar{x}(k) + \bar{y}^T(k)\bar{y}(k) - \gamma^2 v_x^T(k)v_x(k) - \gamma^2 v_y^T(k)v_y(k) + \Delta V(k)\right] \tag{7.39}$$

where n is non-negative integer. Obviously, our goal is to show $J(n) < 0$.

Under the zero initial condition, we have

$$\begin{aligned} J(n) &= \mathrm{E}\sum_{k=0}^{n}\left[\bar{x}^T(k)\bar{x}(k) + \bar{y}^T(k)\bar{y}(k) - \gamma^2 v_x^T(k)v_x(k) - \gamma^2 v_y^T(k)v_y(k) + \Delta V(k)\right] - \mathrm{E}V(n+1) \\ &\le J(n) + \sum_{k=0}^{n}\mathrm{E}(\Delta V(k)) \\ &= \sum_{k=0}^{n}\mathrm{E}\Big\{\xi^T(k)[\bar{\Psi}_{11} + \sigma_\alpha \overline{C}_{01}^T P_{21}\overline{C}_{01} + 2\overline{F}_{01}^T P_{11}\overline{F}_{01}(k) + 2\breve{F}_{01}^T(k)P_{21}\breve{F}_{01}(k) + 2\overline{F}_{11}^T(k)(P_{12}+P_{22})\overline{F}_{11}(k)]\xi(k) \\ &\quad + \eta^T(k)\left[\bar{\Psi}_{22} + \sigma_\beta \overline{C}_{02}^T P_{22}\overline{C}_{02} + 2\overline{F}_{02}^T P_{12}\overline{F}_{02}(k) + 2\breve{F}_{02}^T(k)P_{22}\breve{F}_{02}(k) + 2\overline{F}_{12}^T(k)(P_{11}+P_{21})\overline{F}_{12}(k)\right]\eta(k)\Big\} \end{aligned} \tag{7.40}$$

where

$$\begin{aligned} \xi(k) &= [\xi_0(k), v_x(k)]^T,\ \eta(k) = [\eta_0(k), v_y(k)]^T, \\ \overline{C}_{01} &= \left[\widehat{C}_{01}^T, 0\right],\ \overline{F}_{01} = \left[\widehat{F}_{01}^T(k), 0\right], \\ \breve{F}_{01}(k) &= \left[\breve{F}_{01}^T(k), 0\right],\ \overline{F}_{11}(k) = \left[\breve{F}_{11}^T(k), 0\right], \\ \overline{C}_{02} &= \left[\widehat{C}_{02}^T, 0\right],\ \overline{F}_{02} = \left[\widehat{F}_{02}^T(k), 0\right], \\ \breve{F}_{02}(k) &= \left[\breve{F}_{02}^T(k), 0\right],\ \overline{F}_{12}(k) = \left[\breve{F}_{12}^T(k), 0\right], \\ \bar{\psi}_{11} &= \begin{bmatrix}\psi_{11} & 0 \\ 0 & -\gamma^2 I\end{bmatrix},\ \bar{\psi}_{22} = \begin{bmatrix}\psi_{22} & 0 \\ 0 & -\gamma^2 I\end{bmatrix}. \end{aligned}$$

By (7.40), in order to guarantee $J(n) < 0$, we just need to show

$$\bar{\psi}_{11}+\sigma_\alpha\overline{C}_{01}^T P_{21}\overline{C}_{01}+2\overline{F}_{01}^T P_{11}\overline{F}_{01}(k)+2\breve{F}_{01}^T(k)P_{21}\breve{F}_{01}(k)+2\overline{F}_{11}^T(k)(P_{12}+P_{22})\overline{F}_{11}(k)<0$$
$$\bar{\psi}_{22}+\sigma_\beta\overline{C}_{02}^T P_{22}\overline{C}_{02}+2\overline{F}_{02}^T P_{12}\overline{F}_{02}(k)+2\breve{F}_{02}^T(k)P_{22}\breve{F}_{02}(k)+2\overline{F}_{12}^T(k)(P_{11}+P_{21})\overline{F}_{12}(k)<0 \tag{7.41}$$

which, by Lemma 2, are equivalent to

$$\bar{\psi}_3(k)=\begin{bmatrix}\bar{\psi}_{11} & * \\ \bar{R}_1(k) & T_1\end{bmatrix}<0,\ \bar{\psi}_4(k)=\begin{bmatrix}\bar{\psi}_{22} & * \\ \bar{R}_2(k) & T_2\end{bmatrix}<0 \tag{7.42}$$

where

$$\overline{R}_1(k)=\overline{R}_1+\Delta\overline{R}_1(k)=[R_1,0]+\left[\Delta\overline{R}_1(k),0\right],$$
$$\overline{R}_2(k)=\overline{R}_2+\Delta\overline{R}_2(k)=[R_2,0]+\left[\Delta\overline{R}_2(k),0\right],$$

T_1 and T_2 are defined in Theorem 7.1.

Note that $\bar{\psi}_3(k)$ and $\bar{\psi}_4(k)$ can be rearranged as follows:

$$\bar{\psi}_3(k)=\bar{\psi}_3+\Delta\bar{\psi}_3(k),\bar{\psi}_4(k)=\bar{\psi}_4+\Delta\bar{\psi}_4(k) \tag{7.43}$$

where $\bar{\psi}_3=\begin{bmatrix}\bar{\psi}_{11} & * \\ \overline{R}_1 & T_1\end{bmatrix}<0,\ \bar{\psi}_4=\begin{bmatrix}\bar{\psi}_{22} & * \\ \overline{R}_2 & T_2\end{bmatrix}<0,\quad \Delta\bar{\psi}_3(k)=\begin{bmatrix}0 & * \\ \Delta\overline{R}_1(k) & 0\end{bmatrix}$, $\Delta\bar{\psi}_4(k)=\begin{bmatrix}0 & * \\ \Delta\overline{R}_2(k) & 0\end{bmatrix}$.

Let $\widehat{Q}_1^T=\left[0,\breve{Q}_1^T\right]$, $\widehat{W}_1=\left[\breve{W}_1,0\right]$, $\widehat{Q}_2^T=\left[0,\breve{Q}_2^T\right]$, $\widehat{W}_2=\left[\breve{W}_2,0\right]$, $\breve{Q}_1^T=\left[0,0,\overline{Q}_1^T\right]$, $\breve{W}_1=\left[\overline{W}_1,0,0\right]$, $\breve{Q}_2^T=\left[0,0,\overline{Q}_2^T\right]$, $\breve{W}_2=\left[\overline{W}_2,0,0\right]$.

Using Lemma 1 (i), $\Delta\bar{\psi}_3(k)$ and $\Delta\bar{\psi}_4(k)$ can be rewritten as

$$\begin{aligned}\Delta\bar{\psi}_3(k)&=\widehat{Q}_1F(k)\widehat{W}_1+\widehat{W}_1^TF^T(k)\widehat{Q}_1^T\le\varepsilon_1^{-1}\widehat{Q}_1\widehat{Q}_1^T+\varepsilon_1\widehat{W}_1^T\widehat{W}_1\\ \Delta\bar{\psi}_4(k)&=\widehat{Q}_2F(k)\widehat{W}_2+\widehat{W}_2^TF^T(k)\widehat{Q}_2^T\le\varepsilon_2^{-1}\widehat{Q}_2\widehat{Q}_2^T+\varepsilon_2\widehat{W}_2^T\widehat{W}_2\end{aligned} \tag{7.44}$$

It is implied from (7.43) and (7.44) that

$$\bar{\psi}_3(k)\le\begin{bmatrix}\psi_{11}' & * & * \\ 0 & -\gamma^2I & * \\ R_2 & 0 & T_1\end{bmatrix}+\varepsilon_1^{-1}\widehat{Q}_1\breve{Q}_1^T,\bar{\psi}_4(k)\le\begin{bmatrix}\psi_{22}' & * & * \\ 0 & -\gamma^2I & * \\ R_2 & 0 & T_2\end{bmatrix}+\varepsilon_2^{-1}\widehat{Q}_2\breve{Q}_2^T \tag{7.45}$$

By Lemma 2, the above inequality (7.44) holds if and only if the right-hand side of (7.44) is negative definite, which implies *J(n)* < 0. Letting $n\to\infty$, we have

$$\mathrm{E}\sum_{k=0}^{\infty}\left\{|\bar{x}(k)|^2+|\bar{y}(k)|^2\right\}\leq\gamma^2\mathrm{E}\sum_{k=0}^{\infty}\left\{|v_x(k)|^2+|v_y(k)|^2\right\}$$

which completes the proof of the theorem.

The desired estimator parameters can be determined in the following theorem by solving a set of LMIs.

Theorem 7.3 *Under Assumptions, system* (7.7) *becomes a robust H_∞ state estimator of GRNs* (7.1) *with probabilistic measurement delays* (7.5) *if there exist twelve positive definite matrices X_1, X_2, Y_1, Y_2, P_{11}, P_{12}, P_{21},P_{22}, P_{31}, P_{32}, P_{41}, P_{42}, and three positive constant scalars λ, ε_1, and ε_2 such that the following LMIs hold:*

$$\psi_1=\begin{bmatrix}\psi'_{11} & * & * & * \\ 0 & -\gamma^2 I & * & * \\ R'_1 & 0 & T_1 & * \\ 0 & 0 & Q_1^{-T} & -\varepsilon_1 I\end{bmatrix}<0\ \psi_2=\begin{bmatrix}\psi'_{22} & * & * & * \\ 0 & -\gamma^2 I & * & * \\ R'_2 & 0 & T_2 & * \\ 0 & 0 & Q_2^{-T} & -\varepsilon_2 I\end{bmatrix}<0 \tag{7.46}$$

where

$$R'_1=\begin{bmatrix}-\sqrt{2}P_{11}A & 0 & 0 & 0 \\ -\sqrt{2}(P_{21}A+\alpha_0X_2M-X_1) & -\sqrt{2}X_1 & -\sqrt{2}(1-\alpha_0)X_2M & 0 \\ \sqrt{\sigma_\alpha}X_2M & 0 & \sqrt{\sigma_\alpha}X_2M & 0 \\ 0 & 0 & \sqrt{2}(P_{12}+P_{22})D & 0\end{bmatrix}$$

$$R'_2=\begin{bmatrix}-\sqrt{2}P_{12}C & 0 & 0 & 0 & 0 \\ -\sqrt{2}(P_{22}C+\beta_0Y_2N-Y_1) & -\sqrt{2}Y_1 & -\sqrt{2}(1-\beta_0)Y_2N & 0 & 0 \\ \sqrt{\sigma_\beta}Y_2N & 0 & \sqrt{\sigma_\beta}Y_2N & 0 & 0 \\ 0 & 0 & 0 & \sqrt{2}(P_{11}+P_{21})B & 0\end{bmatrix}$$

and the other variables are the same as defined in Theorem 7.2. *Moreover, the state estimator gain matrices can be designed as follows:*

$$A_x=P_{21}^{-1}X_1,\ B_x=P_{21}^{-1}X_2,\ A_y=P_{22}^{-1}Y_1,\ B_y=P_{22}^{-1}Y_2.$$

Proof The remaining proof for global robust stability is similar to those in the proof of Theorem 7.2. For the sake of simplicity, we omit it here. The proof of Theorem 7.3 is then completed.

7.2.3 Numerical Example

In this section, we present an example to illustrate the effectiveness of the proposed robust state estimator. The dynamics of the repressilator which is a cyclic negative-feedback loop composing of three genes and their corresponding promoters, have been theoretically predicted and experimentally on *Escherichia coli* [19]. Extending the example in [8] by taking into account the transcriptional time delay, norm-bounded uncertainties and stochastic disturbance, assume that system (7.1) with probabilistic measurement delays (7.5) has the following parameters:

$$A = 0.1I\ B = \begin{bmatrix} 0 & 0 & -0.5 \\ -0.5 & 0 & 0 \\ 0 & -0.5 & 0 \end{bmatrix}$$
$$C = D = 0.08I\ M = diag\{0.72, 0.1, 1.3\}$$
$$N = diag\{1.2, 3.1, 0.3\}\ \alpha_0 = 0.2\ \beta_0 = 0.3$$

The parameter uncertainties satisfy Assumption 7.1 with

$$W_1 = W_2 = W_3 = W_4 = 0.1I\ Q = diag\{0.1, 0.3, 0.2\}\ U_1 = 0$$
$$U_2 = diag\{0.65, 0.65, 0.65\}\ H = diag\{0.1, 0.1, 0.1\}$$

The nonlinear regulatory functions $g_i = (\cdot)\ (i = 1, 2)$ are taken as the usually used Hill function, that is, $g(s) = s^2 / 1 + s^2$ The time-varying delays are chosen as $\tau_m = 1$, $\tau_M = 10$, $\sigma_m = 1$, $\sigma_M = 7$. The unknown time-varying matrices $F_i(k)$ (i = 1,2) take values randomly in interval [−1,1]. In this example, the disturbance attenuation level is given as $\gamma = 1$.

By using the LMI toolbox [53], we solve (7.46) and obtain

$$X_1 = 10^5 * diag\{1.4794, 2.0244, 1.2196\}, X_2 = 10^5 * diag\{2.5465, 24.407, 1.1540\}$$
$$Y_1 = 10^5 * diag\{13.955, 8.3399, 8.6990\}, Y_2 = 10^7 * diag\{0.1485, 0.0299, 3.4204\}$$
$$P_{11} = 10^8 * diag\{0.9003, 1.0503, 0.8450\}, P_{12} = 10^8 * diag\{6.1982, 4.7739, 4.9343\}$$
$$P_{21} = 10^5 * diag\{6.3129, 8.4690, 5.1765\}, P_{22} = 10^6 * diag\{5.5071, 2.9641, 3.3308\}$$
$$P_{31} = 10^5 * diag\{2.6201, 3.4870, 2.1430\}, P_{32} = 10^6 * diag\{2.3734, 1.2729, 1.4549\}$$
$$P_{41} = 10^7 * diag\{1.0651, 1.2491, 0.9960\}, P_{42} = 10^7 * diag\{6.0361, 4.6102, 4.7925\}$$
$$\lambda = 8.5642 * 10^{-4},\ \varepsilon_1 = 2.1336 * 10^8,\ \varepsilon_1 = 2.1358 * 10^8$$

According to Theorem 7.3, the state estimator gain matrices can be calculated as follows:

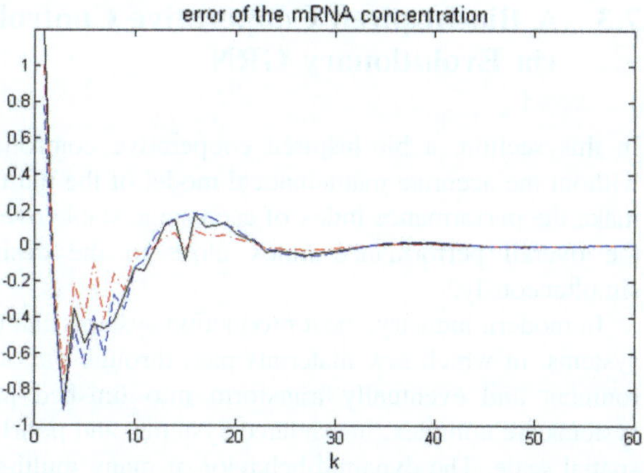

Fig. 7.1 Estimation error $\tilde{x}(k)$

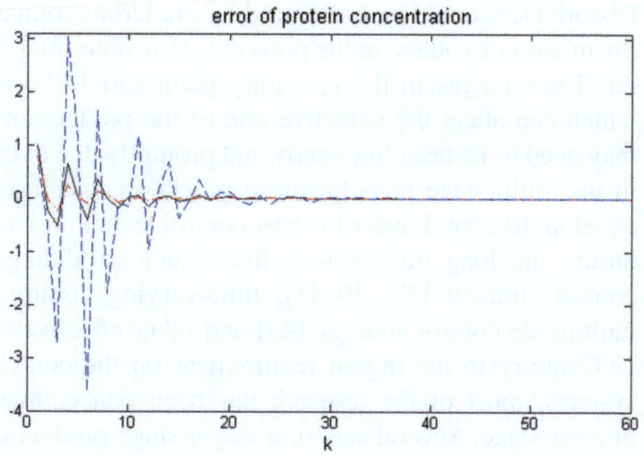

Fig. 7.2 Estimation error $\tilde{y}(k)$

$$A_x = diag\{0.2343, 0.2390, 0.2356\},\ A_y = diag\{0.2534, 0.2814, 0.2612\}$$
$$B_X = diag\{0.4034, 2.8819, 0.2229\},\ B_Y = diag\{0.2696, 0.1010, 10.2691\}$$

In the simulation, the exogenous disturbance inputs are select as $v_x(k) = \sin(6k)\exp(-0.1k)$, $v_y(k) = \cos(2k)\exp(-0.2k)$, $L_x = [1, 0.5, 1.2]^T$, $L_y = [0.7, 0.1, 1.5]^T$ and the initial values are select as $x(0) = [1.2, 1.0, 0.8]$, $y(0) = [1.0, 0.8, 0.7]$. Simulation results are present in Figs. 7.1 and 7.2 where the estimation errors of the concentrations of the mRNAs and proteins are plotted, respectively. The simulation results have confirmed that the designed robust H_∞ state estimator performs very well.

7.3 A Bio-Inspired Cooperative Controller via Evolutionary GRN

In this section, a bio-inspired cooperative controller via evolutionary of GRN, without the accurate mathematical model of the multi-stage system, is presented to make the performance index of each stage stable over the whole study horizon and the overall performance index close to the desired value against vibrations simultaneously.

In modern industry, most production systems can be characterized as multi-stage systems, in which raw materials pass through various processes in different environment and eventually transform into finished products. Most of multi-stage systems are complex, interrelated systems, and poorly understood in both time and spatial scale. The dynamic behavior of many multi-stage systems is heavily influenced by disturbances and, in particular, the changes in the operating point. Disturbances could be introduced by numerous sources, such as load variations, the fluctuation of voltage in the power system, abnormal conditions in machines, and so on. The changes in the operating point should be avoided as much as possible, which can affect the defective rate of the production process. The operating point may need to be reset frequently and promptly due to unpredictable random variation in the multi-stage manufacturing processes. Increasing attention has been paid to develop different kinds of robust control techniques against the vibrations, occurred during the long transmission lines, such as sliding mode control [35, 36], H_∞ control strategy [37, 39–41], time-varying control law [42], the piezo-electric multimode control strategy [43] and other effective controllers [38, 44, 45].

Contrary to the urgent requirement on the control scheme for the multi-stage systems, most of the research has been concentrated on the control of a single process stage. Several sets of a single-stage production system, which is the major component of the multi-stage system, are connected in series. The output of one production stage serves as the input to the subsequent stage. At the level of the multi-stage system, the efficiency and stability of operation could be improved substantially by integrating different distributed control devices and strategies. The control scheme for multi-stage system could maintain the quality levels of not only semi-finished products in one single-stage manufacturing system but also the finished products in the multi-stage system. Dou and Wang [46] took a systematic investigation to analyze the impacts of different parameter variations on system robustness in a multi-stage printing system. The system analysis on the vulnerabilities of the electric power grid and associated communication network was developed in the face of intermitted power generation and uncertain demand [47]. Most recently, Lei et al. developed system-level design optimization methods for electrical drive systems [48].

Recently, many novel control techniques, which do not need the mathematical modeling and require less system information, have attracted extensive attention, such as an iterative learning control [49, 50] and biological inspired control system [51–56]. Liang et al. studied regulating mechanism in neuroendocrine and the endocrine system in the human body respectively and designed a bio-inspired controller against external disturbance [51, 53]. Jin et al. developed a morphogenetic approach using a GRN for adaptive multi-robot pattern generation and formation in changing environment [53–55]. An intelligence-based supervisory control strategy was proposed to adjust the set-points for the distributed controlled grinding system in response to the changes in boundary conditions or the imminent overload condition [56].

7.3.1 Biological Background

Networks in human body, which maintain physical well being, offer fabulous examples of control mechanism with efficiency and accuracy. Studies on networks in human body can contribute to engineering applications. Many bio-inspired controllers show better performances compared with the traditional approaches and have been applied to industrial processes [51, 52].

GRNs are collections of DNA segments in a gene which interact with each other indirectly through their mRNA, protein expression products and other substance. Development of the animal body plan is controlled by large GRNs. GRNs consist largely of the functional linkages among regulatory genes that produce transcription factors, and their target cis-regulatory modules in other regulatory genes, together with genes that express spatially important signaling components [57]. mRNAs serve as a template to synthesize proteins (translation), which accumulate to give the genes particular structural properties. Some proteins can adjust the transcription process in turn. The whole scenario is shown from a system level point of view in Fig. 7.3. The structure of the GRN is inherently hierarchical, that could help us improve the traditional controllers, while not changing the existing controller structure. Theoretical analysis on GRNs have attracted increasing attention, and made significant contributions for understanding both static and dynamic behaviors of biological systems in details [9, 13, 20, 58–61] and references therein.

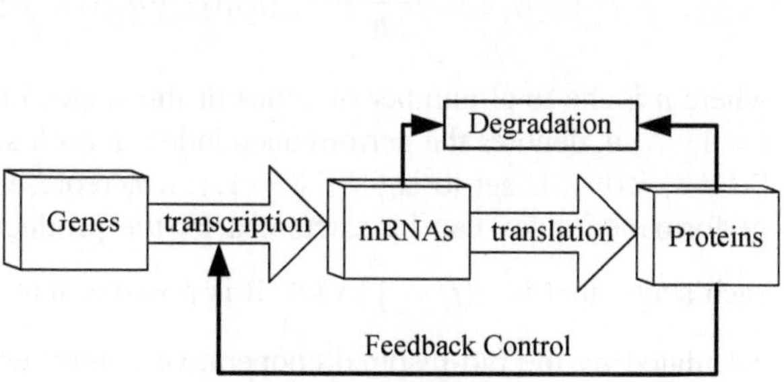

Fig. 7.3 Mechanism of protein synthesis in GRNs

Fig. 7.4 Diagram of the GRN with n genes

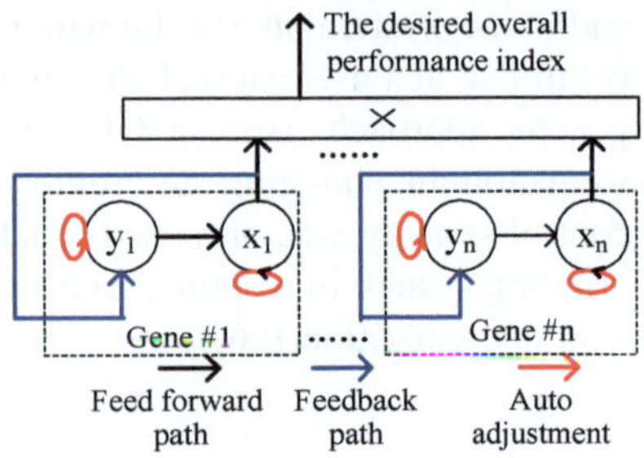

7.3.2 A GRN-Based Controller for Multi-Stage System

To employ the genetic mechanisms in the multi-stage system in manufacturing, it is necessary to establish a metaphor between GRNs and the multi-stage system. In this metaphor, a single gene is mapped to one stage process. There are two types of protein products, namely x and y. Protein type y represents an internal state vector to influence protein type x, which does not have an exact physical meaning that maps to one stage process. Protein type x is mapped to the quality of the product in one stage process. Either protein type x or y can auto-regulate itself. The mechanism of GRNs is shown in Fig. 7.4.

There are n genes in the GRNs. Protein y can be regulated by the difference between the desired and actual overall performance index, i.e. the product of the performance index in each stage process, $x_1 \times \cdots \times x_n$. The performance index is a term that indicates the operating conditions of the machine and the quality of the product.

Based on the structure illustrated in Fig. 7.4, the dynamics of the GRN controller for the multi-stage system can be described by the following differential equations:

$$\begin{aligned} \frac{dx_i(t)}{dt} &= -\alpha x_i(t) + af(y_i(t)) \\ \frac{dy_i(t)}{dt} &= -\beta y_i(t) + bx_i(t), \ 1 \leq i \leq n \end{aligned} \tag{7.47}$$

where n is the total number of genes or the stages in the manufacturing system. x_i, $i = 1, \ldots, n$, denotes the performance index in each stage, which is always positive. So if $x_i < 0$, x_i is set to $|x_i| \cdot y_i$, $i = 1, \ldots, n$, represents internal states. The overall performance index can be calculated by the product of the performance index in each stage, that is, $z(t) = \prod_1^n x_i(t)$. It is possible that the overall performance index calculated by the bio-inspired cooperative controller maybe not appropriate in the manufacturing. Then the control input should be scaled to the fit the situation in the manufacturing according the experience, that is $\tilde{z}(t) = kz(t) = \prod_1^n k_i x_i(t)$. α, a and

β, b are constants with α and β representing the decay rates of protein type x and y, respectively.

Here, $f(y(t))$ is defined to the following sigmoid function:

$$f(y(t)) = \frac{1 - e^{-y(t)}}{1 + e^{-y(t)}}. \tag{7.48}$$

From Eq. (7.47), we can see that the bio-inspired cooperative controller via evolution of gene regulatory network aims to achieve the desired overall performance. The performance index in each stage process is regulated by the internal state vector y. As shown in Fig. 7.4, there are several intricately-interplayed regulation terms: feedback path, feed forward path, and auto regulation (indicated by the arrow). The intricate interplay of different regulations in the bio-inspired cooperative controller makes the overall quality of products be robust to the disturbances. Auto-regulation and coherent regulation are helpful for preventing the solution being local optimum. The problem of being trapped in a local optimum has not been fully resolved in the proposed model. However, the probability of the performance index being trapped in a local optimum can be reduced significantly in this model.

So far, we have introduced the bio-inspired cooperative controller for the multi-stage system. The most important question, whether the bio-inspired controller (7.47) is stable or not, remains to be answered. If the stability of the bio-inspired controller (7.47) is guaranteed, the actual performance index will converge to the desired one. We will address the issue in the next Sect.

1. Stability analysis of the bio-inspired cooperative controller

In this subsection, we will analyze the dynamics of the bio-inspired cooperative controller (7.47) and provide a theoretical proof of the system stability.

Lemma 1 *For the activation sigmoid function defined in Eq.* (7.48), $|f(x)| \leq |x|$ *holds for all* x.

The details for the proof can be referred in the Ref. [53].

Theorem 7.4 *The overall performance index will converge to the desired one and the internal state y(t) in Eq.* (7.47) *will converge to zero, provided that* $\alpha > 0, \beta > 0$, *and* $a + b < 0$.

Proof To proceed with the stability analysis of the Eq. (7.47), we construct the following Lyapunov-Krasovskii function:

$$V(t) = \frac{1}{2}x^2(t) + \frac{1}{2}y^2(t) \tag{7.49}$$

Calculating the difference of *V(t)* along the trajectory of Eq. (7.47), we have

$$\begin{aligned}\frac{dV(t)}{dt} &= x(t)\frac{dx(t)}{dt} + y(t)\frac{dy(t)}{dt}\\ &= x(t)[-\alpha x(t) + af(y(t))] + y(t)[-\beta y(t) + bx(t)]\\ &= -\alpha x^2(t) + ax(t)f(y(t)) - \beta y^2(t) + bx(t)y(t)\\ &(\text{from Lemma 1, we get } |f(y(t))| lt; |y(t)|) \leq -\alpha x^2(t) + a|x(t)||y(t)| - \beta y^2(t) + b|x(t)||y(t)|\\ &= -\alpha x^2(t) + (a+b)|x(t)||y(t)| - \beta y^2(t).\end{aligned} \tag{7.50}$$

If we set

$$\alpha > 0, \beta > 0, a + b < 0 \tag{7.51}$$

we can ensure $dV/dt < 0$. And we could get $x(t) \rightarrow 0$ as $t \rightarrow \infty$.

Since both conditions of the Lypunov functions V, $V > 0$ and $dV/dt < 0$, have been satisfied, we can claim that the bio-inspired cooperative controller (7.47) is stable. It is well known that the system will be stabilized at the points that satisfy $dV/dt < 0$.

2. Parameter selection through multi-objective optimization

The proof of the system stability indicates the bio-inspired controller will be stable under which conditions, but cannot guarantee the controller performance. In this subsection, we will further optimize the system performance by tuning the parameters in the GRN controller model. Two performance indices namely, the absolute value of the error between the desired performance index and the actual one, $|z(t) - z_{desired}|$, and the settling time, T_s, are adopted to evaluate the bio-inspired cooperative controller performance. Settling time is defined as the time required for the controller to settle within a certain percentage, δ, of the steady-state of the response. This can be seen as a multi-objective optimization (MOO) problem, where the objective function is a vector instead of a scalar value. As a consequence, a number of Pareto-optimal solutions should be achieved, no longer one single optimal solution.

In this section, NSGA-II [62], which can find good spread of solutions and convergence near the true Pareto-optimal front in most problems for multi-objective optimization, has been applied to optimize the parameters of the bio-inspired cooperative controller. We use simulated binary crossover (SBX) and polynomial mutation, which could generate the offspring population for real coded NSGA-II. After the offspring population is created, a combined population of the random parent population and the offspring population is formed. Then using the elitist non-dominated sorting algorithm, the solutions with high fitness level are selected as parents for the next generation. As a result, a set of Pareto-optimal solutions that tradeoff between the absolute value of the error and the settling time have been achieved.

Remark: We cannot guarantee that the NSGA-II, which is a stochastic search algorithm, can always find the global Pareto-optimal solutions. Fortunately, our problem is a small optimization problem with several parameters that have been well defined to satisfy the convergence condition. Meanwhile, even if a set of local Pareto-optimal solutions are achieved, it is also sufficient for us in engineering sense.

7.3.3 Application to Multi-Stage Fiber Drafting System

In this subsection, we evaluate the reliability and the efficiency of the proposed bio-inspired cooperative controller for the multi-stage drafting system by performing a set of simulations using Matlab. A four-stage drafting system of fiber process is investigated as an example.

7.3.3.1 Multi-Stage Drafting System

The multi-stage drafting system, a key step in the fiber production, is one of the complex multi-stage manufacturing systems, where the fibers travel longitudinally past a series of rollers in different working environment, e.g. hot air, water, or saturated water vapor [63–65]. After the raw material is formed and extruded from the spinneret, it has essentially zero modulus of elasticity. The speed difference between the adjacent rollers in the drafting system, which is named the draw ratio, stretches the fibers, and then the fiber modulus of elasticity dramatically increases to high values. The motor used to drive the roller can be of any kind for speed regulation, such as direct DC motors, AC motors, and servomotors. If the shaft does

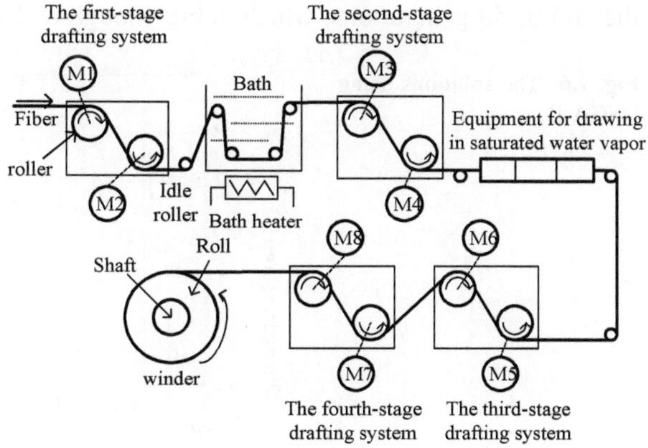

Fig. 7.5 Schematic of a four-stage drafting process in fiber production, where M1, M2,…,M7, and M8 are motors

not have enough stiffness or the gears in the reducer have a strong backlash, the mechanical vibration will appear. This phenomenon can affect the system dynamic response, reduce the accuracy, and limit the working efficiency.

Figure 7.5 depicts the flow diagram of a four-stage drafting process of the carbon fiber precursor manufacturing, in which tows are wound round two rollers in each drawing stage. A moving fiber under the small draw ratio cannot be pulled properly and may slip out of the roller, while the large draw ratio may result in the fiber deformation or even a fiber break. In the case of fiber break, the production line stops, resulting in a waste of time and a lower productivity. Unacceptable levels of the variation in draw ratio can degrade the fiber quality. Stable draw ratio and acceptable speed can improve the fiber structure, enhance fiber properties, and increase productivity. In the application, the draw ratio in each stage is approximately 1–3. The draw ratio grew by degrees at each stage. As a result, the drafting system can gain high velocity and stretches the fiber fully. In this subsection, the draw ratios at each stage are set to 1.2, 1.6, 2.1 and 3.2. As a result, the total draw ratio is set to 12.9024.

7.3.3.2 Multi-Objective Optimization Using NSGA-II

Four parameters in Eq. (7.47), i.e. α, a, β, and b, need to be optimized using NSGA-II under the constraints in Theorem 7.1. The goal of the optimization is to minimize: (1) the absolute value of the error between the desired draw ratio and the actual one and (2) the settling time, while guaranteeing the stability as discussed in the previous subsection.

In the simulation, the overall population of size is set to 100. The crossover probability is 0.9 and the distribution index for SBX is 20. As recommended in [62], the mutation probability is inversely proportional to the number of the decision variables for real-coded NSGA-II which is seven in this case. Therefore, the mutation probability is 0.143. A set of Pareto-optimal solutions can be obtained at the end of 50 generations, which is big enough and sufficient for this optimization

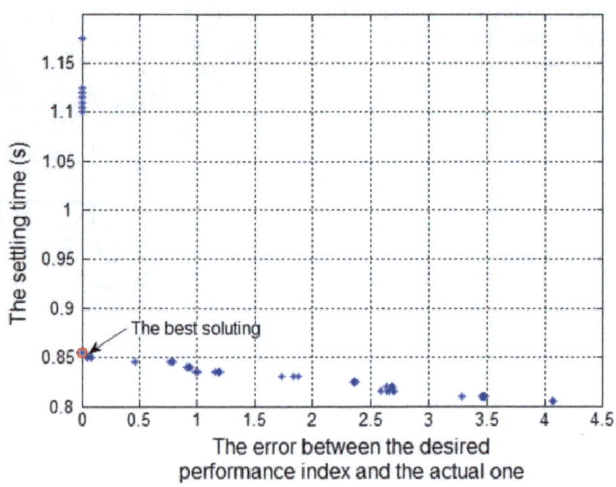

Fig. 7.6 The solutions using NSGA-II

Table 7.1 The parameter of Pareto-optimal solution and their corresponding performance marked in Fig. 7.6

α	β	a	b	$\lvert z(t) - z_{desired} \rvert$	T_s(s)
9.0764	0.0779	−17.2015	−20.0000	0.0018	0.8550

problem. Four parameters α, a, β, and b are randomly initialized between [0, 10], [0, 10], [−20, 0],and [−20, 0] respectively.

The transient constituents of the response have faded. Once the response of the system settles within δ of the steady-state, which is set to 1%, we think that the controller reaches a steady state. 10 independent runs have been formed to optimize the parameters using the NSGA-II. There is minor difference between the Pareto sets from the runs. Therefore, we pick one typical run to present the Pareto-optimal solutions, which are plotted in Fig. 7.6. In the optimization, both the absolute value of the error and the settling time are minimized, subjected to the constraints in Theorem 7.1. The optimized parameters, picked from the best solution, which is indicated by the arrows in Fig. 7.6, and their corresponding performances are listed in Table 7.1. The best solution can provide the parameters for the bio-inspired cooperative controller to follow the desired performance without any difference in the shortest time.

The Pareto-optimal solutions depicted in Fig. 7.6 are helpful to configure the parameters according to their requirements. The solutions can make the bio-inspired cooperative controller follow the desired performance with less difference in shorter time.

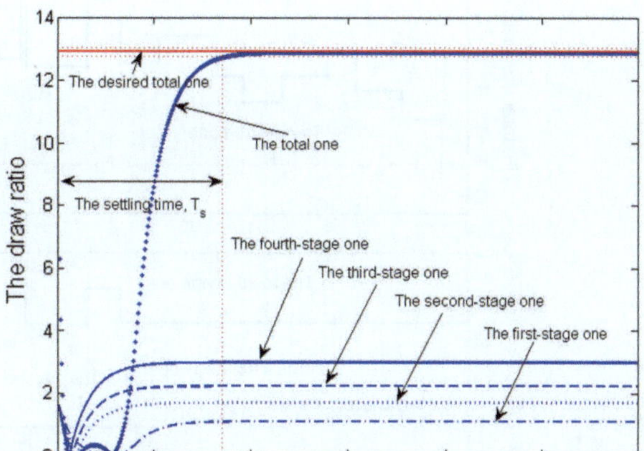

Fig. 7.7 The response of the bio-inspired cooperative controller with different parameter solutions listed in Table 7.1

7.3.3.3 The Performance of the Bio-Inspired Cooperative Controller

The responses of the bio-inspired cooperative controller with the solution in Table 7.1 are shown in Fig. 7.7. In Fig. 7.7, it is obviously that there is no difference between the total controller response and the overall desired performance. For the simplicity, we take the same parameters for each cell, i.e. each stage. As a result, the outputs of the four stages are just same. According to the experience in manufacturing, the draw ratio for each stage calculated by the bio-inspired cooperative controller should be scaled to provide the best working condition. The scales we used here are $k_1 = 1.5793$, $k_2 = 1.1845$, $k_3 = 0.9025$, $k_4 = 0.5922$. The total draw ratio is not need to be scaled, so $k = k_1k_2k_3k_4 = 0.9999$ (due to the round effect in the computer). Generally speaking, the solution picked out in the Fig. 7.6 is the best choice, which can make the controller follow the desired total draw ratio without the steady-state error.

The previous section explained how to configure the parameters using the NSGA-II under the constraints in Theorem 7.1. After optimization, the bio-inspired controller could minimize the absolute value of the difference between the desired total draw ratio and the actual one, and the settling time.

In the following, we are going to check the system performance when the draw ratios vibrate.

7.3.3.4 Robustness to the Vibrations in the Draw Ratio

In this case study, we will evaluate the robustness performance of the proposed bio-inspired cooperative controller to the vibrations in the draw ratio. First, we

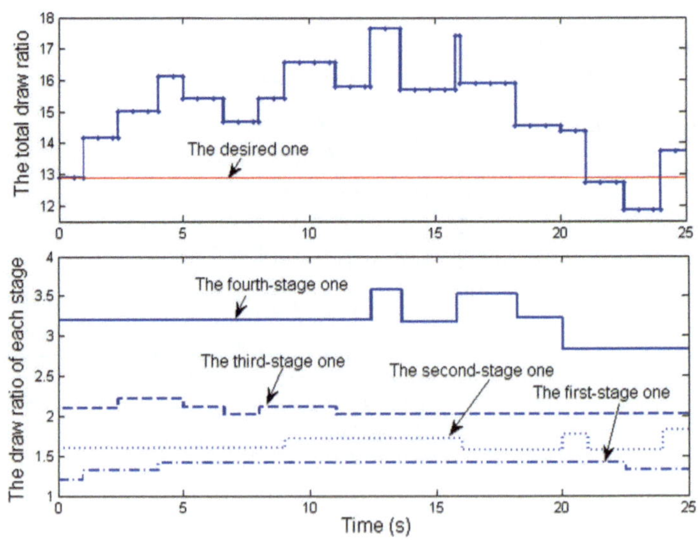

Fig. 7.8 System response without the bio-inspired cooperative controller

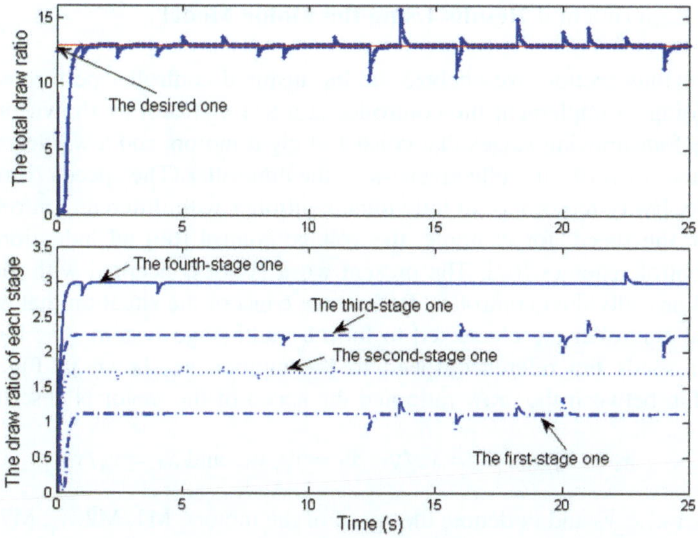

Fig. 7.9 System response of the bio-inspired cooperative controller with the solution (I)

show the system response without the bio-inspired cooperative controller in Fig. 7.8. In Fig. 7.8, the first-stage draw ratio 1.2 increases by 0.12 at time $t = 1$ s and by 0.1 at time $t = 4$ s from 1.32 to 1.42, and decreases from 1.42 to 1.32 at time $t = 24$ s. The second-stage draw ratio increases from 1.6 to 1.72 at time $t = 9$ s, from 1.57 to 1.77 at time $t = 20$ s, and from 1.57 to 1.82 at time $t = 24$ s; the second-stage one decreases from 1.72 to 1.57 at time $t = 16$ s, and from 1.77 to 1.57 at time $t = 21$ s. The third-stage draw ratio increases from 2.1 to 2.22 at time $t = 2.4$ s, and from 2.02 to 2.12 at time $t = 8$ s; the third-stage one decreases by 0.1 at time $t = 5$ s, 6.6 s, 11 s. The fourth-stage draw ratio increases from 3.2 to 3.5 at time $t = 12.4$ s, from 3.1 to 3.5 at time $t = 15.8$ s; the fourth-stage one decreases from 3.5 to 3.1 at time $t = 13.6$ s, from 3.5 to 3.2 at time $t = 18.2$ s, and from 3.2 to 2.8 at time $t = 20$ s. As a result, there is a big vibration in the total draw ratio, which increases from 12.9024 to 17.6625 by 36.89% of 12.9024 during the studying period. The vibrations degrade the fiber quality. Thus, once the vibrations are detected, some compensation in the fiber process is quite desired.

The bio-inspired cooperative controller can eliminate the vibration effect and make the draw ratio stable at the setting value. Once the vibrations affect the draw ratio of each stage, the bio-inspired cooperative controller can drag each draw ratio back to the desired one. The response of the bio-inspired cooperative controller is plotted in Fig. 7.9. The total draw ratio can be stable at the desired points to be 12.9024. Without the controller, the total draw ratio increases 12.9024–17.6625 over the study horizon by 36.89% of the setting value, 12.9024, in Fig. 7.8. The vibration could cause the defective fiber. The bio-inspired cooperative controller can compensate the draw ratio to make it stable at the setting value in Fig. 7.9, as such improves the fiber quality.

7.3.3.5 Experimental Results Using the Motor Model

In the previous section, we analyze the bio-inspired controller performance. Now we are going to implement the controller at a test platform as shown in Fig. 7.3. There are four drawing stages that consist of eight motors and a winder, which are applied here to verify the effectiveness of the controller. The speeds of motors are adjustable. Every motor has an individual controller with different control methods to change the speed, for example, the voltage control [66] nd induction machine torque control schemes [67]. The present work is not concerned with the detailed design of an individual controller. Rather, we concern the situation that arise when individual controller are connected to form a multi-stage system. The bio-inspired controller sends the reference speed to the motors, as shown in Fig. 7.8. The relationship between the draw ratio and the speed of the motor is described by:

$$\bar{x}_1 = v_2/v_1,\ \bar{x}_2 = v_4/v_3,\ \bar{x}_3 = v_6/v_5,\ \text{and}\ \bar{x}_4 = v_8/v_7, \tag{7.52}$$

where v_1, v_2,…, v_7 and v_8 denote the speed of the motors, M1, M2,…, M7, and M8, respectively. $\bar{x}_1$, $\bar{x}_2$, $\bar{x}_3$, and $\bar{x}_4$ are the actual draw ratio in first stage process, second stage process, third stage process, and fourth stage process, respectively. For convenience, the motor M2 is always as fast as the motor M3, i.e. $v_2 = v_3$, the same situation to M4 and M5, M6 and M7, i.e. $v_2 = v_3$, $v_4 = v_5$, $v_6 = v_7$. And the motor M8 is at the same speed with the winder. Obviously, the total draw ratio can be defined as

$$\prod_{i=1}^{n} \bar{x}_i(t) = \frac{v_8}{v_1},\ n = 4 \tag{7.53}$$

In this case, the motor M2 and M3, M4 and M5, M6 and M7, are served as the ladder, which do not contribute to the total draw ratio. The difference in the speeds of the motors M1 and M8 determines the total draw ratio.

First, the bio-inspired cooperative controller calculates the draw ratio for each drawing stage. After the draw ratios are verified by the experience in the

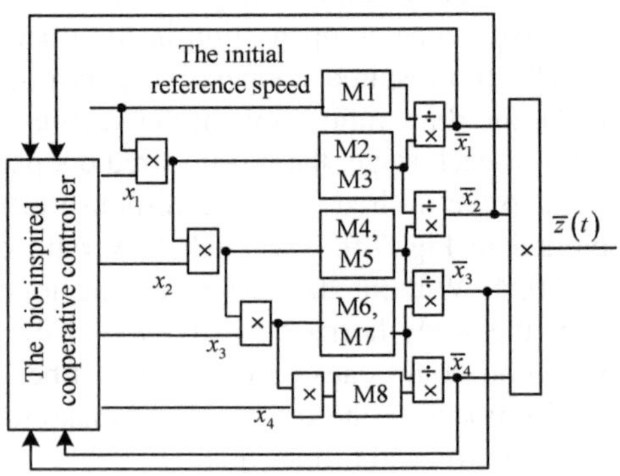

Fig. 7.10 The proposed bio-inspired cooperative control system

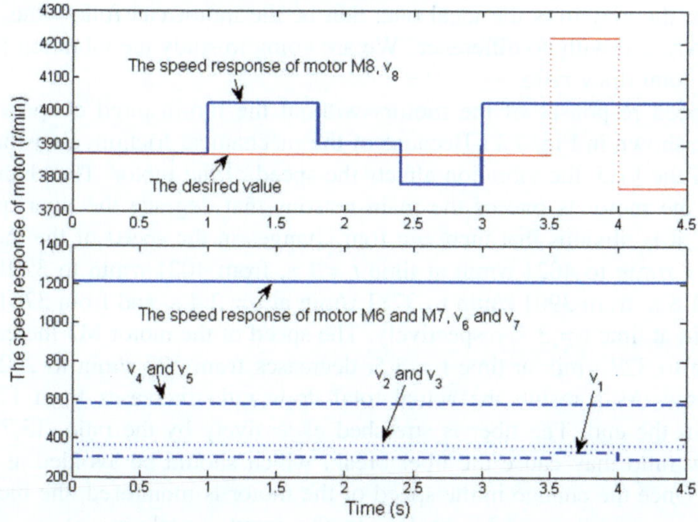

Fig. 7.11 The speed response of the motors without the bio-inspired cooperative controller

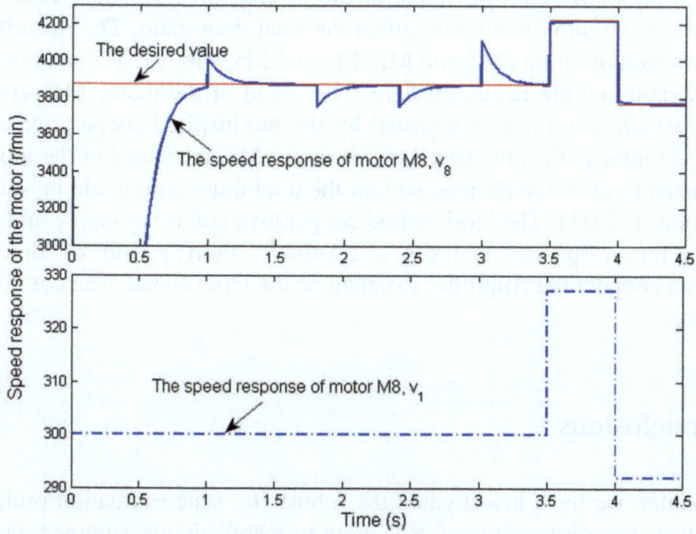

Fig. 7.12 The speed response of the bio-inspired cooperative control system

manufacturing, the draw ratios multiplied the initial reference speed are the desired motor speeds for each stage, as depicted in Fig. 7.10. Then the motors will react according to the input signal. If there is difference between the actual total ratio and the desired one, the bio-inspired cooperative controller will adjust the input signals, x_1, x_2, x_3 and x_4, to eliminate the difference. The individual controller for the motor utilizes different techniques to adjust the speed. Here, the motor utilized to implement the system is the ideal one, that is, the motor can follow the reference very quickly and with no difference. We are going to study the vibration that could affect the total draw ratio.

The speed responses of the motors without the bio-inspired cooperative controller are shown in Fig. 7.11 Because of the mechanical friction, abrasion and the change of the load, the vibration affects the speed of the motor. The change in the speed of the motor is one of the main reasons that degrade the fiber quality. In Fig. 7.11, it is obvious that there are four changes in the speed of the motor, M8, from 3871 r/min to 4021 r/min at time $t = 1$ s, from 4021 r/min to 3901r/min, at time $t = 1.8$ s, from 3901 r/min to 3781 r/min at t = 2.4 s, and from 3781 r/min to 4021 r/min at time t = 3 s, respectively. The speed of the motor M1 increases from 300 r/min to 327 r/min at time t = 3.5, decreases from 327 r/min to 292 r/min at time t = 4 s. As a result, the actual total draw ratio increases from 12.9024 to 13.7705 in the end. The fiber is stretched excessively by the ratio, 13.7705. The high draw ratio may cause the fiber break, which should be avoided as much as possible. Once the change in the speed of the motor is monitored, the bio-inspired cooperative controller would recalculate the input signal, x_1, x_2, x_3 and x_4 to compensate the vibration. After the compensation, the disturbance is eliminated, and the disturbed speed could be back to the desired value, as shown in Fig. 7.12. The total draw ratio could be stable at the desired one, 12.9024. Here, we only concerned the vibration that could affect the total draw ratio. The disturbance that affects the speed of motors M2 and M3, M4 and M5, M6 and M7 can not affect the total draw ratio and are neglected here. The speed of the motor M1 serves as the reference, which could not be adjusted by the bio-inspired cooperative controller. Once the vibrations affect the speed of the motor M1, the speed of the motor M8 is been adjusted to meet the change, so that the total draw ratio could be stable at the desired point, 12.9024. The bio-inspired cooperative control system could provide a desired system response. It takes a relatively short period of time for the bio-inspired cooperative controller to calculate the input signal. The control effect is satisfactory.

7.4 Conclusions

In this chapter, we have investigated the robust H_∞ state estimation problem for a class of stochastic discrete-time GRNs with probabilistic measurement delays. The GRNs under study involve parameter uncertainties, time-varying delays, stochastic disturbances and probabilistic measurement delays. For the admissible uncertainties, effective LMI approaches have been developed to design the estimators such that estimation error dynamics are robustly exponentially stable in the mean square and a prescribed H_∞ disturbance rejection attenuation is guaranteed for the addressed system. Meanwhile, the explicit expression of the desired estimate gains is also shown and a numerical example is given to show the effectiveness of the proposed designed method.

And then, inspired by biological organisms, we propose a bio-inspired cooperative controller via evolution of a GRN for the multi-stage systems. This developmental process is evolved using a multi-objective optimization algorithm to achieve a minimal absolute value of the error between the desired overall performance index and the actual one, and less settling time. Furthermore, a theoretical proof of the system convergence to the desired performance index is also provided. The proposed bio-inspired cooperative controller is implemented with the help of the multi-stage drafting system, a key step in the fiber production. By using the controller, the vibration in the speed of the motor could be eliminated and the total draw ratio may be stable at the desired one. And as a result, there is a decrease in scrap and eventually an improvement of the product quality.

References

1. Wang, T., Ding, Y., Zhang, L., Hao, K.: Robust state estimation for discrete-time stochastic genetic regulatory networks with probabilistic measurement delays. Neurocomputing **111**(3), 1–12 (2013)
2. Ding, Y.S., Wang, T., Jin, Y.C., Ren, L.H., Hao, K.R.: A bio-inspired cooperative controller for multi-stage system via evolutionary gene regulatory network, to be submitted
3. Smolen, P., Baxter, D.A., Byrne, J.H.: Mathematical modeling of gene networks. Neuron **26** (3), 567–580 (2000)
4. Jong, H.D.: Modelling and simulation of genetic regulatory systems: A literature review. J. Comput. Biol. **9**(1), 67–103 (2002)
5. Tomlin, C.J., Axelrod, J.D.: Biology by numbers: mathematical modelling in developmental biology. Nat. Rev. Genet. **8**(5), 331–340 (2007)
6. Schlitt, T., Brazma, A.: Current approaches to gene regulatory network modelling. BMC Bioinform., 8 Suppl **6**(6), S9–S9 (2007)
7. Kobayashi, T., Chen, L., Aihara, K.: Modeling genetic switches with positive feedback loops. J. Theor. Biol. **221**(3), 379–399 (2003)
8. Ma, Q., Xu, S., Zou, Y., Lu, J.: Robust stability for discrete-time stochastic genetic regulatory networks. Nonlinear Anal.-Real. World Appl. **12**(5), 2586–2595 (2011)
9. Li, P., Lam, J.: Disturbance analysis of nonlinear differential equation models of genetic sum regulatory networks. IEEE/ACM Trans. Comput. Biol. Bioinform. **8**(1), 253–259 (2011)
10. Wang, Z., Liao, X., Guo, S., Liu, G.: Stability analysis of genetic regulatory network with time delays and parameter uncertainties. IET Control Theory Appl. **4**(10), 2018–2028 (2010)
11. Pan, W., Wang, Z., Hu, J.: Robust stability of delayed genetic regulatory networks with different sources of uncertainties †. Asian J. Control **13**(5), 645–654 (2011)
12. Ye, Q., Cui, B.: Mean square exponential and robust stability of stochastic discrete-time genetic regulatory networks with uncertainties. Cogn. Neurodyn. **4**(2), 165–176 (2010)
13. Chesi, G.: Robustness analysis of genetic regulatory networks affected by model uncertainty. Automatica **47**(6), 1131–1138 (2011)
14. Yu, W., Lu, J., Chen, G., Duan, Z., Zhou, Q.: Estimating uncertain delayed genetic regulatory networks: an adaptive filtering approach. IEEE Trans. Autom. Control **54**(4), 892–897 (2009)
15. Li, P., Lam, J., Shu, Z.: On the transient and steady-state estimates of interval genetic regulatory networks. IEEE. Trans. Syst. Man. Cybern. B. **40**(2), 336–349 (2010)
16. Cao, J., Ren, F.: Exponential stability of discrete-time genetic regulatory networks with delays. IEEE Trans. Neural Netw. **19**(3), 520–523 (2008)

17. Liang, J., Lam, J., Wang, Z.: State estimation for Markov-type genetic regulatory networks with delays and uncertain mode transition rates. Phys. Lett. A **373**(47), 4328–4337 (2009)
18. Lv, B., Liang, J., Cao, J.: Robust distributed state estimation for genetic regulatory networks with Markovian jumping parameters. Commun. Nonlinear Sci. Numer. Simulat. **16**(10), 4060–4078 (2011)
19. Elowitz, M.B., Leibler, S.: A synthetic oscillatory network of transcriptional regulators. Nature **403**(6767), 335–338 (2000)
20. Chesi, G.: On the steady states of uncertain genetic regulatory networks. IEEE Trans. Syst. Man Cybern. Part A. **42**(4), 1020–1024 (2012)
21. Shen, B., Wang, Z., Liang, J., Liu, X.: Sampled-data H∞ filtering for stochastic genetic regulatory networks. Int. J. Robust Nonlinear Control **21**(15), 1759–1777 (2011)
22. Li, C., Chen, L., Aihara, K.: Stochastic stability of genetic networks with disturbance attenuation. IEEE Trans. Circuits Syst. II **54**(10), 892–896 (2007)
23. Li, X., Rakkiyappan, R.: Delay-dependent global asymptotic stability criteria for stochastic genetic regulatory networks with Markovian jumping parameters. Appl. Math. Model. **36**(4), 1718–1730 (2012)
24. Wu, F.X.: Delay-independent stability of genetic regulatory networks. IEEE Trans. Neural Netw. **22**(11), 1685–1693 (2011)
25. Wang, W., Zhong, S.: Stochastic stability analysis of uncertain genetic regulatory networks with mixed time-varying delays. Neurocomputing **82**, 143–156 (2012)
26. Sun, Y., Feng, G., Cao, J.: Robust stochastic stability analysis of genetic regulatory networks with disturbance attenuation. Neurocomputing **79**, 39–49 (2012)
27. Wang, Z., Lam, J., Wei, G., Fraser, K., Liu, X.: Filtering for nonlinear genetic regulatory networks with stochastic disturbances. IEEE Trans. Autom. Control **53**(10), 2448–2457 (2008)
28. Wang, Z., Liu, Y., Liu, X., Shi, Y.: Robust state estimation for discrete-time stochastic neural networks with probabilistic measurement delays. Neurocomputing **74**(1), 256–264 (2010)
29. Zhang, Y., Xu, S., Zeng, Z.: Novel robust stability criteria of discrete-time stochastic recurrent neural networks with time delay. Neurocomputing **72**(13), 3343–3351 (2009)
30. Liang, H., Zhou, T.: Robust state estimation for uncertain discrete-time stochastic systems with missing measurements. Automatica **47**(7), 1520–1524 (2011)
31. Wang, Z., Ho, D. W., Liu, X.: Variance-constrained filtering for uncertain stochastic systems with missing measurements. IEEE Trans. Autom. Control **48**(7), 1254–1258 (2003)
32. Wang, Z., Yang, F., Ho, D. W., Liu, X.: Robust H∞ filtering for stochastic time-delay systems with missing measurements. IEEE Trans. Signal Process. **54**(7), 2579–2587 (2006)
33. Wei, G., Wang, Z., Shen, B., Li, M.: Probability-dependent gain-scheduled filtering for stochastic systems with missing measurements. IEEE Trans. Circuits Syst. II **58**(11), 753–757 (2011)
34. Ma, L., Da, F., Zhang, K. J.: Exponential H∞ filter design for discrete time-delay stochastic systems with Markovian jump parameters and missing measurements. IEEE Trans. Circuits Syst. I **58**(5), 994–1007 (2011)
35. Chen, C.L., Chang, K.M., Chang, C.M.: Modeling and control of a web-fed machine. Appl. Math. Model. **28**(10), 863–876 (2004)
36. Abjadi, N. R., Soltani, J., Askari, J., Markadeh, G. A.: Nonlinear sliding-mode control of a multi-motor web-winding system without tension sensor. IET Control Theor. Appl. **3**(4), 419–427 (2009)
37. Koç, H., Knittel, D., De Mathelin, M., Abba, G.: Modeling and robust control of winding systems for elastic webs. Econ. Geol. **10**(2), 197–208 (2002)
38. Kang, H.K., Lee, C.W., Shin, K.H., Kim, S.C.: Modeling and matching design of a tension controller using pendulum dancer in roll-to-roll systems. IEEE Trans. Ind. Appl. **47**(4), 1558–1566 (2011)
39. Gassmann, V., Knittel, D., Pagilla, P.R., Bueno, M.A.: Fixed-order tension control in the unwinding section of a web handling system using a pendulum dancer. IEEE Trans. Control Syst. Technol. **20**(1), 173–180 (2012)

40. Kuhm, D., Knittel, D., Bueno, M.A.: Robust control strategies for an electric motor driven accumulator with elastic webs. ISA Trans. **51**(6), 732–742 (2012)
41. Wang, D., Wang, J., Wang, W.: H∞ controller design of networked control systems with Markov packet dropouts. IEEE Trans. Syst. Man Cybern. Syst. **43**(3), 689–697 (2013)
42. Canuto, E., Musso, F.: Embedded model control: Application to web winding. ISA Trans. **46** (3), 379–390 (2007)
43. Li, S.Q., Li, J., Mo, Y.P.: Piezoelectric multimode vibration control for stiffened plate using ADRC-based acceleration compensation. IEEE Trans. Ind. Electron. **61**(12), 6892–6902 (2014)
44. Pagilla, P.R., Siraskar, N.B., Dwivedula, R.V.: Decentralized control of web processing lines. IEEE Trans. Control Syst. Technol. **15**(1), 106–117 (2007)
45. Hur, S.H., Katebi, R., Taylor, A.: Modeling and control of a plastic film manufacturing web process. IEEE. Trans. Ind. Inform. **7**(2), 171–178 (2011)
46. Dou, X., Wang, W.: Robust control of multistage printing systems. Control Eng. Pract. **18**(3), 219–229 (2010)
47. Zio, E., Sansavini, G.: Vulnerability of smart grids with variable generation and consumption: a system of systems perspective. IEEE Trans. Syst. Man Cybern. Syst. **43**(3), 477–487(2013)
48. Lei, G., Wang, T.S., Guo, Y.G., Zhu, J.G., Wang, S.D.: System-level design optimization methods for electrical drive systems: deterministic approach. IEEE Trans. Ind. Electron. **61** (12), 6591–6601 (2014)
49. Ruan, X., Bien, Z.Z., Park, K.H.: Decentralized iterative learning control to large-scale industrial processes for nonrepetitive trajectory tracking. IEEE Trans. Syst. Man Cybern. A Syst. Humans. **38**(1), 238–252 (2008)
50. Taniguchi, S., Kino, H., Ozawa, R., Ishibashi, R., Uemura, M., Kanaoka, K., et al.: Inverse dynamics of human passive motion based on iterative learning control. IEEE Trans. Syst. Man Cybern. A Syst. Humans. **42**(2), 307–315 (2012)
51. Liang, X., Ding, Y.S., Ren, L.H., Hao, K.R., Wang, H.P., Chen, J.J.: A bioinspired multilayered intelligent cooperative controller for stretching process of fiber production. IEEE Trans. Syst. Man Cybern. C. **42**(3), 367–377 (2012)
52. Liang, X., Ding, Y.S., Ren, L.H., Hao, K.R., Jin, Y.L.: Data-driven cooperative intelligent controller based on the endocrine regulation mechanism. IEEE Trans. Control Syst. Technol. **22**(1), 94–101 (2014)
53. Guo, H.L., Meng, Y., Jin, Y.C.: A cellular mechanism for multi-robot construction via evolutionary multi-objective optimization of a gene regulatory network. BioSystems **98**(3), 193–203 (2009)
54. Jin, Y., Guo, H., Yan, M.: A hierarchical gene regulatory network for adaptive multirobot pattern formation. IEEE Trans. Syst. Man Cybern. B Cybern. **42**(3), 805–16 (2012)
55. Meng, Y., Guo, H., Jin, Y.: A morphogenetic approach to flexible and robust shape formation for swarm robotic systems. Robot Auton. Syst. **61**(1), 25–38 (2013)
56. Zhou, P., Chai, T.Y., Sun, J.: Intelligence-based supervisory control for optimal operation of a DCS-controlled grinding system. IEEE Trans. Control Syst. Technol. **21**(1), 162–175 (2013)
57. Coyne, J.A.: Comment on "gene regulatory networks and the evolution of animal body plans". Science **311**(5762), 796–800 (2006)
58. Wang, T., Ding, Y., Zhang, L., Hao, K.: Robust state estimation for discrete-time stochastic genetic regulatory networks with probabilistic measurement delays. Neurocomputing **111**(3), 1–12 (2013)
59. Hori, Y., Kim, T.H., Hara, S.: Existence criteria of periodic oscillations in cyclic gene regulatory networks. Automatica **47**(47), 1203–1209 (2011)
60. Menolascina, F., Bernardo, M.D., Bernardo, D.D.: Analysis, design and implementation of a novel scheme for in-vivo control of synthetic gene regulatory networks. Automatica **47**(6), 1265–1270 (2011)
61. Hasan, S.M.R.: A novel mixed-signal integrated circuit model for DNA-protein regulatory genetic circuits and genetic state machines. IEEE Trans. Circuits Syst. I. **55**(5), 1185–1196 (2008)

62. Deb, K., Pratap, A., Agarwal, S., Meyarivan, T.: A fast and elitist multiobjective genetic algorithm: nsga-ii. IEEE Trans. Evol. Comput. **6**(2), 182–197 (2002)
63. Yusof, N., Ismail, A.F.: Post spinning and pyrolysis processes of polyacrylonitrile (PAN)-based carbon fiber and activated carbon fiber: A review. J. Anal. Appl. Pyrol. **93**(1), 1–13 (2012)
64. Mb Bazbouz, G.S.: Novel mechanism for spinning continuous twisted composite nanofiber yarns. Eur. Polym. J. **44**(1), 1–12 (2008)
65. Ismail, A.F., Rahman, M.A., Mustafa, A., Matsuura, T.: The effect of processing conditions on a polyacrylonitrile fiber produced using a solvent-free free coagulation process. Mater. Sci. Eng. A **485**(1–2), 251–257 (2008)
66. Avila-Montes, J., Campos-Gaona, D., Melgoza VáZquez, E., RodríGuez-RodríGuez, J.R.: A novel compensation scheme based on a virtual air gap variable reactor for ac voltage control. IEEE Trans. Ind. Electron. **61**(61), 6547–6555 (2014)
67. Grcar, B., Stumberger, G., Hofer, A., Cafuta, P.: Im torque control schemes based on stator current vector. IEEE Trans. Ind. Electron. **61**(1), 126–138 (2014)

Chapter 8
Bio-Inspired Optimization Algorithms

For many reasons, one engineering process seeks not only a feasible method, but also the best design in some ways. The process of selecting the best design of an engineering process from some set of available alternatives is called engineering optimization. In this chapter, several bio-inspired optimization algorithms are proposed and applied to engineering process. Firstly, we propose a new synergetic immune clonal selection algorithm (SICSA) to obtain the optimal process parameters [1]. Secondly, a clonal selection algorithm based on a degeneration recognizing method (DR-CSA) is proposed [2] for solving complex engineering multimodal optimization problems. Thirdly, the fruit fly optimization algorithm inspired by immune mechanism and cooperation functional (IFOA) is presented, and used to optimize the parameters in support vector regression (SVR) predictive model for the fiber stretching rate [3]. Finally, we present an improved fruit fly optimization (FOA) algorithm based on the cell communication mechanism (CFOA), and apply it to optimize the controller for pre-oxidation furnaces in carbon fibers production [4].

8.1 Introduction

With the development of modern industrial technologies and computer science, a lot of real-world engineering processes become more and more complicated. Consequently, the optimization process is facing more challenges nowadays. A new research topic of Complex Engineering Optimization (CEO) has risen at the moment and drawn wide attention [5, 6]. Complex engineering optimization usually subjects to various difficulties such as uncertainties, computational expensiveness, system complexity, multiple criteria/objectives/decision variables, and so on [6]. In recent years, a great deal of research effort has been devoted to solving CEO problems [7]. A number of new research areas have emerged including evolutionary optimization in dynamic and uncertain environments, such as surrogate-assisted

Y. Ding et al., *Bio-Inspired Collaborative Intelligent Control and Optimization*, Studies in Systems, Decision and Control 118,
https://doi.org/10.1007/978-981-10-6689-4_8

evolutionary optimization [8], multi- and many-objective optimization [9], large-scale optimization, and integrated control and optimization [10, 11].

Among all different research areas of the CEO problems, multimodal optimization is a specific type of optimization problem which deals with optimization tasks that involve finding all or most of the multiple solutions (as opposed to one single best solution) to a problem [12, 13]. Different from multi-objective optimization, multimodal optimization aims at solving the multiple best solutions of one single objective. Multimodal optimization is quite common in real-word engineering problems, and its solutions always have great practical values. To an engineering optimization task, the knowledge of multiple solutions is usually very helpful. For example, sometimes due to physical (and/or cost) constraints, the best results may not always be realizable. In such a scenario, if multiple solutions (local and global) are known, the implementation can be quickly switched to another solution and still obtain an optimal system performance [13]. Also, multiple solutions could be analyzed to discover the hidden properties or relationships of the system inputs and outputs. Furthermore, multiple solutions could also provide a better flexibility for the engineering process to cooperate with other connected processes. Evolutionary computing techniques play a major role in solving multimodal optimization problems. Evolutionary computing techniques mostly involve metaheuristic optimization algorithms such as Evolutionary Algorithms (EAs), Swarm Intelligence (SI), Differential Evolution (DE), and Artificial Immune System (AIS). These methods have proven to be powerful for global optimization of a wide range of problems. Attempts to solve multimodal optimization problems have been made in all these realms [14–22]. The challenge of using the evolutionary computing techniques for multimodal optimization is that the multimodal problem is against the natural tendency of evolution, which will always converge to the best (or a sub-optimal) solution [14]. To overcome such difficulty, Niching is a technique of finding and preserving multiple stable niches, or favorable parts of the solution space possibly around multiple solutions, so as to prevent convergence to a single solution [15]. Also, Petrowski's clearing method [16] and Goldberg's sharing function approach [17] are two very well studied and respected methods in the Genetic Algorithm (GA) community for multimodal problem. DE based local selection and global selection approaches have also been attempted for solving multimodal problems. A cluster-based DE method for multimodal problems can be seen in [18]. A scatter learning PSO algorithm was proposed for multimodal problems in [19]. In [23], the Clonal Selection Algorithms (CSA) was shown to have a great ability in solving multimodal problems. CSA is capable of locating a larger number of local and global optima solutions. However, CSA is difficult to be applied to some real-world complex engineering problems because the cloning operation in CSA highly increases the evaluation times. In [24], a surrogate-assisted CSA was proposed to improve the computing speed for expensive optimization problems.

There are a lot of bio-inspired optimization algorithms that are applied in practical engineering successfully, such as GA inspired by the genetic science and natural selection [25, 26], PSO algorithm inspired by the simulation of the behavior of birds in nature [27–29], Artificial Bee Colony (ABC) algorithm inspired by the intelligent behavior of honeybee swarm [30], Artificial Immune Algorithm (AIA) inspired by the biological immune system [31–34], and Ant Colony Optimization (ACO) algorithm inspired by the foraging behavior of the real ants [35]. By simulating the foraging behavior of fruit fly swarm, Pan [36] recently proposed a new kind of optimization algorithm called Fruit Fly Optimization Algorithm (FOA) for financial distress model. Li et al. [37] applied the FOA to the forecasting problem of power load in electric power industry. Lin [38] used the FOA to optimize artificial neural network for analysis of service satisfaction in Web auction logistics service. Wang et al. [39] applied the FOA to search the optimal parameters values of the fractional order fuzzy-PID controller. Tu et al. [40] used the FOA to optimize generalized regression neural network for studying on business performance. Li et al. [41] used the FOA to optimize the two parameters for the least squares support vector machine to forecast the annual electricity consumption in China. Chen et al. [42] used the FOA to optimize the grey neural network model to perform satisfaction analysis for e-business service. According to all kinds of applications mentioned above, the FOA seems to be an excellent optimization algorithm. However, there is still an insufficiency in the FOA regarding the search strategy, which is used to produce new candidate solutions of the FOA based on the random information of foregoing solutions.

8.2 A Synergetic Immune Clonal Multi-objective Optimization Method

In this section, first we find out the relationship curves between important product performance and product process parameters based on experimental data, and present a mathematical model of carbon fiber production. And then, a SICSA based multi-objective optimization method to achieve the optimal drawing ratio of six stages is proposed. The SICSA consists of three main operators, including synergetic evolution, clonal operation and non-uniform mutation operation. The results obtained from the proposed method can serve as a reference and a guidance for realistic production of carbon fiber. And the proposed method can also be applied to solve other multi-objective optimization problems.

8.2.1 Drawing Ratio Division of Carbon Fiber Drawing Process

The preparation of PANCF protofilament using dry jet wet spinning method is shown in Fig. 8.1.

Preparation of protofilament is performed in five major steps: polymerization, spinning, solidification, drawing, and oil. Drawing is the necessary means to reduce linear density and improve the strength of protofilament. Drawing ratio is the most

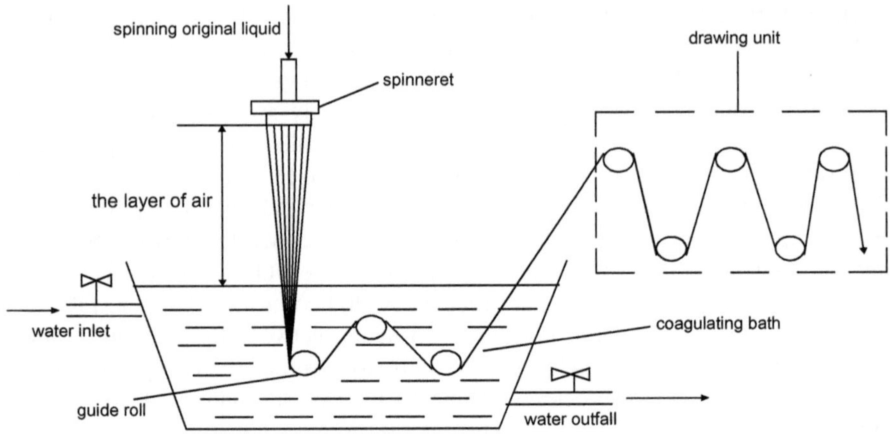

Fig. 8.1 Dry jet wet spinning method

important factor which affects the mechanical properties of protofilament. Prior to breakage, the higher the drawing ratio is, the better the performance of protofilament and carbon fiber will be. Since the drawing ratio within one particular stage is limited, we always adopt multi-stage drawings to meet the requirements. Usually, there are three levels: (1) air drawing and coagulation drawing (first-level drawing); (2) hot water drawing and boiling water drawing (second-level drawing); (3) dry heat drawing and saturated steam drawing (third-level drawing). In this section, we divide the drawing process into six stages according to the chronological order: spinneret drawing ratio r_1, air drawing ratio r_2, dimethylformamide (DMF) coagulation drawing ratio r_3, hot water drawing ratio r_4, boiling water drawing ratio r_5, and third-class drawing ratio r_6.

Excessive drawing will lead to protofilament broken. There is a maximum between drawing ratio and tropism under given conditions. The drawing ratios in different stages must be properly balanced, to avoid filament broken. Therefore, we need to find out the relationship between drawing ratios in various stages. The performance of protofilament can be described by the following function:

$$(\rho_L(i), Tg(i), E_L(i)) = f(r_{1i}, \cdots, r_{ki}, \cdots, r_{ni}) \tag{8.1}$$

where

$\rho_L(i)$ linear density of the ith experiment;
$Tg(i)$ strength of the ith experiment;
$E_L(i)$ breaking elongation of the ith experiment;
r_{ki} drawing ratio of the kth stage of the ith experiment.

The ultimate goal is to determine the optimal drawing ratios in different stages to minimize the linear density, maximize the strength and the breaking elongation ratio.

The relationship between drawing ratio of various stages and performance of protofilament can be estimated by constructing a model using experiment data. In this work, the following fitting function is used:

$$s_i = f(r_i) = a \cdot (r_i^b) \cdot \exp\left(\frac{c}{r_i}\right),\ r_i \geq 1.0 \tag{8.2}$$

where, r_i is the drawing ratio, s_i is the performance parameter of the ith experiment. We need to determine the constants a, b and c. The fitting function can be expressed using the following form:

$$\log(s_i) = \log(a) + b \times \log(r_i) + \frac{c}{r_i} \tag{8.3}$$

We can take N ($N \geq 3$) groups of experiment data (r_i, s_i) in proper order to compose N-linear equations, and solve the model parameters (a, b, c) via the least square method.

The total drawing ratio is obtained by multiplying the drawing ratios all stages $r_{sumi} = \prod_{k=1}^{n} r_{ki}$. As the influence of the drawing ratios in various on the properties of protofilament are very similar, we can apply the same model $s_k = f(r_k)$ for different stages. Due to different roles of each drawing stage, the weight w_k of every drawing stage is different from each other. The final performance of the protofilament is the weighted sum of each stage:

$$x_k = \sum_{k=1}^{6} (w_k \times f(r_k)) \tag{8.4}$$

8.2.2 *Optimal Division of Drawing Ratio with the SICSA*

As mentioned above, a high-quality carbon fiber should have a small linear density, high strength, and a high breaking elongation ratio. In this section, we propose a stochastic global optimization algorithm, termed SICSA to optimize the drawing ratios.

The immune algorithm converges fast and is less likely to get trapped in local optimums. The clonal selection has found to be the antibody which can recognize the antigen will thus be amplified and selected to keep down by immune system. However, the antibodies which can not recognize the antigen have no chance to be selected and amplified. Based on the above mechanisms, clonal selection is able to maintain optimal gene in the current generation and prepare for mutation to maintain a high diversity of population.

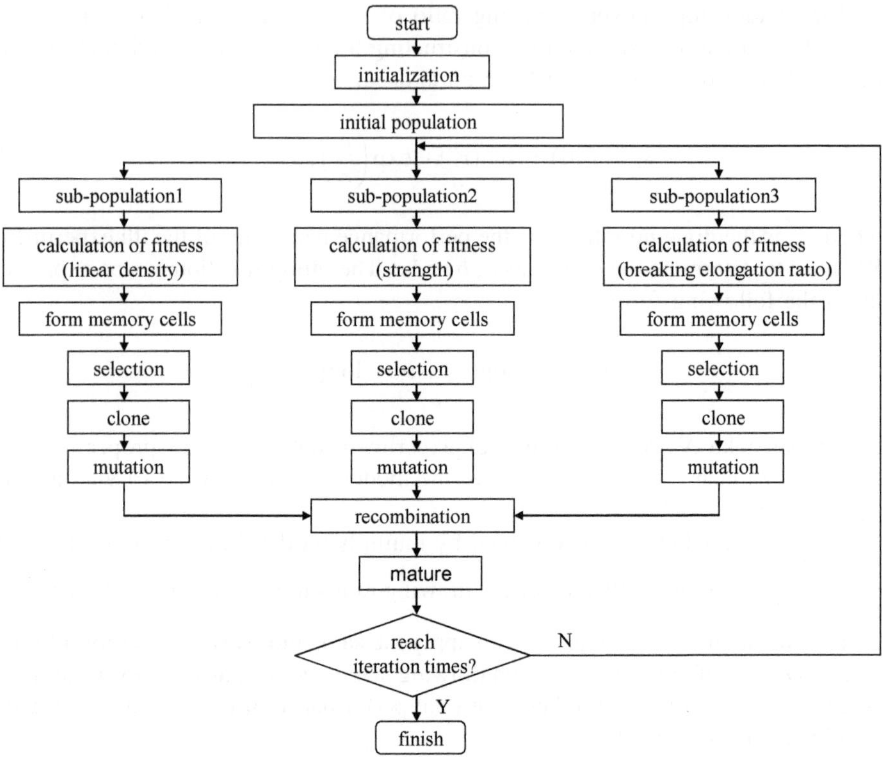

Fig. 8.2 Flow chart of SICSA

Steps of the SICSA are as shown in Fig. 8.2.

Step 1. Initialize population size, iteration times, weights in the fitness function, crossover rate p_c, mutation rate p_m;

Step 2. Generate the initial population randomly. The population contains a number of individuals, each of which represents a candidate drawing ratio division scheme of six drawing stages;

Step 3. Equally divide the current population into three sub-populations. Three objective functions are defined, namely, minimization of linear density, maximization of the strength and breaking elongation ratio;

Step 4. For every sub-population, form their own memory cells. If it is the first generation, we initialize a new memory cell group, otherwise we use the better one to replace the existing one;

Step 5. For every sub-population, select several best individuals according to their own fitness function;

Step 6. For every sub-population, clone their own best individuals. The expansion number is an increasing function of fitness, if the fitness becomes higher, the expansion number becomes bigger;

Step 7. Do overclocking mutation in every sub-population;

Step 8. Recombine the three new sub-populations to a complete a new population;
Step 9. Produce mature antibodies in the new population;
Step 10. If the maximum iteration number is not reached, go to Step 2;
Step 11. Stop. The final solution is the optimized drawing ratios distribute scheme

The main immune evolution operators of the SICSA are as follows:

(1) ***Synergetic evolution***

Collaboration and competition are the two common ways of synergetic evolution. It is a kind of synergetic to keep dynamic balance of population. In this section, we divide all individuals of the population into three sub-populations of same population size, each optimizing one of the three objectives. After selection operation, clonal expansion and overclocking mutation are applied independently in each sub-population, the sub-populations are combined into one new population, then crossover operation is performed in this population to obtain the next population. This approach is termed as the "division-parallel-recombine" mode, which is repeated to solve the three-objective optimization problem we have here.

(2) ***Clonal expansion***

Clonal expansion decides which antibodies will be selected to be amplified in the population. In tradition, the sole criterion of selection operation is the fitness value. In this section, we propose some new criteria to guide the selection based on immune evolution algorithm. Every individual x in population can be selected to be cloned as $clonals(x)$ copies:

$$clonals(x) = \mathrm{int}\left[m_{clonal} \cdot Aff_{Ab\sim Ag} \cdot Aff_{Ab\sim Ab} + n_{clonal}\right] \tag{8.5}$$

In which, $Aff_{Ab\sim Ag}$ indicates the affinity between antibody and antigen, $Aff_{Ab\sim Ab}$ indicates the concentration of antibodies, m_{clonal} and n_{clonal} are two reference clonal expansion coefficients, m_{clonal} is generally greater than one, and n_{clonal} is generally less than one. We consider $Aff_{Ab\sim Ag}$ and $Aff_{Ab\sim Ab}$ in the meanwhile, because they can balance the convergence and diversity in the evolutionary process, thereby overcoming the premature convergence and improving its global search ability.

(3) ***Overclocking mutation***

Overclocking mutation ensures the diversity. Generally, mutation rate p_m is determined by the concentration of antibodies $Aff_{Ab\sim Ab}$. $Aff_{Ab\sim Ab}$ means the crowdedness of the individuals in their field. In this section, we definite p_m as follow:

$$p_m = 1/(m_{mut} \cdot Aff_{Ab\sim Ab}(x) + 1) \tag{8.6}$$

In which, m_{mut} is a reference mutation coefficient, and generally greater than one. p_m can give up the individuals that neither meet the constraints nor in the feasible solution space, and guarantee the rest individuals meet the constraints. In that case, overclocking mutation is a random variation, which keeps diversity base on the concentration of antibodies.

Table 8.1 Part of the experiment data

No.	r_1	r_2	r_3	r_4	r_5	r_6	ρ_L (tex)	Tg (CN/d)	E_L (%)
1	2.5	1.5	1.0	1.0	1.0	1.5	5.4	3.6	24.3
2	2.5	1.5	1.0	1.0	1.2	1.5	4.2	4.7	17.9
3	2.5	1.5	1.0	1.0	1.5	1.5	3.6	5.4	15.5
4	2.5	1.5	1.0	1.0	1.7	1.5	3.0	6.2	14.7
5	2.5	1.5	1.0	1.0	2.0	1.5	2.5	7.0	13.6
6	2.5	1.5	1.0	1.0	1.0	1.5	5.4	3.9	34.0
7	2.5	1.5	1.0	1.2	1.0	1.5	4.2	5.2	19.9
8	2.5	1.5	1.0	1.5	1.0	1.5	3.6	5.9	17.5
9	2.5	1.5	1.0	1.7	1.0	1.5	3.0	6.3	16.3
10	2.5	1.5	1.0	2.0	1.0	1.5	2.5	6.6	13.4

8.2.3 Result and Discussion

The relationship between drawing ratio of various stages and performance of protofilament can be estimated using experiment data we collected, part of them are shown in Table 8.1.

Take the relationship between total drawing ratio and performances of protofilament for example, the fitting curves from the data are showed in Fig. 8.3.

The relationship between drawing ratio (r_i) and linear density (s_{1i}) can be derived from Fig. 8.3(a) as follows:

$$s_{1i} = f_1(r_i) = 94.72 r_i^{-1.37} \exp\left(\frac{-3.05}{r_i}\right) \tag{8.7}$$

Similarly, the relationship between drawing ratio (r_i) and strength (s_{2i}) can be derived from Fig. 8.3(a):

$$s_{2i} = f_2(r_i) = 34.48 r_i^{-0.38} \exp\left(\frac{-8.54}{r_i}\right) \tag{8.8}$$

The relationship between drawing ratio (r_i) and breaking elongation ratio (s_{3i}) can be derived from Fig. 8.3(b):

$$s_{3i} = f_3(r_i) = 0.37 r_i^{0.99} \exp\left(\frac{14.28}{r_i}\right) \tag{8.9}$$

The importance of each drawing stage is assessed by material experts based on different roles of drawing stages to determine the weight for each stage. We structure the judgement matrix by Satty scaling as follows:

Fig. 8.3 The relationship between drawing ratio and performance (linear density, strength and breaking elongation ratio) fitting curves

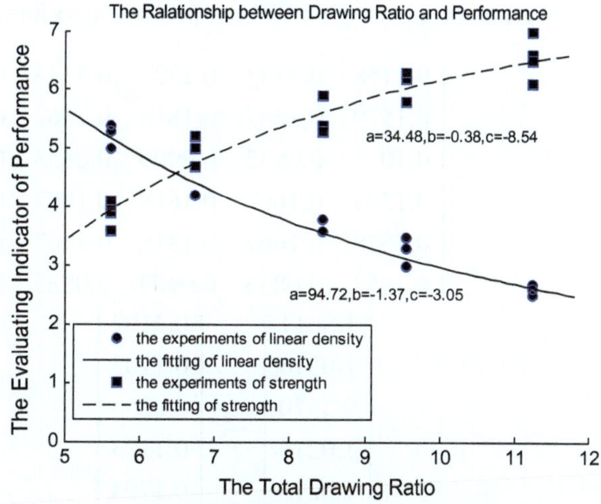

(a) Linear density and strength

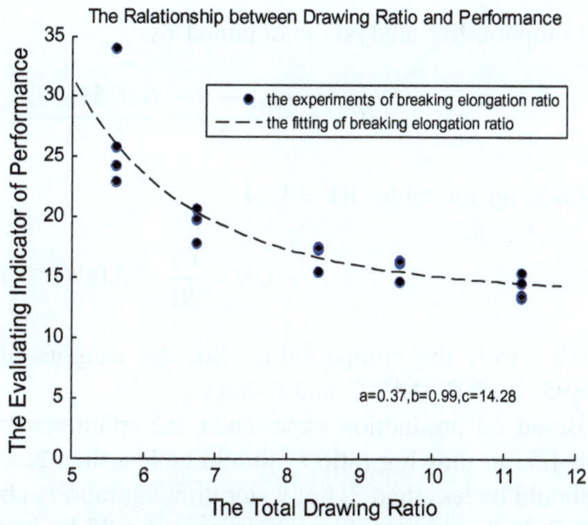

(b) Breaking elongation ratio

	r_1	r_2	r_3	r_4	r_5	r_6
r_1	1	2	3	2	2	3
r_2	1/2	1	2	1	1	2
r_3	1/3	1/2	1	1/2	1/2	1
r_4	1/2	1	2	1	1	2
r_5	1/2	1	2	1	1	2
r_6	1/3	1/2	1	1/2	1/2	1

We obtain the weights by sum-product algorithm:

$$\begin{bmatrix} 0.3158 & 0.3333 & 0.2727 & 0.3333 & 0.3333 & 0.2727 \\ 0.1579 & 0.1667 & 0.1818 & 0.1667 & 0.1667 & 0.1818 \\ 0.1053 & 0.0833 & 0.0909 & 0.0833 & 0.0833 & 0.0909 \\ 0.1579 & 0.1667 & 0.1818 & 0.1667 & 0.1667 & 0.1818 \\ 0.1579 & 0.1667 & 0.1818 & 0.1667 & 0.1667 & 0.1818 \\ 0.1053 & 0.0833 & 0.0909 & 0.0833 & 0.0833 & 0.0909 \end{bmatrix}$$

$$\rightarrow \begin{bmatrix} 1.8611 \\ 1.0216 \\ 0.5370 \\ 1.0216 \\ 1.0216 \\ 0.5370 \end{bmatrix} \rightarrow \begin{bmatrix} 0.3102 \\ 0.1703 \\ 0.0895 \\ 0.1703 \\ 0.1703 \\ 0.0895 \end{bmatrix}$$

Compatibility analysis is obtained by:

$$\mathrm{CI} = \frac{\lambda_{\max} - n}{n - 1} = \frac{6.0251 - 6}{6 - 1} = 0.0050$$

Look up the table: RI = 1.24

$$CR = \frac{CI}{RI} = 0.004 < 0.1$$

CR meets the compatibility. So, the weights of $r_1 \sim r_6$ are 0.3102, 0.1703, 0.0895, 0.1703, 0.1703 and 0.0895.

Based on production experience, the spinneret drawing ratio r_1 should be less than 4.5, air drawing ratio r_2 should be less than 2, DMF coagulation drawing ratio r_3 should be less than 2, hot water drawing ratio r_4 should be less than 6 and greater than 2, boiling water drawing ratio r_5 should be less than 1.3, third-class drawing ratio r_6 should be less than 2.5. Taking all these constraints into account, the mathematical model of carbon fiber drawing ratio division can be determined as follows:

$$\begin{cases} \min 0.3102 \times 94.72 r_i^{-1.37} \exp\left(\frac{-3.05}{r_i}\right) + 0.1703 \times 94.72 r_2^{-1.37} \exp\left(\frac{-3.05}{r_2}\right) \\ \quad + 0.1895 \times 94.72 r_3^{-1.37} \exp\left(\frac{-3.05}{r_3}\right) + 0.1703 \times 94.72 r_4^{-1.37} \exp\left(\frac{-3.05}{r_4}\right) \\ \quad + 0.1703 \times 94.72 r_5^{-1.37} \exp\left(\frac{-3.05}{r_5}\right) + 0.1895 \times 94.72 r_6^{-1.37} \exp\left(\frac{-3.05}{r_6}\right) \\ \max 0.3102 \times 34.48 r_i^{-0.38} \exp\left(\frac{-8.45}{r_i}\right) + 0.1703 \times 34.48 r_2^{-0.38} \exp\left(\frac{-8.54}{r_2}\right) \\ \quad + 0.1895 \times 34.48 r_3^{-0.38} \exp\left(\frac{-8.45}{r_3}\right) + 0.1703 \times 34.48 r_4^{-0.38} \exp\left(\frac{-8.54}{r_4}\right) \\ \quad + 0.1703 \times 34.48 r_5^{-0.38} \exp\left(\frac{-8.45}{r_5}\right) + 0.1895 \times 34.48 r_6^{-0.38} \exp\left(\frac{-8.54}{r_6}\right) \\ \max 0.3102 \times 0.37 r_i^{0.99} \exp\left(\frac{14.28}{r_i}\right) + 0.1703 \times 0.37 r_2^{0.99} \exp\left(\frac{14.28}{r_2}\right) \\ \quad + 0.1895 \times 0.37 r_3^{0.99} \exp\left(\frac{14.28}{r_3}\right) + 0.1703 \times 0.37 r_4^{0.99} \exp\left(\frac{14.28}{r_4}\right) \\ \quad + 0.1703 \times 0.37 r_5^{0.99} \exp\left(\frac{14.28}{r_5}\right) + 0.1895 \times 0.37 r_6^{0.99} \exp\left(\frac{14.28}{r_6}\right) \\ 1 \le r_1 \le 4.5 \\ 1 \le r_2 \le 2 \\ 1 \le r_3 \le 2 \\ 2 \le r_4 \le 6 \\ 1 \le r_5 \le 1.3 \\ 1 \le r_6 \le 2.5 \end{cases} \tag{8.10}$$

The above mathematical model indicates that optimization of carbon fiber drawing process is an inherently multi-objective optimal problem, which will be solved using SICSA We set the population size as 120, the iteration times as 50, the coefficient of fitness as 0.7, and the crossover rate p_c as 0.7, adopt the non-unity mutation, and use different mutation rates p_m. The optimized results are shown in Fig. 8.4.

Figure 8.4 indicates that the SICSA has a fast convergence and good ability to maintain the diversity. We can also observe that the optimal solution of the first objective function, the second objective function, the third objective function and the sum of the three objective functions have converged in 50 iteration times, which demonstrate the effectiveness of the SICSA. The obtained optimal drawing ratio allocation scheme is: the spinneret drawing ratio r_1 is 4.5, air drawing ratio r_2 is 2, DMF coagulation drawing ratio r_3 is 2, hot water drawing ratio r_4 is 4, boiling water drawing ratio r_5 is 1.3, and third-class drawing ratio r_6 is 2.5.

In order to fully examine the performance of the SICSA, we empirically compare the performance of a genetic algorithm (GA), an immune algorithm (IA) with that of the SICSA. The parameters of the GA and the IA are set the same with those of the SICSA, the optimal solutions of the three algorithms are as shown in Table 8.2. Figure 8.5 shows the mean value of objective functions of three algorithms.

We can see from Fig. 8.5 that only the SICSA convergences and finds the optimal solution in 50 iterations. Furthermore, SICSA obtains the best objective

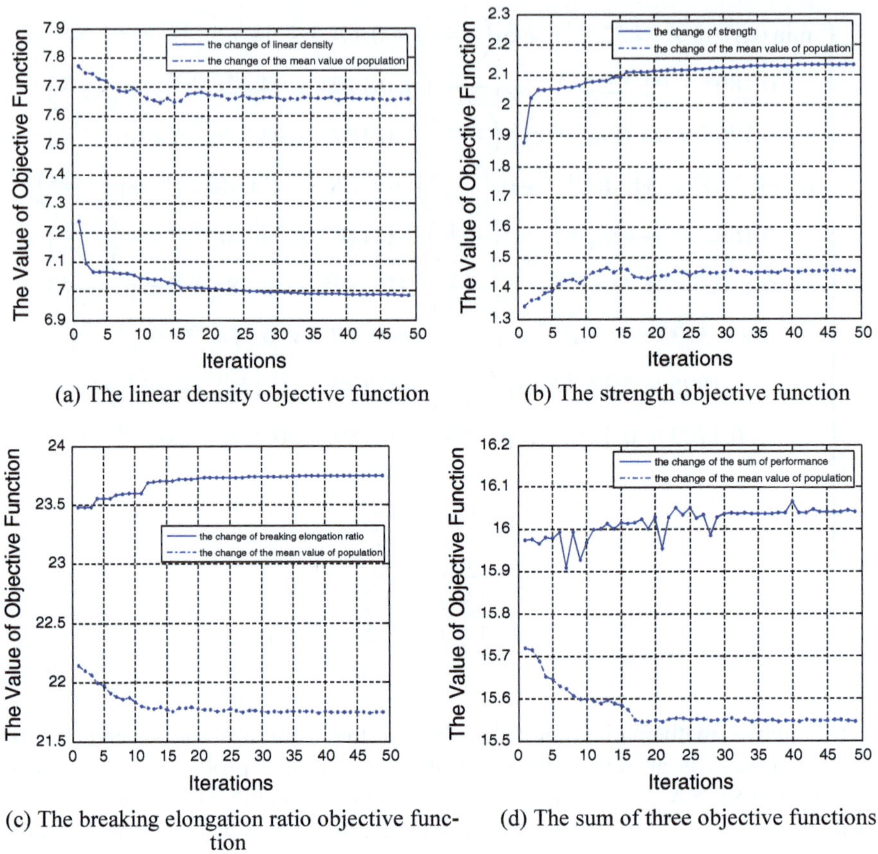

(a) The linear density objective function

(b) The strength objective function

(c) The breaking elongation ratio objective function

(d) The sum of three objective functions

Fig. 8.4 The trace of optimal solution and performance

Table 8.2 drawing ratio distribution schemes of SICSA, IA and GA

Algorithms	r_1	r_2	r_3	r_4	r_5	r_6
SICSA	4.5	2.0	2.0	4.0	1.3	2.5
IA	4.5	2.0	2.0	6.0	1.3	2.5
GA	4.4	2.0	2.0	2.0	1.3	2.0

values among the three algorithms. It confirms the fast convergence and global search capability of the SICSA. Table 8.3 shows optimization results of three algorithms over 100 independent runs. We can include from these results that GA performs similarly to IA. We can also observe that SICSA has achieved a slightly larger linear density than the IA and the GA a slightly smaller strength than the IA

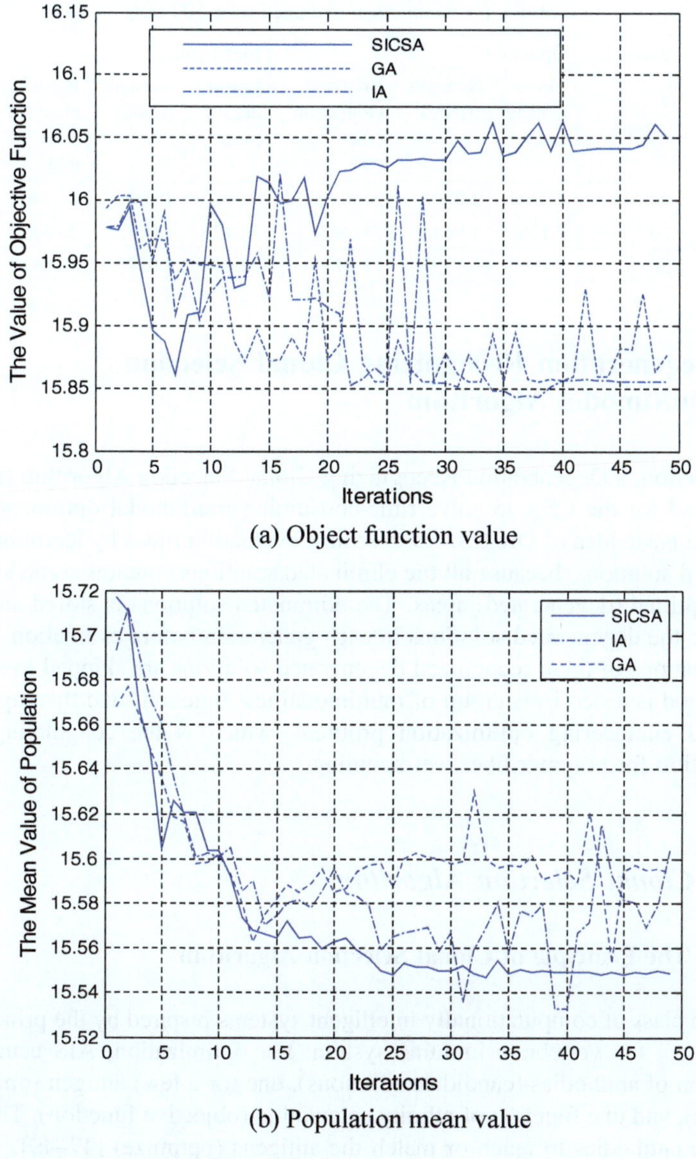

Fig. 8.5 Comparison of the SICSA, the IA and the GA

and the GA. Most importantly, however, SICSA has achieved a much bigger breaking elongation ratio than IA and GA. As a whole, the parameters obtained by SICSA will lead to higher-quality carbon fiber. In addition, the computation time SICSA takes is acceptable in practice.

Table 8.3 Optimization results for comparison averaged over 100 runs

Algorithm	Iterations	Optimum			Mean value			Time (s)
		Linear density (tex)	Strength (CN/d)	Breaking elongation ratio (%)	Linear density (tex)	Strength (CN/d)	Breaking elongation ratio (%)	
SICSA	50	7.3394	1.7802	21.6255	7.2572	1.8632	21.4406	57.93
IA	50	7.1566	1.9645	21.0879	7.0436	2.0787	20.8452	9.12
GA	50	7.1812	1.9397	21.1454	7.0648	2.0576	20.8878	2.01

8.3 Degeneration Recognizing Clonal Selection Multimodal Algorithm

In this section, a Degeneration Recognizing Clonal Selection Algorithm (DR-CSA) is designed for the CSA to solve time-consuming multimodal optimization problems. The basic idea of DR-CSA is to reduce evaluation times by learning from the eliminated solutions, because all the eliminated solutions contains some knowledge of non-optimal (degenerated) areas. The eliminated solutions is stored and used to recognize the degenerated solutions in each generation before evaluation. And then the evaluations of these recognized degenerated solutions are skipped to save time. The method is tested by a group of multimodal test functions and then applied to a real-word engineering optimization problem, which is the coagulating process optimization for polymer fiber wet spinning.

8.3.1 Clonal Selection Algorithm

8.3.1.1 The Principle of Clonal Selection Algorithm

AIS are a class of computationally intelligent systems inspired by the principles and processes of the vertebrate immune system. For optimization, AIS usually has a population of antibodies (candidate solutions), one (or a few) antigen (optimization objective), and one function of affinity calculation (objective function). The attempt is for the antibodies to reach or match the antigens (optimize) [47–49].

In AIS family, CSA is a class of algorithms inspired by the clonal selection theory of acquired immunity that explains how B and T lymphocytes improve their response to antigens over time called affinity maturation. These algorithms focus on the Darwinian attributes of the theory where selection is inspired by the affinity of antigen-antibody interactions, reproduction is inspired by cell division, and variation is inspired by somatic hyper-mutation.

According to the clonal selection theory, only cells that are capable of recognizing an antigen will reproduce and be maintained as memory cells; differentiation

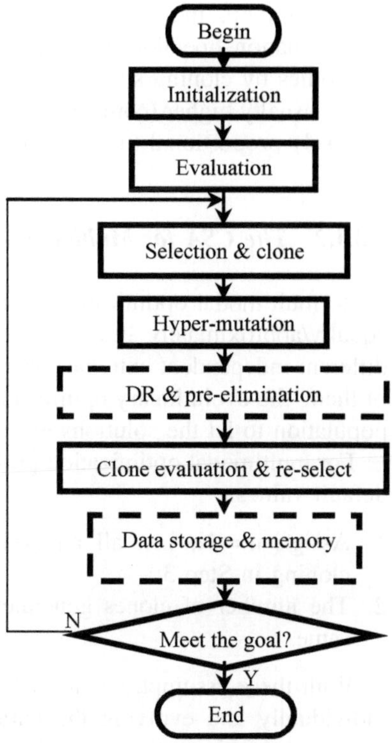

Fig. 8.6 Flowchart of the CSA and DR-CSA

happens along with proliferation; diverse antibody patterns are formed by accelerated somatic mutation [50]. Based on such phenomenon, in the CSA method, antibodies with higher affinity are selected and cloned, and the clones are hyper-mutated and then re-selected again.

Figure 8.6 (the solid line part) presents the block diagram of the computational procedure for the CSA. And these general steps shown in Fig. 8.6 of the CSA to solve optimization problems are defined as follows:

Step 1. Initialization: randomly initialize a population N of antibodies.
Step 2. Evaluation: determine the affinity of each antibody of the population.
Step 3. Selection and cloning: select a number (n) of the highest affinity antibodies and generate clones independently and proportionally to their affinities.
Step 4. Hyper-mutation: generating a population of matured clones. The higher the affinity, the smaller the mutation rate.
Step 5. Clone evaluation and re-selection: determine the affinity of the matured clones in relation to antigen. Select the antibody with the highest affinity from the matured clones and form the new population N.
Step 6. Repeat Steps 2-5 until termination criterion is met.

To sum up, the main characteristics of the CSA are: (1) Instead of using recombination operators (such as crossover in GA), the CSA generates the new antibodies by cloning so that all the antibodies are independent; (2) The mutation rate is usually higher (compares with other evolutionary computing techniques) and inversely proportional to the affinity.

8.3.1.2 *The CSA for Multimodal Problem*

As to multimodal optimization problem, the objective is to find all or most of the equally/approximately best solutions. The multiple best solutions can locate in different independent extremes or a continuous extremal area. In order to locate all of them, the evolutionary optimization algorithms need to evolve all or most of the population to let the solutions evenly distributed on the optimal peaks/areas.

For multimodal optimization problems, two parameters of the CSA may assume default values:

1. Assign n = N, i.e., all antibodies from the population will be selected for cloning in Step 3.
2. The number of clones generated for each of the N antibodies should be the same.

With these assumptions, the CSA is capable of exploring the fitness landscape individually and evolving the entire population simultaneously. The population tends to spread on the optimal peaks/areas. As a result, the CSA is a feasible method for solving multimodal optimization problems.

However, the CSA for multimodal optimization clones all of the population ($n = N$) on every iteration. And in order to explore the fitness landscape, the clone amount is usually more than a few. So on each generation, there is a large population that requires a large amount of affinity evaluations. For example, if the antibody population is 100 and the clone amount is 10, there will be 10^3 times of affinity evaluation on every iteration. If the iteration times is 100, then there will be a 10^5 times of evaluation in total. If one evaluation of some engineering model takes a few seconds, then the overall optimization time for such task will take days to finish. Thus, for CSA to be applicable to complex engineering optimization problems, certain computing speed improvement is necessary.

8.3.2 Degeneration Recognizing Clonal Selection Algorithm

8.3.2.1 General Idea of Degeneration Recognizing Method

There are few ways to improve the computing speed of the CSA, such as the surrogate-assisted or hybrid methods. The methods used may vary, but the principle is mainly the same. This basic principle is to reduce the times of direct calculation of the affinity via the mechanism model (evaluation), because the evaluation is the major reason of the time cost.

The intention here is to design a fast and simple method which aims at reducing the times of evaluation as well. The focus is on the knowledge that contained in the eliminated antibodies which is usually overlooked by other methods. The selection operation of each generation selects part of the population and eliminates the rest. We noticed that the selected ones may or may not be the final optimal solutions, but the eliminated ones are definitely not the optimal solutions. These eliminated antibodies along with theirs mark of "eliminated" are of some useful information.

To be specific, first, one can be sure that an antibody, which was eliminated once before due to poor affinity, should be eliminated again if it reappeared, its evaluation can be skipped. Therefore, use a database to store all of the eliminated antibodies and match every new antibody with this database to see if it is a once-eliminated antibody before its evaluation should reduce some evaluation times. As a most preliminary idea, we can foresee that such method could avoid a few times of model calculation and save a few time. However, it is not efficient enough on timesaving because the probability of an exactly same antibody's reappearing is very low.

To further develop the method (assuming the problem is continuous and differentiable), if one antibody is eliminated before, all other antibodies located inside a very small neighbourhood of this antibody should be eliminated. This determination is made under the assumption that two very close antibodies have approximately same affinities. For most of the real-word engineering problems, this assumption is acceptable because real-word problems are usually continuous and smooth. Then, if the size of the neighbourhood is properly chosen, a considerable amount of calculation times would be avoided. In this section, a new generated antibody (by clone and hyper-mutation) falling into a small neighbourhood of a once-eliminated antibody is called ***degeneration***. This antibody is called ***degenerated antibody*** and the small neighbourhood is called ***degenerated area***. Once a degenerated antibody is recognized, it will be eliminated immediately. This method is called Degeneration Recognizing Clonal Selection lgorithm.

The surrogate-assisted kind of method usually uses a faster approximation to replace some of the evaluations, it is effective in saving evaluation time but the approximation may affect the accuracy of the result. However, the optimization accuracy of the proposed DR-CSA method is as accurate as the CSA. This is due to the fact that all the approximations in the DR-CSA are used on the soon-to-be-eliminated antibodies, while all the selected antibodies were evaluated.

To sum up, there are three reasons to believe that the proposed DR-CSA is feasible:

1. Since the clone operation of the CSA creates a huge population, which means the eliminated population is also huge. Even though each eliminated antibody only has a very small neighbourhood (degenerated area), when the amount of

the eliminated antibody is great, the combined degenerated area will be large enough to make a difference.

2. Due to the hyper-mutation process, which is a great help for exploring the function landscape, a large number of degenerated solutions is created. So this pre-elimination method has a relatively large probability in recognizing the degenerated solutions so that the calculation time can be obviously saved.
3. The accuracy of the multimodal optimization results is not affected by the proposed degeneration recognizing method.

8.3.2.2 DR-CSA Procedure and Computational Complexity Analysis

For the DR-CSA, two more steps are added into the regular CSA, show in Fig. 8.6 (dashed line part). One is degeneration recognizing and pre-elimination, and the other one is degeneration database storage and management.

For the degeneration recognizing and pre-elimination step, every new matured clone is strategically compared with the database. The recognized degenerated antibodies will be eliminated immediately. The pseudo-code of this step is shown as follow:

Algorithm: Degeneration Recognizing and Pre-Elimination
Input: new population (*newCnds*); degeneration database (*dgBase*), which is made of degeneration antibodies (*dgBase.points*) and their respective degeneration neighbourhood radius (*dgBase.radius*).
Output: non-degenerated antibodies (*Cnds*).

```
n_c = the number of newCnds
n_d = the number of dgBase.points
for j = 1 to n_c
   dgMark = 0;
   for i = 1 to n_d
      distance = the Euclidean distance between newCnds(j) and
                 dgBase.points(i)
      if distance < dgBase.radius(i)
         dgMark = 1;
         break;
      end if
   end for
   if dgMark = 0
      Store newCnds(j) into Cnds
   end if
end for
return Cnds
```

After this degeneration recognizing and pre-elimination step, the final result (*Cnds*) is the rest of the population that survive pre-elimination. The next steps are the standard CSA evaluation and selection. And then, the selected part of the *Cnds* becomes the new generation and the rest (eliminated) part becomes the input (*elmCnds*) of the following step:

Algorithm: Degeneration Database Storage & Management
Input: eliminated antibodies (*elmCnds*); degeneration database (*dgBase*).
Output: renewed degeneration database (*dgBase*).

```
n_e = the number of elmCnds
for i = 1 to n_e
    radius = the degeneration neighbourhood radius of antibody
             elmCnds(i), detailed calculation method is introduced in the
             next segment
    Store elmCnds(i) into dgBase.points
    Store radius into dgBase.radius
end for
return dgBase
```

As shown, the database expands on each iteration by accumulating eliminated solutions. On large-scale problems, some proper database management can be applied to prevent overflow. For example, the data in the database with the smallest neighbourhood can be deleted.

According to the code structure introduced above, the computational complexities of the two algorithms that are added for the DR-CSA are $O(n_c n_d)$ and $O(n_e)$ respectively. Obviously, the computational complexity is increased comparing to the regular CSA. In Degeneration Recognizing and Pre-Elimination Algorithm, n_c is the population size which is a fixed number. n_d is the number of all the eliminated antibodies (pre-eliminated antibodies are not included) and keeps increasing along with iteration times. In Degeneration Database Storage and Management Algorithm, the n_e is the number of the eliminated antibodies in each iteration, it is the increment of n_d in each iteration. n_e is relative small and stable with certain randomness.

As a result, besides evaluation, the major computational expensiveness of the DR-CSA is dominated by n_d and the computation time is also increasing along with iteration. In complex engineering optimization, only when the evaluation time cost is much higher than the DR-CSA algorithm itself, the proposed method is able to reduce considerable computation time. The increasing pattern of n_d is monitored and plotted in the following experiments.

8.3.2.3 Degeneration Neighbourhood Size Calculation Method

The determination of the neighbourhood size is important in the DR-CSA. If the neighbourhood size is too small, the probability of a new antibody's falling into this neighbourhood will become too small to effectively save time. On the other hand, if the neighbourhood size is too large, the actual optimal peaks/areas may accidentally be covered by some degenerated antibody's neighbourhood, causing the algorithm unable to locate these optimal peaks/areas. In this segment, three ways of calculating a proper neighbourhood size is introduced by using affinity as guidance.

Assuming that the optimization objective is y_{obj} and the definition of the affinity f is

$$f = |F(x) - y_{obj}| \tag{8.11}$$

where $F(\cdot)$ is the complex engineering model and $\boldsymbol{x}$ is the engineering input variables. The multimodal optimization task is to achieve the $\boldsymbol{x}$ that minimize f.

Such optimization task is commonly appeared in the engineering optimization problem. For example, when the optimization objective is a certain property index of the product/process and expected to be stable around a set value, the set value is the y_{obj} here.

Let r_{nbh} be the radius (defined by Euclidean distance) of the neighbourhood area for an antibody with affinity f. Three different ways of neighbourhood radius calculation method are proposed as follow:

1) *Fixed size method*

This method used a fixed value for the neighbourhood size, which means the radius of the neighbourhood is a constant. The calculation formula can be written as,

$$\begin{cases} r_{nbh} = \varepsilon, & f \geq \delta \\ r_{nbh} = 0, & f < \delta \end{cases} \tag{8.12}$$

where ε is the set value for the neighbourhood radius. δ is a threshold for the affinity. The threshold is used to protect the optimal peaks/areas. A simple diagrammatic sketch of the neighbourhood size relating to affinity is shown in Fig. 8.7(a).

2) *Linear increasing method*

This method is a size-varied method. r_{nbh} is increasing along with the affinity. The method is defined as,

$$\begin{cases} r_{nbh} = p * f, & f \geq \delta \\ r_{nbh} = 0, & f < \delta \end{cases} \tag{8.13}$$

where p is parameter of the increased gradient of the r_{nbh}. A diagrammatic sketch of this method is shown in Fig. 8.7(b).

3) *Non-linear increasing method*

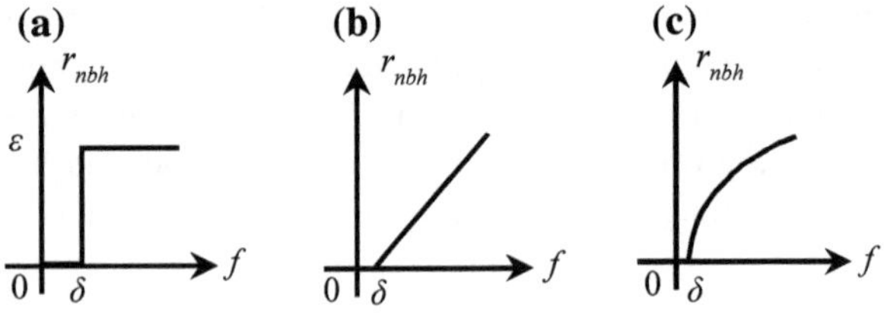

Fig. 8.7 Sketch of three neighbourhood size methods

This method is a development of method 2. It is defined as,

$$\begin{cases} r_{nbh} = p * f^{\tau}, & f \geq \delta \\ r_{nbh} = 0, & f < \delta \end{cases} \tag{8.14}$$

The parameter τ is an exponent. $\tau < 1$ is necessary. With the exponent smaller than 1, r_{nbh} increases faster when f is small, but won't be too large when f is big. A simple diagrammatic sketch of this method is shown in Fig. 8.7(c).

If the optimization objective value is not known in advance for the optimization problem, these neighbourhood size calculation methods need some further improvement because the range of the affinity is unknown. The improvement method can be a recalculation of the affinity. For example, a normalization within all the known affinity (the range that have been found) would solve the problem. This can be a further improvement for the DR-CSA in the future research.

8.3.3 Experimental Studies

8.3.3.1 Regular CSA Multimodal Optimization Experiments

In complex engineering optimization problems, there may be more than one optimal solution for an objective, and sometimes the optimum can even be continuous area. So, the test function that adopted first is an altered Schaffer's F6 function. The Schaffer's F6 function is

$$f(x, y) = 0.5 + \frac{\sin^2\left(\sqrt{x^2 + y^2}\right) - 0.5}{\left[1 + 0.001(x^2 + y^2)\right]^2} \tag{8.15}$$

The term $0.001(x^2 + y^2)$ is set to zero so that the fluctuation of the function is not dampened. Thus, there are multiple optimal (maximum) areas of the function (see Fig. 8.8). The feasible region is $-4 \leq x, y \leq 4$. The amount of the initial solutions is 100.

First, a pair of contrast experiment is conducted between the traditional CSA and the traditional GA. These two methods are both used to optimize the test function. In order to observe clearly, the optimization objective is set as: $y_{obj} = 1$. The parameters of these two evolutionary algorithms are: antibody/ individual population N = 100; number of the best solutions to be clone n = 100; number of clone M = 10; iteration/generation times I, G = 30; binary coding length 22; crossover probability is 0.50; mutation rate is 0.01; hyper-mutation rate is 0.10.

The optimization results are shown in Figs. 8.8 and 8.9. Figure 8.8 shows the entire population of the 30th iteration of the traditional CSA. Apparently, the CSA is able to locate all the maximum areas and spread the solutions evenly on these areas. However the traditional GA is not able to evenly spread them on all of the

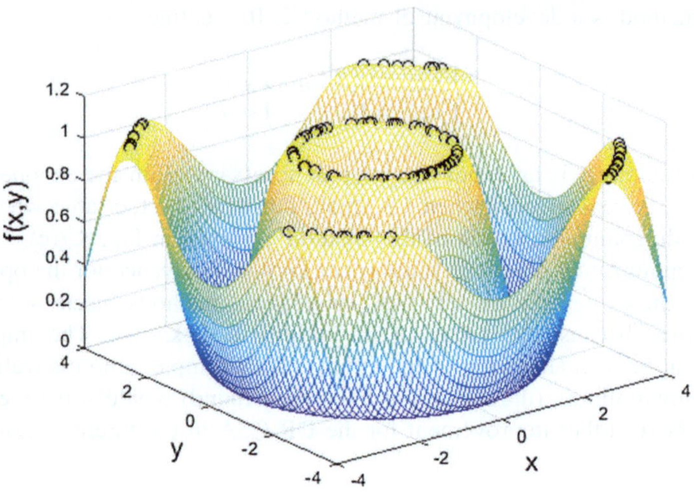

Fig. 8.8 Results of the 30th iteration of the CSA experiment

Fig. 8.9 Maximum and mean results of the GA and CSA experiments

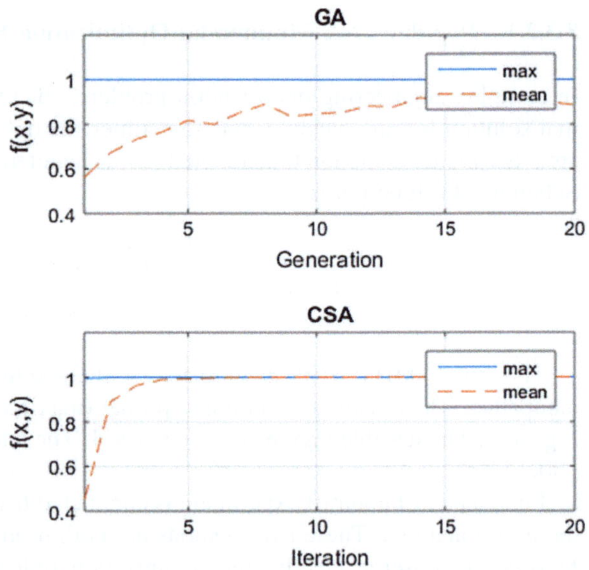

fittest areas like CSA. The explanation for this phenomenon is that the traditional GA always selects the individuals with better fitness. If one fittest individual appeared, it will be kept in the population till the end. And due to the crossover operation, this fittest individual will start affecting other individuals. As a result, the entire population will converge towards this fittest individual. But for the traditional CSA, each antibody is independently evolving. Since $n = N$, each antibody has its own clones. The selection is always among the origin with its own clones respectively. So, each antibody has the opportunity to evolve independently so that the entire population does not converge while evolving. As a result, the entire population is randomly located on the best-affinity areas at the end.

Figure 8.9 gives the maximum and mean values of the entire population of the first 20 generations/iterations. Both of the traditional GA and traditional CSA can reach the maximum solution rapidly (within 3 generations) and GA maybe faster. However, since the traditional GA only focuses on selecting the best solutions among the population, the mean value of the population has a certain randomness. But as to the traditional CSA, the entire population is evolving and gradually reaching the maximum solutions. So the mean value of the population also tends to reach the maximum.

From these results we can see that the traditional CSA is a favorable method for solving multimodal optimization problem. However, in this experiment, the affinity evaluation that CSA carried out is ten times more than GA in order to obtain multimodal solutions. When it comes to complex time-consuming engineering problem, the large amount of evaluation times will cause a severe problem.

8.3.3.2 Experiments of the DR-CSA

After proving the effectiveness of the CSA on multimodal optimization problem, a set of test experiments of the proposed DR-CSA is conducted. Meanwhile, the three neighbourhood size calculation methods are tested and compared with each other.

The test function is the same as the former experiment, so is the optimization objective and the CSA parameters. For the following experiments, to reveal the DR-CSA's abilities of solution-evolving and time-saving, the termination criterion for iteration is set as $f_{mean} < 10^{-4}$. Three sets of parameters for the three neighbourhood size calculation methods are listed in Table 8.4. To be fair, these parameters are roughly chosen.

In order to observe how the three methods perform, a set of the same 100 random initial antibodies is given to all three experiments. A quarter of the initial antibodies (dots) and their degenerated areas (shadows) are shown in Fig. 8.10. The degenerated area size is calculated by the three methods respectively after a pre-evaluation of the initial antibodies' affinities. In the following DR-CSA steps, any antibody that falls into these shadows will be regarded as degenerated.

Figure 8.10 shows that the three different methods have distinctive performances. First of all, all three methods have a few antibodies which do not have any neighbourhood because of δ. It means that these antibodies have very good

Table 8.4 Parameters for three neighbourhood size methods

Fixed Size Method	Linear Increasing Method	Non-linear Increasing Method
$\varepsilon = 0.1$	$p = 0.3$	$p = 0.04$
$\delta = 0.01$	$\delta = 0.005$	$\tau = 0.50$
		$\delta = 0.001$

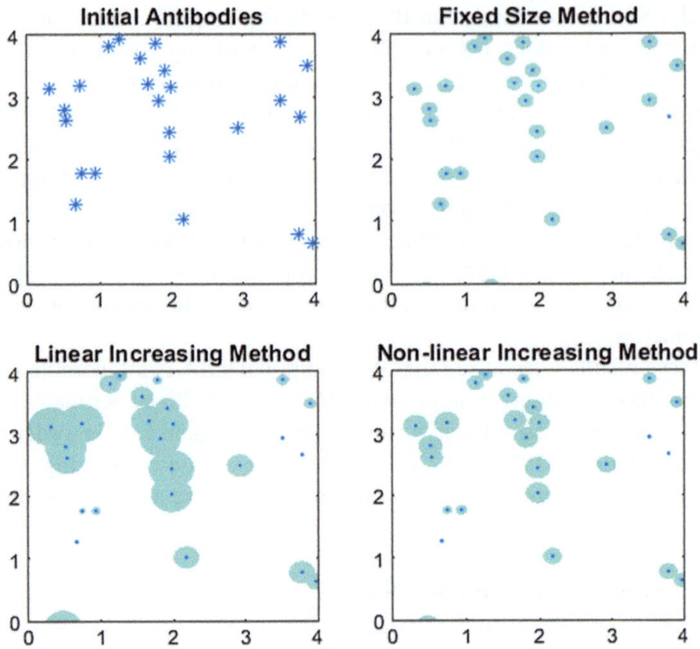

Fig. 8.10 Initial antibodies and initial degenerated areas of the three methods

affinities. Any other antibodies which are near these antibodies may also have very good affinities so that evaluation is required to ensure the method's accuracy. And for all the other antibodies which have an obvious degenerated neighbourhood, their neighbourhood sizes are various. For the fixed size method, as its name, all the sizes are identical. And for the linear increasing method, some antibodies have an unusually large neighbourhood because these antibodies' affinity is relatively bad. And for the non-linear increasing method, the neighbourhood sizes are more appropriate. Its neighbourhood sizes are not too big nor too small, within a reasonable range.

After initialization, as the iteration continues, more and more antibodies will be eliminated and then enter the degeneration database. Consequently, the degenerated areas will gradually expand so that the probability of a degenerated antibody being recognized will gradually increase. As a result, the DR-CSA method will be more and more efficient as the iteration continues.

In order to evaluate the timesaving ability of the DR-CSA, an evaluating indicator is introduced. Since the DR-CSA is aimed to save time by avoiding a number of evaluations, two evaluating indexes for timesaving are defined as

$$\Omega \text{ or } \omega = \frac{\mathrm{M}_{\text{actual}}}{\mathrm{M}_{\text{original}}} \cdot 100\% \tag{8.16}$$

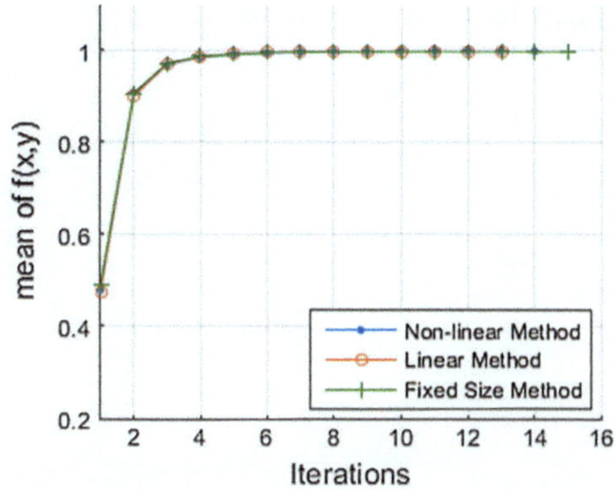

Fig. 8.11 Test function mean value of three methods

where M_{actual} is the number of actual evaluation times using the DR-CSA, and $M_{original}$ is the number of the evaluation times using the traditional CSA. So Ω and ω here represents the percentage of the time cost of the DR-CSA over the time cost of the traditional CSA. But the difference is, Ω stands for the overall time cost of all iterations together, and ω stands for the time cost of each single iteration. Therefore, a value of Ω and a curve of ω can both help understanding the method's performance of timesaving.

Meanwhile, the database usage, the mean value of the entire population, and the iteration times to reach the target are also monitored to show the optimization performance. Due to the randomness in heuristic algorithm, all of the following experiment results are an average result of 10 times repetition of the same experiment. As shown in Fig. 8.11, the curves of the population's mean value of the three methods are very similar. And the iteration times are also very close to each other (around 13–15 times). As a result, the three different methods do not affect the method's accuracy and effectiveness.

However, when it comes to the time cost and database usage, the three different methods have some obvious differences. Table 8.5 provides the index Ω of the three methods. For timesaving, under the listed parameters, the fixed size method is better than the linear method and the non-linear method is better than the fixed size method.

Figure 8.12 shows the curves of index ω. For the initial antibodies, the entire population needs to be calculated because there is no antibody in the eliminated database yet. So the time cost for the first iteration is always 100%. Then, after the first selection, a lot of antibodies are eliminated and stored to recognize degenerated

Table 8.5 Experimental results of the three neighbourhood size methods

Methods	Iterations times	Database usage	Ω
Fixed Size	14.9	2913.3	50.13%
Linear	13.0	3882.3	61.81%
Non-linear	13.8	3745.4	38.98%

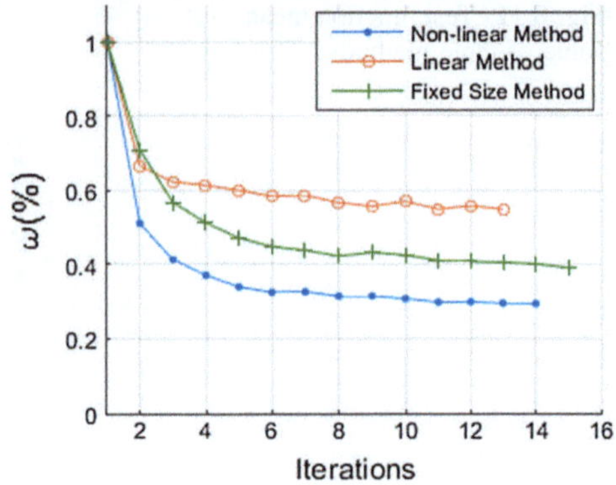

Fig. 8.12 Curves of index ω of three methods

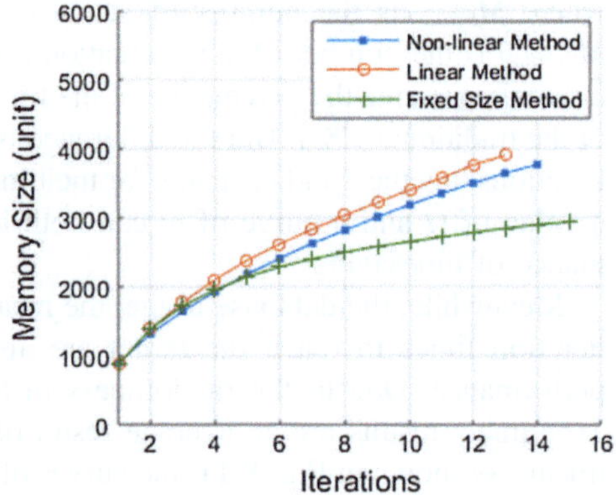

Fig. 8.13 Database usage of three methods

antibodies, so the time cost is decreasing. From Fig. 8.12, we can see that the time cost is dropping remarkably within the first few iterations. And then the time cost is changing slowly with certain randomness. This is because, after a few iterations, a big part of the population are gathering around the best-affinity areas and protected from pre-elimination by δ. And when most of the population are gathering in such best areas, the mean value of this iteration is close to the optimum. As a result, an inverse similarity of Figs. 8.12 and 8.11 can be observed. According to Ω and ω, under the chosen parameters, the non-linear increasing method has a better performance on timesaving.

And looking at the final database usage in Table 8.5 and its changing curve in Fig. 8.13, the data storage is also increasing fast at first, and then gradually slowing down.

The fixed sized method's data storage is smaller than the other two methods. This is because the neighbourhood size of the fixed size method is not continuously changing starting from 0. So it doesn't have tiny neighbourhood cases and the amount of the data storage is relatively small. And the other two methods, the non-linear method is better than the linear method because the neighbourhood size

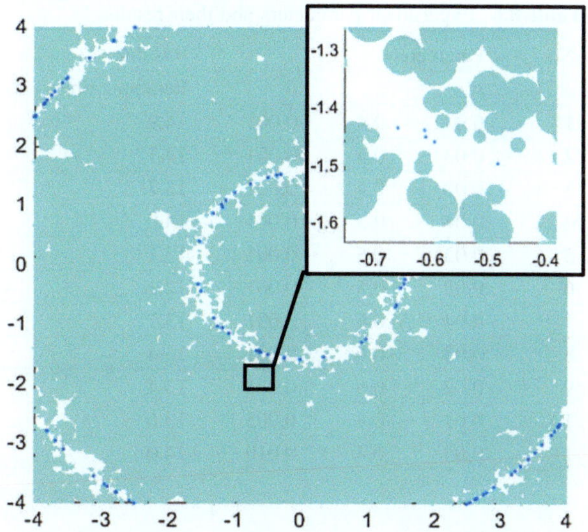

Fig. 8.14 Degenerated area of the 13th iteration of non-linear method

of the non-linear method is increasing faster than the linear method when close to 0. So the non-linear method has less tiny neighbourhood cases than linear method.

Another way to observe the DR-CSA's performance is to see if the degenerated area can gradually expand and cover most of the not-best-affinity areas and not covering the best-affinity areas. Therefore Fig. 8.14 gives the degenerated area of the 13th iteration in non-linear increasing method experiment and a partial enlarged detail. As shown, the shadows cover almost all of the degenerated areas and does not cover the optimal areas. As a result, the method is proved to be valid.

8.3.3.3 Experiments of the Non-linear Increasing Method

The neighbourhood size determination method is important for the DR-CSA, and its parameters affect the method's performance. In this segment, a set of test experiments is conducted to show how the method is affected by these parameters. And the method chosen for this experiment is the non-linear increasing method.

For the non-linear increasing method, there are three parameters, i.e., p, τ, and δ.. The parameter values chosen for the experiment are listed on the left side of Table 8.6. And the experiments results are listed on the right side. Each result is an average of 10 times of the same experiment.

First of all, all the experiments reach the termination criterion around 12–15 iterations. The accuracy of the optimization is not much affected by the parameters. Then, several comparison curves of ω are selected to form plots that demonstrate the influences of the parameters in the method.

In Fig. 8.15, the three curves (experiment 1, 2, and 3) have the same τ and δ but different p. According to the definition of the method, p is the general proportion of

Table 8.6 Experiment parameters and their results

No.	Parameters			Results		
	p	τ	δ	Iterations times	Database usage	Ω
1	0.04	0.4	0.001	14.9	3106.0	27.96%
2	0.03	0.4	0.001	14.7	3833.7	34.86%
3	0.02	0.4	0.001	13.7	4915.3	46.76%
4	0.04	0.5	0.001	14.0	3782.9	38.73%
5	0.03	0.5	0.001	13.1	4286.2	45.77%
6	0.02	0.5	0.001	12.3	5068.3	56.00%
7	0.04	0.6	0.001	13.7	4099.4	46.14%
8	0.03	0.6	0.001	13.4	4589.0	51.52%
9	0.02	0.6	0.001	13.4	5089.5	59.98%
10	0.04	0.4	0.005	13.6	2759.9	32.93%
11	0.04	0.4	0.010	14.0	2548.4	35.10%

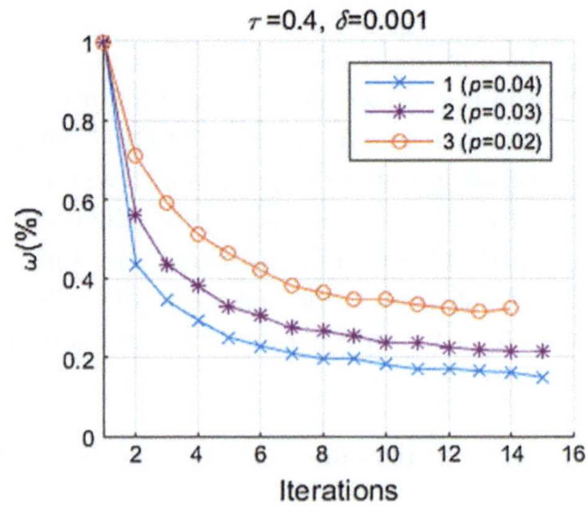

Fig. 8.15 Comparison curves of experiment 1, 2, and 3

the neighbourhood size so it determines the average neighbourhood size. As shown in the Fig. 8.15, the time cost decreases with an increasing p.

In Fig. 8.16, the three curves (experiment 3, 6, and 9) have the same p and δ but different τ. When p and δ are the same, a smaller τ corresponds to a lower time cost. That is because as a less-than-one exponential term, the smaller τ is, the faster the neighbourhood size grows when the affinity is close to zero. However, unlike p, τ only affects part of the neighbourhood size (when affinity is small). In fact, the main function of τ is to prevent the neighbourhood size from becoming too large and causing a failure of the algorithm.

In Fig. 8.17 the three curves (experiment 1, 10, and 11) have the same p and τ but different δ. δ protects the optimal areas by keeping a zero neighbourhood zone around the optimal areas. So the performances of the first few iterations is not affected by δ because the population average is not close to optimum. The three

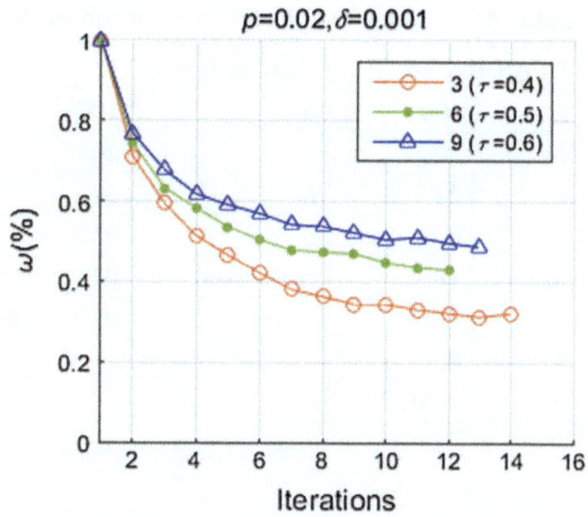

Fig. 8.16 Comparison curves of experiment 3, 6, and 9

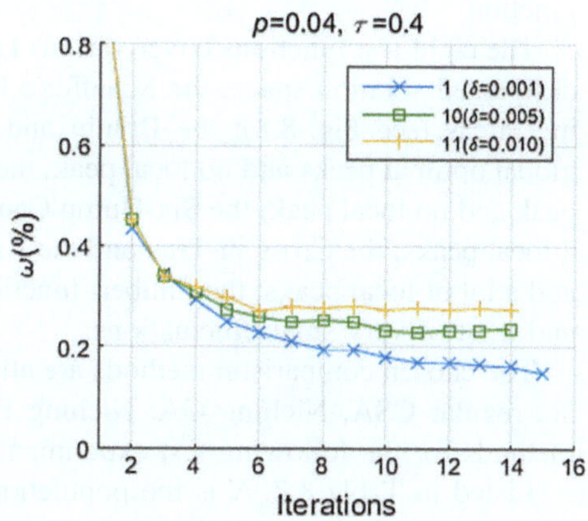

Fig. 8.17 Comparison curves of experiment 1, 10, and 11

curves are mostly overlapped. But after that, most of the population is evolved and entered the optimal nearby area, the δ starts to affect the curves. As shown in Fig. 8.17 and Table 8.6, a relatively smaller δ helps improving the computation speed but bigger δ helps decreasing data storage.

At last, by comparing all the experiment results in Table 8.6, it is pleasant to find that when parameters are properly chosen, the time saving of the DR-CSA method is more than 82%.

8.3.3.4 Comparing Experiments on Multiple Multimodal Functions

In order to prove the effectiveness of the proposed DR-CSA method, the method was tested on eight different benchmark multimodal functions, comparing with several well-developed heuristic optimization methods. All of the chosen multimodal functions are frequently used as test function. They are Schaffer's F6

Table 8.7 Parameters of each comparison method

CSA/DR-CSA		Niching GA		Niching PSO		opt-aiNet	
N	100	N	100	N	100	N	100
I	30	I	200	I	200	I	300
L	22	L	22	w_{max}	1	–	–
μ_h	0.1	μ	0.05	w_{min}	0	–	–
M	10	cross-over	0.9	c_1	2	M	10
p	0.03		$0.01 \sim 0.1$	m	0.1	τ	0.1
τ	0.5			n	3	d	0.4
δ	0.001	–	–	–	–	β	50

function, Branin function, Himmelblau function, Rosenbrock function, Six-Hump Camel Back function, Cross-in-Tray function, Shubert function, and Rastrigin function.

The eight test functions cover various kinds of multimodal functions. Within the designated solution space, the Schaffer's F6 function contains several global optimal areas (see Fig. 8.8); the Branin and the Himmelblau function have several global optimal peaks and no local peak; the Rosenbrock function has one flat global peak and no local peak; the Six-Hump Camel Back function has 2 global peaks and 4 local peaks; the Cross-in-Tray and the Rastrigin function have a few global peaks and a lot of local peaks; the Shubert function has 2 global peaks, a few local peaks, and a lot of very small fluctuations.

The chosen comparison methods are all multimodal optimization methods. They are regular CSA, Niching GA, Niching PSO, and opt-aiNet. Parameters of these methods for the following test experiments are listing in Table 8.7.

Listed in Table 8.7, N is the population size, I is the maximum iteration (generation), L is the binary code length, M is the clone amount, μ is the mutation rate, μ_h is the hyper-mutation rate, and all other parameters are with each method itself. The distance parameter of Niching GA is chosen within 0.01–0.1 according to function's range. All of the parameters are chosen carefully so that all of the methods perform well.

In the neighbourhood size calculation of the DR-CSA, in order to employ the same parameters (p, τ, and δ) to different test functions, this part carries out an additional normalization of the function and the solution space. But normalization is not necessary if the parameters can chose freely. The experiments calculated the best result error (the error between experimental and theoretical best), mean value error (the error between population mean and theoretical best), the standard deviation of the final population, and the number of evaluation times.

Amount these results, the number of evaluation times reflects the computation speed of the optimization method. Besides the numerical results, the multimodal performance (including the number of the optimal peaks/areas that found and the distribution of the solutions) of each method is observed and recorded. As a computation speed improving method, these results are the most important results.

To evaluate the evaluation times, besides the termination criterion of mean value error $< 1e^{-4}$, another termination criterion which is the maximum iteration times is added. This criterion is adopted because some optimization methods will converge before satisfying the first criterion, so without a second criterion of maximum iteration times the computation will run forever and the judgment of its computation speed will not be fair. The results of the comparison test experiments are listed in Table 8.8. Every listed result is an average value of the same experiment repeating ten times.

As shown in Table 8.8, the computation speed (the number of fitness evaluation times) of the DR-CSA is clearly better than all other methods. The best result appeared in the Rastrigin function experiment, the evaluation times of the DR-CSA is only **11.5%** of opt-aiNet and **18%** of the regular CSA. While the computation speed is highly improved, all the other optimization results are also comparable with other methods.

When look at the multimodal optimization performance, the DR-CSA is the same as the regular CSA. They are able to locate all of the global optimal results and if the optimal result is an area (like Schaffer's F6 function), these two methods can evenly (with randomness) distribute the results on this area. If the function has suboptimal peaks which are almost as good as the global peaks, the two methods can also locate such suboptimal peaks, such as the Shubert function and the Rastrigin function. The multimodal optimization ability of Niching GA is relatively weaker than other methods. If there are many peaks in the function, the result of Niching GA may miss several of the peaks. Besides, the distribution of the final population on the peaks of Niching GA is usually not even. The peak-locating ability of Niching PSO is not very stable and the result distribution is not even either. The opt-aiNet method has an excessive peak-locating ability, it can locate all of the peaks in the function, including global and local peaks and even the very small fluctuations. If there are no local peaks in the function, such as the Himmelblau function, this opt-aiNet method has an excellent result. But if the function has many very small peaks, such as the Shubert function, this method's result will be very unsatisfactory.

When look at the result of standard deviation, the Niching GA usually has the smallest standard deviation due to its convergence property. However the strong convergence affects the distribution ability in multimodal optimization problem. Considering both the standard deviation and the distribution performance, the DR-CSA and the CSA have a better overall performance. Also, in most cases, the DR-CSA has favorable best result and mean result, the same as the CSA, comparing to all other methods. In conclusion, the proposed DR-CSA is able to highly improve the optimization speed while maintaining a favorable optimization result.

Table 8.8 Experiment results of different optimization methods on multimodal benchmark functions

Optimization Function	Objective	Algorithm	Error of Best Result	Error of Mean Result	Standard Deviation	Evaluation times	Located(Real) Peaks
Schaffer's F6 x, y∈[−4,4]	maximum (1)	DR-CSA	3.98e-11	6.91e-5	2.77e-4	6344	5 (5)
		CSA	1.92e-10	5.30e-5	2.14e-4	16400	5 (5)
		Niching GA	2.54e-13	7.28e-5	2.78e-4	18830	5 (5)
		Niching PSO	0	2.95e-2	9.59e-2	20400	5 (5)
		opt-aiNet	8.39e-14	8.04e-5	3.68e-4	20737.3	5 (5)
Branin x,y∈[−10,10]	minimum (0.39789)	DR-CSA	3.23e-6	0.4466	1.0605	11537	3 (3)
		CSA	3.67e-6	0.4187	1.1174	30000	3 (3)
		Niching GA	5.56e-5	0.2059	0.1479	20100	2.8 (3)
		Niching PSO	4.42e-3	1.6047	1.2145	20400	3.9 (3)
		opt-aiNet	4.19e-6	0.5770	0.5026	63777	3 (3)
Himmelblau x, y∈[−6,6]	minimum (0)	DR-CSA	5.47e-6	0.8392	1.7482	14877	4 (4)
		CSA	1.04e-5	0.7981	1.6689	30000	4 (4)
		Niching GA	1.70e-4	0.7273	0.5303	20100	3.6 (4)
		Niching PSO	1.02e-2	3.5099	3.4065	20400	4 (4)
		opt-aiNet	1.39e-6	9.44e-5	8.71e-5	30094.4	4 (4)
Rosenbrock x∈[−2,2] y∈[−1,3]	minimum (0)	DR-CSA	2.56e-4	0.7689	1.0482	11539	1 (1)
		CSA	4.85e-4	0.8049	1.0614	30000	1 (1)
		Niching GA	3.07e-4	0.6198	0.5402	20100	1 (1)
		Niching PSO	1.89e-3	1.2478	1.7092	20400	1 (1)
		opt-aiNet	8.39e-7	0.2787	0.5846	33510	1 (1)
Six-Hump Camel Back x∈[−2,2] y∈[−1,1]	minimum (-1.0316)	DR-CSA	6.65e-6	0.0227	0.0328	9094	2 (2)
		CSA	6.51e-6	0.0218	0.0336	30000	2 (2)
		Niching GA	2.75e-5	0.0266	0.0183	20100	2 (2)

(continued)

Table 8.8 (continued)

Optimization Function	Objective	Algorithm	Error of Best Result	Error of Mean Result	Standard Deviation	Evaluation times	Located(Real) Peaks
		Niching PSO	4.74e-4	0.4116	0.3475	20400	3.9 (2)
		opt-aiNet	2.84e-5	0.5244	0.9400	29032.4	6 (2)
Cross-in-Tray $x, y \in [-4,4]$	minimum (−2.0626)	DR-CSA	1.07e-6	9.06e-4	2.14e-3	11205	4 (4)
		CSA	1.04e-6	8.81e-4	1.97e-3	30000	4 (4)
		Niching GA	1.86e-6	2.62e-4	1.82e-4	20100	3.8 (4)
		Niching PSO	1.66e-6	2.14e-2	2.97e-2	20400	4.7 (4)
		opt-aiNet	1.87e-6	7.91e-2	9.76e-2	32342.8	13.1 (4)
Shubert $x, y \in [-2,2]$	minimum (−186.73)	DR-CSA	1.54e-4	13.03	24.99	6561	4.7 (2)
		CSA	2.01e-4	13.04	25.38	30000	4.8 (2)
		Niching GA	2.15e-3	13.55	10.15	20100	1.7 (2)
		Niching PSO	9.06e-2	112.17	51.64	20400	13.4 (3)
		opt-aiNet	6.02e-4	130.76	51.09	21645	31.4 (2)
Rastrigin $x, y \in [-5,5]$	minimum (0)	DR-CSA	1.39e-5	2.2656	2.2853	5403	18.5 (1)
		CSA	6.75e-5	2.0638	1.9986	30000	17.3 (1)
		Niching GA	1.32e-7	2.1829	0.9085	20100	9 (1)
		Niching PSO	0.4797	6.5484	3.7553	20400	27.2 (1)
		opt-aiNet	2.95e-4	9.4255	5.5570	46925.4	58.9 (1)

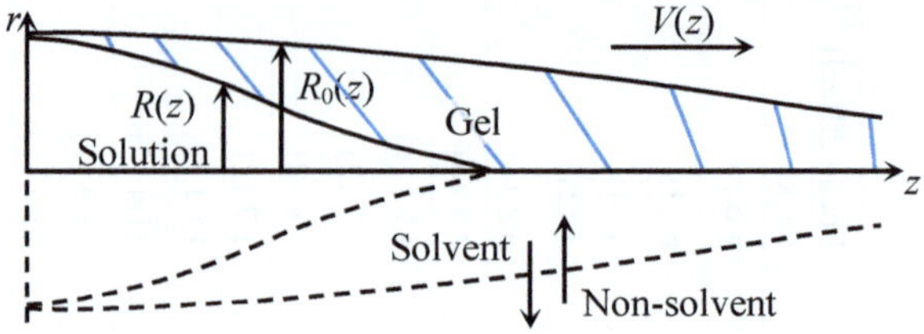

Fig. 8.18 Enlarged diagrammatic sketch of polymer fiber's sectional view

8.3.4 Wet Spinning Coagulating Process Modeling and Optimization

8.3.4.1 Wet Spinning Coagulating Model

Wet spinning is one of the most traditional producing methods of polymer fibers. This process is used for polymers that need to be dissolved in a solvent to be spun. It is the key processes during the whole fiber production [51, 52].

A spinneret is submerged in a chemical bath where the dissolved polymer is extruded. The polymer solution is a mix of polymer, solvent, and non-solvent (coagulant). The bath liquid contains mostly the coagulant for the polymer fibers to precipitate and solidify. A diagrammatic sketch of a polymer fiber's sectional view is shown in Fig. 8.18. The z axis is the direction from the spinneret to the take-up wheel along bath. The r axis is the direction from the fiber center to the outside.

$V(z)$ is the velocity at each point of the fiber. $R_0(z)$ is the outer radius of the fiber. As the polymer solution gradually solidifies, the gel phase grows from the outside to the center. $R(z)$ is the radius of solution-gel interface. The solvent and non-solvent are diffusing between the fiber and the bath as the polymer moves inside the bath. The concentrations of solvent and non-solvent inside the fiber are $C_S(r, z)$ and $C_{NS}(r, z)$.

By studying the physical relationships between the variables and process parameters, the mechanism of the coagulating process can be divided into four major parts: (1) double diffusion; (2) phase inversion; (3) mass balance; (4) force balance. Details of the model please see [53–56]. By combing all four sub-models, the coagulating model is written as: (Fig. 8.19)

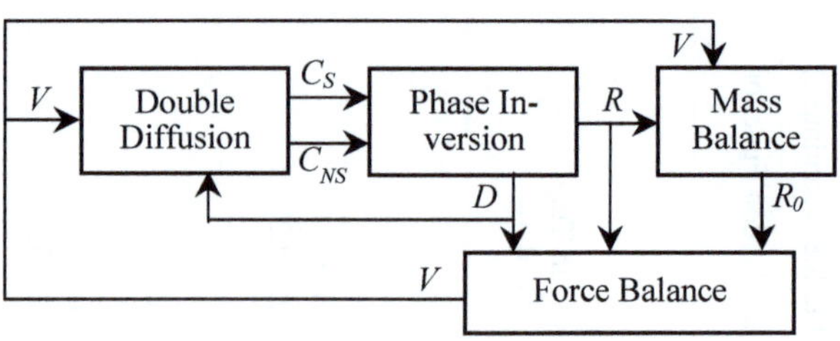

Fig. 8.19 Structure block of coagulating mechenism model

$$\begin{cases} V(z)\frac{\partial C_S}{\partial z} = \frac{1}{r}\frac{\partial\left(rD_S\frac{\partial C_S}{\partial r}\right)}{\partial r} \\ V(z)\frac{\partial C_{NS}}{\partial z} = \frac{1}{r}\frac{\partial\left(rD_{NS}\frac{\partial C_{NS}}{\partial r}\right)}{\partial r} \\ R(z) = f(C_S(r,z), C_{NS}(r,z)) \\ R_0(z) = \sqrt{\frac{G}{\pi\rho_g V(z)} - \frac{(\rho_s-\rho_g)R^2}{\rho_g}} \\ V'(z) = \frac{F_e - G(V(z)-V_0) - 2\pi\int_0^x P_{xs}Rdx - \pi\sigma(R_0(0)-R_0(z))}{\mu_s\pi R^2(z) + \mu_g\pi\left(R_0^2(z)-R^2(z)\right)} \end{cases} \tag{8.17}$$

8.3.4.2 Numerical Computation Method for Wet Spinning Coagulating Model

The equations in wet spinning coagulating model need to be solved numerically at the same time. It is tricky when the two PDEs and one ODE need to be combined and calculated simultaneously.

One way to combine PDE with ODE is to adopt the Method of Lines (MOL). The basic idea of the MOL is to replace the spatial (boundary-value) derivatives in the PDE with algebraic approximations. Once this is done, the spatial derivatives are no longer stated explicitly in terms of the spatial independent variables. Thus, in effect, only the initial-value variable, typically time in a physical problem, remains. In other words, with only one remaining independent variable, there is a system of ODEs that approximate the original PDE [57].

Via adopting MOL, the two PDEs are converted into two sets of ODEs. And then the other equations and programs can be combined together. The details of the MOL method to change PDE into ODE can be found in [57].

8.3.4.3 Model Numerical Computation Results

The polymer wet spinning coagulating system that is studied in this section is the PAN-DMSO-water system. Which means in the process the polymer is polyacrylonitrile, the solvent is dimethyl sulphoxide, and the non-solvent is water. According to [53–58], coefficients of this PAN-DMSO-water system are adopted. The numerical computation of the model is conducted by MATLAB (Version R2014b).

Some of the model computation results are shown in Figs. 8.20 and 8.21. Figure 8.20 shows the concentration curves of the solvent and the non-solvent inside the fiber from the beginning of the bath to the end. The different curves are the different layer-positions inside the fiber relating to different radiuses. The solvent inside the fiber is gradually reducing, and the outer layer (lower curves) is reducing faster than the inner layer (upper curves). At the same time, the non-solvent is gradually increasing insider the fiber. The outer layer is increasing faster than the inner layer. Figure 8.21 shows the results of fiber radius and the

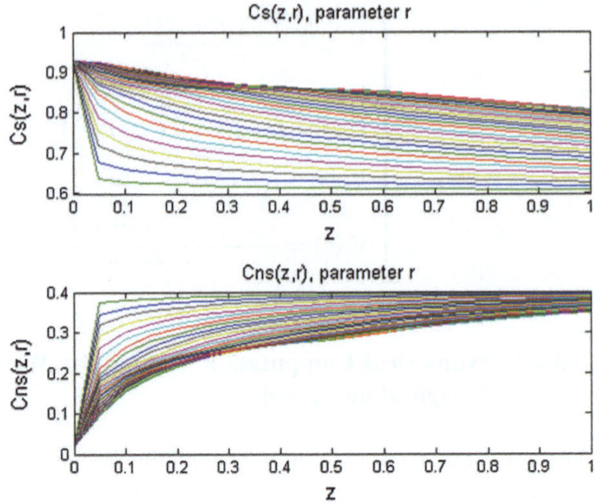

Fig. 8.20 Concentration curves of solvent and non-solvent

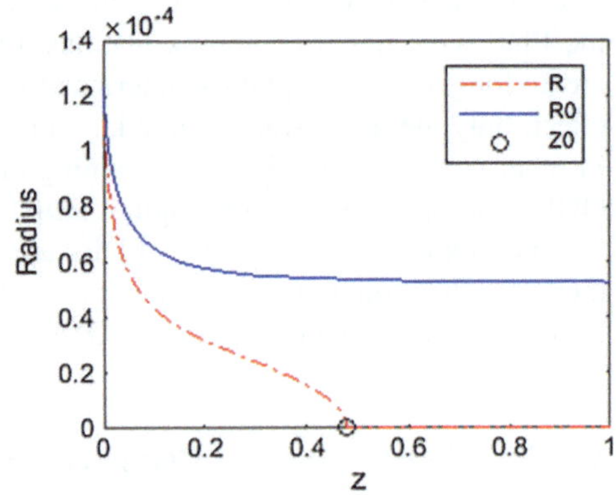

Fig. 8.21 Curves of fiber radius and solution-gel interface radius

solution-gel interface radius. As shown in the figure, after the point $z = 0.48$ the fiber is fully gelled. The point Z_0 is named ***fully gelled point***.

The computer environment of this experiment is 3.40 GHz Intel Core i7-3770 CPU, 8.00 GB RAM, and 64-bit operating system. The overall time cost of one complete computation is 2.93 s. If an optimization evaluates this model thousands of times, the overall time cost would be days.

8.3.4.4 Coagulating Process Optimization Based on DR-CSA

By observing and studying the model outcomes, the fully gelled point Z_0 appears to be a significant result. The outcome of this point is the result of all the process parameters such as solvent diffusivity, non-solvent diffusivity, polymer viscosity, bath concentration, fiber concentration, fiber extrusion speed, polymer solution flow rate, take up speed, spinneret orifice diameter, bath length, and so on. At the same time, this fully gelled point is related to all kinds of fiber performances. It directly reflects the gelation degree. And the gelation degree can reflect the fiber's qualities such as strength, evenness, viscoelasticity, elongation at break, et al. As a result, this fully gelled point can be a good choice as an optimization objective.

Instead of maximizing or minimizing the objective, the optimization objective of the fully gelled point should be a desired value, i.e. $Z_0 = Z_{set}$. But eventually, it can be converted into a minimize problem as $Z_0 - Z_{set} = 0$. And when the focus is only on the output of Z_0, the system can be regarded as a multiple-input single-output (MISO) system. By a few test runs of this MISO system, it is realized that the solution of a desired output is not unique. There will be multiple different combinations of inputs that have the same output. So this optimization task is a multimodal optimization problem. Also, it is expected that its solutions are located on one or more extremal intervals instead of independent extremes. The optimization result should be a set of equally best solutions for the technicians to choose.

Due to the effectiveness of the DR-CSA in solving time-consuming multimodal optimization problem. It is adopted for the coagulating process optimization.

Within all the influencing factors of the wet spinning coagulating process mentioned above, several factors are considered as constants because they are impossible or hard to change. These factors include: specifications of the spinneret, specifications of the bath, physical properties of the polymer, solvent and non-solvent, parameters of diffusion, Reynolds number, surface tension, and so on. And then, some other controllable factors are chosen as the optimization input variables (decision variables). The chosen input variables are:

1. Bath liquid proportion: includes solvent concentration C_{S_bath}, non-solvent concentration C_{NS_bath}. Since $C_{S\_bath} + C_{NS\_bath} = 100\%$, only one variable needs to be used as an input factor. In this experiment, the C_{S_bath} is chosen.
2. Original polymer solution proportion: includes polymer concentration $C_{polymer}$, solvent concentration C_{S_solu}, and non-solvent concentration C_{NS_solu}. Since $C_{polymer} + C_{S\_solu} + C_{NS\_solu} = 100\%$ and $C_{polymer}$ is a constant, C_{S_solu} is chosen to be the input that represents this factor.
3. Fiber extrusion speed V_0 and polymer solution flow rate G. Since $V_0 = G/(\rho\pi R^2)$, only G is chosen as a input.
4. Take up speed V_{end} .

With the inputs all selected, the optimization objective for this experiment is set as $Z_{set} = 0.70\ (m)$. And the computing parameters are: antibody population $N = 100$; number of the best solutions to be clone $n = 100$; number of clone $M = 10$; iteration times $I = 15$; binary coding length is 22; hyper-mutation rate is 0.10; parameters for non-linear neighbourhood size increasing method: $p = 0.03$, $\tau = 0.6$, and $\delta = 0.001$. Both of the CSA and DR-CSA are applied.

The max and mean values of the optimization population are shown in Fig. 8.22, and the curve of timesaving index ω is shown in Fig. 8.23. As shown in the figures,

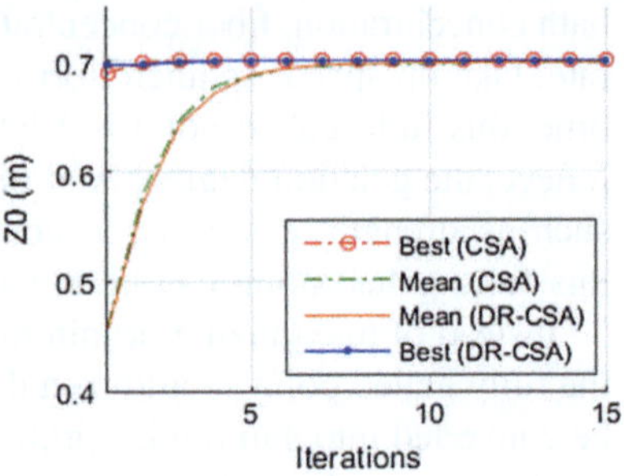

Fig. 8.22 Results of max and mean values of Z_O optimization

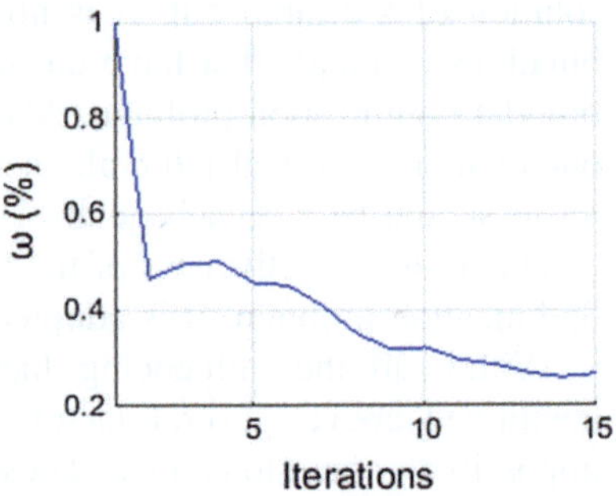

Fig. 8.23 Timesaving index changing curve of Z_O optimization

both of the CSA and the proposed DR-CSA is able to fulfill the task of optimizing this time-consuming wet spinning coagulating model. DR-CSA and CSA have the approximately same performance on accuracy.

However, the DR- improvement results of DR-CSA are: the overall time costs index $\Omega = \mathbf{39.60\%}$; the final database usage is **5025** units. Which means the overall running time of DR-CSA is only about 40% of the traditional CSA, and it can be smaller if the parameters are optimal.

The first 10 best solutions are listed in Table 8.9. The optimization statistical results are listed in Table 8.10.

Table 8.9 First 10 best solutions of z_O optimization

C_{S_solu} (%)	C_{S_bath} (%)	G (kg/s)	V_{end} (m/s)	Z_O (m)
93.6	68.2	5.8e-07	0.048	0.70
93.7	58.4	6.6e-07	0.060	0.70
92.9	55.1	6.8e-07	0.068	0.70
93.1	65.6	6.9e-07	0.053	0.70
93.9	65.4	5.2e-07	0.051	0.70
93.7	64.9	5.1e-07	0.052	0.70
93.6	58.2	7.2e-07	0.060	0.70
92.6	60.2	6.6e-07	0.063	0.70
94.0	68.4	6.6e-07	0.046	0.70
93.3	60.8	6.1e-07	0.059	0.70

Table 8.10 Statistical results of z_0 optimization

	Error of Best	*Error of Mean*	*Std*	*Evaluate Times*	*Running Time (s)*
CSA	9.06e-7	4.51e-4	8.34e-4	15000	20185.6
DR-CSA	8.53e-7	4.94e-4	9.62e-4	5936	8129.2

8.4 A Dynamic SVR-ARMA Model with Improved Fruit Fly Algorithm

In this section, we study the fiber stretching process which plays a key role in the process of fiber production. Based on the history data from the mechanism model of the fiber stretching process, the prediction SVR-ARX model is established between the input data and output data. The fruit fly optimization algorithm (FOA) is used to optimize the parameters in SVR-ARX model. Furthermore, to circumvent the FOA easily fall into the local optimum and no communication relations among the generation population, IFOA is proposed, which can be optimized the parameters in SVR-ARX model. And since, compared the output values between the soft sensor IFOA-SVR-ARX method and the actual fiber stretching system, a series of the error values are obtained. This error values can be considered as an ARMA time series. It reflects the relationship between the current output and the previous output from IFOA-SVR-ARX predict model. Meanwhile, The ARMA model is to achieve a transform from the static model to dynamic model, and reach the dynamic modification for the static model.

8.4.1 The Fiber Stretching Process

In practical fiber production process, the stretching unit is often organized into several small parts under different working environments because these environments can exert different influence on fibers, e.g., in hot air, water, or high-pressure vapour [59, 60]. A single stretching process on the fiber production is as shown in Fig. 8.24. In Fig. 8.24, two stretching units and a bath between them is illustrated. The equipment in the single stretching unit is also provided in detail. The raw strand is derived from the coagulating bath, and filaments are uniaxial stretched into the solidified filaments. Winding wound around the two rollers, there have different rotational speeds. Hence a careful correction and proof reading before the final publication is very necessary.

Table 8.11 gives the parameter and their steady state values of the fiber stretching process. Focusing on the stretching process, the steady state is defined as

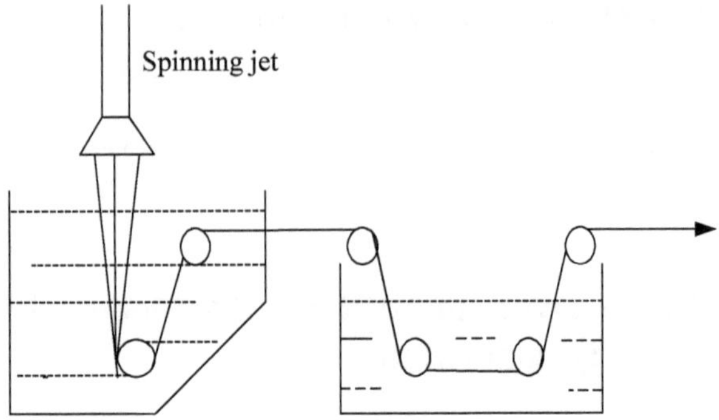

Fig. 8.24 A single stretching process of the fiber production

Table 8.11 The model parameters and their steady state values of the fiber stretching process

Parameters	Steady state values
Tension F	Output1
Velocity V	Output2
Voltage U	Input
Modulus of elasticity E	211.1 N/tex
Radius of Motor roller R	0.1 m
Moment of inertia Ja	0.02 N.M.s^2
Length of span between two driving rollers L	6.09 m
Coefficient of friction roller f	0.685
Filaments cross-sectional area A	1.13 μm^2
Ratio between driving Motor and roller n	1

follows. For any spinning line points, the change of polymer is in the "steady state" and continuous state refers to any point of the polymer through a spinning production line with a constant, and its parameter values do not change over time. That is, velocity, temperature and other parameters in the whole spinning line points have different locations. However, variation between different points is continuous, and their parameters do not change over time in the fixed position.

$$\begin{aligned} J_a\frac{dV(t)}{dt} &= -fV(t) + (F(t) - F_0(t))R^2 + nuR \\ L\frac{dF(t)}{dt} &= EA(V(t) - V_0) + F_0(t)V_0 - F(t)V(t) \end{aligned} \tag{8.18}$$

8.4.2 *The Predictive Model*

8.4.2.1 The SVR-ARX Model

(1) SVR

Assumed a group of the sample sets $\{(x_i, y_i)\}_{i=1}^{m}$, where $x_i \in R^n$ represent the input vector, $y_i \in R$ represent the corresponding output variable of x_i, m is the number of the given sample data. The nonlinear function $\varphi(x)$ is a mapping between the $x_i \in R^n$ and $y_i \in R$. Then, the regression function $f(x)$ can be expressed as follows:

$$f(x) = \langle w * \varphi(x) + b \rangle \tag{8.19}$$

where w is the weighting vector in the feature space, b is the bias variables. According to the minimizing empirical risk principle, the solved problem of the Eq. (8.19) can be equivalent to solving the Eq. (8.19) :

$$\frac{1}{2}\|w\|^2 + C\sum_{k=1}^{n} g(x_k, y_k, f(x_k)) \tag{8.20}$$

where C is a regularized constant, the function $g(x_k, y_k, f(x_k))$ denotes ε insensitive loss function and defines as Eq. (8.21) :

$$g(x, y, f(x)) = \begin{cases} 0 & |y - f(x)| < \varepsilon \\ |y - f(x)| - \varepsilon & else \end{cases} \tag{8.21}$$

The slack variables $\xi = (\xi_1, \quad \xi_1^*, \quad \cdots, \quad \xi_n, \quad \xi_n^*)^T$ will be introduced into the Eq. (8.21), and we obtain a convex quadratic programming with respect to the original problem of the ε-SVR:

$$\begin{aligned} \min_{w,b,\xi^{(*)}} \quad & \frac{1}{2}\|w\|^2 + C\sum_{k=1}^{n}(\xi_k + \xi_k^*) \\ s.t. \quad & f(x_k) - y_k \le \varepsilon + \xi_k \\ & y_k - f(x_k) \le \varepsilon + \xi_k^* \\ & \xi_k, \xi_k^* \ge 0 \quad k = 1, \cdots, n. \end{aligned} \tag{8.22}$$

A linear separable problem is also a convex quadratic programming problem. To solve convex quadratic programming problem, the Lagrange Multiplier is introduced in the Eq. (8.22). Then, the constrained optimization problem can be transformed to an unconstrained dual space optimization problem. Furthermore, to deduce the dual problem of the original space, Lagrange function is introduced as Eq. (8.23).

$$\begin{aligned}L(w,b,\xi,\alpha,\beta) = \frac{1}{2}\|w\|^2 + C\sum_{k=1}^{n}(\xi_k+\xi_k^*) - \sum_{k=1}^{n}(\beta_k\xi_k+\beta_k^*\xi_k^*) \\ - \sum_{k=1}^{n}\alpha_i(\varepsilon+\xi_k+y_k-f(x_k)) \\ - \sum_{k=1}^{n}\alpha_i^*(\varepsilon+\xi_k^*+y_k-f(x_k))\end{aligned} \tag{8.23}$$

where $\alpha = (\alpha_1, \alpha_1^*, \cdots, \alpha_n, \alpha_n^*)^T$ and $\beta = (\beta_1, \beta_1^*, \cdots, \beta_n, \beta_n^*)^T$ are lagrange multipliers vector. Finally, the optimization problem can be written as follows:

$$\begin{aligned}\min_{\alpha,\beta} \quad & -\tfrac{1}{2}\sum_{i,k=1}^{n}(\alpha_i^*-\alpha_i)(\alpha_k^*-\alpha_k)(\varphi(x_i)\varphi(x_k)) \\ & -\varepsilon\sum_{k=1}^{n}(\alpha_k^*+\alpha_k)+y_k\sum_{k=1}^{n}(\alpha_k^*-\alpha_k)\end{aligned} \tag{8.24}$$

with the constraints

$$\begin{aligned}&\sum_{k=1}^{n}(\alpha_k^*-\alpha_k)=0 \\ &C-\alpha_k-\beta_k=0 \quad k=1,\cdots,n \\ &\alpha_k,\alpha_k^*\geq 0,\beta_k,\beta_k^*\geq 0 \quad k=1,\cdots,n\end{aligned}$$

According to the Eq. (8.24), the nonlinear regression problem can be transformed into maximizing problem in regard of solving w with the constraints.

where

$$w = \sum_{k=1}^{n}(\alpha_k^*-\alpha_k)\varphi(x_k) \tag{8.25}$$

$$f(x) = \sum_{k=1}^{n}(\alpha_k^*-\alpha_k)\left(\phi(x_k)^T,\varphi(x)\right)+b \tag{8.26}$$

On the basis of the Mercer theorem [51], the Kernel function $K(x_k, x)$ is defined by:

$$K(x_k,x) = \left(\varphi(x_k)^T,\varphi(x)\right) \tag{8.27}$$

The effectiveness of SVR depends on the selection of kernel, the kernel's parameters, and soft margin parameter C. The Gaussian Kernel function RBF is selected as kernel function in this section, considering that it has excellent properties of the nonlinear forecasting performance among the linear kernel function, the polynomial kernel function, the sigmoid kernel function [61].

$$K(x_k, x) = \exp\left(-\frac{\|x - x_k\|^2}{\sigma^2}\right) \tag{8.28}$$

where the σ is the width of the Gaussian RBF kernel function. Then, we will substitute the expression (8.27) into the Eq. (8.26), Eq. (8.29) is obtained as follows:

$$f(x) = \sum_{k=1}^{n} (a_k^* - a_k)K(x_k, x) + b \tag{8.29}$$

Finally, the SVR function can be expressed as Eq. (8.29).

(2) The structure of the SVR-ARX model

To consider the effect of the historical input-output data in the fiber stretching process, we combine the ARX model with SVM model, the specific combination method is written as follows:

We will consider the input-output data before k time as input vector of the SVM model at k time. Based on the history input-output data of the fiber stretching process, the SVR-ARX model is considered as predict model. The Eq. (8.30) represents the structure of the ARX model.

$$y(k) = \sum_{i=1}^{p} a_i y(k-i) + \sum_{j=1}^{q} b_j u(k-j) + e(k) \tag{8.30}$$

where the k is the discrete time, $u(k)$ is the system input at k time, $y(k)$ denotes the system output at k time. $u(k-j)$ is the j -th input before $u(k)$. $y(k-i)$ is the i-th output before $y(k)$. $e(k)$ is the Gaussian white noise. $a_i, i = 1, \cdots, p$ and $b_j, j = 1, \cdots, q$ denote the coefficient of the $y(k-i)$ and $u(k-j)$ respectively. p, q represent the order of the input variables $u(k)$, and output variables $y(k)$.

$$x = \left[y_{k-1}, \cdots, y_{k-p}, u_{k-1}, \cdots, u_{k-q}\right] \tag{8.31}$$

Then, the predict output of the SVR-ARX model is obtained by the Eq. (8.32).

$$\hat{y}_k(x) = f(x) = f\left[y_{k-1}, \cdots, y_{k-p}, u_{k-1}, \cdots, u_{k-q}\right] \tag{8.32}$$

where the $f(x)$ is a SVR function.

8.4.2.2 IFOA

(1) FOA

Based on simulate behavior of fruit fly find out the food, an intelligent swarm algorithm FOA is proposed [62]. At present, the FOA has been applied in the different fields [63–65]. It is an interactive evolutionary computation method. Firstly, considering the fruit fly smell and visual be better than the other species, they use the smell to locate the food source, and then, rely on the visual to locate the exact location of the food. Finally, the information interaction realize among the populations. The specific implement steps of the FOA are listed as follows.

Step 1. Parameters initialization

The initial fruit fly swarm position X_0, Y_0, the evolution generation number, and the swarm size are fixed.

Step 2. Position initialization

For each fruit fly in the swarm, initialize its position. According to the random flight direction and the distance for searching food, the fruit fly location is given.

$$X_i = X_0 + Rand \tag{8.33}$$

$$Y_i = Y_0 + Rand \tag{8.34}$$

Step 3. Determine the current distance $Dist_i$ of the fruit fly, and calculates the smell function S_i as follows

$$Dist_i = \sqrt{X_i^2 + Y_i^2} \tag{8.35}$$

$$S_i = 1/Dist_i \tag{8.36}$$

Step 4. Calculate the fitness function

The fitness function g_i of the fruit fly location can be expressed;

$$g_i = g(S_i) \tag{8.37}$$

Step 5. Update strategy

The fruit fly with best fitness function among the fruit fly swarm is found using the following formula;

$$[g_b, p] = \min(g_i) \tag{8.38}$$

$$g = g_b \tag{8.39}$$

$$X_0 = X(p)\ X_0 = X(p) \tag{8.40}$$

$$Y_0 = Y(p) \tag{8.41}$$

Step 6. Iterative optimization

Judge the termination condition is satisfied. If not, go to Step 2; otherwise, finish the loop.

(2) The flowchart of the IFOA

In the FOA optimization process, there exist two problems: one is the parameters premature convergence; the other one is the poor quality of the final solution. In the FOA individual update process, there exist some defects as follows:

(1) All of the individuals have the same roles in a generation population.
(2) There is no interact information among the individuals.
(3) In a generation population, all of the individuals have the same update strategy and optimization methods.

To solve these problems, the immune system is introduced in this section. The immune system is a kind of the system with biological structures, and it can protect the organism from the dangers of the disease. It mainly consists of the immune recognition, response (primary and secondary response), surveillance, memory, learning mechanism [66]. The details of immune memory mechanism used in search process of IFOA. The food is seen as an antigen. The fruit fly individuals are recorded as the antibodies. Here, the antigen and antibody affinity is judged by using the fitness function values. The relationship between the fitness function values and affinity is opposite, namely the lower fitness value, the higher affinity, and otherwise reverse. At the end of each generation optimization, the individuals with the high fitness values are retained. These part individuals play a leading role in the next generation of the optimization process. The specific measures are adopted as follows:

(1) To retain the excellent individuals corresponding to the best fitness function values in the optimization process of the previous generation population, the immune memory mechanism is introduced in the iterative optimization process.
(2) The whole swarm is separated by two parts. One part populations are the excellent individuals, which have been retained in the previous generation. These individuals are denoted 'elite individuals', and they are used for guidance the search way in the next generation populations. The other part individuals search method is followed by the best individual in the previous generation.
(3) The 'elite individuals' have their own independent search ways, and the rest individuals is defined using another search way.

The search process is repeated and by doing so it is hoped. This search way can enhance cooperation among the individuals and highlight an important role of the 'elite individuals' in the previous generation. It can also promote communication

Fig. 8.25 The flowchart of the IFOA

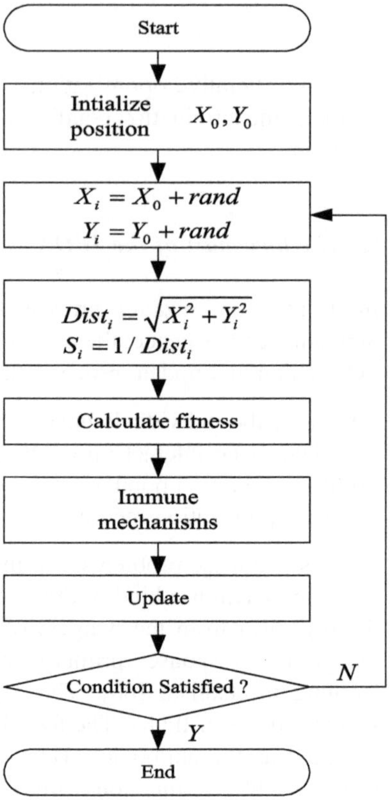

between the individuals in the same generation and next generation, and accelerate the optimization speed in the search process.

The flowchart of the IFOA is as shown in Fig. 8.25.

8.4.3 IFOA-SVR-ARX-ARMA Model

The time series analysis method is a statistical method, which could use the history data to reveal the variables trend in the system inborn structure and character [67]. As well known, the ARMA model is a time series model, and it belongs to a linear model. It consists of two parts, AR model part and MA model part. The aim of the ARMA model is to modify the predict error, according to the finite length of the observed data, from the IFOA-SVR model, and achieve the minimum value of predict error.

AR model prediction method is based on a linear combination of past observations output values and present interference values to predict the future trend of the system data. The mathematical formula for the AR model is expressed as follows:

$$y_k = \varphi_1 y_{k-1} + \ldots + \varphi_p y_{k-p} + e_k \quad (8.42)$$

where $\varphi_i, i = 1, \ldots, p$ are the coefficients of the formula, p is the order of the AR model, and the random variable e_k is the Gaussian white noise.

MA prediction model is based on the observations values of past and current interference values to predict the trend of the data. The mathematical formula for the MA model is written as follows:

$$y_k = e_k - \theta_1 e_{k-1} - \ldots - \theta_q e_{k-q} \quad (8.43)$$

where the $\theta_j, j = 1, \ldots, q$ are the unknown coefficient of the model, and q is the order of the MA model, and e_k is also a white noise error terms.

$$y_k = \varphi_1 y_{k-1} + \ldots + \varphi_p y_{k-p} + e_k - \theta_1 e_{k-1} - \ldots - \theta_q e_{k-q} \quad (8.44)$$

where y_k is the stationary, $\varphi(z)$ and $\theta(z)$ represent the polynomial as follows

$$\varphi(z) = 1 - \varphi_1 z^{-1} - \varphi_2 z^{-2} - \ldots - \varphi_p z^{-p}$$
$$\theta(z) = 1 - \theta_1 z^{-1} - \theta_2 z^{-2} - \ldots - \theta_q z^{-q}$$

where z^m represents the lag operator, $z^m y_k = y_{k-m}$, $z^m a_k = a_{k-m}$, $z^m c = c$, c is a constant. $\varphi(z)$, $\theta(z)$ represent the operator polynomial.

The model is usually referred to as the $ARMA(p,q)$, and can be rewritten by:

$$\varphi(z) y_k = \theta(z) e_k \quad (8.45)$$

We can use the $ARMA(p,q)$ model to approximate the random system, the basic theory of the linear operator on the Hibert space has been proved. To estimate the coefficients φ_i, θ_j, firstly, the order of the ARMA model (p,q) should be determined. Obviously, on one hand, with the increase of the model order, the predictive accuracy of the model is gradually increased; on the other hand, the complexity of the ARMA model is dramatically improved. In the practical engineering process, there is a difficult problem between the high prediction accuracy and low complexity for the fiber stretching system. The AIC (Akaike's Information Criterion) is used to determine the order of the ARMA model, which is proposed by (Akaike 1974) [68]. The $A(s)$ function is defined as follows:

$$A(s) = \ln \sigma^2 + \frac{2s}{N} \quad (8.46)$$

where the σ^2 is a Gaussian white noise, N represents the sizes of the observed sample, s denotes the number of the model parameters ($s = p + q + 1$). The range of the model order p, q can be limited in the interval $(0, N/3]$ or $(0, \ln N]$. Then, the matlab 'armax' function can be used to fix the coefficients of the ARMA model. The flowcharts of the proposed predict method is shown in Fig. 8.26. In Fig.8.26

Fig. 8.26 The figure of IFOA-SVR-ARX-ARMA

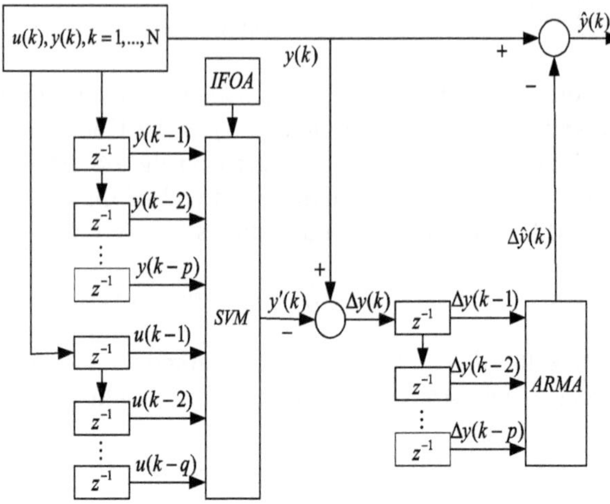

the $u(k)$ is the system input at k time, $y(k)$ denotes the system output at k time. All of these data $(u(k), y(k)), k = 1, \ldots N$ are generated by the model of the fiber stretching process (1).

The specific steps of modeling correspoding to the Fig. 8.26. can be written as follows:

- Given the fiber stretching process at k time input $u(k)$, the model (1) is used to generate correspongding to the output $y(k)$.
- Delay operator z^{-1} is used to obtain the input and output data before $k$time $u(k-1), \ldots, u(k-q), y(k-1), \ldots, y(k-p)$, repspectively.
- There is a nonlinear relation in input and output of the fiber stretching process. The SVM-ARX model is selected as the nonlinear ARX model.
- Using the proposed IFOA to optimize parameter C, σ in the SVM-ARX model.
- The prediction value $y'(k)$ of the SVM-ARX model at k time is obtained. The error time series $\Delta y(k)$ can be expressed as $\Delta y(k) = y(k) - y'(k)$.
- Considered the ARMA model to build a time series with the $\Delta y(k)$, the predict value $\Delta \hat{y}(k)$ is obtained. Then, the final system perdiction value at k time is expressed as $\hat{y}(k) = y(k) - \Delta \hat{y}(k)$.

8.4.4 Simulation Results

8.4.4.1 The Prediction Results of the IFOA-SVR-ARX

The results of the prediction model IFOA-SVR-ARX are simulated under assumed of the prediction model order $p = q = 2$. The others order of the model are discussed by using the same method with the order $p = q = 2$. In this section, the whole experiment process can be separated into two parts independent experiments. The first part, to validate robustness and prediction accuracy of the proposed prediction model, the three groups experiment are carried out at the number of the test data (20, 15, 10), and the corresponding simulation results are listed in Figs. 8.27, 8.28, 8.29 and 8.30, respectively. The second part, using the last two

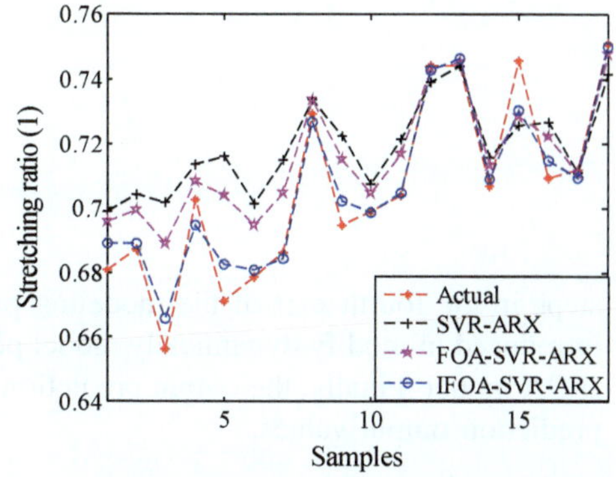

Fig. 8.27 The results of the prediction

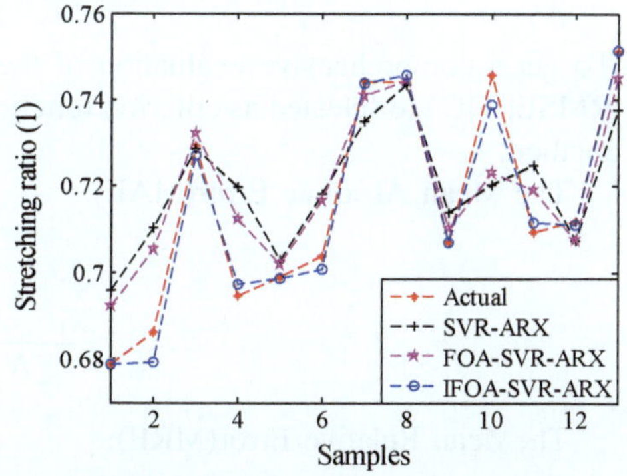

Fig. 8.28 The results of the prediction

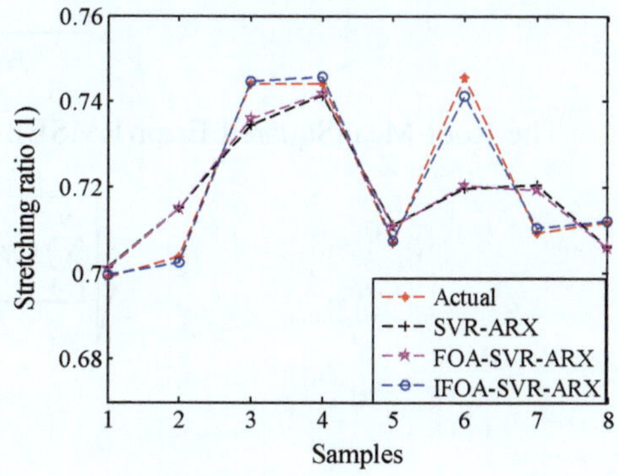

Fig. 8.29 The results of the prediction

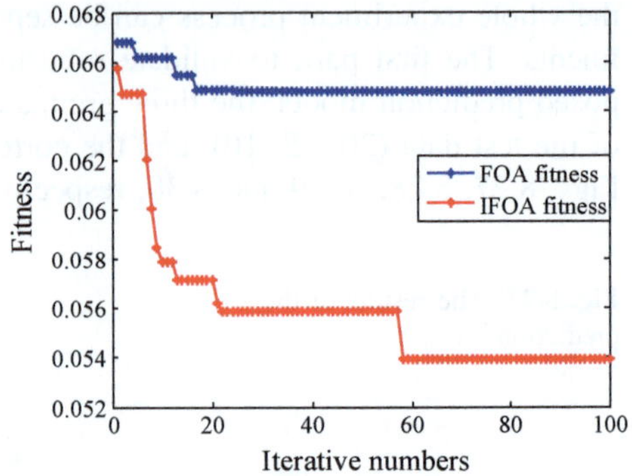

Fig. 8.30 The fitness function compared in FOA and IFOA

steps in the fourth part of the modeling process, the ARMA time series model is introduced to modify dynamically model prediction error, which is generated by the IFOA model. Finally, the output prediction results are obtained as the whole system prediction output values.

8.4.4.2 *The Results of the Error Analysis*

To get a comprehensive evaluation of the identification results, the MAE, MRE, RMSE, TIC are selected as criterion functions to judge the accuracy of the proposed method.

The Mean Absolute Error(MAE):

$$\frac{\sum_{i=1}^{n} |y_i - \hat{y}_i|}{N} \tag{8.47}$$

TheMean Relative Error(MRE):

$$\frac{\sum_{i=1}^{N} |y_i - \hat{y}_i|}{y_i} \tag{8.48}$$

The Root MeanSquared Error(RMSE):

$$\sqrt{\frac{\sum_{i=1}^{N} (y_i - \hat{y}_i)^2}{N}} \tag{8.49}$$

Theil's Iequality Coefficient(TIC):

$$\frac{\sqrt{\sum_{i=1}^{N}(y_i-\hat{y}_i)^2}}{\sqrt{\sum_{i=1}^{N}y_i^2}+\sqrt{\sum_{i=1}^{N}\hat{y}_i^2}} \tag{8.50}$$

Under the different error criterion functions MAE, MRE, RMSE, TIC, Table 8.12 gives the results at the 20 groups of the test data among the predict method of the SVR-ARX, FOA-SVR-ARX, IFOA-SVR- ARX. From Table 8.12, in terms of MAE we can see that the IFOA- SVR-ARX achieves a decrease of the 61.48%, 72.35% respectively compared to the method of FOA-SVR-ARX and SVR-ARX, MRE with a decrease of 62.15%, 72.76% respectively, RMSE with a decrease of 59.35%, 70.28% respectively, and TIC with a decrease of 58.72%, 69.80% respectively. Table 8.13 lists the results at the 15 groups of the test data. From Table 8.13, in terms of MAE, the IFOA- SVR-ARX achieves a decrease of the 80.19%, 83.06% respectively compared to the method of FOA-SVR-ARX and SVR-ARX, MRE with a decrease of 80.54%, 83.33%, RMSE with a decrease of 76.15%, 79.47%, and TIC with a decrease of 75.82%, 79.05%. Table 8.14 shows the results at the 10 groups of the test data. In terms of MAE, the IFOA- SVR-ARX achieves a decrease of the 84.44%, 85.57% respectively compared to the method of FOA-SVR-ARX and SVR-ARX, MRE with a decrease of 84.55%, 85.71%, RMSE with a decrease of 83.19%, 84.03%, and TIC with a decrease of 83.33%, 84.34% respectively. Comparing to the results in Tables 8.12, 8.13, and 8.14, we can see that the results of the predict is related to the number of the test data, and with the decrease of the test sample numbers, the prediction accuracy is gradually increase.

Table 8.12 The predict error comparison among SVR-ARX 、FOA-SVR-ARX 、IFOA-SVR-ARX

	SVR-ARX	FOA-SVR-ARX	IFOA-SVR-ARX
MAE	0.0170	0.0122	0.0047
MRE	0.0246	0.0177	0.0067
RMSE	0.0212	0.0155	0.0063
TIC	0.0149	0.0109	0.0045

Table 8.13 The predict error comparison among SVR-ARX 、FOA-SVR-ARX 、IFOA-SVR-ARX

	SVR-ARX	FOA-SVR-ARX	IFOA-SVR-ARX
MAE	0.0124	0.0106	0.0021
MRE	0.0174	0.0149	0.0029
RMSE	0.0151	0.0130	0.0031
TIC	0.0105	0.0091	0.0022

Table 8.14 The predict error comparison among SVR-ARX 、FOA-SVR-ARX 、IFOA-SVR-ARX

	SVR-ARX	FOA-SVR-ARX	IFOA-SVR-ARX
MAE	0.0097	0.0090	0.0014
MRE	0.0133	0.0123	0.0019
RMSE	0.0119	0.0113	0.0019
TIC	0.0083	0.0078	0.0013

8.4.4.3 The Error for the ARMA Model

In the following experiments, the perdiction output data from the IFOA-SVR-ARX model are collected to bulid the ARMA model. Each data corresponding production stretch ratio. Based on these static prediction output data $y'(k)$, the ARMA model is used to validate the predict output through the dynamic way. The specific information about ARMA modeling can be found in Fig. 8.26 (last two steps). For the ARMA prediction model, the parameters are set as follows: noise $e_t \sim N(0,1)$, the number of the observed data $N = 30$, the range of the model order p, q can be limited in the interval $(0, \ln N]$, respectively.

Before the ARMA model predictions, we should identify the ARMA model order. According to formula (8.46), the different orders of the ARMA model corresponding the AIC values can be shown in Table 8-XV., we can see that the AIC minimum value is −4.03, corresponding the model order $p = 1, q = 3$, respectively.

$$\varphi(z) = 1 - 0.8996z^{-1}$$
$$\theta(z) = 1 + 0.3544z^{-1} - 0.2982z^{-2} - 0.8826z^{-3}$$

The above formula $\varphi(z)$, $\theta(z)$ are substituted into formula (28). Consequently, the ARMA model parameters and orders can be fixed as follows:

$\Delta y_k = 0.8996\Delta y_{k-1} + e_k + 0.3544z^{-1} - 0.2982z^{-2} - 0.8826z^{-3}$. Based on the obtained the ARMA model, we use the 18 data as the test samples set to validate the perdictive performance of the model. Table 8.15 gives the compared results

Table 8.15 The aic values under the different model orders

p	0	0	0	1	1
q	1	2	3	0	1
AIC	−2.46	−3.30	−3.50	−3.69	−3.76
p	1	1	2	2	2
q	2	3	0	1	2
AIC	−3.75	−4.03	−3.73	−3.52	−3.81
p	2	3	3	3	3
q	3	0	1	2	3
AIC	−3.99	−3.83	−3.88	−4.01	−3.95

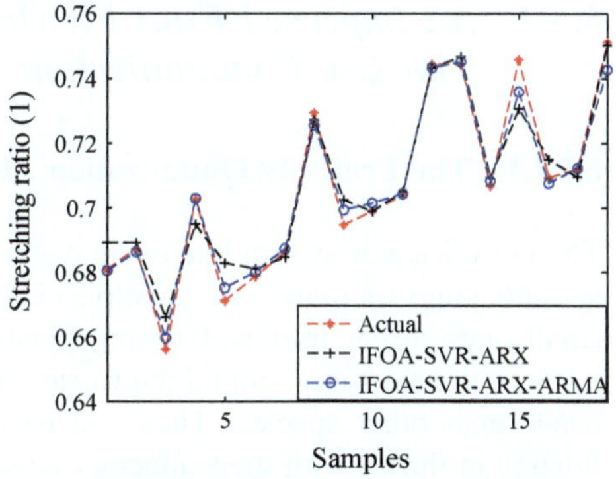

Fig. 8.31 The results of the IFOA-SVR-ARMA prediction

Table 8.16 The predict error compared between IFOA-SVR and IFOA-SVR-ARMA model

	IFOA-SVR-ARX	IFOA-SVR-ARX-ARMA
MAE	0.0047	0.0026
MRE	0.0067	0.0037
RMSE	0.0063	0.0038
TIC	0.0045	0.0027

between IFOA-SVR-ARX and IFOA-SVR-ARX-ARMA perdiction method. From Table 8.16, in terms of MAE, the IFOA-SVR-ARX-ARMA achieves a decrease of the 44.68%, MRE with a decrease of 44.78%, RMSE with a decrease of 39.68%, and TIC with a decrease of 40.00%. Figure 8.31 shows the tested results with test set. From Fig. 8.31, it can be observed that the IFOA-SVR-ARX-ARMA model approximates the actual value of stretching ratio higher than the IFOA-SVR-ARX model that one.

8.5 An Improved Fruit Fly Optimization Algorithm Inspired from Cell Communication Mechanism

In this section, we present an improved Fruit Fly Optimization Algorithm (FOA) based on the cell communication mechanism (CFOA). The CFOA is used to optimize the controller for pre-oxidation furnaces in carbon fibers production. With the proposed approach, we can control the temperature of pre-oxidation reactions, and reduce not only wasted time but also energy consumption.

8.5.1 *An Improved Fruit Fly Optimization Algorithm with Cell Communication*

8.5.1.1 The Fruit Fly Optimization Algorithm

The FOA is a new swarm intelligent method based on fruit fly's foraging behaviors, and it belongs to a kind of interactive evolutionary computation. Fruit flies are very small ones to eat fruit and rotting plants, which widely exist in temperate and tropical climate zones around the world. Fruit flies have visual and olfactory senses better than other species. They can easily make good search of various odors floating in the air with their olfactory organ, or even smell the food sources 40 km away from them. Then, they would fly to the food by their sensitive vision. The food finding process made by the fruit fly can be summarized into the following steps: (1) Firstly, smelling the food source by olfactory organ, and flying towards that location; (2) Secondly, getting close to the food location by their sensitive visions; (3) At last, other fruit flies' flocking location and towards that direction. Figure 8.32 [42] shows the food searching iterative process of fruit fly.

According to the food finding characteristics of fruit fly swarm, the FOA can be divided into seven steps as follows:

Step 1. Parameters initialization.

The main parameters of the FOA are the total evolution number, the population size pop, and the initial fruit fly swarm location (X_0, Y_0).

Step 2. Population initialization.

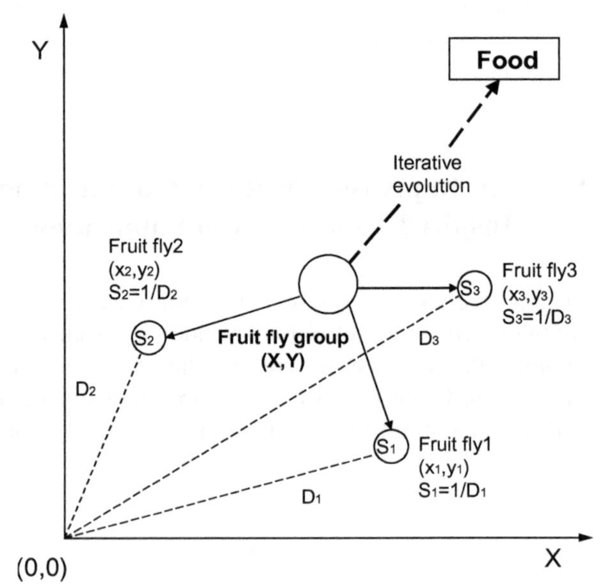

Fig. 8.32 Food searching iterative process of fruit flies

$$X_i = X_0 + rand \quad (8.51)$$

$$Y_i = Y_0 + rand \quad (8.52)$$

Step 3. Computation of distance (D_i) and smell (S_i)

$$D_i = \sqrt{X_i^2 + Y_i^2} \quad (8.53)$$

$$S_i = 1/D_i \quad (8.54)$$

Step 4. Computation of the fitness function (f_i).

$$f_i = f(S_i) \quad (8.55)$$

Step 5. Find out the minimum individual fruit fly with the best fitness function (f_b) among the fruit fly swarm.

$$[bestX \quad bestindex] = \min(f(S_i)) \quad (8.56)$$

Step 6. Selection operation.

Keep the best fitness function value and coordinates (X_b, Y_b). Then, the fruit fly swarm flies towards that location with the best fitness function value by using vision.

$$f_b = bestX \quad (8.57)$$

$$X_b = X(bestindex) \quad (8.58)$$

$$Y_b = Y(bestindex) \quad (8.59)$$

Step 7. Judge if the stopping condition is satisfied. If not, go to Step 2; otherwise, stop the circulation.

8.5.1.2 The Basic Characteristic of the Cell Communication

Cell communication refers to a message through a medium to another cell and interacts with corresponding receptors of target cells, and after that a series of physiological changes taking place in the interior of cells. Indeed, cells can not survive in isolation, and survive by receiving and processing information from the external environment, whether that information pertains to the availability of change and nutrients in temperature. Cell to cell communication is essential for coordination of cellular events in multi-cellular systems [69]. Intercellular communication has three main ways: gap junction, cell recognition and chemical communication.

Gap junction: narrow water-filled channels that connect the cytoplasm of adjacent epithelial cells, as well as of some other types of cells.

Cell recognition: mutual recognition among cells, usually with specific complementary interaction among their respective surface molecules or membrane glycoproteins.

Chemical communication: cells secrete some chemicals (such as hormone) to the external cells, as a signal molecule in target cells, adjust their function, specific and can be divided into four forms: (a) endocrine, (b) paracrine, (c) synaptic, (d) autocrine.

Based on the three main ways of cell communication, three evolutionary strategies are established. Especially, the gap junction, the cell recognition, and the chemical communication in the mode of cell communication can be used in the iterative and evolutionary process, respectively.

8.5.1.3 The CFOA Based on the Cell Bommunication Mechanism

Although the FOA has been applied in several fields, it still suffers with some degree of premature convergence and poor quality of solution, especially in solving higher dimension multimodal objective functions. To overcome the shortcomings, we present the CFOA by combining the cell communication mechanism. The CFOA applies the modified update formula to maintain the population diversity and enhances the convergence velocity and precision by incorporating the information of the global worst, mean and best solutions into the search strategy to improve the exploitation. The CFOA has three major steps: (1) randomly generate a fruit fly swarm's initial position; (2) get the smell concentrations at positions of each and every fruit fly; and (3) assign each fruit fly a direction and distance for its movement to look for food with the modified update formula. The main steps are described as follows and their implementation procedure is illustrated in Fig. 8.33.

According to the food searching characteristics of fruit flies, the CFOA can be divided into eleven steps as follows:

Step 1. Parameters initialization.

The main parameters of the CFOA are the total evolution number, the population size pop, and the initial fruit fly swarm location.

Step 2. Population initialization.

Give the random flight direction and the distance for food searching of a fruit fly by using olfactory model (1)–(2).

Step 3. Computation of the D_i and the S_i from Eqs. (8.53)–(8.54).

Step 4. Computation of the f_i by Eq. (8.55).

Step 5. Computation of the mean fitness function.

$$f_{man} = \frac{\sum_{i=1}^{n} f(S_i)}{n} \tag{8.60}$$

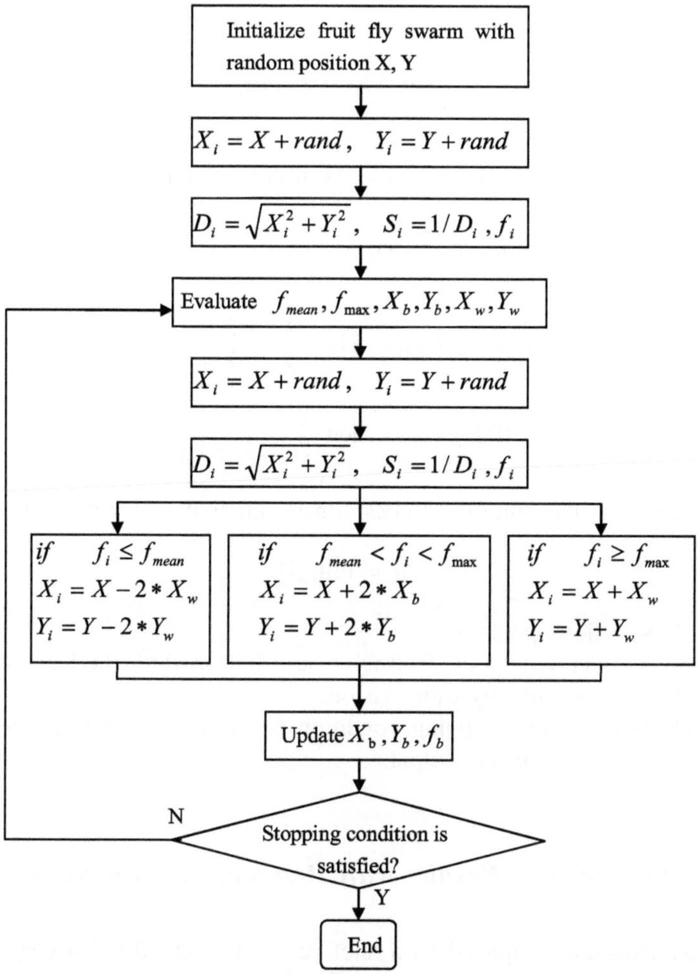

Fig. 8.33 The implement procedure of the CFOA

Step 6. Find out the minimum individual fruit fly with the f_b among the fruit fly swarm using Eqs. (8.56)–(8.59).

Step 7. Find out the maximal individual fruit fly with the worst fitness function ($f_{\max}$) among the fruit fly swarm.

$$[worstX \quad worstindex] = \max(f(S_i)) \tag{8.61}$$

$$f_{\max} = worstX \tag{8.62}$$

$$X_w = X(worstindex) \tag{8.63}$$

$$Y_w = Y(worstindex) \tag{8.64}$$

Step 8. Update strategy.
Check each fruit fly's position using the following formulas:

$$\text{If}(f_i \le f_{mean}) \text{ Then } \begin{array}{l} X_i = X - 2 * X_w \\ Y_i = Y - 2 * Y_w \end{array} \tag{8.65}$$

$$\text{If}(f_{mean} < f_i < f_{\max}) \text{ Then } \begin{array}{l} X_i = X + 2 * X_b \\ Y_i = Y + 2 * Y_b \end{array} \tag{8.66}$$

$$\text{If}(f_i \ge f_{\max}) \text{ Then } \begin{array}{l} X_i = X + X_w \\ Y_i = Y + Y_w \end{array} \tag{8.67}$$

Step 9. Find out the fruit fly with best fitness function among the fruit fly swarm:

$$f_b = \min(f_i) \tag{8.68}$$

Step 10. Selection operation.
Keep X_b, Y_b, and f_b, then the fruit fly swarm flies towards the location with the best fitness function value by using vision.

Step 11. Judge if the stopping condition is satisfied. If not, go to enter the circulation; otherwise, get the output.

8.5.2 Experimental Results with Benchmark Functions

Our experiments are completed on a Intel Pentium Dual, CPU 2.4 GHZ, Personal Computer (PC) with 3 GB RAM under Windows XP. All simulations are implemented by using MATLAB_7.9 program.

8.5.2.1 Parameter Settings

In all the experiments, the values of the common parameters are chosen to be the same for all algorithms. The total evolution number is 100, the initial range of X is [−600, 600], population size is 20 for all functions. The PSO algorithm to be compared is given below: social and cognitive components are both set to 2.0 [70]. Inertia weight, which influences the velocity of the particle in the next iteration, is set to be 0.6. Each of the experiments is repeated 10 times independently, and the reported results are average, best, worst and standard deviations of the statistical experimental data.

Table 8.17 Benchmark functions definition

Name	Definition
Schaffer	$f_1(x) = 0.5 + \left[\left(\sin\sqrt{\sum_{i=1}^{n} x_i^2}\right)^2 - 0.5\right]^2 \Big/ \left(1 + 0.001\left(\sum_{i=1}^{n} x_i^2\right)\right)^2$
Rosenbrock	$f_2(x) = \sum_{i=1}^{n}\left(100(x_i^2 - x_{i+1})^2 + (x_i - 1)^2\right)$
Sphere	$f_3(x) = \sum_{i=1}^{n} x_i^2$
Griewank	$f_4(x) = \frac{1}{4000}\sum_{i=1}^{n} x_i^2 - \prod_{i=1}^{n}\cos\left(\frac{x_i}{\sqrt{i}}\right) + 1$
Rastrigin	$f_5(x) = \sum_{i=1}^{n}\left[x_i^2 - 10\cos(2\pi x_i) + 10\right]$
Ackley	$f_6(x) = 20 + e - 20\exp\left(-0.2\sqrt{\frac{1}{n}\sum_{i=1}^{n} x_i^2}\right) - \exp\left(\frac{1}{n}\sum_{i=1}^{n}\cos 2\pi x_i\right)$

8.5.2.2 Experimental Results for Benchmark Functions

In the field of computational optimization, it is common to test different algorithms using plenty of test functions, especially when the test functions include various kinds of problems such as regular, irregular, separable, non-separable unimodal and multimodal. Multimodal functions which have more than one local optimal value are used to verify the ability of algorithms escaping from local minima. If an algorithm is poor that it can not be efficiently completed the search process during global searching, it gets stuck at the local optimal value. The tested benchmark functions are numbered as f_1 to f_6 and tabulated in Table 8.17.

The experiments are tested on three groups of benchmark functions with dimension (D) of 5, 30 and 50. The increment in the dimension of function increases the difficulty. Hence, if an algorithm can find the global optimal values for many dimensions, it will be easily copied to the practical engineering problems. The performance of the CFOA is compared with that of the PSO and the FOA in terms of (1) the best result, (2) the worst result, (3) the average result and (4) the standard deviation (Std. Dev) result. Initial range, formulation, dimensions and the characteristics of these problems are listed in Tables 8.18–8.20.

(1) Experiment 1

In Experiment 1, the performance of the CFOA has been compared to other optimization algorithms: the PSO and the FOA. Functions in the set are with low dimension. The worst, best, mean of best and the standard deviations of the function values found after 10 runs are presented in Table 8.18. From the results in this table, it is clear that the CFOA is precise more on six functions. This is because the CFOA is a new strategy by incorporating the information of global best, worst and mean solutions

Table 8.18 Results for 5d functions

Function	C	Range	Comparison	FOA	CFOA	PSO
f_1	MN	[−600,600]	Best	4.35e-004	0	0.0049
			Worst	5.83e-004	0	0.0190
			Average	4.99e-004	0	0.0119
			Std.Dev	5.07e-005	0	0.0074
f_2	UN	[−600,600]	Best	3.9574	4.35e-005	0.4423
			Worst	5.3111	0.5691	4.6508
			Average	4.3386	0.1271	1.9483
			Std.Dev	0.5844	0.1972	1.5192
f_3	US	[−600,600]	Best	4.24e-004	0	6.49e-007
			Worst	5.34e-004	2.57e-142	1.76e-004
			Average	4.71e-004	2.57e-143	7.41e-005
			Std.Dev	3.15e-005	8.12e-143	5.93e-005
f_4	MN	[−600,600]	Best	9.61e-005	0	5.12e-007
			Worst	0.0141	2.6024e-013	2.61e-004
			Average	0.0015	2.6024e-014	3.98e-005
			Std.Dev	0.0044	8.2294e-014	7.93e-005
f_5	MS	[−600,600]	Best	0.0873	0	1.0681
			Worst	19.748	0	3.9807
			Average	4.5809	0	2.7500
			Std.Dev	7.4937	0	0.8968
f_6	MN	[−600,600]	Best	0.0413	1.09e-010	0.0071
			Worst	1.2502	0.0047	0.1863
			Average	0.1642	5.28e-004	0.0342
			Std.Dev	0.3816	0.0015	0.0539

devised for optimal exploitation of local information. None of the three algorithms have found the optimal value on Rosenbrock and Ackley functions, but the result of the CFOA algorithm is better than the other two. On Schaffer, Griewank and Rastrigin functions the CFOA have produced the optima. For space limitation, here we just present the convergence curves of the PSO, the FOA and the CFOA algorithms to show the progresses of the mean of best function values presented in Fig. 8.34.

In order to make this clear, the interpretation of the convergence curves of the CFOA are described separately, Fig. 8.34 (b) is for Fig. 8.34 (a), Fig. 8.34 (e) is for Fig. 8.34 (d), Fig. 8.34 (g) is for Fig. 8.34 (f), Fig. 8.34 (i) is for Fig. 8.34 (h), respectively.

(2) Experiment 2

In Experiment 2, the performance of the CFOA algorithm has been compared with other optimization algorithms: the PSO and the FOA. From Table 8.19, for three algorithms, on Rosenbrock and Ackley functions none of algorithms have found the optimal value,

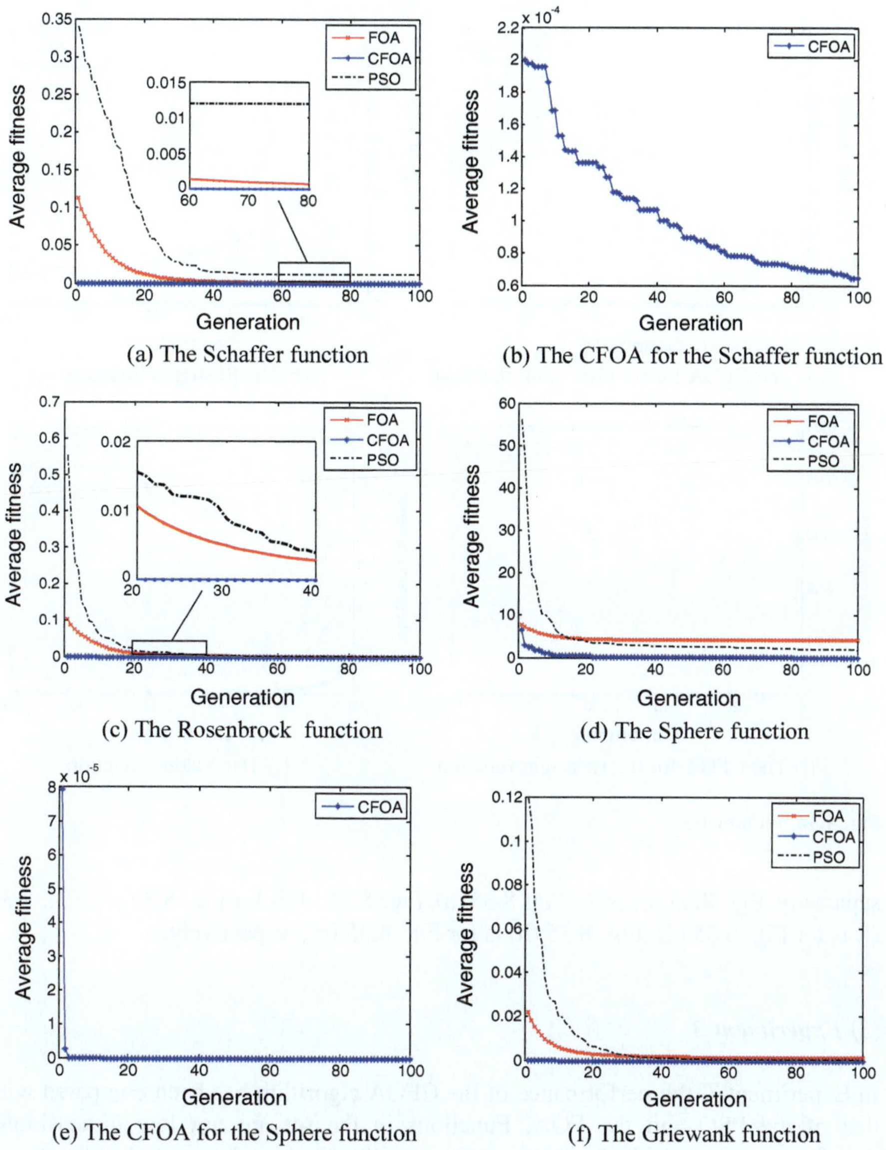

(a) The Schaffer function (b) The CFOA for the Schaffer function

(c) The Rosenbrock function (d) The Sphere function

(e) The CFOA for the Sphere function (f) The Griewank function

Fig. 8.34 Convergence graphs of the mean of best function values

but the result of the CFOA is better than the other two. On Schaffer, Sphere, Griewank and Rastrigin functions the CFOA have produced the optima. The CFOA shows better robustness on Schaffer and Rastrigin functions. For space limitation, here we just present the convergence curves of the PSO, the FOA and the CFOA to show the progresses of the mean of best function values presented in Fig. 8.35. In order to make this clear, the interpretation of the convergence curves of the CFOA are described

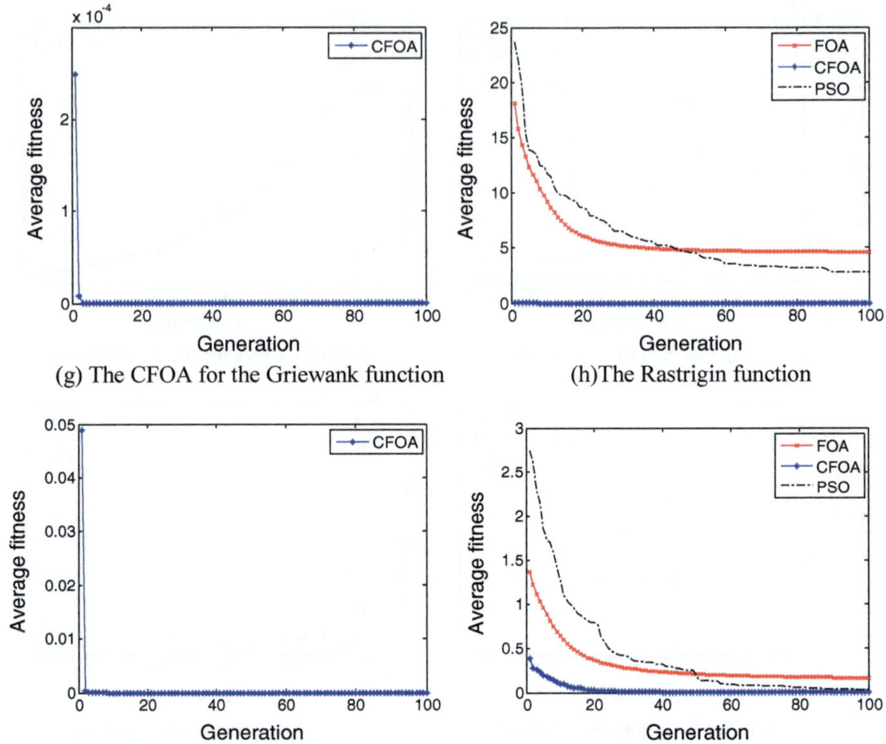

(g) The CFOA for the Griewank function

(h)The Rastrigin function

(i) The CFOA for the Rastrigin function

(j) The Ackley function

Fig. 8.34 (continued)

separately, Fig. 8.35 (b) is for Fig. 8.35 (a), Fig. 8.35 (d) is for Fig. 8.35 (c), Fig. 8.35 (f) is for Fig. 8.35 (e), Fig. 8.35 (h) is for Fig. 8.35 (g), respectively.

(3) Experiment 3

In Experiment 3, the performance of the CFOA algorithm has been compared with that of the PSO and the FOA. Functions in the set are not low dimensional, therefore, non-separable functions are more difficult than the separable functions. The dimensionality of the search space is an important problem [71]. In some functions, the global minimum value is very small when compared to whole search space or is very close to the local ones (Schaffer). As for multimodal functions, if the algorithm cannot explore the search space effectively and cannot keep up the direction changes in the functions having narrow curving valley (Rosenbrock), it fails in these kinds of problems. From Table 8.20, for three algorithms, on Rosenbrock and Ackley functions none of the algorithms have produced the optimal value but the result of the CFOA algorithm is better than the other two. On

Table 8.19 Results for 30d functions

Function	C	Range	Comparison	FOA	CFOA	PSO
f_1	MN	[−600,600]	Best	0.0031	0	0.0049
			Worst	0.6099	0	0.0190
			Average	0.1602	0	0.0176
			Std.Dev	0.2539	0	0.0045
f_2	UN	[−600,600]	Best	28.707	7.90e-05	88.072
			Worst	108.70	0.4244	166.19
			Average	72.133	0.0811	136.16
			Std.Dev	26.257	0.1359	26.393
f_3	US	[−600,600]	Best	0.0029	0	0.4305
			Worst	1.5799	1.7617e-141	1.7187
			Average	5.13e-001	1.7617e-142	1.1396
			Std.Dev	0.5270	5.5710e-142	0.4359
f_4	MN	[−600,600]	Best	1.92e-004	0	0.0444
			Worst	0.0588	7.40e-003	0.1181
			Average	0.0210	7.44e-004	0.0703
			Std.Dev	0.0236	0.0024	0.0212
f_5	MS	[−600,600]	Best	0.6323	0	67.382
			Worst	128.24	0	140.63
			Average	91.073	0	99.362
			Std.Dev	49.057	0	27.050
f_6	MN	[−600,600]	Best	0.0446	−8.8818e-016	1.6693
			Worst	1.8229	0.5775	2.3261
			Average	0.9374	0.1225	1.9970
			Std.Dev	0.7885	0.2328	0.3511

Schaffer, Sphere, Griewank and Rastrigin functions the CFOA has produced the optima. In order to make this clear, the interpretation of the convergence curves of the CFOA are described separately, Fig. 8.36 (c) is for Fig. 8.36 (b), Fig. 8.36 (e) is for Fig. 8.36 (d), respectively.

8.5.3 The CFOA to Optimize the Controller in Carbon Fibers Production

8.5.3.1 Pre-oxidation Furnace Model for PANCF

Pre-oxidation is to pass the precursor tow through a furnace divided into several zones with increasing temperature gradient at 190–300 °C in air [72], followed by a pre-carbonization and a carbonization [73]. A pre-carbonization treatment usually between 300–700 °C is performed prior to carbonization. To avoid thermal shock of

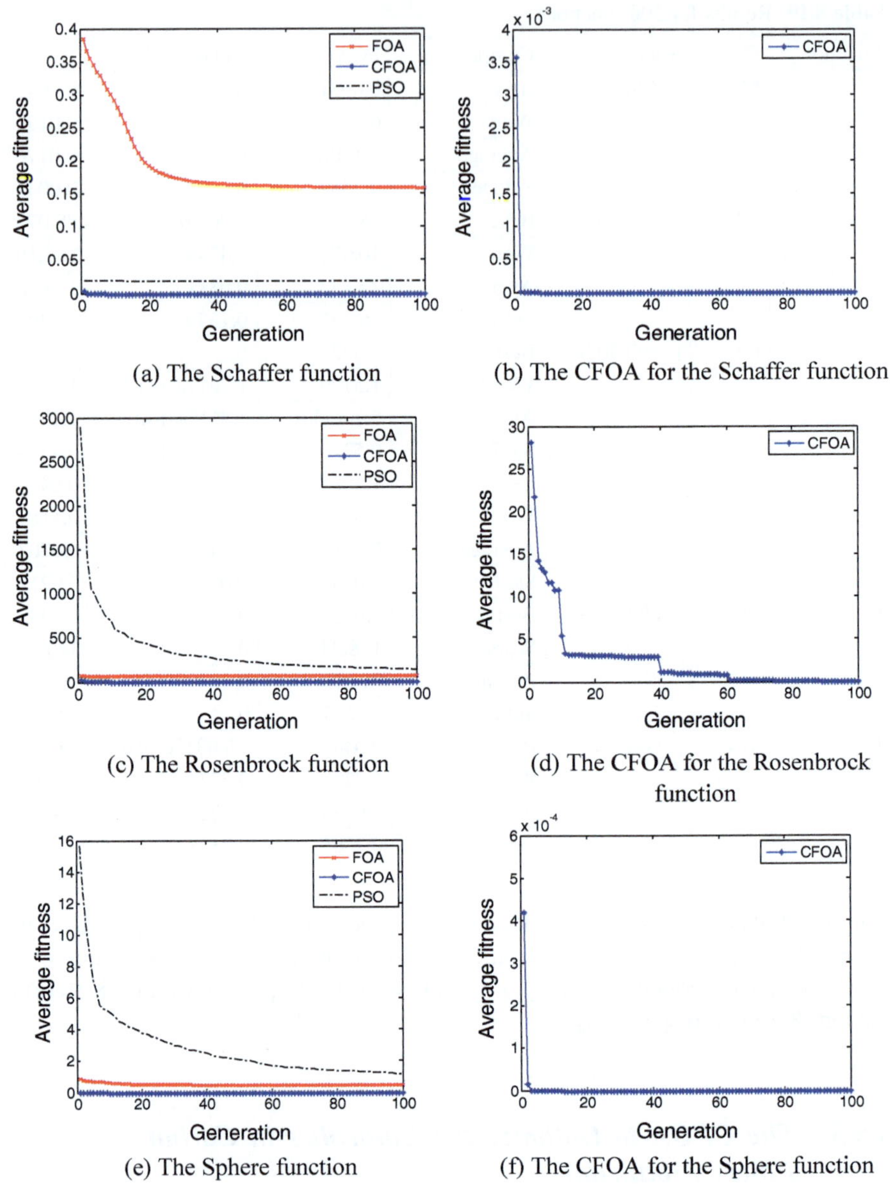

Fig. 8.35 Convergence graphs of the mean of the best function values

fibers, low temperature heat treatment is applied. Carbonization is carried out in an inert atmosphere at the highest temperature of 1400–1600 °C and a total resident for a few minutes in oxygen-free atmosphere is applied, during which non-carbon elements are driven off and the high-strength inorganic carbonaceous fibrous materials are produced at last [74].

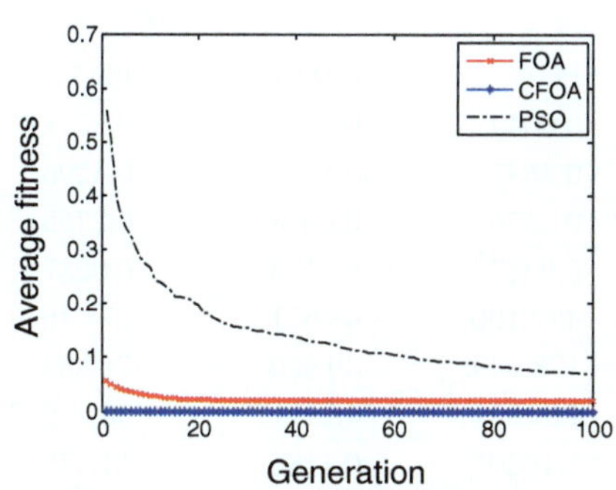

(g) The Griewank function

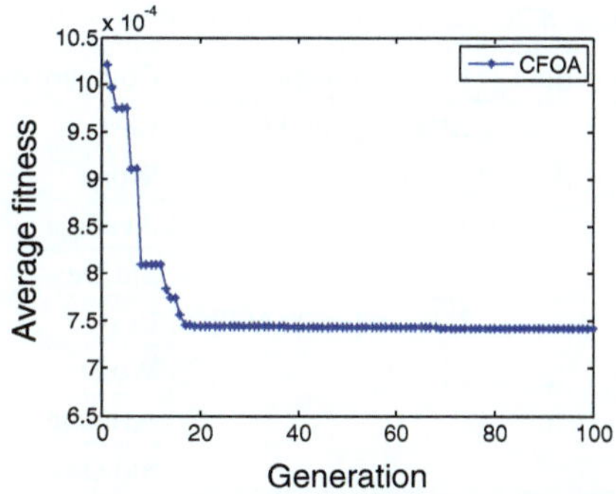

(h) The CFOA for the Griewank function

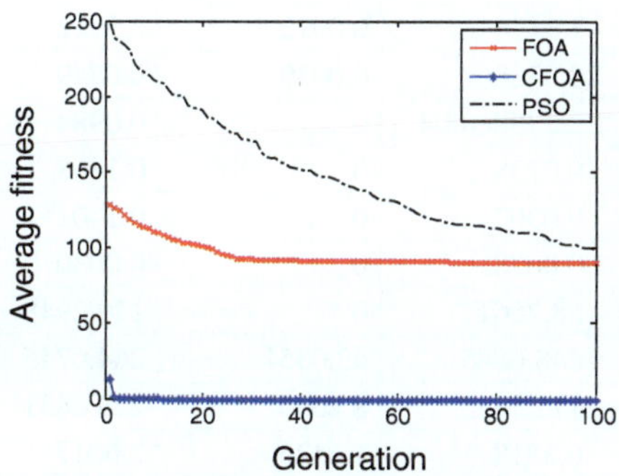

(i) The Rastrigin function

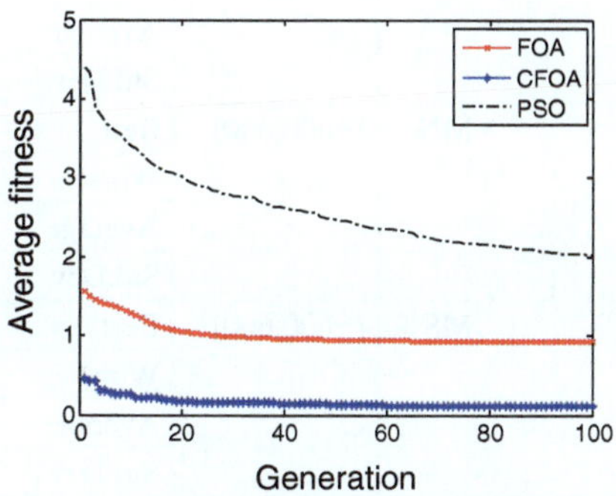

(j) The Ackley function

Fig. 8.35 (continued)

PAN precursor fibers are conducted continuously through pre-oxidation furnaces with 10 temperature zones in air at an initial feeding speed of 0.4 m/min, and then the pre-oxidized fibers are conducted through one low temperature carbonization furnace and one high temperature carbonization furnace. Through such a two-step process, carbon fibers can be obtained finally, and the detailed process is shown in Fig. 8.37. The detailed temperatures of the 10 zones are programmed as 190-200-210-220-230-240-250-260-260-270 °C, as shown in Table 8.21.

8.5.3.2 Optimal Control Model of Pre-oxidation Furnace

The diagram of the Smith-predictor control system by combining with the CFOA is shown in Fig. 8.38, where *r* is the desired input, *e* is the system error between the desired output and the actual output, *u* is the control force, *y* is the actual output. We wish to properly design a set of PI gains such that the system output response satisfies certain specifications. In the CFOA-based Smith-predictor control system, let $\Theta = [\theta_1, \theta_2] = [\mathrm{K_p}, \mathrm{K_i}]$ be a parameter vector or a fruit fly.

Table 8.20 Results for 50d functions

Function	C	Range	Comparison	FOA	CFOA	PSO
f_1	MN	[−600,600]	Best	0.0049	0	0.2221
			Worst	0.8832	0.6784	0.3296
			Average	0.2576	0.0678	0.2755
			Std.Dev	0.0387	0.2145	0.0037
f_2	UN	[−600,600]	Best	48.5100	48.6213	274.4630
			Worst	178.0495	49.000	769.8354
			Average	119.8091	48.9517	508.4547
			Std.Dev	3.1920	0.1186	14.2781
f_3	US	[−600,600]	Best	0.8342	0	2.6038
			Worst	1.9000	0.0123	6.1062
			Average	1.3174	0.0012	3.5712
			Std.Dev	0.0388	0.0039	0.0949
f_4	MN	[−600,600]	Best	2.2758e-004	0	0.0984
			Worst	0.0738	0	0.1868
			Average	0.0307	0	0.1401
			Std.Dev	0.0028	0	0.0030
f_5	MS	[−600,600]	Best	58.7561	0	179.2948
			Worst	248.6845	43.0351	264.6745
			Average	172.1425	8.2006	225.2831
			Std.Dev	6.3818	1.6426	2.6017
f_6	MN	[−600,600]	Best	0.0458	-8.8818e-016	2.3402
			Worst	1.5043	1.3538	2.9950
			Average	0.8789	0.1354	2.5752
			Std.Dev	0.0687	0.4281	0.0209

The convergence of the CFOA toward the global optimal solution is guided by the objective function. Hence it should be properly defined before the CFOA is executed. The objective function is defined by the integral of the error (*IE*) as

$$IE = \int_0^{T_i} (t|r(t) - y(t)|)dt = \int_0^{T_i} t|e(t)|dt \tag{8.69}$$

where T_i is the time of integration. This function will be minimized by using the CFOA to obtain the optimal PI gains.

Pre-oxidization furnace is approximately a first-order system with time lag given as

$$G(s) = \frac{K}{Ts + 1} e^{-\tau s} \tag{8.70}$$

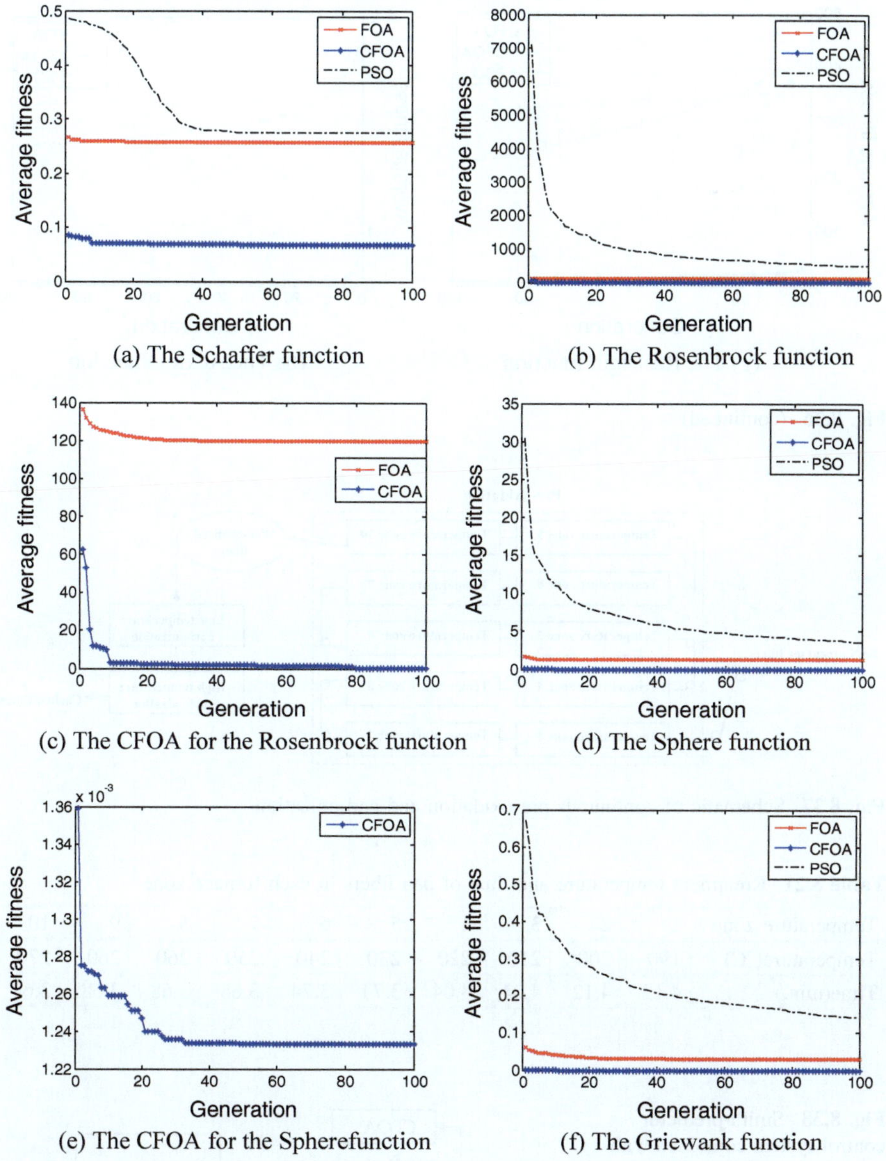

(a) The Schaffer function (b) The Rosenbrock function

(c) The CFOA for the Rosenbrock function (d) The Sphere function

(e) The CFOA for the Spherefunction (f) The Griewank function

Fig. 8.36 Convergence graphs of the mean of best function values

where K is the process gain, τ is time delay, T is time constant. The control model for pre-oxidization furnace with substituting the actual system parameters can be written as

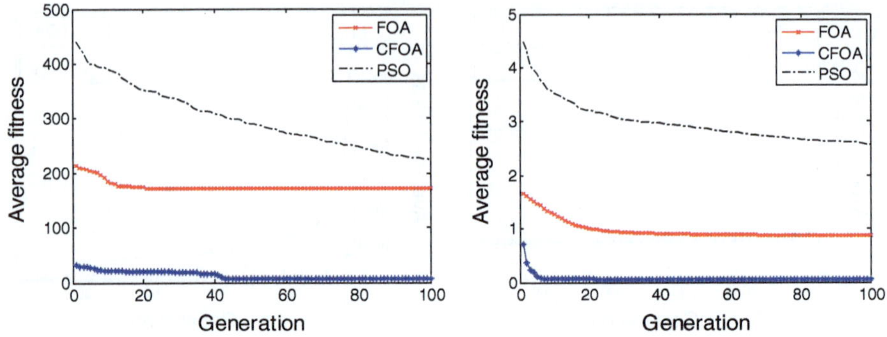

(g) The Rastrigin function (h) The Ackley function

Fig. 8.36 (continued)

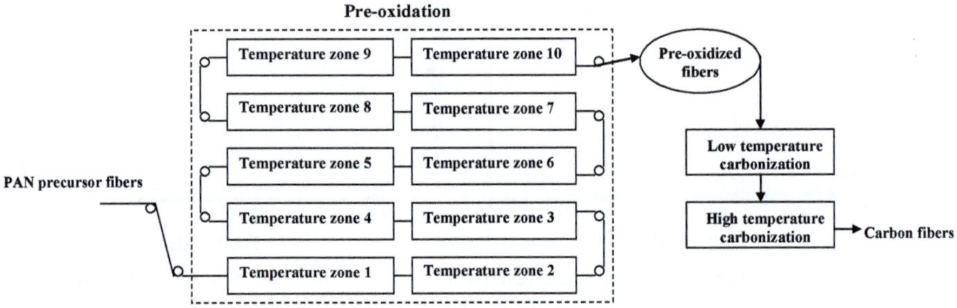

Fig. 8.37 Schematic of continuous pre-oxidation and carbonization

Table 8.21 Rreatment temperature and time of pan fibers in each furnace zone

Temperature zone	1	2	3	4	5	6	7	8	9	10
Temperature(°C)	190	200	210	220	230	240	250	260	260	270
Time(min.)	4.12	4.12	4.04	4.04	3.74	3.74	3.68	3.68	3.68	3.68

Fig. 8.38 Smith-predictor control system combined with the CFOA

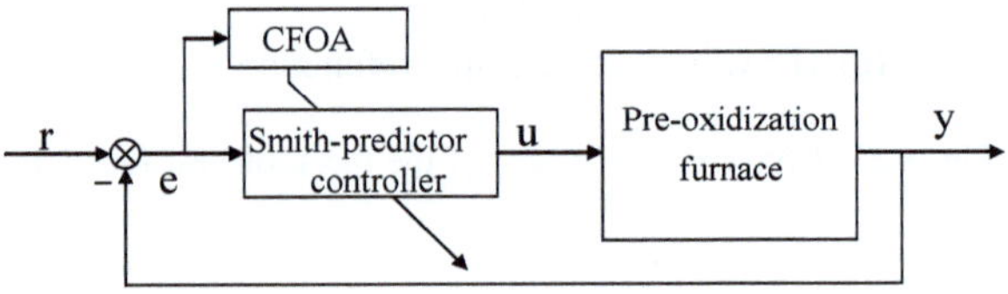

$$G(s) = \frac{50}{60s + 1} e^{-30s} \tag{8.71}$$

The Smith-predictor controller can eliminate the effect of time delay, which is extensively used in process industry. So, the Smith-predictor controller is used in pre-oxidization furnace. Parameter search interval $[\theta_{\min}, \theta_{\max}] = [0, 300]$, the population size is 20, the number of iterations (generations) is 50. Search for a set of optimal PI control gains for a class of pre-oxidization furnace of Eq. (8.71) via the CFOA such that the objective function *IE* of Eq. (8.69) is minimized. The complete steps for designing the CFOA-based Smith-predictor controller system are summarized as follows:

1: Begin

2: Initialize fruit fly swarm with random position X_0, $Y_0 \in [0, 300]$

3: Give the random direction and distance for search of food, set G=0

$$X_i = X_0 + rand, \quad Y_i = Y_0 + rand$$

4: The smell concentration judgment value (S_i) is the reciprocal of the distance of the food location to the origin (D_i)

$$D_i = \sqrt{X_i^2 + Y_i^2}, \quad S_i = 1/D_i, \quad S_i = [K_p, K_i]$$

5: Evaluate the corresponding objective function IE_i of every fruit fly and record each fruit fly's best previous position(X_b, Y_b), worst position(X_w, Y_w), the mean of objective function(IE_{mean}), and the worst of objective function($IE_{\max}$).

6: While $G \le G_{\max}$ do

7: $X_i = X_0 + rand, \quad Y_i = Y_0 + rand$

8: $D_i = \sqrt{X_i^2 + Y_i^2}, \quad S_i = 1/D_i, \quad S_i = [K_p, K_i]$

9: Update each fruit fly's position using the following formulas:

If($IE_i \le IE_{mean}$)

$$X_i = X - 2 * X_w$$
$$Y_i = Y - 2 * Y_w$$

end

If($IE_{mean} < IE_i < IE_{\max}$)

$$X_i = X + 2 * X_b$$
$$Y_i = Y + 2 * Y_b$$

end

If($IE_i \ge IE_{\max}$)

$$X_i = X + X_w$$
$$Y_i = Y + Y_w$$

end

10: Update X_b, Y_b, IE_b

11: End

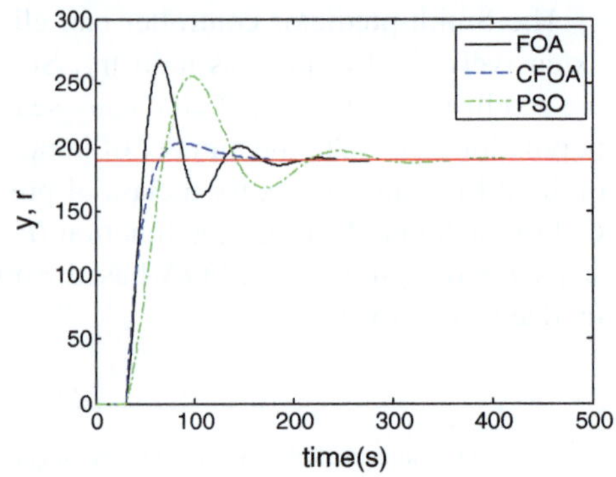

Fig. 8.39 Comparison of control effectiveness among the FOA, the PSO, and the CFOA

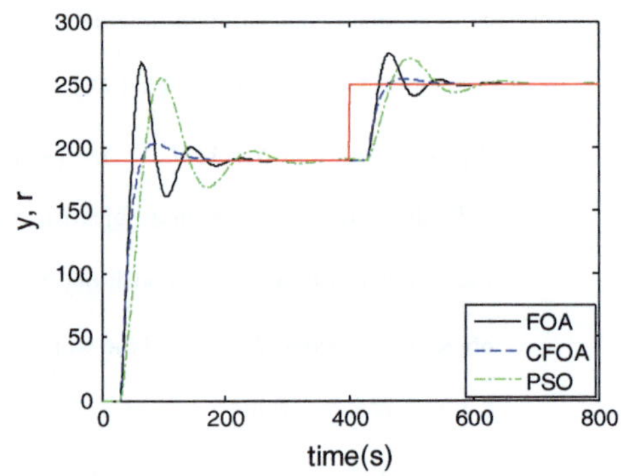

Fig. 8.40 Responses with temperature set-point changing

8.5.3.3 Simulation Results

Usually, we use the same transfer function model for all of the 10 zones. So, taking Zone 1 as an example, we can obtain satisfactory control performance of the system (21) by the Smith-predictor controller, as shown in Fig. 8.39. We compare the control effectiveness of the FOA, the PSO and the CFOA to obtain the optimal PI gains. The Fig. 8.40 is the contrast effectiveness of the influence on Zone 1's temperature of pre-oxidization furnace with the level changing.

We compare the control effectiveness of the CFOA with that of the FOA and the PSO optimal control schemes. Figure 8.39 shows the performance of the three types of optimization algorithms, the simulation results show that the CFOA-based Smith-predictor controller is very sensitive toward the set points. Figure 8.40 is the contrast effectiveness of the influence on the pre-oxidization furnace temperature with the sudden changing. The goal is to test whether the optimization control algorithm has the ability to quickly return to its normal working status after the sudden changing. We change the set point of the temperature at the 400th sampling period. From Fig. 8.40, we can see that the Smith-predictor controller based on the CFOA to optimize PI gains, and we get the system response curve with faster rise

Table 8.22 Control parameters

Control Algorithm	K_p	K_i
PSO	0.0162	0.0024
FOA	0.04	0.008
CFOA	0.0786	0.0026

time, small overshoot, reduced settling time. The detailed parameters are as shown in Table 8.22.

8.6 Conclusions

In this chapter, first, we propose a global optimization algorithm SICSA based on concept of co-operative evolution and clonal selection theory. The SICSA adopts the "division-parallel-recombine" mode to take the three different objectives into account, complemented with clonal expansion and overclocking mutation to perform balanced explorative and exploitative search. We also compare the SICSA with the IA and the GA under the same experimental setup. Our results show that the SICSA exhibits the best performance in precision and convergence time. The obtained parameters can be used to guide to parameter settings in real production of carbon fiber.

And then, a degeneration recognizing clonal selection algorithm is proposed for time-consuming multimodal optimization problems. The basic idea of this method is to utilize the eliminated solutions to identify and eliminate the degenerated solutions before evaluation operation. As a result, a number of evaluation times is avoided so that the computing time is reduced. The method is tested by a group of test function experiments and then applied to a real-world engineering problem, namely, the optimization of the wet spinning coagulating process. From the test experiments, it can be concluded that the proposed DR-CSA method is as efficient as the traditional CSA in solving multimodal problems. Moreover, the DR-CSA is a lot faster in solving this problem. As a result, the DR-CSA is able to save a considerable computing time, benefiting from recognizing and pre-eliminating these high-probability degenerated solutions.

Third, we consider the prediction problems for the fiber stretching process. The fiber stretching ratio is selected as the performance indices to appraisal the quality of the final fiber products. The SVR model is taken as the foundation of the prediction model, and an ARX model to consider as the SVR model input. To optimize the parameter in the SVR-ARX, the immune mechanism improved FOA method is used to enlarge the searching scope of the solution area. Furthermore, under the different error criterion functions, the simulation experiment is divided into two parts: one part simulation results show that the IFOA optimize method have the better performance than the conditional FOA optimize method. The other part simulation result demonstrates that the introduced ARMA model can improve predictive accuracy.

At last, an improved CFOA algorithm is proposed and its performance is compared with the standard version of the FOA and the PSO. In experimental functions, we use the same maximum iteration number and the population number for all problems. The reason is that we assume the algorithms of other methods do not know much about the recommended values of these parameters for their problems to be optimized. The results from a set of numerical benchmark functions show that the CFOA outperforms the FOA in most of the experiments. Meanwhile, the CFOA is applied to optimize the controller for pre-oxidation furnaces in carbon fibers production. Simulation results demonstrate the CFOA can achieve the optimal PI gains of the Smith-predictor controller, as such overcome the shortcomings of poor adaptability of the FOA.

References

1. Chen, J.J.: Ding*, Y.-S., Jin, Y.-C., Hao, K.-R.: A synergetic immune clonal selection algorithm based multi-objective optimization method for carbon fiber drawing process. Fibers and Polymers **14**(10), 1722–1730 (2013)
2. N. Xu, Y. S. Ding, L. H. Ren, and K. R. Hao, Degeneration recognizing clonal selection algorithm for multimodal optimization, IEEE Transactions on Cybernetic, 2017, in press
3. F. Guo, L. H. Ren, Y. C. Jin, and Y. S. Ding, A dynamic SVR-ARMA model with improved fruit fly algorithm for nonlinear fiber stretching process, Natural Computing, 2017, in press
4. C. C Xiao, K. R. Hao, and Y. S. Ding, An improved fruit fly optimization algorithm inspired from cell communication mechanism: Taking the pre-oxidation process of carbon fiber production as an example, Mathematical Problems in Engineering, Vol. 2015 (2015)
5. Rao, S.S.: Engineering optimization: theory and practice. John Wiley & Sons (2009)
6. Chai, T.Y., Jin, Y.C., Sendhoff, B.: Evolutionary complex engineering optimization: opportunities and challenges. IEEE. Comut Intell. M **8**(3), 12–15 (2013)
7. Jin, Y.C., Branke, J.: Evolutionary optimization in uncertain environments—A survey. IEEE. T. Evolut. Comput **9**(3), 303–317 (2005)
8. Jin, Y.C.: Surrogate-assisted evolutionary computation: Recent advances and future challenges. Swarm and Evolutionary Computation **1**(2), 61–70 (2011)
9. Marler, R.T., Arora, J.S.: Survey of multi-objective optimization methods for engineering. Struct. Multidiscip. O **26**(6), 369–395 (2004)
10. Chai, T.Y.: Challenges of optimal control for plant-wide production processes in terms of control and optimization theories. ACTA Automatica Sinica **35**(6), 641–649 (2009)
11. Liang, X., Ding, Y.S., Wang, Z.D., Hao, K.R., Hone, K., Wang, H.P.: Bidirectional optimization of the melting spinning process. IEEE Transactions on Cybernetics **44**(2), 240–251 (2014)
12. Woldemariam, K.M., Yen, G.G.: Vaccine-enhanced artificial immune system for multimodal function optimization. IEEE Transactions on Cybernetics **40**(1), 218–228 (2010)
13. Das, S., Maity, S., Qu, B.Y., Suganthan, P.N.: Real-parameter evolutionary multimodal optimization—A survey of the state-of-the-art [J]. Swarm and Evolutionary Computation **1** (2), 71–88 (2011)
14. Roy, S., Islam, S.M., Das, S., Ghosh, S.: Multimodal optimization by artificial weed colonies enhanced with localized group search optimizers. Appl. Soft Comput. **13**(1), 27–46 (2013)

15. Deb, K.: An efficient constraint handling method for genetic algorithms. Computer. Method. Appl. M **186**(2), 311–338 (2000)
16. Pétrowski, A.: A clearing procedure as a niching method for genetic algorithms. Proceedings of IEEE International Conference on Evolutionary Computation 798–803 (1996)
17. Goldberg, D.E., Richardson, J.: Genetic algorithms with sharing for multimodal function optimization [C]. Proceedings of the Second International Conference on Genetic Algorithms, Hillsdale, NJ: Lawrence Erlbaum, Genetic algorithms and their applications: 41–49 (1987)
18. Gao, W., Yen, G.G., Liu, S.: A cluster-based differential evolution with self-adaptive strategy for multimodal optimization. IEEE Transactions on Cybernetics **44**(8), 1314–1327 (2014)
19. Ren, Z.G., Zhang, A.M., Wen, C.Y., Feng, Z.R.: A Scatter learning particle swarm optimization algorithm for multimodal problems. IEEE Transactions on Cybernetics **44**(7), 1127–1140 (2014)
20. Yang, P., Tang, K.: Lu X. Improving estimation of distribution algorithm on multimodal problems by detecting promising areas . IEEE Transactions on Cybernetics, **45**(8): 1438–1449 (2014)
21. Zhang, G., Li, Y.: A memetic algorithm for global optimization of multimodal nonseparable problems. IEEE Transactions on Cybernetics **46**(6), 1375–1387 (2015)
22. Hui, S., Suganthan, P.N.: Ensemble and arithmetic recombination-based speciation differential evolution for multimodal optimization. IEEE Transactions on Cybernetics **46**(1), 64–74 (2015)
23. De, Castro, L.N., Von, Zuben, F.J.: Learning and optimization using the clonal selection principle. IEEE. T. Evolut. Comput, **6**(3): 239–251 (2002)
24. Bernardino, H.S., Barbosa, H.J.C., Fonseca, L.G.: Surrogate-assisted clonal selection algorithms for expensive optimization problems. Evol. Intel. **4**(2), 81–97 (2011)
25. Chtioui, Y., Bertrand, D., Barba, D., Science Food Agric. J.: "Feature selection by a genetic algorithm application to seed discrimination by artificial vision," **76**(1), 77–86 (1998)
26. Tang, K.S., Man, K.F., Kwong, S., He, Q.: Genetic algorithms and their applications. IEEE. Signal. Proc. Mag. **13**(6), 22–37 (1996)
27. Kennedy, J., Eberhart, R.: Particle swarm optimization. Proceedings of IEEE International Conference on Neural Networks. **4**, 1942–1948 (1995)
28. Gao, L., Hailu, A.: Comprehensive Learning Particle Swarm Optimizer for Constrained Mixed-Variable Optimization Problems. Int. J. Comput. Int. Sys. **3**(6), 832–842 (2010)
29. Hu, Y.-F., Ding, Y.-S., Hao, K.-R., Han, H., Ren, L.-H.: An immune orthogonal learning particle swarm optimization algorithm for routing recovery of wireless sensor networks with mobile sink. Int. J. Syst. Sci. **45**(3), 337–350 (2014)
30. Karaboga, D., Basturk, B.: A powerful and efficient algorithm for numerical function optimization: artificial bee colony (ABC) algorithm. J. Global Optim. **39**(3), 459–471 (2007)
31. Hu, Z.H.: A container multimodal transportation scheduling approach based on immune affinity model for emergency relief. Expert Syst. Appl. **38**(3), 2632–2639 (2011)
32. Xu, N., Ding, Y.-S., Hao, K.-R.: Immunological mechanism inspired iterative learning control. Neurocomputing. **145**(5), 392–401 (2014)
33. Chen, J.J., Ding, Y.-S., Jin, Y.-C., Hao, K.-R.: A synergetic immune clonal selection algorithm based multi-objective optimization method for carbon fiber drawing process. Fibers and Polymers. **14**(10), 1722–1730 (2013)
34. Han, H., Ding, Y.-S., Hao, K.-R., Liang, X.: An evolutionary particle filter with the immune genetic algorithm for intelligent video target tracking. Comput. Math Appl. **62**(7), 2685–2695 (2011)
35. Manue, L., Thomas, S.: The automatic design of multi-objective ant colony optimization algorithms. IEEE. T. Evolut. Comput **16**(6), 861–875 (2012)
36. Pan, W.T.:"A new fruit fly optimization algorithm: taking the financial distress model as an example," Knowl-Based Syst. **26**(2), 69–74 (2012)
37. Li, H., Guo, S., Li, C., Sun, J.: "A hybrid annual power load forecasting model based on generalized regression neural network with fruit fly optimization algorithm," Knowl-Based Syst. **37**, 378–387 (2013)

38. Lin, S.M.: Analysis of service satisfaction in web auction logistics service using a combination of Fruit fly optimization algorithm and general regression neural network. Neural Comput. Appl. **22**(3–4), 783–791 (2013)
39. Wang, S., Yan, B.:Fruit fly optimization algorithm based fractional order fuzzy-PID controller for electronic throttle, Nonlinear. Dynam, 73(1–2), 611–619 (2013)
40. Tu, C.S., Chang, K., Chen, K., Lu, H.A.: A study on business performance with the combination of Z-score and FOAGRNN hybrid model. Afr. J. Bus. Manag. **7**, 7788–7798 (2012)
41. Li, H., Guo, S., Zhao, H. R., S, C. B., Wang, B.: Annual electric load forecasting by a least squares support vector machine with a fruit fly optimization algorithm, Energies. **5**(11), 4430–4445 (2012)
42. Chen, P.W., Lin, W.Y., Huang, T.H., Pan, W.T.: Using fruit fly optimization algorithm optimized grey model neural network to perform satisfaction analysis for e-business service. Applied Mathematics and Information Sciences. **7**(2), 459–465 (2013)
43. Wang, Y., Li, B., Weise, T., et al.: Self-adaptive learning based particle swarm optimization. Inform. Sciences **181**(20), 4515–4538 (2011)
44. Sabat, S.L., Ali, L., Udgata, S.K.: Integrated learning particle swarm optimizer for global optimization. Appl. Soft Comput. **11**(1), 574–584 (2011)
45. Karaboga, D., Basturk, B.: On the performance of artificial bee colony (ABC) algorithm, Appl. Soft. Comput. **8**(1), 687–697 (2008)
46. Karaboga, D., Akay, B.: A comparative study of artificial bee colony algorithm. Appl. Math. Comput. **214**(1), 108–132 (2009)
47. Dasgupta, D.: Advances in artificial immune systems. IEEE. Comput. Intell. M **1**(4), 40–49 (2006)
48. Hofmeyr, S.A., Forrest, S.: Architecture for an artificial immune system. Evol. Comput. **8**(4), 443–473 (2000)
49. Zhang, W., Yen, G.G., He, Z.: Constrained optimization via artificial immune system. IEEE Transactions on Cybernetics. **44**(2), 185–198 (2014)
50. Luh, G.C., Chueh, C.H.: Multi-modal topological optimization of structure using immune algorithm. Computer. Method. Appl. M **193**(36), 4035–4055 (2004)
51. Ding, Y.S., Liang, X., Hao, K.R., Wang, H.P.: An intelligent cooperative decoupling controller for coagulation bath in polyacrylonitrile carbon fiber production. IEEE Trans. Control Syst. Technol. **21**(2), 467–479 (2013)
52. Xiao, L., Ding, Y.S., Wang, Z.D., et al.: Bidirectional optimization of the melting spinning process. IEEE Transactions on Cybernetics. **44**(2), 240–251 (2014)
53. Ding, Y.S., Xu, N., Ren, L.H., Hao, K.R.: Data-driven neuroendocrine ultrashort feedback-based cooperative control system [J]. IEEE. T. Contr. Syst. T **23**(3), 1205–1212 (2015)
54. Kalabin, A.L., Pakshver, E.A.: Simulation of kinetics and heat and mass transfer for the spinning of chemical fibers from polymer solutions. Theor. Found. Chem. Eng. **31**(6), 520–525 (1997)
55. Walczak, Z.K.: Processes of fiber formation. Elsevier. (2002)
56. Xu, N., Ding, Y.S., Schiesser, W.E., Kothare, M.V.: Mathematical simulation of wet spinning coagulation process: dynamic modeling and numerical results. AIChE Journal, in press. (2016)
57. Vande, Wouwer: A., Saucez, P., Schiesser, W.E.: Simulation of distributed parameter systems using a Matlab-based method of lines toolbox: chemical engineering applications. Ind. Eng. Chem. Res. **43**(14), 3469–3477 (2004)
58. Sano, Y.: Drying behavior of acetate filament in dry spinning. Dry. Technol. **19**(7), 1335–1359 (2001)
59. Liang, X., Ding, Y.S., Ren, L.H., Hao, K.R.: A Bioinspired multilayered intelligent cooperative controller for stretching process of fiber production. IEEE Transactions on Systems, Man, and Cybernetics-Part C: Applications and Reviews. **42**(3), 367–377 (2012)

60. Liang, X., Ding, Y.S., Wang, Z.D., Hao, K.R., Hone, K., Wang, H.P.: Bidirectional optimization of the melting spinning process. IEEE Transactions on Cybernetics. **44**(2), 240–251 (2014)
61. Keerthi, S.S., Lin, C.J.: Asymptotic behaviors of support vector machines with Gaussian kernel. Neural Comput. **15**, 1667–1689 (2003)
62. Pan, W.T.: A new fruit fly optimization algorithm: Taking the financial distress model as an example. Knowl.-Based Syst. **26**, 69–74 (2012)
63. Li, H., Guo, S., Li, C., Sun, J.: A hybrid annual power load forecasting model based on generalized regression neural network with fruit fly optimization algorithm. Knowl.-Based Syst. **37**(1), 378–387 (2013)
64. Lin, S.M.: Analysis of service satisfaction in web auction logistics service using a combination of Fruit fly optimization algorithm and general regression neural network. Neural Comput. Appl. **22**(3), 783–791 (2013)
65. Chen, P.W., Lin, W.Y., Huang, T.H., Pan, W.T.: Using fruit fly optimization algorithm optimized grey model neural network to perform satisfaction analysis for e-business service [J]. Applied mathematics and Information Science. **7**, 459–465 (2013)
66. Xu, N., Ding, Y.S., Hao, K.R.: Immunological mechanism inspired iterative learning control. Neurocomputing **145**(5), 392–401 (2014)
67. Zhang, G.P.: Time series forecasting using a hybrid ARIMA and neural network model. Neurocomputing. **50**, 159–175 (2003)
68. Akaike, H.: A new look at the statistical model identification. IEEE Trans. Autom. Control **19**, 716–723 (1974)
69. Ahmed, K.A., Xiang, J.: Mechanisms of cellular communication through intercellular protein transfer. J. Cell Mol. Med. **15**(7), 1458–1473 (2011)
70. Liu, B., Wang, L., Jin, Y.H., Tang, F., Huang, D.X.: Improved particle swarm optimization combined with chaos. Chaos, Solitons Fractals **25**(5), 1261–1271 (2005)
71. Boyer, D.O., Martfnez, C.H., Pedrajas, N.G.: A crossover operator for evolutionary algorithms based on population features. J. Artif. Intell. Res. **24**, 1–48 (2005)
72. Jing, M., Wang, C.G., Wang, Y.X., et al.: Effects of Preoxidation and Carbonization Technologies on Tensile Strength of PAN-Based Carbon Fiber. J. Appl. Polym. Sci. **108**(2), 1259–1264 (2008)
73. Rahaman, M.S.A., Ismail, A.F., Mustafa, A.: A review of heat treatment on polyacrylonitrile fiber. Polym. Degrad. Stabil. **92**(8), 1421–1432 (2007)
74. Ji, M.X., Wang, C.G., Bai, Y.J., et al.: Structural evolution of polyacrylonitrile precursor fibers during preoxidation and carbonization. Polym. Bull. **59**(4), 527–536 (2007)

Chapter 9
Bio-Inspired Bi-Directional Optimization Algorithms

In this chapter, based on different biological mechanisms, some bi-directional optimization methods are proposed. Firstly, a bi-directional optimization method based on an immune-enhanced neural network is introduced [1]. Then, a hybrid approach of genetic algorithm (GA) and improved particle swarm optimization (IPSO) is proposed to construct the radial basis function neural network (RNN) [2]. Next, a bi-directional prediction approach based on neural networks and multi-objective evolutionary algorithm is developed [3]. At last, a bi-directional prediction model based on a support vector machine (SVM) and improved particle swarm optimization algorithm (SVM-IPSO) is created [4].

9.1 Introduction

The rapid development of artificial intelligence (AI) has provided the industry with a powerful tool for analyzing its process and products. Due to the comparatively slow executing speed, most of the AI-based approaches are still taken to build offline systems for estimation and prediction of the reasonability of manufacturing and overall quality of the final products, but the effort to bring them to practical online monitoring and control never stops [5–8]. Another trend is that more and more researchers have turned their attention from application of a single AI approach to combination of several ones, e.g. the artificial neural network (ANN), fuzzy system (FS) and genetic algorithm (GA), and usually such an approach can be proved more effective for the highly nonlinear systems, such as the textile production. ANN has proved to be an excellent adaptive method with dark-box operating performance, powerful study and generalized ability to deal with modeling the dynamic process for process control. Kadi [9] used the ANN to predict mechanical modeling for fiber-reinforced composite materials. For the ANN with its variants, Kadi et al. introduced it into the performance estimation of the fiber-reinforced composites. Arafeh et al. combined the fuzzy mechanism with neural network to form an

Y. Ding et al., *Bio-Inspired Collaborative Intelligent Control and Optimization*, Studies in Systems, Decision and Control 118,
https://doi.org/10.1007/978-981-10-6689-4_9

intelligent approach for the material processing [10]. Yu et al. proposed a fuzzy neural network (FNN) that simplified the network structure and feature selection of the classical ANN. This approach can then be applied with a reasonable rule set to conduct the fabric selection among different fabric specimen [11]. Liu et al. introduced another approach, the firing-strength transform matrix to the adaptive neuro-fuzzy inference system (ANFIS) and then took it as a predicator for the trim-beam numbers in the textile manufacturing [12]. As to the upper level of problem solving, namely, the methodology, Deng et al. proposed a series of intelligent decision support tools consisting of different AI methods (or concepts and frameworks) among which different methods take charge of different tasks in a whole plan, and a case of multifunctional textile material design is then taken to verify the effectiveness of such a guiding plan [13]. Yang et al. proposed a hybrid method using back-propagation neural network (BPNN), genetic algorithm (GA) and simulated annealing algorithm to determine the optimal mixing ratio of different components in special types of fibers [14]. Yang et al. proposed an improved GA to subtract unnecessary elements from a large feature set so that the classification for foreign fibers can be realized more efficiently and effectively [15]. There are also some other aspects of the AI, e.g. the artificial immune mechanism, have great potential to be applied to the optimization of the manufacturing process of fiber [16] or used to broaden the application scope of special textiles [17]. Topology of ANN is not compact enough which leads to more training time for learning. Then radial basis function (RBF) neural network (RNN) rises in artificial intelligence filed. Du et al. [18] investigated the center selection of multi-output RNN. Roy et al. [19] investigated the learning theory of the RNN. Hong et al. [20] presented a novel topology of the RNN, referred to as the boundary value constraints. Huang et al. [21] investigated the function approximation of the RNN. Because of its robustness and fast training rate, RNN becomes the most suitable choice for real time signal processing of an unknown nonlinear dynamical system. Qiao et al. [22] proposed a self-organizing RNN to model and identify nonlinear dynamical systems. Wang et al. [23] presented a self-correcting RNN based forecasting method for adapting changing conditions.

Neural networks and multi-objective evolutionary algorithm are effective in solving black box problems, and both algorithms have been applied on the industrial problems successfully for many years. There are some reviews for different neural networks [24–28]. Multi-objective evolutionary algorithm is an area of multiple criteria decision making, which is used to make an optimal decision to be taken in presence of trade-offs between two or more conflicting objectives. It has been successfully used in pattern recognition, adaptive control and prediction problems [29, 30]. Since every single algorithm has its own shortcomings, combining several artificial intelligence (AI) approaches is a new study trend in resent years. They can fully promote their respective superiorities while avoid their respective defects to enhance their optimizing effects. In the literatures, particle swarm optimization (PSO) is a popular algorithm because of its easy implementation procedure and high performance. Hu et al. [31] proposed an immune cooperative PSO algorithm for fault-tolerant routing optimization in heterogeneous wireless sensor networks. Alireza [32] introduced an adaptive mutation mechanism and a dynamic inertia weight to the PSO to estimate the unknown parameters in nonlinear systems. Behrang et al. [33] used the PSO to predict monthly average daily global solar radiation on horizontal surface for different regions of Iran. Mahor et al. [34] presented a PSO with self adaptive inertia weight to determine the optimal short term generation schedule of real operated cascaded hydroelectric system located at Narmada River in Madhya Pradesh, India. Tang et al. [35] proposed a novel PSO combined with a modified simulated binary crossover (MSBX). Luitel et al. [36] combined quantum principles to the PSO and neural networks to optimize multiple-input-multiple-output learning systems. Vasumathi et al. [37] added the PSO to adaptive neural network for harmonic isolation and reduction. Oh et al. [38] designed a polynomial RNN which optimized by the PSO for handling high-dimensional pattern classification problems. Huang et al. [39] combined orthogonal least-squares and enhanced PSO to RNN for real-time power dispatch of practical Taiwan Power Company systems. Li et al. [40] proposed a hybrid algorithm with modified PSO, simulated annealing and RNN to predict the product specifications melt index.

The training procedure for ANN models is not only time consuming but it is also possible to get trapped in local minima and subjectively in selecting the model architecture [41]. SVM is a relatively new artificial intelligence technique which is increasingly being applied to geotechnical problems and has yielded encouraging results [42, 43]. SVM implements the structural risk minimization principle rather than the empirical risk minimization principle implemented by most traditional ANN models [44, 45]. Based on this principle, SVM achieves an optimum network structure; meanwhile, it can lower the global error of the model. It raises the generalization capability of the model, which is more prominent in small-sample learning. SVM have found a wide application in the fields of pattern recognition, bio-informatics, and nonlinear regression estimation problems. Particularly, support vector regression (SVR) is an extension of SVM. Wang et al. [46] proposed a hybrid load forecasting model combining differential evolution algorithm and SVR. Gilan et al. [47] developed a hybrid SVR and PSO model to predict the compressive and rapid chloride penetration test (RCPT) results of concretes containing metakaolin. Hong [48] developed a hybrid SVR and chaotic particle swarm optimization model to predict electric load. Kang et al. [49] developed swarm intelligence approaches to optimize power flow problems in power networks. Liang et al. [50] developed an adaptive PSO method based on clustering to solve multimodal function optimization problems.

9.2 Bi-Directional Optimization with an Immune-Enhanced Neural Network

In this section, a set of optimization problems for both the manufacturing processes of staple and filament are raised and generalized, through which we try to find a reasonable way to solve by applying a hybrid intelligent approach. This approach is established on the neural network and the artificial immune system (AIS). The neural network is used for building a black-box model for the bi-directional relationship between production parameters and final quality indices of the textile product, and the AIS is further introduced to make the approximation of the neural network model more precisely. Based on the practical production data, a software platform for the realization of the proposed model is established, and the bi-directional simulation and assessment can therefore be conducted on it. With the help of the software platform, the proposed model can further be applied to the practical textile manufacturing process.

9.2.1 *Bi-Directional Optimization for Melting Spinning Process*

9.2.1.1 The Basic Structure of the Melting Spinning Process

A general melting spinning system for filament can be depicted as in Fig. 9.1. It mainly consists of a melting transportation system and a spinning system. The main task of the melting transportation system is to convert the raw materials for spinning to liquid with a predefined viscosity and temperature. The liquid can then be shaped with ease, and the detailed spinning procedure can be accomplished by the following spinning system [51, 52].

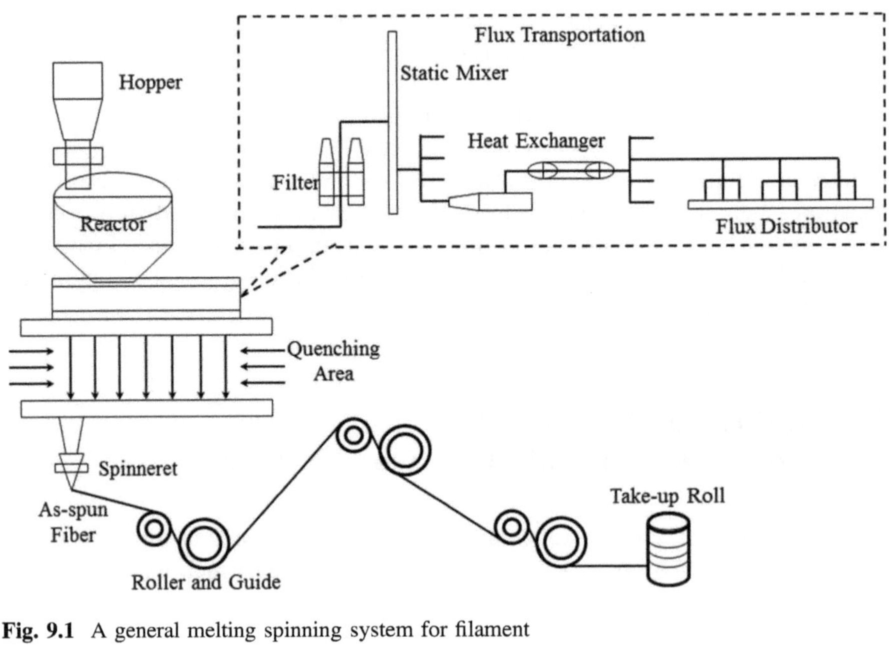

Fig. 9.1 A general melting spinning system for filament

At the very beginning of the process, the raw materials for spinning, no matter what its original form is, are firstly molten down to solution, and then transported to a group of spinnerets through branched tube systems. The dense spinning solution is then intruded through the holes on bottom of the spinnerets and enters the blowing area. In the blowing area, the air blowing from different directions helps to solidify the solution streams so that the so called “as-spun” fibers can be formed. The “as-spun” fibers can be stretched according to a designated stretching ratio in the following procedures. Different types of filaments ask for different combinations of equipment and special raw materials, but the basic processing procedures are similar. As requested by various specifications of the final products, more procedures may be added to the whole production line. The melting spinning technology for the staple is similar to that for the filament to a large extent, and the main difference is that a cutting procedure is added in so its final product would be short fibers, compared to the long fiber of filament.

9.2.1.2 Key Parameters and Quality Indices

The determination of the parameters for spinning is one of the most critical parts on optimizing the manufacturing process [53, 54]. Since the whole production line is a combination of several complex sections, some features can be summarized during the selection of these parameters. (1) There are numerous variables to be determined in a single processing section, and these variables should be coordinated harmoniously so that such a single section can work properly. (2) Different sections have different parameters. This indicates the whole production is under a specific condition based on a reasonable combination of each section. So the parameters for all the sections should be taken into consideration together to build such a foundation. (3) The working status of the spinning line may vary with time, which results in redetermination of process configuration. On most occasions, the changes for better parameters are inevitable. Moreover, the challenge brought by the great nonlinear nature is that the shift of parameter value cannot be judged by purely applying a constant trend or fixed schemes, e.g. the rising of heating temperature will cause either lower spinning speed afterwards or higher, which may not be explained by a simple mathematical model. So all the factors in the whole spinning process (including the key parameters) should be put together to make a thorough analysis, and a correctly modified plan against the change of working status can be generated. Plus, the intensity of changes is also a great factor against the processing fluctuation. Inappropriate modification of parameters may lead worse results rather than pulling the production back to normal.

All the features about the parameters in the production as above imply a challenging task to determine the most critical ones and how to tune them when changes occur.

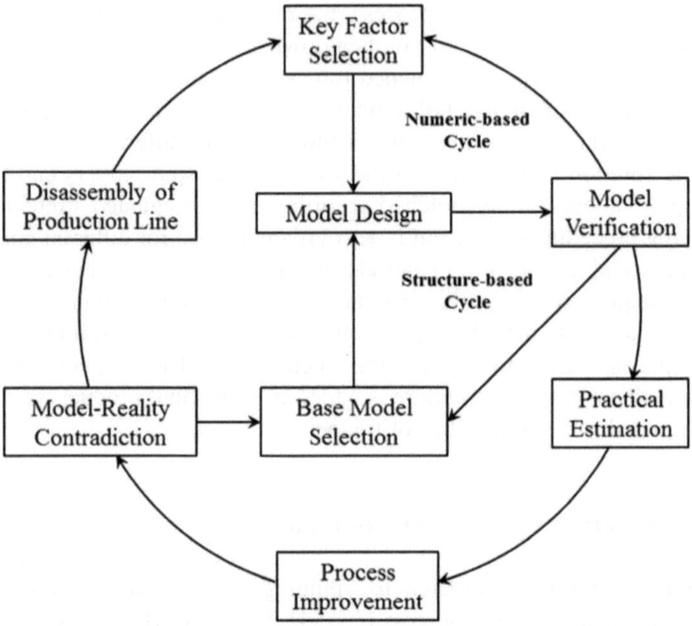

Fig. 9.2 General workflow of the spinning process optimization

9.2.1.3 The General Optimization Paradigm for the Spinning Process

The melting spinning process is a large complex system with numerous variables coupling with each other. Its complex mechanism raises the difficulty level of conducting an accurate model to demonstrate its detailed dynamic behaviors, which consequently brings more challenges to the process analysis and optimization. When approaching a practical problem with multiple key points, people always tend to apply the "divide and conquer" strategy, and so do they with the optimization of the textile spinning system. Figure 9.2 illustrates a workflow diagram of a complete process for modeling the textile spinning production with its application. The whole process starts from a disassembly of the spinning line, and then the key factors that play critical roles can be extracted for further inspection. With the aid of computation techniques, one or more models can be selected as candidates for modeling the spinning process. The model that fits for the actual process is generated out of the framework (base models) and the data source (key factors). After that, a series of experiments should be elaborately designed so that the performance of the selected model can be quantized for comparison. It may also be necessary to roll back to the previous steps for picking up better model basis or influential factors. The verification of models is followed by the practical value estimation whose responsibility is to put the model and results with the actual fiber product data together, and some advice may appear to make the process optimized during this procedure. The optimization may include addition/remove/modification of the existing devices, or better valuation for process configurations. This may probably lead to a distraction between the proposed model in theory and the process itself. In this case, another round of model designing and test will start. During the whole process, the output (or results) of one procedure can be used as the input (or source) of its subsequent procedure(s).

For the candidate models of the spinning process, a rigid mathematical deduction is usually accompanied by numerous equations with plenty of parameters and assumptions as conclusion. However, these achievements always ask for more care if the working status of the spinning procedure changes even a little [55, 56]. A reasonable alternative to solve such problem is to build a black-box-like model that only the input and output of the target system are concerned while the internal details can be ignored, or more precisely, be modeled by the black-box itself. Usually, researchers take the process parameters and the quality indices as the input and the output of the model, respectively. By applying some nonlinear methods, the quality of the final textile product related to the selected parameters can be calculated [57–59]. But such model still has limitations. First of all, the data of the manufacturing process required for building and verifying the model may not be acquired sufficiently, because the parameters of a fixed manufacturing process for a specific textile product are mostly stable (or just fluctuates within a tiny scope). It cannot provide a large range for fully analyzing the behavior of the whole process. Meanwhile, a simple input-output model is only responsible for acquiring the predicted properties of textile products, which can therefore be regarded as a static analysis. It does not have the ability to tell the on-site operators how the process parameters should be tuned to improve the product quality, while this may be more significant for the practical textile manufacturing. As a result, a reverse optimizing procedure from indices of final products to the process configuration is required.

9.2.1.4 The Bi-Directional Intelligent Optimization

The basic idea of building a bi-directional optimization model rather than a common one as above is derived from the actual needs of textile manufacturing itself. It may be useful and sufficient for scientists and researchers in laboratories to implement a forward model that shows the relation from process to products. But what the industrial experts, analysts and operators in the spinning workshop need is a tool that helps them find the possible sources of problems that deteriorate the quality and how to eliminate them. Unlike the lab version model, such requirement just asks for a backward one to make the problems above solved. This is also the main motivation for looking for a bi-directional spinning model and related methodology. Based on the situation of the textile industry, an effective model that qualifies for practical application in the textile production should meet the requirements as below,

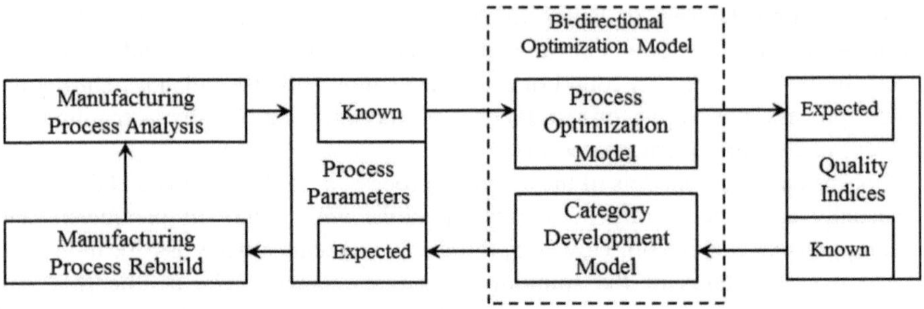

Fig. 9.3 The modeling processes consisting of a forward path and a reverse path

1. The relation between the process configuration (e.g. the parameters) and the quality indices of final product should be simulated so that the possible changes of product quality are foreseeable. This is actually the main task that the contemporary models should finish.
2. The quality indices of the final textile product should have the ability to be taken as a source to get process configurations with the proposed model. By this means, the on-site operators can take advantage of such model to develop new types of textile productions with desired qualities. The actual production line can then be reconfigured with the deducted set of parameters. The possible waste that may come with unsupervised change of process configuration can also be prevented.

These two points as above actually reveals a pair of modeling processes consisting of a forward path and a reverse path, which is shown in Fig. 9.3. In the forward path, the process parameters are taken as input to gain the corresponding quality indices of the final product through a nonlinear model (which may be covered by a black-box type mechanism). In the reverse path, on the other hand, these parameters are regarded as the destination that can be calculated by another model with the quality indices as its source. Both the models can be those with strong nonlinear characteristics to match the features of production, and self-learning capability to generalize knowledge from varieties of in-process data. In the textile industry, the forward approach here can be called "process optimization" for it can help to discover the relation between the changes of production configuration and the subsequent product variations. The reverse approach can be called "category development" for its ability to propose a reasonable set of parameters that lead the textile quality to a known desired level. This would be of special significance for some textile factories and engineers because the idea how to produce something with expected performance is their kernel consideration, which may be more meaningful than just grasping the relationship between production and product.

9.2.2 The Immune-Enhanced Neural Network for the Melting Spinning Process

9.2.2.1 The Neural Network for the Spinning Process

As a useful tool, varieties types of neural networks (NNs) with their derivations have been applied to system analysis and optimization [60]. An appropriate selection of a NN model depends on the characteristics of the target system and the type of data set involved (this may be the most critical point for the selection). The spinning process is a classical system with a huge number of process data collected in a sequential but discrete sampling time series. So the corresponding NN for modeling this process should have a good master on discrete data. The radial basis function (RBF) NN is featured by its ability on approximating discrete system [61], which is therefore picked here to play as the foundation of the optimization model. The basic structure of the RBFNN for the melting spinning process is shown in Fig. 9.4. The process data are received by an input layer, and then transformed by a hidden layer with tunable connecting weights and kernel functions. The output layer is responsible for generating results for the whole RBFNN. For the analysis and optimization of the spinning process, two types of data can be taken as the input data, namely, the parameters of different sections and the final quality indices of fibers. Their detailed roles depend on which direction of the analysis should be made. For the process optimization (the forward path), the parameters of sections on the production line are selected as the input data, and the desired output is the

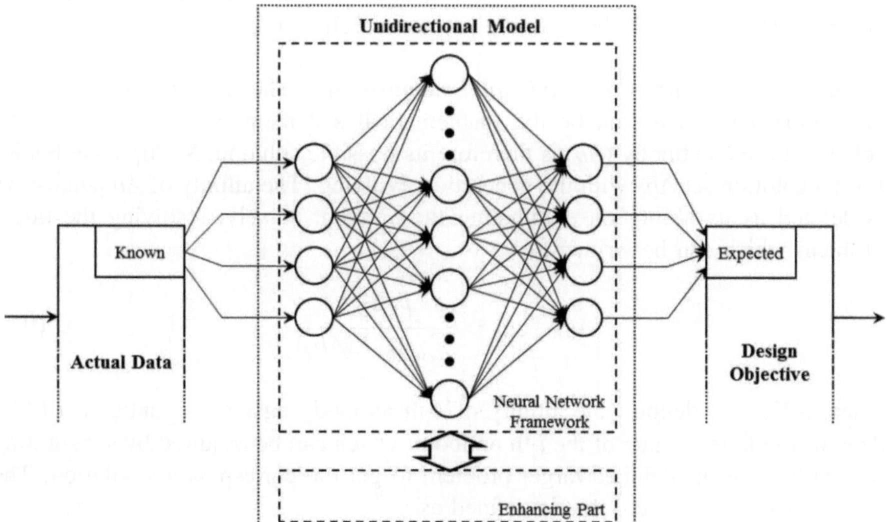

Fig. 9.4 RBFNN -based optimizing model for the melting spinning process

predicted quality indices. For the category development, the expected quality indices should be treated as the data source for optimization. The in-process parameters can be calculated by the corresponding model, hoping that such parameters can lead to a satisfactory production of fibers. The basic structure of the RBFNN can be shared by both the procedures above, but the specific configuration of the models may vary according to different requirements.

9.2.2.2 The Immune-Enhanced Optimization

The RBFNN based optimization model could bring an approximation to the spinning process, but its accuracy may not be satisfied. The input data collected from the fiber production can be treated as independent points in a continuous data space, but there are no apparent clues indicating how the other points, e.g. other production configurations are located in such a space. This feature will make the RBFNN model achieve wrong results. Moreover, the data set for building the optimization model is limited in practice, but the test data set (or more practically, the real production data) is unlimited theoretically and the trend of the data cannot be well predicted. Such a contradiction between the idealized model and the reality may further deteriorate the accuracy of the proposed model in the long run. To solve these problems, the artificial immune mechanism is introduced to strengthen the robustness of the RBFNN framework by tuning the weights and parameters of the kernel functions of the nodes in the RBFNN. The algorithm can be demonstrated as in Fig. 9.5a. For the optimization of the spinning process, each part of the algorithm should have its own specific meaning. A detailed procedure for optimizing the spinning process with the artificial immune mechanism can be given by the steps as follows. All the spinning process-specified procedures are as shown in Fig. 9.5b.

Definition of antigen, antibody, and memory cells. The antigen (*Ag*) in an artificial immune system can be the mathematical statement of the problem to be solved, and the antibody (*Ab*) is therefore its possible solution. Multiple antibodies form a solution set *Abs* with the generation evolving. The affinity of *Ab* against *Ag* is defined as its ability on eliminating the antigen, namely, resolving the target problem which can be written as

$$Aff_{Ab_x \sim Ag} = \frac{F(Ab_x)}{\sum_{y \in Abs} F(Ab_y)}, \tag{9.1}$$

where $Aff_{Ab_x \sim Ag}$ denotes the affinity of x-th antibody against the antigen, $F(Ab_i)$ denotes the fitness value of the i-th antibody which can be acquired by substituting the antibody to its detailed target problem to get the corresponding solution. The affinity between different *Abs* is defined as

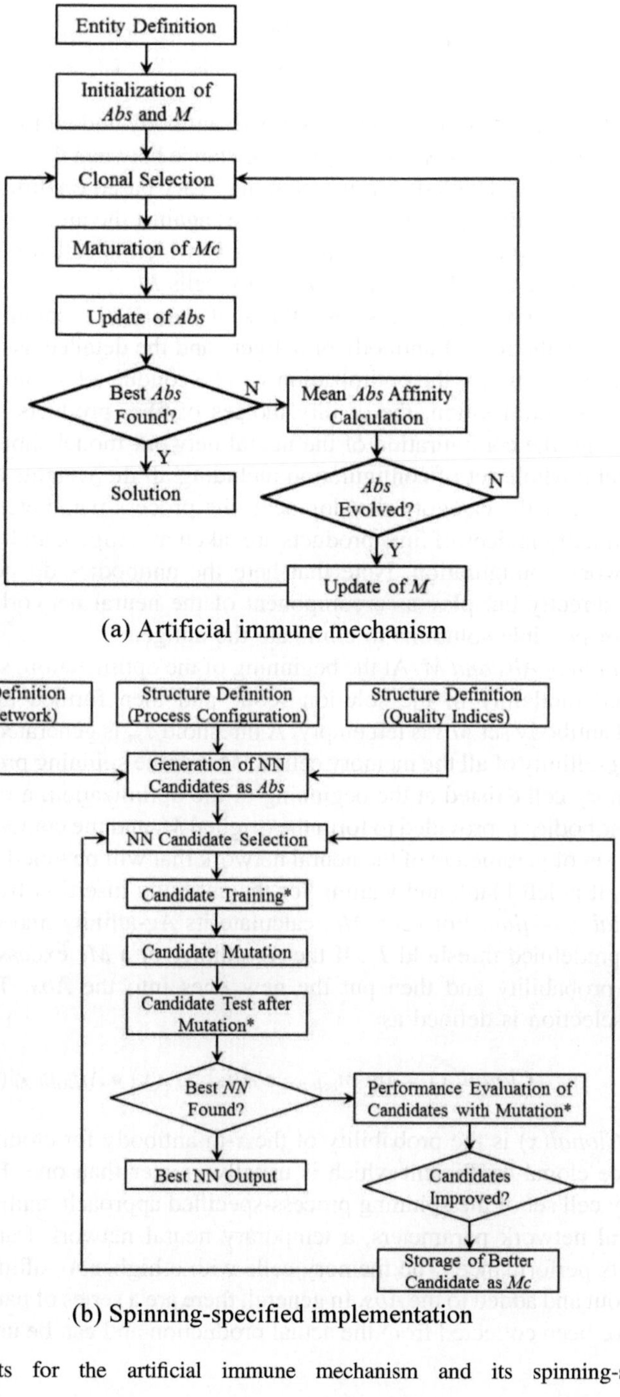

Fig. 9.5 Flow charts for the artificial immune mechanism and its spinning-specified implementation

$$Aff_{Ab_x \sim Ab}(x) = \frac{\min_{y \in (Abs - \{x\})}(\|x - y\|)}{\max_{y,z \in Abs}(\|y - z\|) + 1}, \tag{9.2}$$

where $Aff_{Ab_x \sim Ab}(x)$ is the affinity of the x-th antibody and all the other antibodies in a certain antibody set Abs, $\|x - y\|$ is the distance between the x-th antibody and the y-th antibody (the detailed measurement may vary based on different requirements). The superior antibodies with higher affinities against the antigen or other antibodies can be picked up and stored in another set M which is called the memory cell set, and the antibodies in M are called memory cells Mc.

As to the spinning process, two types of data in the spinning process can be taken to play the role of antibody or antigen, and the detailed assignment is decided by the direction where the optimization will be conducted as shown in Fig. 9.3. For the process optimization, the quality indices of fiber products should be taken as antigen, and the configuration of the neural network model should be the antibody (note that a whole set of configuration including all the weights and bias is a single antibody). For the category development, the process parameters that may lead to certain quality indices of final products are taken as antigen, and the antibody is still the network configuration. Note that here the antibodies do not function on the antigen directly but play as a component of the neural network which is the real source of possible solutions to eliminate the antigen.

Creation of Abs and M. At the beginning of the optimization, some antibodies are generated randomly in the solution scope and then formed an original M. The original antibody set Abs is left empty. A threshold T_m is generated by calculating the mean Ag-affinity of all the memory cells in M. For the spinning process, since there is no memory cell existed at the beginning of the optimization, a randomly generated set of antibodies is provided to form the original M, and the contents of each memory cell is a set of parameters of the neural network that will be tuned. As to the antibody set Abs, it is left blank and waiting for the antibody insertion from M.

Clonal selection. For each Mc, calculate its Ag-affinity and compare the result with a predefined threshold T_s. If the Ag-affinity of a Mc excesses T_s, clone it at a certain probability and then put the new ones into the Abs. The probability for clonal selection is defined as

$$Clonal(x) = \mathrm{int}[m_{rclonal} \bullet Aff_{Ab \sim Ag}(x) \bullet Aff_{Ab \sim Ab}(x)], \tag{9.3}$$

where $Clonal(x)$ is the probability of the x-th antibody for cloning and $m_{rclonal}$ is a reference clonal coefficient which is usually greater than one. For each Mc in the memory cell set of the spinning process-specified approach, namely, a candidate set of neural network parameters, a temporary neural network framework is built to verify its performance. The memory cells with a higher Ag-affinity over T_s will be picked out and added to the Abs. In general, there are a series of parameter-index pairs that have been collected from the actual production and can be used for verification.

Meanwhile, a potential qualified neural network should be capable to simulate the behavior of the real system on most occasions. So the *Ag*-affinity of the memory cell should be generated based on tests with all the data.

Maturation of Mc with high affinity. For the *Mc* that is selected to join the *Abs*, make a mutation with a certain probability on its contents (e.g., by changing several parameters in the whole set on a random basis whose probability is predefined by field experience) and put the mutated individuals into the current *Abs*. The mutation probability is defined as

$$P_{mut}(x) = \frac{1}{Aff_{Ab \sim Ab(x)} + 1}, \tag{9.4}$$

where $P_{mut}(x)$ denotes the probability for mutation. Equation (9.4) indicates that the antibody with a higher *Ab*-affinity will acquire more chances for mutation so that all the antibodies can spread in the solution scope as widely as possible.

Update the Abs. The *Ag*-affinity of each antibody in the current *Abs* should be tested, and those with lower affinities are removed. Meanwhile, check if there is any antibody whose *Ag*-affinity has been high enough to match the antigen, namely, whether the solution with a certain precision has been found. The value of such a threshold depends on the specific parameters of a certain network. A large threshold will call the algorithm for much time to get to the optimum solution, and a small one may keep the algorithm stuck in an infinite loop. If a solution is found, end the optimization and exit, or the following steps will proceed. For the memory cell that consists of network configuration, the mutation is implemented by introducing a tiny fluctuation to the parameters it represents, which would result in different results on the network based on it. The memory cells both with and without mutation are added to the *Abs*, and then all the antibodies in the *Abs* are substituted to the network structure to verify its present performance represented by *Ag*-affinity. Those with inferior performance will be removed from the current *Abs*. Meanwhile, the best antibody will also be checked to determine whether the optimal solution, namely, a neural network that satisfactorily matches the actual spinning system has been found, and in that case the optimization will terminate.

Cycling. Calculate the mean affinity in the current *Abs*, and then compare it with the predefined threshold. If the mean affinity is lower than the threshold which means the mutation makes no quality improvement to the current antibodies, turn back to *Clonal selection* step.

Update the memory cells. Compared the *Ag*-affinity of the antibodies in the *Abs* with those in the current *M*, and insert the individual antibodies with a higher affinity than the current memory cells into *M* to become new memory cells. Adjust the threshold for the memory cells for the new *M*, and remove those *Mc* whose affinity is under the new threshold from *M* so that the mean affinity of the *M* can be increased generation after generation.

9.2.3 Application and Results

9.2.3.1 Experiment Design

An industrial fiber production line (1.56dtex cotton-type polyester staple fiber, semi-open outer quenching applied) in one of the leading Chinese textile manufacturer is taken as the optimizing target by applying the proposed immune-NN based bi-directional optimization approach. It is a classical polymer manufacturing process whose various requirements for the fiber products should be met by tuning the in-process parameters. A large amount of real-time data has been accumulated by technicians for further analysis, which is beneficial for applying the intelligent optimizing methods.

1. Data collection

A RBFNN-based optimizing mechanism generally requires a series of input and output data for training and testing, and these data should cover a wide range of values so that the trained model with the related RBFNN approximates the practical plant as precisely as possible. But such requirement cannot always be satisfied in the analysis of the spinning system. As a practical industrial process with large quantities of products, the manufacturing configuration should always be kept stable, or at least generally unchanged for a long time. So the room for adjusting the parameters and the corresponding quality indices is limited. To eliminate the disadvantage of such characteristic of the spinning system, more running data must be acquired to guarantee the optimizing model can cover a larger solution scope, which could consequently improve the accuracy of the model. In the following experiments, 200 groups of input and output data from the practical spinning process are collected for building the optimizing model. Each group consists of process parameters and the corresponding production quality indices. Note that the roles of these two parts of data depend on what kind of optimization needs to be made. For the process optimization, the process parameters are taken as the input, and the quality indices of fibers are taken as the output. For the category development, however, their roles should be swapped. The detailed information about the data collected is listed in Table 9.1, e.g. the categories and the approximate ranges of the data. Before the formal experiments, all the data are randomly arranged and then taken to train the proposed model, and a pure NN model for comparison is also trained with the same data set. The reason for the random arrangement of data before training is to guarantee the trained model with better adaptation. Consequently, the model itself does not need to make an extra extrapolation which may decrease the accuracy.

2. Process design

The basic configuration of the production line including the process and devices for conducting the experiments is listed in Table 9.2. Here four process parameters, namely, the spinning velocity (SV), the spinning temperature (ST), the quenching

Table 9.1 Process configuration and collected data

Item	Unit	Range
Spinning Velocity (SV)	$m \times min^{-1}$	1000–1197
Spinning Temperature (ST)	°C	280–299
Quenching Velocity (QV)	$m \times min^{-1}$	100–139
Quenching Temperature (QT)	°C	20–24
EYS	1	196.29–237.78
EYSCV	1	5.46–10.04
DT	1	5.82–6.81
DE	1	20.94–24.05

Table 9.2 Basic configuration of the spinning process and devices

Category	Item	Value
Product category	Fineness/*dtex*	1.56
	Post-drawing ratio	3.6523
Spinning parameters	Spinning velocity/$m \times min^{-1}$	Variable
	Spinning temperature/°C	Variable
	Characteristic viscosity/$dl \times g^{-1}$	0.63
	Quenching velocity/$m \times min^{-1}$	Variable
	Quenching temperature/°C	Variable
Device parameters	Non-quenching gap height/cm	6
	Number of spinneret orifice	3064
	Diameter of spinneret orifice/cm	0.0022
	Pump mass throughput/$g \times (min \times hole)^{-1}$	0.0097
Performance indices	EYS1.5	Variable
	EYSCV	Variable
	DT	Variable
	DE	Variable

velocity (QV), and the quenching temperature (QT) are taken as the major factors of the spinning process, and the performance indices, namely the elongation corresponding to 1.5 times the yielding stress (EYS1.5) and its coefficient of variance (EYSCV), breaking tenacity (DT), and the ability of elongation (DE), are represented by four major considerations. The basic production configuration, e.g. the fineness and the post-drawing ratio is kept unchanged, and the device parameters are also stable due to the practical foundation of such simulation. The whole procedure consists of three parts as below.

Preliminary verification. An optimizing model with a pure NN as its core is introduced, and one of the important quality indices of the spinning process, the EYS1.5 is taken to verify the ability of approximation of such model. 150 groups of

Table 9.3 Parameters of the models applied

Item	Value
Number of input neurons	4
Number of output neurons	4
Number of hidden neurons	6
Size of memory cell set (*M*)	5
Size of antibody set (*Abs*)	20
T_s	0.25
$m_{rclonal}$	10
Maximum cycling time	50

actual data (sorted by spinning speed ascending) are taken for training the model and the remaining 50 groups act as the test set. The aim for doing a preliminary experiment is to reveal the possible drawback of an optimizing approach with conventional AI methods that have been widely applied to the spinning process, and therefore the necessity for modification can be easily concluded.

Process optimization. The process optimization is conducted by applying the process parameters as known knowledge and the quality indices of the fiber products as the expected results. The proposed intelligent model and the pure NN-based model are taken for experiment, and their respective results can be compared with the actual quality indices collected from the real production. The objective of this part is to explore the dynamics of the process parameters and their corresponding influence, e.g. a minor modification on the final quality of the fiber products. Practically, such experiment can also help to verify the ability of the proposed model on generating products with a series of specifically tuned producing environments so that the time-consuming and high-cost online test can be avoided. The parameters of the models applied are listed in Table 9.3.

Category development. Opposite to the process optimization part, the category development turns the input and the output around, namely, acquiring the process configuration through known fiber quality indices. The approaches applied here are the same as those in the process optimization. The implementation in such a reverse direction is similar to the process optimization simulation technically, but may have greater significance for the industrial manufacturers. The aim for conducting this part is to determine whether the proposed model has the ability to provide reasonable configuration under the guidance of the final products. With the proposed intelligent mechanism, it is easy to design and make fiber products with desired qualities, and the cost for testing different combinations of manufacturing parameters can be cut off in a great amount. The process and device parameters for this experiment are the same as those for the process optimization instance.

All the experiments above take actions through a Microsoft .NET framework-based software specifically designed for connecting and optimizing the spinning manufacturing process. Both the optimizing approaches including the proposed intelligent method and the NN model are coded. This software provides the possibility that the proposed AI-based approach can be applied to the practical manufacturing without

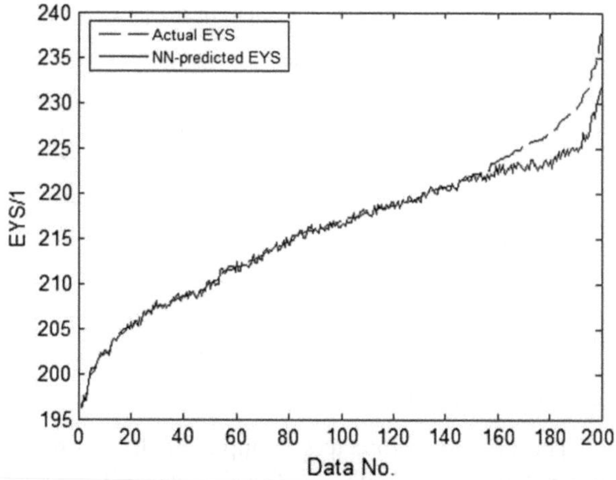

Fig. 9.6 Response of the EYS1.5 against variation of spinning velocity using the RBFNN

support of the laboratory (or even computerized) environment. As a result, the proposed algorithm can be implemented by an experienced programmer (who builds the codes of the algorithm) along with a control engineer (who connects the software with the field devices to get running data, meanwhile managing to send instructions downward to establish optimization). So the expectation that the academic achievements involved can be served as a powerful tool for the on-site engineers in the spinning factories can be realized, which is the very goal for this work.

9.2.3.2 Results and Analysis

1. Preliminary verification

Figure 9.6 shows the response of the EYS1.5 (shown as "EYS" in figures for short) for a model based on the RBFNN, in which the EYS1.5 is given in ascending order. The RBFNN-based model is firstly trained with the training set of 150 groups of the original data, and then tested with both the training set and the test set (50 groups). It can be observed that the RBFNN-based model approximates the actual value of EYS1.5 satisfactorily for the first 150 groups of original data (the training set). But the absolute error starts to increase when it comes to the test set. Noted that although there're 150 groups of data which cover three fourth of the curve being taken for training, their corresponding errors of them still exist. For the test dataset (covers the last one fourth part of the curve), the error between the model generated value and the actual value of the EYS1.5 increases gradually. By comparing the actual value and the calculated one, the proportion between them can be acquired as a tiny decimal, which indicates this model should be compensated by multiplying a small coefficient to its original result to get a better performance. But such a procedure is not theoretically proved so that it cannot be introduced to the current model as a reasonable correction. The preliminary experiment shows that the RBFNN-based optimizing model may not be capable to stimulate the spinning

Table 9.4 Process parameters as input of the process optimization

Fineness/*dtex*	SV/m × min^{-1}	ST/°C	QV/m × min^{-1}	QT/°C	Notes
1.56	1100	289	23.5	125	Base
	1110	293	24.5	130	
	1120	297	25.5	135	
	1130	301	26.5	140	
	1140	305	27.5	145	

system for acceptable performance, especially for the uncovered data, which therefore asks for further modification.

2. Process optimization

The process parameters as input of the optimization are listed in Table 9.4, in which the row marked with "Base" means it's the basic configuration for experiment. One out of four major parameters is selected to make changes while keeping others stable and its value is tuned for four times. So the whole experiment process can be separated to four smaller independent experiments and the responses of the product indices are provided as shown in Fig. 9.7a–d, respectively. It can be observed that the EYS1.5 rises with the increment of spinning velocity, spinning temperature or quenching velocity, but drops as the quenching temperature becomes higher. For the EYSCV and DT, a higher spinning velocity leads to a lower index, and the results for tuning the spinning temperature, quenching velocity and quenching temperature are similar. For the DE, the trend is just opposite to that of the DT.

As to the performance of optimization, both the models have the ability to follow the fluctuation of the process parameters as input, and the difference lies that in the proposed algorithm has smaller absolute errors. The introduction of the AIS broads the searching scope of the optimizing model so that the model can "see" those solutions that may have been excluded during the training process of the original NN (the configuration of the network has actually been fixed once the training process finishes, so the result for a specific input is known without ability for possible exploration). In the view of numeric calculation and the inherent feature of the NN, the errors of both the models can be acceptable. Note that a minor fluctuation of the process parameter, however, may result in substantial changes in quality indices, so a comparative small error should still be an advantage.

3. Category development

In the category development experiment, five sets of quality indices are selected as the input of the reverse models, compared to those for the production development. These quality indices are collected from the practical manufacturing process, covering the whole scope that the quality may possibly fluctuate and the required process configuration for each set is known. Their detailed values of these sets are listed in Table 9.5, and the row marked with "Base" indicates this set is also used for training of the NNs included in the models. By applying the proposed intelligent optimizing model, the possible process configuration that contains four

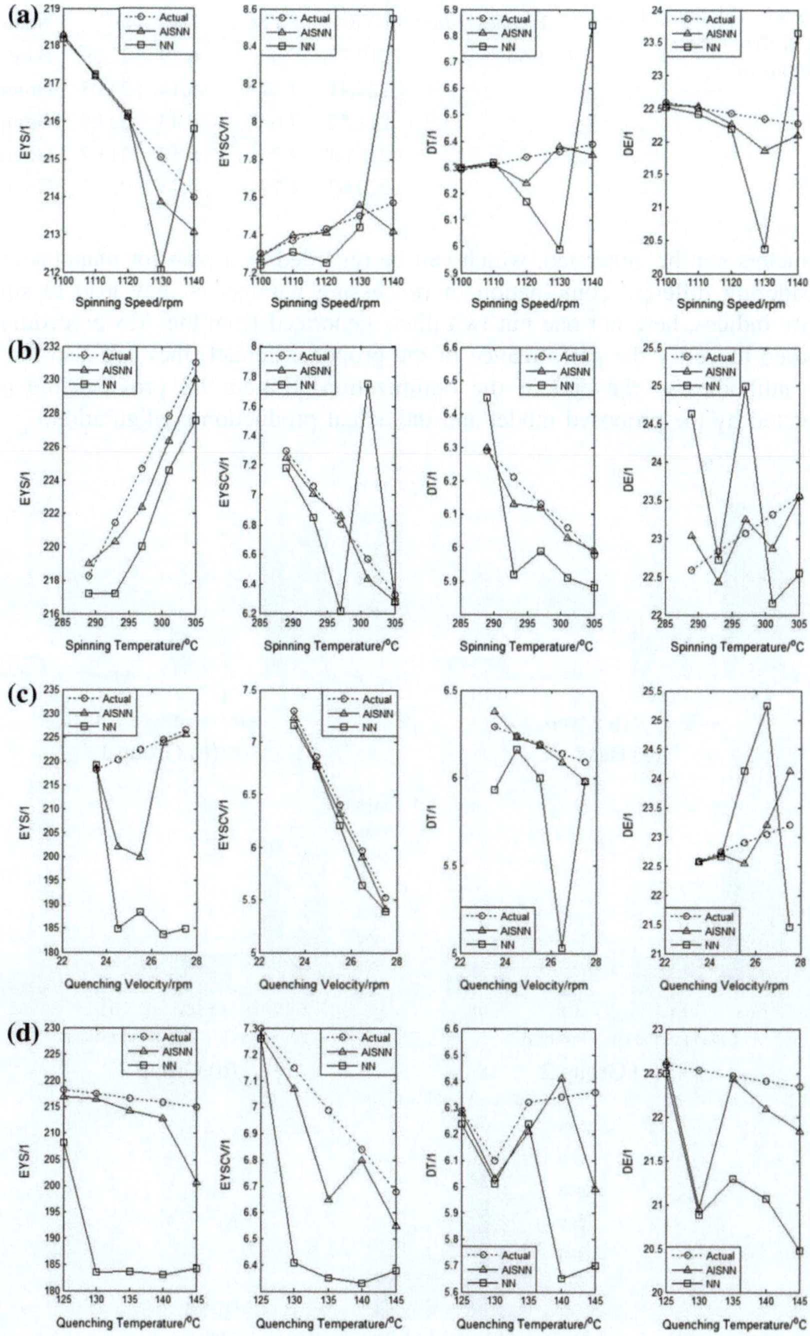

Fig. 9.7 Responses of quality indices against variations of different process parameters. **a** Response of quality indices against variations of spinning velocity. **b** Response of quality indices against variations of spinning temperature. **c** Response of quality indices against variations of quenching velocity. **d** Response of quality indices against variations of quenching temperature

Table 9.5 Quality indices as input of the category development

Fineness/*dtex*	EYS	EYSCV	DT	DE	Notes
1.56	218.27	7.3	6.29	22.59	Base
	224.41	7.42	6.14	23.05	Group 1
	211.52	7.6	6.45	22.08	Group 2
	206.04	8.23	6.58	21.67	Group 3
	220.67	8.79	6.23	22.77	Group 4

parameters can be generated, which can be regarded as a plan for manufacturing. Considering different combination of processing parameters may lead to similar quality indices, here not one but two plans generated from the AIS procedure are recorded to verify the performance of the proposed model (they are also the two best antibodies at the end of the optimization). Figure 9.8 provides the plans generated by the proposed model and the actual production configuration.

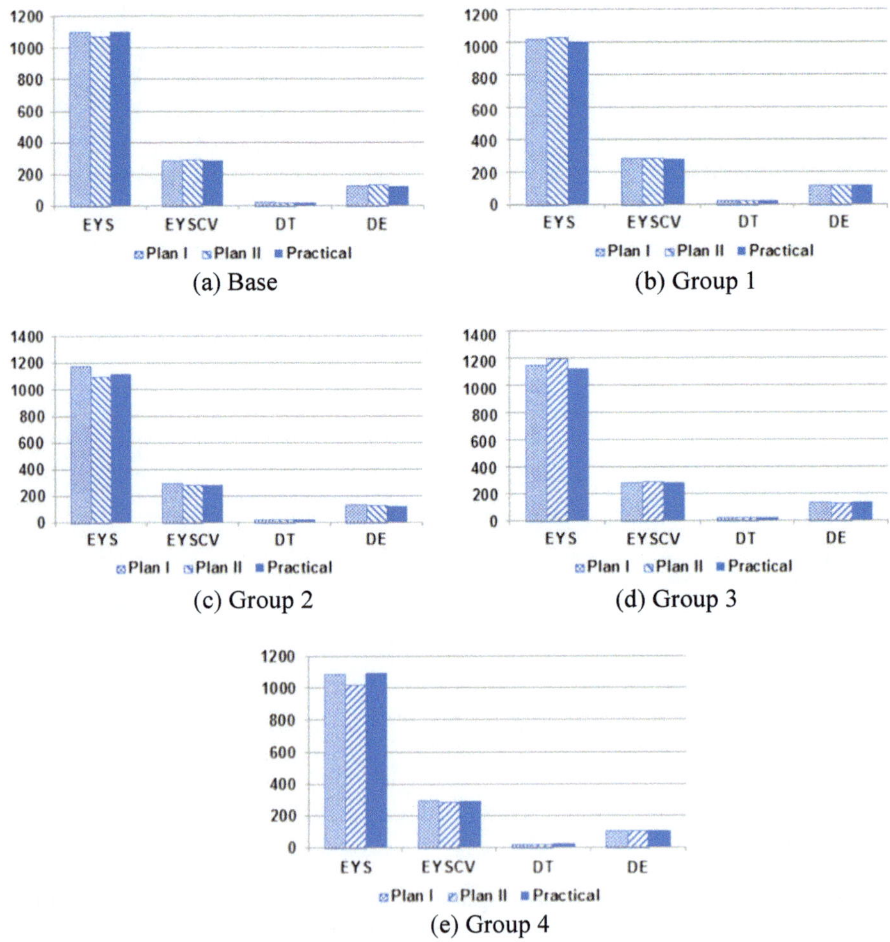

Fig. 9.8 Optimal plans generated under different quality requirements

Table 9.6 Quality verification of the generated plans and real ones with errors

	Indices	Actual value	Plan I	Error/%	Plan II	Error/%
Base	EYS1.5	218.27	218.27	0	217.31	0.44
	EYSCV	7.3	7.3	0	7.00	4.11
	DT	6.29	6.29	0	6.00	4.61
	DE	22.59	22.59	0	22.99	1.77
Group 1	EYS1.5	224.41	224.35	0.03	223.86	0.25
	EYSCV	7.42	7.28	1.92	7.09	4.45
	DT	6.14	6.14	0	6.15	0.16
	DE	23.05	23.04	0.04	23.01	0.17
Group 2	EYS1.5	211.52	211.14	0.18	212	0.23
	EYSCV	7.6	7.81	2.76	7.33	3.55
	DT	6.45	6.46	0.16	6.44	0.16
	DE	22.08	22.05	0.14	22.12	0.18
Group 3	EYS1.5	206.04	205.71	0.16	205.05	0.48
	EYSCV	8.23	7.98	3.04	8.22	0.12
	DT	6.58	6.59	0.15	6.6	0.30
	DE	21.67	21.65	0.09	21.6	0.32
Group 4	EYS1.5	220.67	222.53	0.84	225.28	2.09
	EYSCV	8.79	8.48	3.53	8.55	2.73
	DT	6.23	6.18	0.80	6.12	1.77
	DE	22.77	22.91	0.61	23.11	1.49

The verification for these plans is a little different from the procedure in the production development simulation. Since a different combination of processing parameters may contribute to similar products, the performance of these plans cannot be summarized only through comparing their contents with the actual configuration but should be put back into the actual manufacturing process to see what kind of product can be produced, and the difference between plans can therefore be verified by comparing their final products. Table 9.6 shows the production quality with the generated plans and the real ones, and the mean errors of the two plans for each index are also provided.

9.3 The Bi-Directional Optimization by Neural Network with a GA-IPSO Hybrid Algorithm

In this section, we propose a neural network model with a GA-IPSO hybrid algorithm for bi-directional optimization of PAN-based carbon fiber. We use the nearest neighbor-clustering algorithm (NNCA) to decide the hidden layer nodes number of neural network, the GA-IPSO hybrid algorithm is used to tune the parameters in the network. By this model, for one direction, we predict the carbon fiber properties; for the other direction, we obtain produce scheme for new type carbon fiber.

9.3.1 The PANCF Production and Its Bi-Directional Optimization

Design new types carbon fiber with superior properties, is a urgent task for engineers and scholars. In the past, we always make a production scheme of new type carbon fiber through former experiences, then we produce this kind of carbon fiber according to the proposed experiential scheme by testing production line. After obtaining carbon fiber from the practical experiment, we put the carbon fiber to physical and chemical instruments to test its property indices, compare these indices to expect value and get the disparity, adjust several produce parameters in proposed scheme, produce carbon fiber according to adjusted scheme by testing production line again, repeat executing produce, test and adjust process until we gain the carbon fiber which is satisfactory to us. This method wastes time, energy and cost. The most terrible thing is that, if we do not choose the appropriate parameters of control devices, it will result in oscillation and unstable of the whole system, which brings in severe loss. Therefore, it is a better choice to carry out computer simulation and can analyze and synthesize all the antecedent produce schemes, avoid one-sided decision made by workers from their personal experience and interests, and give out a more scientific and proper control parameters form an optimal production scheme.

In this section, we propose a bi-directional optimization model of carbon fiber production process based on RNN. We adopt the nearest neighbor-clustering algorithm to select a suitable set of centers of the network, and introduce a GA-IPSO hybrid algorithm to tune the parameters of the network. When we take production parameters as input and properties indices as output, this model can monitor production parameters in real time and predict properties indices online. When we take properties indices as input and production parameters as output, this model can be viewed as a designer of developing new type carbon fiber. All these bi-directional optimization functions are as shown in Fig. 9.9.

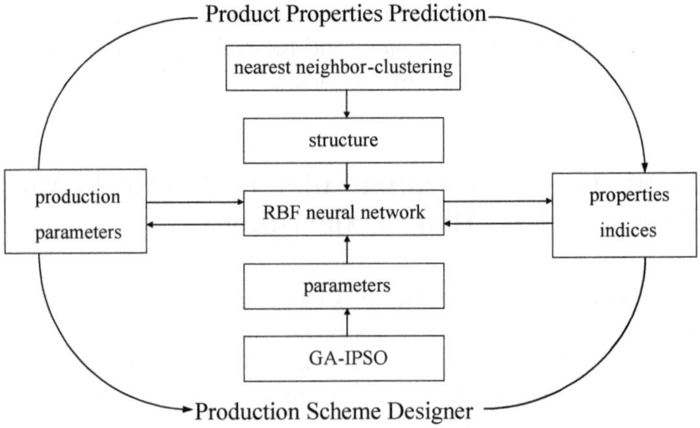

Fig. 9.9 The bi-directional optimization model of carbon fiber production process

9.3.2 The GA-IPSO Hybrid Algorithm for Neural Network

9.3.2.1 RBF Neural Network

The RNN is a typical supervised learning feed-forward neural network which is put forward by Moody and Darken in 1989 [62]. It is a three-layer network composed by input layer, hidden layer and output layer. The input layer just receives and transforms input signal to hidden layer; the hidden layer adapts radial basis functions as activation function; and the output layer always gives the final responses through simple linear functions.

The common radial basis functions are listed as follows:

$$f(x) = \exp^{-(x/\sigma)^2} \tag{9.5}$$

$$f(x) = 1/(\sigma^2 + x^2)^\alpha, \quad \alpha > 0, \tag{9.6}$$

$$f(x) = (a^2 + x^2)^\beta, \quad \alpha < \beta < 1 \tag{9.7}$$

As Gaussian function is the outstanding representative among them, we use it as activation function of hidden layer in this section:

$$u_j = G(\|x_d - c_j\|) = \exp\left(-\frac{\|x_d - c_j\|^2}{2\sigma_j^2}\right), j = 1, 2, \ldots, N_h \tag{9.8}$$

where u_j is the output of node j in hidden layer, $x_d = (x_1, x_2, x_3, \ldots, x_{Ni})^T$ is the input vector, N_i is the number of input nodes, c_j is the j-th center node in the hidden layer, σ_j is the width of the j-th center node, N_h is the number of hidden nodes, and $\|\cdot\|$ denotes Euclidean distance.

The output is formed by a linear combination of the hidden layer responses

$$y_i = \sum_{j=1}^{N_h} \omega_{ij} u_j - \theta = W_i^T U, \; i = 1, 2, \ldots, N_O, \tag{9.9}$$

where $W_i = (\omega_{i1}, \omega_{i2}, \ldots, \omega_{iNh} - \theta)^T$ represents the linking weight of the node in hidden layer and the node in output layer, $U = (u_1, u_2, \ldots, u_{Nh}, 1)^T$, N_O is the number of output nodes.

First of all, we should confirm the structure of RNN, the number of input nodes N_{i} and the number of output nodes N_O are decided by production parameters and product properties indices respectively, according to their optimization directions. The crux is fixing the number of hidden nodes N_h. N_h influences the capability of network directly. If the network structure is redundant, it will occupy mass software and hardware while training and testing, even leads to the network over fit, obtains excellent performance when it is training while shows horrible results during testing. If the network structure is too small, the network can not stand for global information. At present, there is not an uniform method to determine N_h. Generally, N_h is selected by trail and error refer to experience expression given by

$$N_h = \sqrt{N_\text{i} + N_\text{O}} + \xi, \ \xi = 1, 2, \ldots 10, \tag{9.10}$$

In this work, we adopt the nearest neighbor-clustering algorithm to acquire N_h. The NNCA is firstly presented in 1973 [63]. This algorithm is an online self-adaption cluster algorithm, and it does not need to fix N_h first. It collects data, constructs structure and trains network at the same time, which is called online-learning. It displays as a gradual increase method, which starts from a simple network and gradually increases the number of hidden nodes until to fulfill the purpose. While clustering is finished, the optimal network is formed. The nearest neighbor-clustering algorithm can be shown in Fig. 9.10.

The specific procedures of the nearest neighbor-clustering algorithm are shown below.

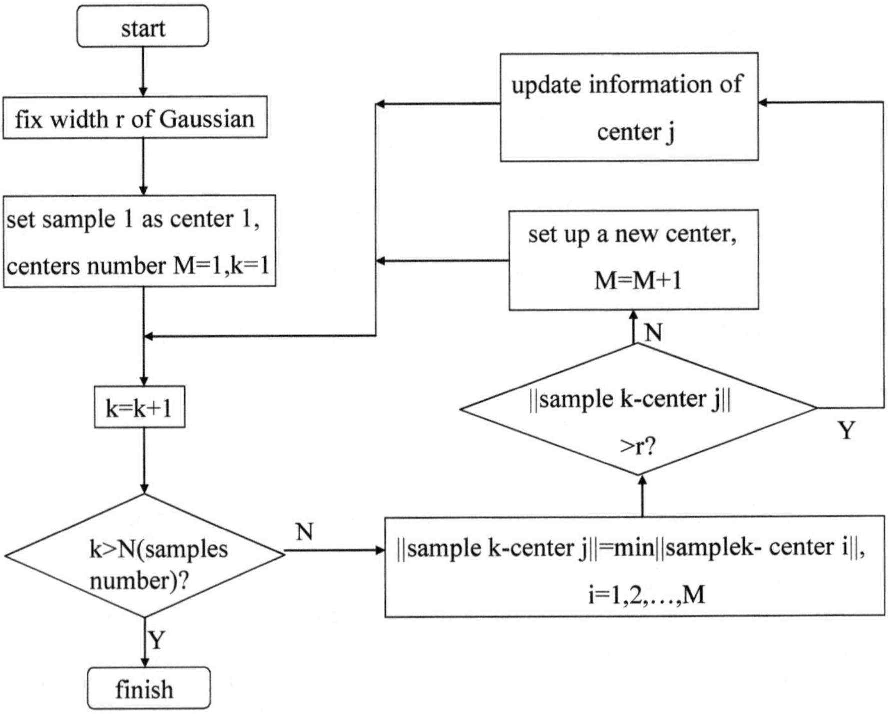

Fig. 9.10 The nearest neighbor-clustering algorithm

Step 1: Choose a proper width r of Gaussian function, here it means σ, define a vector A(l) to store the sum of output vectors of each class, and also define a counter $B(l)$ to calculate the numbers of samples belong to each class, in which l means how many classes there exists.

Step 2: Begin from the first sample (x_{1d}, y_1), set up x_{1d} as the first cluster center, sign it as follows: $c_1 = x_{1d}$, $A(1) = y_1$, $B(1) = 1$. At this time, an original RNN is formed. So far, this RNN includes only one node in hidden layer, c_1 is the center of this node, the weight vector from this node to output layer is $w_1 = A(1)/B(1)$.

Step 3: Consider the second sample (x_{2d}, y_2), calculate the distance between x_{2d} and c_1 and mark it as $\|x_{2d}\text{-}c_1\|$. If $\|x_{2d}\text{-}c_1\| \le r$, c_1 pertains to the nearest neighbor-cluster of x_{2d}, then $A(1) = y_1 + y_2$, $B(1) = 2$, $w_1 = A(1)/B(1)$. If $\|x_{2d}\text{-}c_1\|>r$, x_{2d} can be viewed as another cluster center, then $c_2 = x_{2d}$, $A(2) = y_2$, $B(2) = 1$. Add a new node to hidden layer of existing RNN, the weight vector from this node to output layer is $w_2 = A(2)/B(2)$.

Step 4: Think about the samples one by one. Assuming that the nodes number of hidden layer is M when we take the k-th sample (x_{kd}, y_k) $(k = 3, 4,\ldots, N)$ into account, which means the number of cluster centers is M, the centers are $c_1, c_2, \ldots, c_M$, respectively. At this situation, calculate distances between x_{kd} and these cluster centers which described as $\|x_{kd}\text{-}c_i\|$, $i = 1, 2, \ldots, M$. Set up $\|x_{kd}\text{-}c_j\|$ as the minimal distance, in this way, c_j is the nearest neighbor cluster of x_{kd}. If $\|x_{kd}\text{-}c_j\| \le r$, c_j pertains to the nearest neighbor-cluster of x_{kd}, then $A(j) = A(j) + y_k$, $B(j) = B(j) + 1$. When $i \neq j$, $i = 1, 2, \ldots, M$, keep A(i) and B(i) as the same as before, weight vectors from hidden layer to output layer are $w_i = A(i)/B(i)$, $i = 1, 2, \ldots, M$. If $\|x_{kd}\text{-}c_j\|>r$, x_{kd} becomes a new cluster center, then $c_{M+1} = x_{kd}$, $M = M+1$, $A(M) = y_k$, $B(M) = 1$, A(i) and B(i) will not change $(i = 1, 2, \ldots, M-1)$, add the M-th node to hidden layer of RNN which is built in last step, the weight vector from it to output layer is $w_M = A(M)/B(M)$.

Step 5: Do step 4 again and again until the last sample has been regarded, finish it.

After executing the nearest neighbor-clustering algorithm, we obtain the number of centers and the parameters of RNN which means the position and width of centers, weights from hidden nodes to output nodes at the same time. However, in this section, we just use the nearest neighbor-clustering algorithm to determine the number of hidden nodes, which is the number of centers. We will tune the position and width of centers, weights from hidden nodes to output nodes by proposed GA-IPSO which is described in Sect. 9.3.2. But the parameters of RNN we gained from the nearest neighbor-clustering algorithm will be taken as initial particle swarm range.

9.3.2.2 GA-IPSO Hybrid Algorithm

1. **Basic PSO Algorithm**

PSO algorithm is an efficient population-based stochastic optimization technique inspired by social behavior of bird flocks or fish schools which is originally developed by Kennedy and Eberhart in 1995 [64]. Since PSO requires short computer time and less memory, it has been successfully applied to wide range optimization applications.

The principle of PSO is: For a *D*-dimension problem, it maintains a population of particles distribute uniformly around a *D*-dimension search space at first. Suppose the size of the swarm is *m*, the population is represented as $x = (x_1, x_2, \ldots, x_m)^T$. Each particle in this population indicates a potential solution to the problem. A particle is recorded from two aspects of its position and velocity, the position is represented as $x_{id} = (x_{i1}, x_{i2}, \ldots, x_{iD})$, the velocity is represented as $v_{id} = (v_{i1}, v_{i2}, \ldots, v_{iD})$. The position is associated with the fitness. The best fitness the particle has achieved is considered as its current best position, marked it as $p_{id} = (p_{i1}, p_{i2}, \ldots, p_{iD})$. The best position in the whole population is recorded as $p_{gd} = (p_{g1}, p_{g2}, \ldots, p_{gD})$. In every generation, each particle updates its status in the light of three criterions: (1) its own inertia; (2) the cognitive information p_{id}; (3) the social information p_{gd}. These particles in the population tend to cluster together by updating their status until the optimum solution is obtained at last.

For the *t* iteration, the velocity and position of each particle are updated in the following manners:

$$v_{id}^{t+1} = wv_{id}^{t} + c_1 r_1 (p_{id}^{t} - x_{id}^{t}) + c_2 r_2 (p_{gd}^{t} - x_{gd}^{t}) \tag{9.11}$$

$$x_{id}^{t+1} = (x_{id}^{t} + v_{id}^{t+1}) \tag{9.12}$$

where *w* is the inertia weight, which decides influence on the next generation of the particle current velocity. c_1 and c_2 are acceleration constants, called the cognitive and social parameter, respectively. r_1 and r_2 are uniform random numbers between 0 and 1.

Commonly, $c_1 = c_2 = 2.05$, $w_{max} = 0.9$, $w_{min} = 0.4$, for *T* is the maximal number of iterations, *w* of every iteration is calculated according to this formula:

$$w = w_{max} - \frac{t}{T}(w_{max} - w_{min}) \tag{9.13}$$

2. **GA-IPSO Hybrid Algorithm**

Furthermore, although PSO has good convergent property, if the initial population of particles can not effectively cover the whole region, the population diversity will be greatly reduced after some iterations, as such the particles will converge easily to a local optimum. Aim to avoid the limitations and enhance the superiorities, due to PSO utilizes ω, p_{id} and p_{gd} to modify the current search point to prevent the particles flying in the same direction, in this section we employ dynamic-adjust w and introduce a penalty factor (θ_1, θ_2) for coordinating the guides of p_{id} and p_{gd}. In that case, the formulas (9.11) can be rewrite as:

$$v_{id}^{t+1} = w^t v_{id}^t + \theta_1^t c_1 r_1 (p_{id}^t - x_{id}^t) + \theta_2^t c_2 r_2 (p_{gd}^t - x_{id}^t) \tag{9.14}$$

1. Particle and its fitness

The parameters of RNN needed to be tuned are positions and widths of hidden centers, weights between hidden layer and output layer. A group of parameters is a particle. Since numbers of input, hidden, output layer are N_i, N_h, N_0, the dimension of a particle is given by

$$D = (N_i + 1 + N_0) \times N_h \tag{9.15}$$

Mark the output of RNN as y, the expect output as $\hat{y}$. The goal is to minimum the error between y and $\hat{y}$. So the fitness function of particles is given by

$$f = \frac{1}{2n} \sum_{i=1}^{n} |y_i - \hat{y}_i|^2 \tag{9.16}$$

2. Dynamic-adjust inertia weight w

Inertia weight w expresses the search length of the PSO. When w is a large value, it is good for global optimum search owing to it can easily get out of local optimum; when w is a small value, it is good for local optimum search and accelerate convergence. Although w in formula (9.13) is not a constant, it can not be called dynamic-adjust, it is just a kind of linear decrease. The linear decrease method has weak inspire ability of search direction though it is convenient to realize. Here, we dynamic adjust w based on the change of the smoothness [65] of the object function. The definition of smoothness is shown below:

$$a^t = \frac{1}{m} \sum_{i=1}^{m} |f(x_{id}^t) - f(x_{\min d}^t)| \tag{9.17}$$

where $f(x_{id}^t)$ is the fitness value of the i-th particle on iteration t, $f(x_{\min d}^t)$ is the fitness value of the best particle on iteration t, these are expressed below:

$$f(x_{id}^t) = f(x_{i1}^t, x_{i2}^t, \ldots, x_{iD}^t) \tag{9.18}$$

$$f(x_{\min d}^t) = \min_{i=1,2,\ldots m} f(x_{id}^t) \tag{9.19}$$

On every iteration, w changes with a^t like this:

$$w^t = 1.1^{-\lambda} w^{t-1}, \quad t = 2, \ldots, T \tag{9.20}$$

$$\lambda = \begin{cases} 1, & a^t - a^{t-1} > 0 \\ 0, & a^t - a^{t-1} = 0 \\ -1, & a^t - a^{t-1} < 0 \end{cases} \tag{9.21}$$

When $a^t - a^{t-1} > 0$, it indicates this iteration is emanative, the search step length should be decreased for precisely search in range around extremum; when $a^t - a^{t-1} < 0$, it indicates this iteration is convergent, the search step length should be increased for accelerating search for extremum.

3. Cognitive parameter c_1 and social parameter c_2

In GA-IPSO, we also change cognitive parameter c_1 and social parameter c_2 linearly as follows:

$$c_1 = c_{1\min} + \frac{t}{T}(c_{1\max} - c_{1\min}) \tag{9.22}$$

$$c_2 = c_{2\max} - \frac{t}{T}(c_{2\max} - c_{2\min}) \tag{9.23}$$

4. enalty factor(θ_1, θ_2)

After calculating resent fitness of a particle on each iteration, we change the position of the particle a little, then calculate the fitness of the new position. If the new fitness is smaller than the resent fitness, it means the current p_{id} and p_{gd} can not give out direction guidance precisely, then we compare the new fitness to the best fitness of this particle and the best fitness of the best particle in the population respectively to check out which one shows a bad answer. When the new fitness is smaller than the best fitness of this particle, it means the best position of this particle does not affect search direction correctly, then its influence should be decreased as a penalty, on the contrast, the cognitive influence should be increase. The mechanism is the same to the social influence expressed by the best position of the best particle in the population. The mathematical expression is showed below:

$$x^t_{newd} = x^t_{id} - 0.5rand(), \quad t = 2, \ldots, T, \tag{9.24}$$

$$\theta^t_1 = 1.1^{-\alpha} \theta^{t-1}_1 \tag{9.25}$$

$$\alpha = \begin{cases} 1, & (f(x^t_{newd}) - f(p^t_{id}) < 0) \cap (f(x^t_{newd}) - f(x^t_{id}) < 0) \\ 0, & (f(x^t_{newd}) - f(p^t_{id}) = 0) \cap (f(x^t_{newd}) - f(x^t_{id}) < 0) \cup (f(x^t_{newd}) - f(x^t_{id}) \geq 0) \\ -1, & (f(x^t_{newd}) - f(p^t_{id}) > 0) \cap (f(x^t_{newd}) - f(x^t_{id}) < 0) \end{cases} \tag{9.26}$$

$$\theta^t_2 = 1.1^{-\beta} \theta^{t-1}_1 \tag{9.27}$$

$$\beta=\begin{cases} 1, & (f(x_{newd}^t)-f(p_{gd}^t)<0)\cap(f(x_{newd}^t)-f(x_{id}^t)<0) \\ 0, & (f(x_{newd}^t)-f(p_{gd}^t)=0)\cap(f(x_{newd}^t)-f(x_{id}^t)<0)\cup(f(x_{newd}^t)-f(x_{id}^t)\geq 0) \\ -1, & (f(x_{newd}^t)-f(p_{gd}^t)>0)\cap(f(x_{newd}^t)-f(x_{id}^t)<0) \end{cases} \tag{9.28}$$

5. The GA-IPSO

With the iteration increasing, the population diversity will be reduced. Sometimes stagnation happens, it displays that v almost be zero. For GA is famous for its cross and mutation operations, it can produce new particles, rebirth the whole population, escape from the above terrible situation. So, we combine GA to the IPSO, and adopt p_{id} and p_{gd} to execute cross and mutation operations, which means the new particles keeping the advantage of being guided by current p_{id} and p_{gd}. The mechanism is showed below:

$$\text{cross}\quad v_{idc}^{t+1}=\eta p_{id}^t+(1-\eta)p_{gd}^t=(v_{i1c}^{t+1},v_{i2d}^{t+1},\ldots,v_{iD}^{t+1}) \tag{9.29}$$

$$\text{mutation}\quad dm=\lfloor 1+(D-1)rand()\rfloor \tag{9.30}$$

$$v_{id}^{t+1}=(v_{i1c}^{t+1},\ldots,v_{idmd}^{t+1}rand(),\ldots,v_{iD}^{t+1}) \tag{9.31}$$

where η is a cross coefficient between 0 and 1, formula (9.30) is used to select which dimension will be mutation, $\lfloor\rfloor$ indicates getting an integer downward.

The flowchart of GA-IPSO is indicated in Fig. 9.11.

The specific procedures of the GA-IPSO are shown below:

Step 1: Set iteration as t = 1. Initiate the parameters such as ω^1, θ_1^1, θ_2^1, $c_{1\min}$, $c_{1\max}$, $c_{2\min}$, $c_{2\max}$, η, maximum T, swarm size M, error_goal.
Step 2: Initiation the particle swarm according to the range given by the nearest neighbor-clustering algorithm.
Step 3: Calculate the fitness of each particle in the population.
Step 4: Set up p_{id} and p_{gd} of the first iteration.
Step 5: Set up x_{id} and v_{id} of each particle for next iteration.
Step 6: Update iteration $t = t + 1$.
Step 7: Calculate the fitness of each particle in the new population.
Step 8: Update p_{id} and p_{gd}.
Step 9: Change the position of each particle in this population a little, calculate the new fitness, update θ_1^1 and θ_2^1 according to formula (9.24)–(9.28).
Step 10: Update x_{id} and v_{id} of each particle according to (9.17)–(9.23), (9.14) and (9.12).
Step 11: If v = 0, execute GA, cross and mutation v_{id} according to formula (9.29)–(9.31).
Step 12: Do Step 6–12 until fitness satisfy the error_goal or t reaches T.

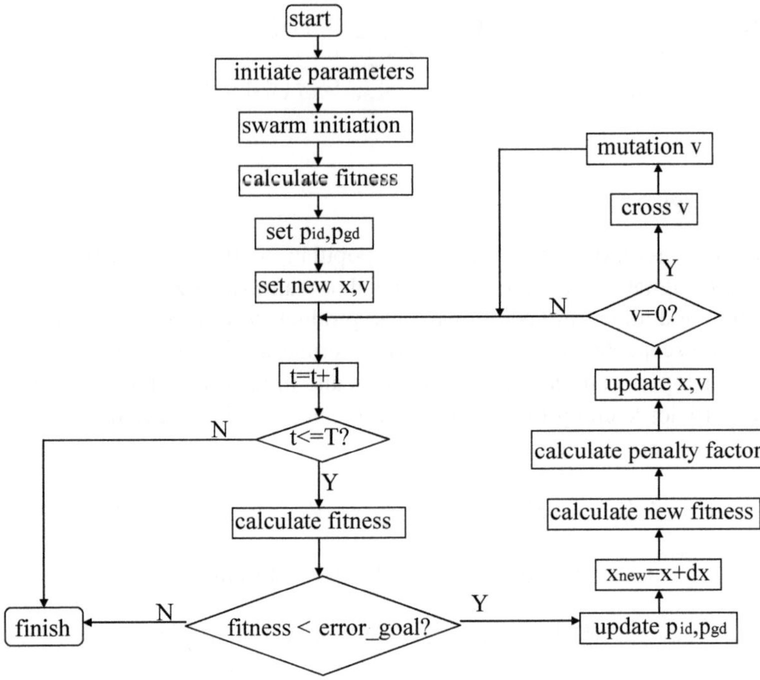

Fig. 9.11 The flowchart of GA-IPSO

9.3.3 Simulation and Results

9.3.3.1 Data Preparation

Aim to test and demonstrate the performance of the model we proposed, we used these data collected and organized from experiments in Table 9.7.

In these data, viscosity average molecular weight, conversion ratio, solid content, spinning jet drawing ratio, coagulating bath temperature and total drawing ratio are production parameters, strength and structure parameters are property indices. These specific production parameters and property indices can relatively fully represent the whole manufacture process and product performance of PAN-based carbon fiber. In the bi-directional optimization process, we take the first 45 groups of samples as training dataset, and the last 5 groups of samples as test dataset. We also make comparison between conventional RNN, basic PSO-RNN and the proposed method, observe their performance in different ways which are defined as the following respectively:

$$\text{Mean absolute error}: \quad MAE = \frac{1}{n}\sum_{i=1}^{n}\left|y_i - \hat{y}_i\right| \tag{9.32}$$

Table 9.7 Experiment data of carbon fiber production process

No	Viscosity average molecular weight (10^4)	Conversion ratio (%)	Solid content (%)	Spinning jet drawing ratio (%)	Coagulating bath temperature (°C)	Total drawing ratio	Strength (CN/d)	Structure parameter
1	8.9	94.5	20.8	−50.3	14	6.33	4.08	14.82
2	6.3	91.0	20.0	−59.7	15	5.89	3.23	12.63
3	11.6	92.0	20.4	−50.5	14	6.03	3.76	13.24
4	8.8	94.8	21.8	−63.4	13	6.65	4.17	17.24
5	7.0	81.8	17.9	−63.4	15	6.32	3.99	15.14
6	8.2	85.5	21.7	−59.5	15	5.49	4.58	16.61
7	7.2	89.8	19.5	−53.1	13	5.88	3.64	15.49
8	8.9	82.5	17.5	−56.8	19	6.38	4.07	17.57
9	8.0	83.4	18.6	−62.1	17	5.72	3.18	15.48
10	11.7	90.6	17.9	−53.8	16	6.47	3.22	12.10
11	11.5	82.8	18.7	−64.8	17	5.79	3.27	12.73
12	6.3	95.1	19.6	−54.9	16	6.37	4.36	17.18
13	10.4	98.6	20.2	−68.3	17	6.41	3.99	14.91
14	7.6	93.1	19.7	−55.4	16	5.88	3.38	17.07
15	8.5	84.6	22.3	−65.3	18	5.04	3.99	13.26
16	9.3	89.8	20.1	−53.8	16	5.66	3.30	15.31
17	11.7	79.4	22.7	−55.8	19	5.85	3.11	15.78
18	8.5	96.9	20.8	−51.8	14	5.54	4.70	12.19
19	11.9	96.4	22.7	−61.5	13	5.39	4.12	15.69
20	7.8	95.0	18.4	−63.7	13	6.64	4.86	14.17
21	10.2	82.7	21.1	−60.9	13	5.86	4.39	12.30
22	10.0	90.1	18.7	−58.5	15	6.78	4.17	14.94

(continued)

Table 9.7 (continued)

No	Viscosity average molecular weight (10^4)	Conversion ratio (%)	Solid content (%)	Spinning jet drawing ratio (%)	Coagulating bath temperature (°C)	Total drawing ratio	Strength (CN/d)	Structure parameter
23	9.2	77.5	21.0	−62.9	16	5.78	4.63	13.16
24	10.2	86.4	21.2	−63.0	15	6.54	4.76	12.74
25	10.0	83.9	17.4	−63.6	18	5.79	4.98	13.23
26	7.1	80.6	18.5	−62.7	17	6.62	3.00	12.88
27	6.8	80.9	18.3	−68.9	18	6.51	4.73	13.13
28	12.0	86.3	21.0	−54.2	19	5.75	4.23	12.26
29	7.0	79.1	22.1	−64.2	19	5.43	4.98	15.81
30	6.2	90.2	19.1	−54.7	14	6.58	4.06	13.69
31	9.4	87.4	21.7	−52.4	13	6.90	3.96	15.23
32	11.3	92.3	21.1	−62.1	17	5.66	4.60	16.17
33	10.0	92.4	17.0	−59.0	13	6.34	3.46	14.99
34	7.1	91.0	20.6	−59.2	16	5.88	4.00	15.21
35	8.2	77.7	19.3	−63.2	16	6.67	4.80	14.67
36	8.8	78.5	22.5	−65.4	19	6.54	4.15	12.74
37	11.9	84.0	17.0	−57.0	16	5.33	4.69	14.94
38	6.9	88.7	19.8	−63.2	15	6.72	4.48	17.12
39	11.1	91.4	19.5	−58.3	17	6.98	4.17	17.24
40	9.9	86.0	19.8	−66.8	18	6.03	3.49	13.62
41	8.3	95.0	21.6	−66.7	16	6.77	4.33	13.25
42	7.1	92.8	18.9	−55.1	15	6.18	3.17	15.39
43	8.6	98.3	21.7	−62.3	14	5.31	4.25	15.84
44	8.9	88.7	19.8	−61.6	17	5.40	4.32	14.50

(continued)

Table 9.7 (continued)

No	Viscosity average molecular weight (10^4)	Conversion ratio (%)	Solid content (%)	Spinning jet drawing ratio (%)	Coagulating bath temperature (°C)	Total drawing ratio	Strength (CN/d)	Structure parameter
45	6.7	84.2	17.2	−60.8	14	5.81	4.46	13.24
46	11.0	98.0	22.7	−74.6	12	6.89	3.90	12.82
47	9.5	92.2	20.3	−50.5	17	5.92	3.91	13.20
48	7.7	78.2	17.2	−63.4	19	6.17	3.99	15.19
49	9.5	79.3	18.1	−67.4	13	6.50	4.78	17.69
50	7.4	90.4	21.3	−55.3	18	6.65	4.96	12.49

$$\text{Mean relative error}: \quad MRE = \frac{1}{n}\sum_{i=1}^{n}\left|\frac{y_i - \hat{y}_i}{y_i}\right| \tag{9.33}$$

$$\text{Root mean square error}: \quad RMSE = \sqrt{\frac{1}{n}\sum_{i=1}^{n}(y_i - \hat{y}_i)^2} \tag{9.34}$$

Theil's Inequality Coefficient:

$$TIC = \frac{\sqrt{\sum_{i=1}^{n}(y_i - \hat{y}_i)^2}}{\sqrt{\sum_{i=1}^{n}(y_i)^2} + \sqrt{\sum_{i=1}^{n}(\hat{y}_i)^2}} \tag{9.35}$$

9.3.3.2 Properties Prediction

For properties prediction, the production parameters are input, the properties indices are output, N_i is 6, N_O is 2. Set r = 1, we obtain $N_h = 6$ by applying the nearest neighbor-clustering algorithm (NNCA), all the parameters information are list in Table 9.8.

Based on Table 9.8, an initial RNN model is developed. We consider the range of initial swarm for both proposed method and basic PSO-RNN by referencing Table 9.8. We set $w^1 = \theta_1^1 = \theta_2^1 = 0.5$, $c_{1\min} = c_{2\min} = 0.4$, $c_{1\max} = c_{2\max} = 0.9$, $\eta = 0.3$, error_goal = 0.01, maximum iteration T is 5000, swarm size is 20 for

Table 9.8 Parameters list of rnn got from nnca

Position of hidden layer centers					
c_1	c_2	c_3	c_4	c_5	c_6
0.4664	0.1387	0.7297	0.4038	0.9834	0.9819
0.8050	0.2038	1.0000	0.3347	0.8927	0.3094
0.6609	0.1556	0.5499	0.9139	1.0000	0
1.0000	0.2958	0.0298	0.1932	0.3967	0.6396
0.1667	0.3333	0.6667	0.8333	0	0.5000
0.6652	0.6600	0.7054	0	0.1829	0.1522
Width of hidden layer centers					
σ_1	σ_2	σ_3	σ_4	σ_5	σ_6
1.0000	1.0000	1.0000	1.0000	1.0000	1.0000
Weights of output layer					
W_1	W_2	W_3	W_4	W_5	W_6
0.4340	0.5585	0.6185	0.6064	0.6339	0.8537
0.5112	0.4440	0.4670	0.3342	0.4585	0.5194

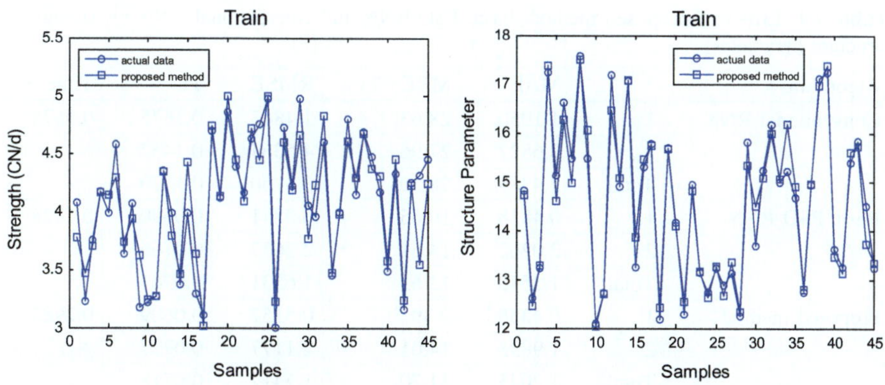

(a) strength (b) structure parameter

Fig. 9.12 Training results of the proposed GA-IPSO- RNN model

proposed method in this section, set $w_{\min} = 0.4$, $w_{\max} = 0.9$ for basic PSO-RNN, and the other parameters of the basic PSO-RNN are agreed with the proposed method.

By employing the proposed GA-IPSO algorithm to optimize the parameters of RNN, we gain the best parameters for the RNN. The training accuracy and the agreement level of the proposed method are shown in Fig. 9.12.

From Fig. 9.12, the prediction values of training dataset have a high level of agreement with the actual data, it can be seen directly that the proposed model has an accuracy and good tracking performance which reflects the model's outstanding universality.

The model's prediction accuracy also can be reflected from its performance on the test dataset. We show the prediction results of the test dataset of the proposed method, basic PSO-RNN and conventional RNN in Fig. 9.13 to compare their

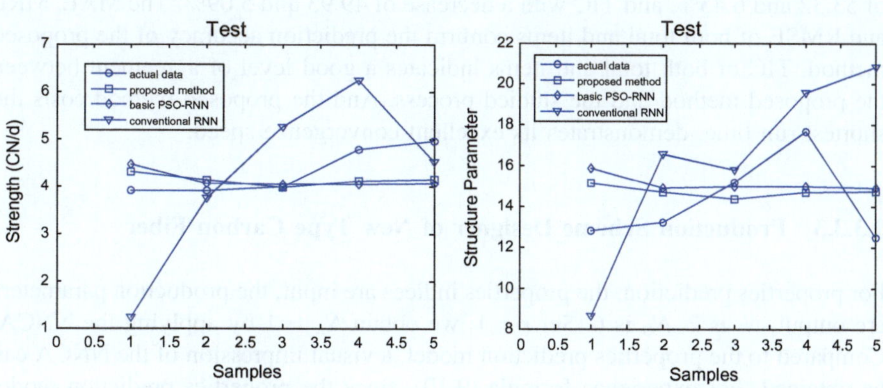

(a) Strength (b) Structure parameter

Fig. 9.13 Performance of proposed method, basic PSO-RNN and conventional RNN

Table 9.9 Errors of proposed method, basic PSO-RNN and conventional RNN (1. strength, 2. structure parameter)

Algorithms		MAE	MRE (%)	RMSE	TI	Time (s)
Conventional RNN	1	1.1950	28.63	1.4843	0.1675	0.9575
	2	3.6827	27.96	4.5364	0.1452	
	Total	2.4389	28.30	3.3750	0.1470	
Basic PSO-RNN	1	0.4818	10.65	0.5841	0.0690	0.7428
	2	2.0262	14.61	2.2637	0.0766	
	Total	1.2540	12.63	1.6531	0.0761	
Proposed method	1	0.4258	9.39	0.5157	0.0609	0.2985
	2	1.9833	14.01	2.1177	0.0727	
	Total	1.2045	11.70	1.5412	0.0718	

performance. It can be seen obviously that, the curve of the proposed GA-IPSO-RNN model is most closed to the actual data curve, which reveals it has the best prediction accuracy and reliability among the three models. The basic PSO-RNN displays worse performance, and the conventional RNN is the worst one with big errors. The detailed numeral comparison of MAE, MRE, RMSE and TIC is also listed in Table 9.9.

From data in Table 9.9, we can see that, in terms of MAE, the proposed GA-IPSO-RNN achieves with a decrease of approximately 50.61 and 3.95% compared to the other two models, respectively. The same information can be obtained from the other items such as MRE with a decrease of 58.66 and 7.36%, RMSE with a decrease of 54.33 and 6.77%, and TIC with a decrease of 51.16 and 5.65%. Similar results are observed in terms of strength and structure parameter. For strength, MAE with a decrease of 64.37 and 11.62%, MRE with a decrease of 67.20 and 11.83%, RMSE with a decrease of 65.26 and 11.71%, and TIC with a decrease of 63.64 and 11.74%. For structure parameter, MAE with a decrease of 46.15 and 2.12%, MRE with a decrease of 49.89 and 4.11%, RMSE with a decrease of 53.32 and 6.45%, and TIC with a decrease of 49.93 and 5.09%. The MAE, MRE and RMSE of both total and items confirm the prediction accuracy of the proposed method. TIC of both total and items indicates a good level of agreement between the proposed method and the studied process. And the proposed method costs the shortest run time, demonstrates its excellent convergence speed.

9.3.3.3 Production Scheme Designer of New Type Carbon Fiber

For properties prediction, the properties indices are input, the production parameters are output, N_i is 2, N_O is 6. Set $r = 1$, we obtain $N_h = 1$ by applying the NNCA. Compared to the properties prediction model, a visual impression of the NNCA can be obtained. For experience formula (9.10), since the properties prediction model and production scheme design model have the same sum of N_i and N_O, they should almost have the same hidden nodes number, however, they have quite different N_h

Table 9.10 The parameters value of the rnn got from the nnca

Position of hidden layer centers	Weights of output layer
c_1	W_1
0.5453	0.4766
0.4971	0.5011
Width of hidden layer centers	0.5117
σ_1	0.4849
1.0000	0.4630
	0.5397

by the NNCA, which indicates its good performance of simplify the RNN structure. And once there add a new sample, it can be treated online as a training sample for RNN at once, which will improve the prediction accuracy of the model. All the parameters are list in Table 9.10.

Based on Table 9.10, an initial RNN model is developed. We consider the range of initial swarm for both proposed method and basic PSO-RNN by referencing Table 9.10. We set $w^1 = \theta_1^1 = \theta_2^1 = 0.5$, $c_{1\min} = c_{2\min} = 0.4$, $c_{1\max} = c_{2\max} = 0.9$, $\eta = 0.3$, error_goal = 0.01, maximum iteration T is 5000, swarm size is 20 for proposed method in this section, set $w_{\min} = 0.4$, $w_{\max} = 0.9$ for basic PSO-RNN, and the other parameters of the basic PSO-RNN are agreed with the proposed method.

By employing the proposed GA-IPSO algorithm to optimize the parameters of the RNN and gain the best parameters. The training accuracy and the agreement level of the proposed method are shown in Fig. 9.14. From Fig. 9.14, the prediction values of training dataset have a high level of agreement with the actual data, it can be seen directly that the proposed model has an accuracy and good tracking performance which reflect the model's outstanding universality.

The model's prediction accuracy can also be reflected from its performance on the test dataset. We show the prediction results of the test dataset of the proposed method, basic PSO-RNN and conventional RNN in Fig. 9.15 to compare their performance. It can be seen obviously that, the curve of the proposed GA-IPSO-RNN model is most closed to the actual data curve, which reveals it has the best prediction accuracy and reliability among the three models. The conventional RNN displays the worst performance, we just compare the proposed method and the basic PSO-RNN as shown in Fig. 9.16. The detailed comparison of MAE, MRE, RMSE and TIC is also listed in Table 9.11.

From data in Table 9.11 we can see that, in terms of MAE, the proposed GA-IPSO-RNN achieve, with a decrease of approximately 98.59 and 13.71% compared to the other two models, respectively. The same information can be obtained from the other items such as MRE with a decrease of 98.61 and 19.75%, RMSE with a decrease of 99.34 and 11.20%, and TIC with a decrease of 93.41 and 10.79%. Similar results are observed in terms of viscosity average molecular weight, conversion ratio, solid content, spinning jet drawing ratio, coagulating bath temperature and total drawing ratio. The MAE, MRE and RMSE of both total and

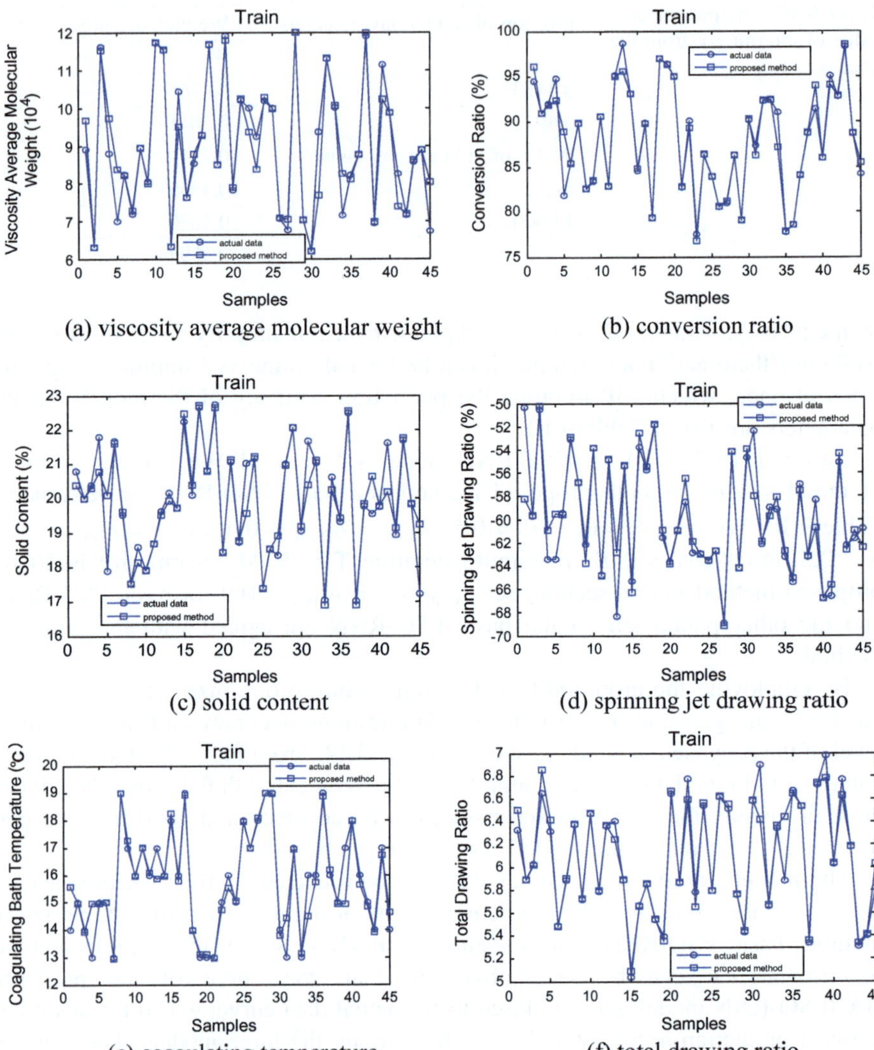

(a) viscosity average molecular weight (b) conversion ratio

(c) solid content (d) spinning jet drawing ratio

(e) coagulating temperature (f) total drawing ratio

Fig. 9.14 Training results of the proposed GA-IPSO-RNN model

items confirm the prediction accuracy of the proposed method. TIC of both total and items indicates a good level of agreement between the proposed method and the studied process. As the curves shown in Fig. 9.15, the data list in Table 9.11 also indicate quite unacceptable errors of conventional RNN, it verifies the proposed method's effective optimization of RNN's parameters indirectly. And the proposed model costs the shortest run time, demonstrates its excellent convergence.

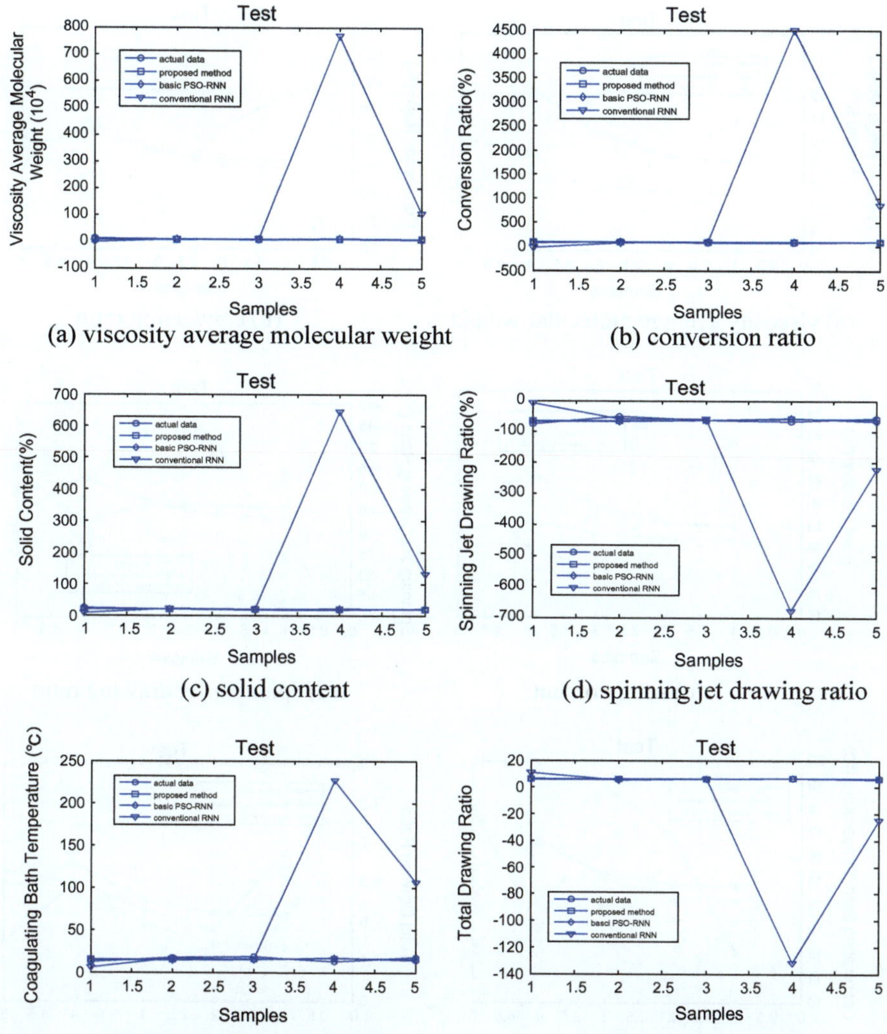

Fig. 9.15 Performance of proposed method, basic PSO-RNN and conventional RNN

9.4 Performance Prediction of Differential Fibers with a Bi-Directional Optimization Approach

Based on neural networks and multi-objective evolutionary algorithm, this section develops an approach to predict both fiber performance and production parameters, and its final target is to optimize the fiber production. This approach only depends

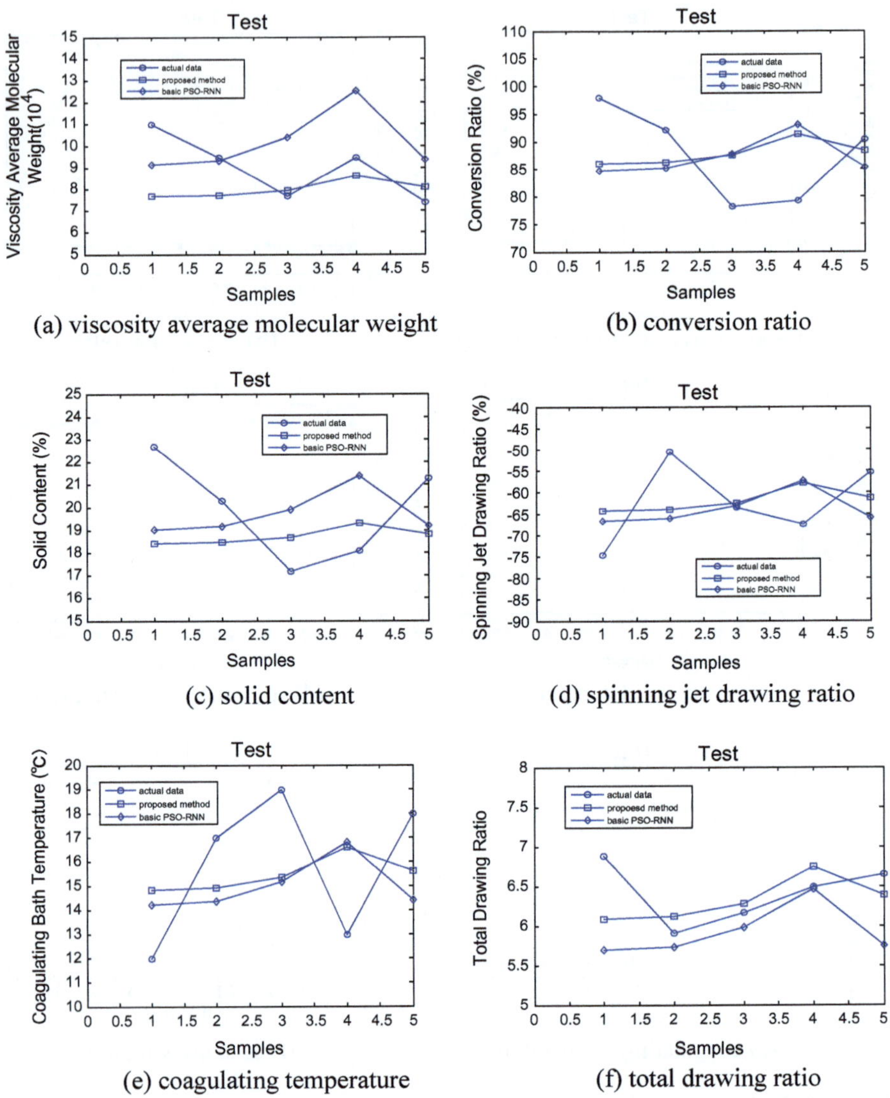

Fig. 9.16 Performance comparison of proposed method and basic PSO-RNN

on the production parameters and their corresponding fiber performance, so it remains unaffected when the production modes or the equipment conditions are changed. The prediction in this section consisted of the bi-directional prediction process, which includes the prediction of the fiber performance by the production parameters and the prediction of the production parameters by the fiber performance. If these two prediction processes are regarded as two irrelevant parts, they

Table 9.11 Errors of proposed method, basic PSO-RNN and conventional RNN

Algorithms		MAE	MRE (%)	RMSE	TIC	Time (s)
Conventional RNN	1	173.1400	1876.58	341.7052	0.9612	1.7296
	2	1062.8000	1310.66	2004.3000	0.9391	
	3	151.6600	816.09	285.3062	0.9051	
	4	171.6800	264.56	285.1956	0.7416	
	5	62.6800	443.07	103.5493	0.8054	
	6	34.8800	533.97	63.3625	0.9510	
	total	276.1367	874.16	847.6972	0.9285	
Basic PSO-RNN	1	1.9652	22.73	2.2049	0.1138	0.4065
	2	9.7604	11.30	10.3417	0.0590	
	3	2.5932	13.19	2.7487	0.0690	
	4	8.9433	15.27	10.2110	0.0806	
	5	3.2098	20.66	3.2757	0.1054	
	6	0.4920	7.38	0.6731	0.0544	
	total	4.4940	15.09	6.2559	0.0686	
Proposed method	1	1.3996	14.37	1.7597	0.1026	0.1464
	2	8.2660	9.61	9.1148	0.0518	
	3	2.2693	11.04	2.5107	0.0647	
	4	8.0976	13.47	9.1473	0.0733	
	5	2.9085	19.20	2.9776	0.0944	
	6	0.3260	4.94	0.4054	0.0317	
	Total	3.8778	12.11	5.5555	0.0612	

Note 1. viscosity average molecular weight; 2. conversion ratio; 3. solid content; 4. spinning jet drawing ratio; 5. coagulating bath temperature; 6. total drawing ratio

can be solved by many kinds of algorithms, but there maybe be some problems. Because, the forward prediction is based on the production process so it is a forward process in nature which means it can be seen as an independent prediction. However, the backward reasoning is complex, because it is a reverse process in nature. For example, different production parameters can achieve same fiber performance, and this situation will affect the backward reasoning results, which means that one way prediction is not enough to solve this backward reasoning. In a word, comparing with the forward prediction which can be solved by different algorithms simply, the backward reasoning is too complex to be realized by the same algorithms as the forward prediction because of the missing data, the conflict data, the interaction influences and other problems. In order to avoid those problems, the forward prediction and the backward reasoning are designed together in the bi-directional prediction approach in this section, and in another word, the prediction approach adopted in this section is a bi-directional prediction approach whose backward reasoning is based on the forward prediction. In order to solve the accuracy problem, through experiments and data analysis, this section uses hybrid intelligent algorithms, including particle swarm optimization algorithm with

K-means algorithm to optimize the clustering and to increase the accuracy of the forward prediction. The backward reasoning as mentioned before is based on the forward prediction, and it can be seen as a multi-objective evolutionary problem. This section uses multi-objective evolutionary algorithm, the clustering results and the forward prediction to achieve the backward reasoning results.

9.4.1 Fiber Production and Bi-Directional Prediction Optimization

9.4.1.1 Fiber Production Process

Fiber production is a complex production line which generally consists of two systems, the melting transportation system and the spinning system. The melting transportation system is to convert the fiber materials to liquid and accomplished by the following spinning system. The spinning system includes a blowing area and a stretching process. The blowing area is to help the liquid to solidify the streams which called 'as-spun' fibers [51, 52], and then the 'as-spun' fibers can be stretched in the stretching process according to the stretching ratio. Although different fibers ask for different combinations of equipments and materials, the basic production processes are similar. The production process of polyester staple fiber is depicted as Fig. 9.17.

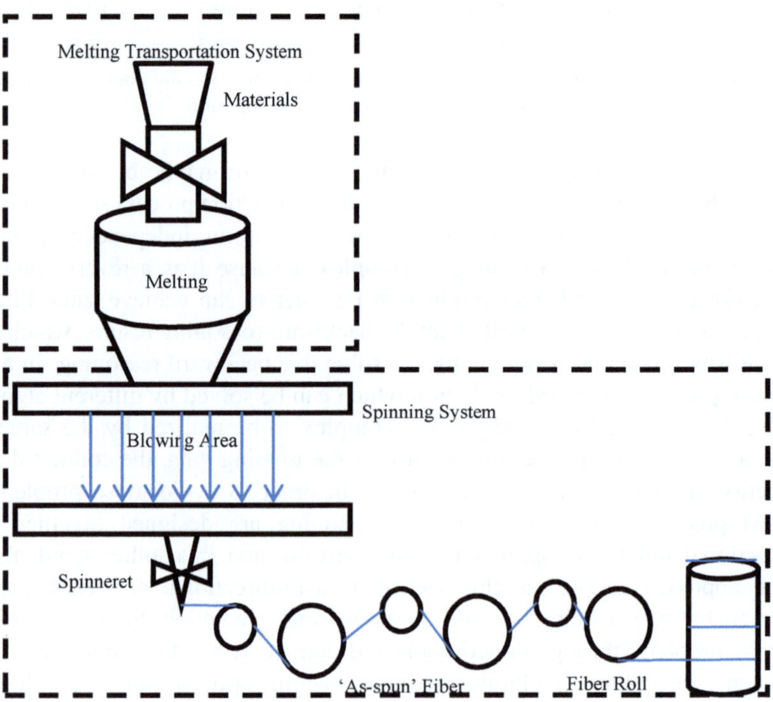

Fig. 9.17 The process of fiber production

In order to product fibers with good performance, the optimization of fiber production always depends on the adjusting and controlling of production parameters. Because the production line consists of several processes, so there are numerous production parameters needed to be determined in every process. In addition, all the processes have the interaction influences to each other. The traditional approaches which follow the experience are not accurate enough to solve the problems above, and the mathematic models are not adaptable enough to solve differential fibers. So, if we can find a way to achieve the bi-directional prediction results with hybrid intelligent algorithms, the production parameters can be determined and the fiber production can be optimized.

First of all, the selections of the key production parameters and fiber performance are the foundation of this section. At the beginning of the production, the materials will be melted into liquid according to a predefined viscosity and temperature, this process belongs to chemical categories, so it is not taken into account in this section. After this process, the materials will enter the blowing area and the stretching process. These two processes are the most important ones in this section, so all the parameters of these processes are considered into the bi-directional prediction approach.

In the bi-directional prediction approach, the prediction of fiber performance by production parameters is a forward prediction process and the prediction of production parameters by fiber performance is a backward reasoning process.

9.4.1.2 Fiber Production of Bi-Directional Prediction Approach

As mentioned above, the bi-directional prediction approach consists of the forward prediction and the backward reasoning.

1. **Forward prediction**

Because the fiber production is a positive process, so the forward prediction is a positive process too. It stands to reason that when the parameters of fiber production are changed, its corresponding fiber performance will be changed accordingly. In addition, the production is a long term process, which means that the predetermined production parameters usually need not be changed in once production. Because of the production condition and environment condition, there must be a small disturbance between the real production parameters and the presupposed production parameters. In other words, the real measured production parameters of the production at the same time are not the same, but they are similar. So if the production parameters are similar, there must be so little difference between the fiber performances. Meanwhile, though the production parameters of ever production are unstable, their fiber performance must all have incredibly small variations with each other, which means they all have the character of aggregation.

From the above analysis, the relationship between the production parameters and the fiber performance can be assumed as a many-to-many problem. Because there is a character of aggregation in the production parameters and their corresponding fiber performance, so this many-to-many problem can be simplified to a one-to-one problem by clustering algorithm. In order to solve this one-to-one problem, this section developed the neural networks based on clustering algorithm to fulfill the function of the forward prediction.

2. **Backward reasoning**

The prediction of production parameters by fiber performance is a reverse process, so the prediction process can be seen as a backward reasoning. Comparing with the forward prediction process, the backward reasoning process is more complex. Traditional prediction approaches will meet some problems as follows. First, the variety of fiber performance is more inadequate than the production parameters. Secondly, because the production has several processes and every process of the whole production is not independent of each other, so it is possible that the similar fiber performances can sometimes be achieved with different production parameters, which means that the backward reasoning cannot be simplified to be a one to one problem, it is a one-to-many problem. The one stands for the fiber performance and the many stands for the production parameters. Therefore, traditional neural networks based on clustering algorithm, which is used in the forward prediction process, are not suitable for the backward reasoning.

In order to solve the above problems, the backward reasoning consists of not only itself but also the clustering results and the forward prediction. First, we use the clustering results as the foundation of the backward reasoning. Since the clustering results are based on the production parameters, it can keep the variety of fiber performance without changing its structure and avoid the influences of the missing and conflict data. Secondly, in order to solve the one to many problem, the multi-objective evolutionary algorithm is used in it to find the optimal answer in the many, and the forward prediction are used to calculate the objectives.

Figure 9.18 is the overall chart of the bi-directional prediction approach.

9.4.1.3 Design and Improvement of the Bi-Directional Prediction Approach

The design and improvement of the bi-directional prediction approach will be divided into three parts, the clustering process, the forward prediction and the backward reasoning.

1. **Design of clustering process**

As mentioned above, the bi-directional prediction is composed of the forward prediction and the backward reasoning. The backward reasoning depends on the forward prediction and the forward prediction is based on the positive clustering result, so the clustering result is the most important point in the bi-directional prediction approach.

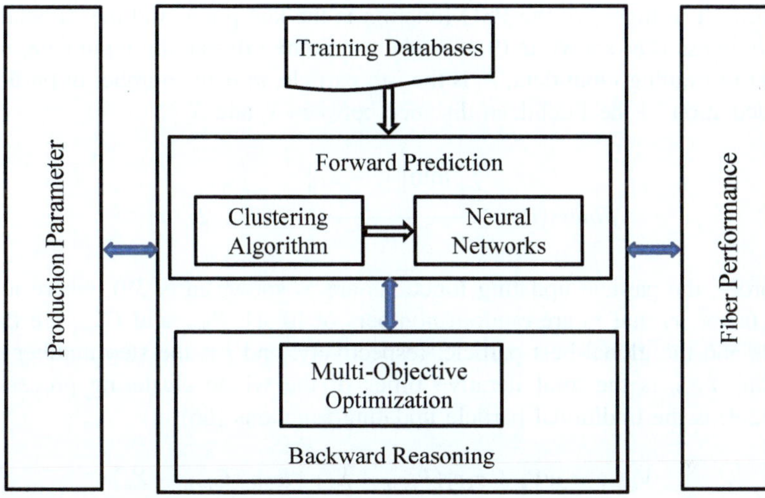

Fig. 9.18 The overall chart of the bi-directional prediction approach

Clustering algorithm is an important tool for data analysis and an unsupervised classification algorithm, which can classify the unlabeled data automatically. Up to now, a lot of algorithms have been proposed and applied in the clustering problems, but they all have a common problem that the results is the locally optimal solution. In order to solve this problem, this section uses some hybrid intelligent algorithms to optimize tradition clustering algorithms. Comparing with other clustering algorithms, the particle swam optimization algorithm (PSO) with its multifarious particle can get the global optimal solution more efficiently [66, 67], and the K-means algorithm is good at optimizing the local optimal solution [68, 69].

Firstly, the whole clustering process will be operated by every kind of fiber performance, which can optimize the clustering results. All the training data are pre-processed to be the same order as (9.36), this normalize approach can avoid the effects from the difference among different data kinds and increase the clustering efficiency and accuracy. Relieve the preprocess effect is by (9.37), where X^* is the initial data, and X is the processed data.

$$X = \frac{X - \min(X^*)}{\max(X^*) - \min(X^*)} \tag{9.36}$$

$$X = X \times (\max(X^*) - \min(X^*)) + \min(X^*) \tag{9.37}$$

Secondly, we use the PSO function and multifarious particles to find the preliminary result. The selection of initial particles of the PSO function is in the training data randomly, and the initial particles include the input data and the output data of every center. The selection in training data can avoid the generation of the invalid centers, and the including of the output data can avoid the error of the data mutations. The fitness of the PSO function is the key point, and the calculation of the *i-th* fitness is as shown in (9.38), where x_j is the j-th training input data, n is the number of training input data, X_i is the i-th particle, m is the number of particles. is the calculation of the Euclidean distance between x_j and X_i.

$$fitness(i) = \frac{\sum_{j=1}^{n} \min\left\|x_j - X_i\right\|^2}{n}, i = 1, 2 \ldots m \tag{9.38}$$

Thirdly, the particle updating functions are as shown in (9.39), where w is the speed factor, r_1 and r_2 are random numbers of [0, 1], P_{best} and G_{best} are the best particle and the global best particle, respectively, and t is the step number of this updating, n_{step} is the total iterative times of the whole clustering process. The Eq. (9.39) is the traditional particle updating functions [66].

$$\begin{aligned} V_{k+1} &= wV_k + c_1 r_1 (P_{best} - X_k) + c_2 r_2 (G_{best} - X_k) \\ w &= w_{\max} - (w_{\max} - w_{\min}) t / n_{step} \\ X_{k+1} &= X_k + V_{k+1} \end{aligned} \tag{9.39}$$

Through iterative analysis and calculation, all the particles will be directed to the best result, which means most of the particles will be similar to each other. When the similarity of particles reaches a certain high degree, the best particle is more near to the global optimal solution. But all the initial particles are different in centers' order to each other, so the similarity is difficult to judge by the particles themselves. However the fitness of every particle stands for both the cluster accuracy and its particle, so the similarity of the fitness can stand for the similarity of the particles. Then the K-means function is added to increase the accuracy of the optimal solutions and to optimize the clustering result.

In addition, mutation operator is used to make the particles far from the local optimal solution. Because mutation operator can lead the particles both to the better and to the worse, so the mutation operator is used on the particles except the optimal particle. Then the optimal particle can keep optimizing only.

Finally, we calculate the output centers according to the input centers and its training output data as shown in (9.40), where Y_j is the j-th output center, X_j is the j-th input center, x_i is the training input data which are clustered into the j-th input center, y_i is the corresponding training output data of x_i, n is the number of training data in the j-th center. The numerator of the coefficient calculation is the exponent of the distance between input data and input center, and its denominator is the total of all the distance factors. This can not only keep the inverse proportion between the coefficient and the distance, but also avoid the effect from the coefficients' order of magnitude. In this function, the closer the input data is to the input center, the bigger coefficient the corresponding output data will be weighted by, which means it can decrease the effect from some data which has the similar production parameters and the performance with large difference.

$$Y_j = \frac{\sum_{i=1}^{n} y_i \exp\left[-\frac{\|x_i - X_j\|^2}{2}\right]}{\sum_{i=1}^{n} \exp\left[-\frac{\|x_i - X_j\|^2}{2}\right]} \tag{9.40}$$

The steps of the clustering algorithm are as follows.

Step 1: Pre-process the training data by (9.36).

Step 2: Select the initial particles randomly in the training data and calculate the initial fitness by (9.38) of all the particles and the initial P_{best} and G_{best} with their particle as the local and global optimal one.

Step 3: Update the particles by (9.39) and make sure that all velocities of particles are less than V_{max}.

Step 4: Calculate the fitness of all the particles by (9.38) and update P_{best} and its local optimal particle. If $P_{best} < G_{best}$, then $G_{best} = P_{best}$, and update its globally optimal particle.

Step 5: Regard the P_{best} particle as the optimal particle, other particles have the mutation operator. The rate of mutation operator is 10%, and the mutations will change into another particle selected in the training data randomly.

Step 6: If all the steps are over, or two thirds of the fitness values are similar to the global best particle's, then use the K-means function to optimize the global best particle by 3 times, else return to Step 3.

Step 7: Calculate the output centers as (9.40) by the global best particle and output the clustering result. During the calculation, if there is no training data in one center, then this center will be deleted from the final clustering result.

2. **Design and improvement of the neural networks**

Traditional neural networks can do prediction, but this prediction is point prediction which conveys little information about the prediction accuracy [25, 26]. Radial basis function neural networks based on Gaussian function are used here to optimize the prediction accuracy and avoid the point prediction [70, 71]. There are some traditional difficulties in this kind of algorithm, the calculation of σ and the calculation of weight [9, 24].

Firstly, the value of σ in the radial basis function neural networks shows the width of each center, and the final fitness value of each clustering result shows the average distance of all the centers. So this section uses the final fitness value of each clustering result to calculate the value of σ by (9.41), where m is the kind number of fiber performance, and the factor 2 enlarges the width of every center by the average distances as the maximum distance.

$$\sigma(i) = 2 \times fitness(i),\ i = 1, 2 \ldots m \tag{9.41}$$

Secondly, the traditional function to calculate the weight of every output center relies on the value of σ. Although the calculation of σ is improved and simplified, in this section, it is an average value. In order to reduce the prediction errors, this section also improves the function of weight value, which can assure that the total value of all the weights is 1 and the prediction value is within the range. The traditional Gaussian function is as shown in (9.42), while the improved Gaussian function is as shown in (9.43), where w_{ij} is the j-th output center's weight of the i-th output kind, X is the input data, C_{ij} is the j-th input center of the i-th output kind, σ is a constant, m is number of output kind, and n is the clustering amount.

$$w_{ij} = \exp\left[-\frac{(X - C_{ij})^2}{2\sigma^2}\right], i = 1, 2\ldots m, \ j = 1, 2\ldots n \tag{9.42}$$

$$w_{ij} = \frac{\exp\left[-\frac{(X-C_{ij})^2}{2\sigma^2}\right]}{\sum_{j=1}^{N}\left[-\frac{(X-C_{ij})^2}{2\sigma^2}\right]}, i = 1, 2\ldots m, \ j = 1, 2\ldots n \tag{9.43}$$

The steps of the clustering algorithm are as follows.

Step 1: Preprocess the input production parameters by (9.36).
Step 2: Calculate the σ by (9.40) and the weights by (9.42).
Step 3: Add up the production of every center and its weight.
Step 4: Relieve the preprocess effect by (9.37) and get the prediction result of fiber performance.

3. **Design and improvement of the multi-objective evolutionary algorithm**

The backward reasoning is a one-to-many problem, in other words, there are more than one production parameter set can achieve the target fiber performance. As mentioned above, the foundation of the backward reasoning is the positive clustering result based on output classes. So every clustering result can be seen as a subset and every local optimal solution can be seen as a sub-goal. Multi-objective evolutionary algorithm is a good tool to solve this kind of problem and find the globally optimal solution [72, 77]. The NSGA-II [72, 73] algorithm is one of the hottest algorithms to solve this kind of problems in recent years.

The backward reasoning process based on multi-objective evolutionary algorithm is shown as follows. Figure 9.19 is the overall chart of the algorithm above.

Step 1: According to the input data in backward reasoning, select the high similarity output centers and their corresponding input centers in the positive clustering results by the classes of the output data and save the deviations between the input data and the centers. All of the corresponding input centers are candidate solutions.

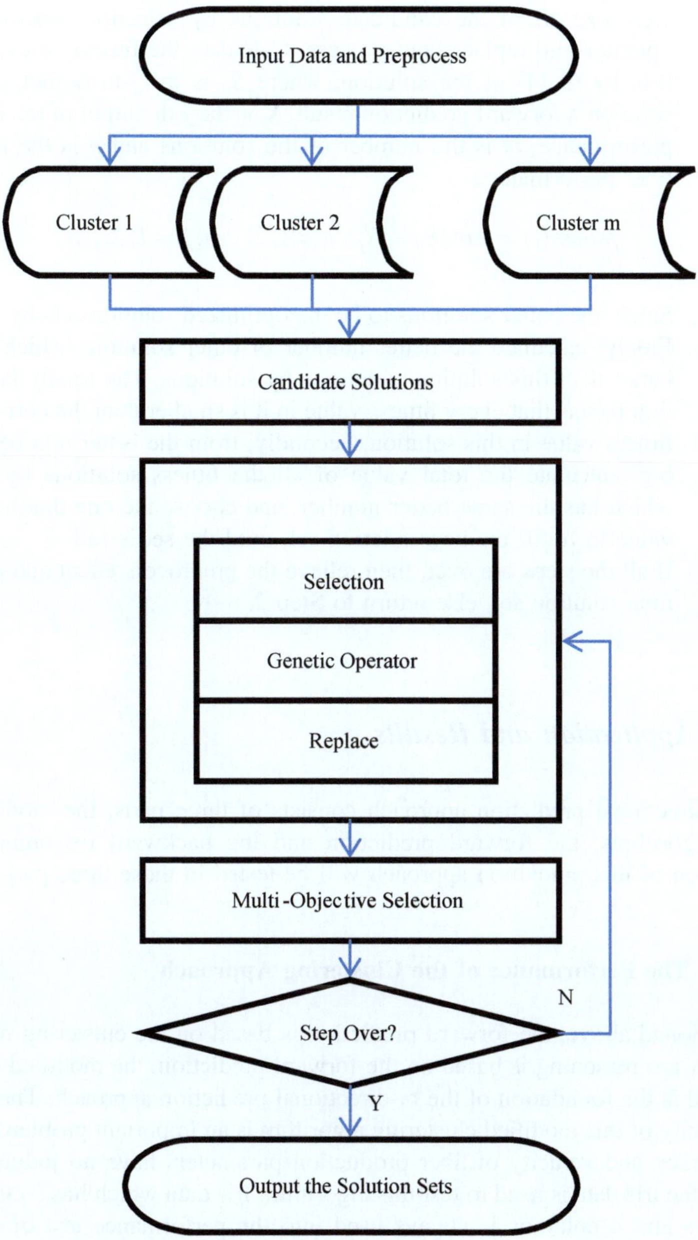

Fig. 9.19 The overall chart of backward reasoning

Step 2: Optimize all of the candidate solutions by selection process, genetic operator and replacement process. Calculate the fitness of every objection by (9.44) of the solution, where S_{ij} is the j-th output of the i-th solution's forward prediction result, X_j is the j-th output of the input fiber performance, m is the number of the solutions and n is the number of fiber performance.

$$fitness(j) = abs(S_{ij} - X_j), i = 1, 2...m, j = 1, 2...n \tag{9.44}$$

Step 3: Select the better solutions to be the optimized solution sets by two rules. Firstly, calculate the better number of other solutions which is totally better than this solution, solutions by solutions. The totally better solution means that every fitness value in it is smaller than the corresponding fitness value in this solution. Secondly, from the better number small to big, calculate the total value of all the fitness solutions by solutions which has the same better number, and choose the one that has smaller value to fulfill the new solution set, until the set is full.

Step 4: If all the steps are over, then relieve the preprocess effect and output the final solution set, else return to Step 2.

9.4.2 *Application and Results*

The bi-directional prediction approach consists of three parts, the modified clustering algorithms, the forward prediction and the backward reasoning. So the application of this prediction approach will be tested in those three parts.

9.4.2.1 The Performance of the Clustering Approach

As mentioned above, the forward prediction is based on the clustering results and the backward reasoning is based on the forward prediction, the modified clustering algorithm is the foundation of the bi-directional prediction approach. The accuracy and veracity of this modified clustering algorithm is an important problem. Because the accuracy and veracity of fiber production parameters have no judgment standard, so the iris data is used to test this algorithm. Iris data which has 3 classes, 150 data rows and 4 columns is always used into the performance test of clustering algorithm.

Table 9.12 lists the parameters of the algorithm applied in this section and Table 9.13 is the error list and the fitness of the proposed algorithm, traditional PSO algorithm and K-means algorithm. According to the results, the proposed algorithm

Table 9.12 Parameters of the algorithm applied

Item	Value
Learning factor (C_1,C_2)	1.5
Speed factor (w_{max})	0.9
Speed factor (w_{min})	0.4
Velocity maximum	0.5
Maximum step	100
Classify number	3
Particle number	10
Size of training input data	(150 × 4)
Size of training output data	(150 × 1)
Size of input centers	(3 × 4×1)
Size of output centers	(3 × 1)

Table 9.13 Clustering results of different algorithms

	PSO	K-means	This section
Error	15	17	6
Final fitness	0.67	0.53	0.27

has an accuracy and veracity at high level, and the clustering result by this algorithm can also optimize the prediction accuracy.

9.4.2.2 The Prediction Results of Polyester Staple Fiber

1. **Parameters selection**

As mentioned before, the most important sections of a fiber production is the blowing area and the spinning system, and the material melting is assumed to be predetermined in this section. The experiments are based on a 1.56dtex cotton-type polyester staple fiber with fully-closed blowing area fiber production line. Through many productions with different production parameters, there are a large number of production parameters and their corresponding fiber performances. Table 9.14 is a parameters list and value ranges of this fiber productions. In Table 9.15, the major variable production parameters are the spinning velocity (SV), the spinning temperature (ST), the blowing velocity (QV) and the blowing temperature (QT). In addition, the major variable fiber performance is the elongation corresponding to 1.5 times the yielding stress (EYS1.5), the coefficient of variance (EYSCV), the breaking tenacity (DT), and the ability of elongation (DE).

2. **Design of the bi-directional prediction approach**

All of the practical fiber data are divided into two parts. About 3000 of them are selected as training data and others are the testing data. Table 9.15 lists the parameters setting of different parts of the bi-directional prediction approach.

Table 9.14 Clustering results of different algorithms

	Item	Value
Fiber category	Fineness (*dtex*)	1.56
	Post-drawing ratio	3.6523
Equipment parameters	Non-quenching gap height (cm)	6
	Number of spinneret orifice	3064
	Diameter of spinneret orifice (cm)	0.0022
	Pump mass throughput (g/min · hole)	0.0097
Spinning parameters	Spinning velocity (m/min)	1000–1197
	Spinning temperature (°C)	280–299
	Characteristic viscosity (dl/g)	0.63
Blowing parameters	Blowing velocity (m/min)	100–139
	Blowing temperature (°C)	20–24
Fiber performance	EYS1.5	196.29–237.78
	EYSCV	5.46–10.04
	DT	5.82–6.81
	DE	20.94–24.05

Table 9.15 Parameters of the bi-directional prediction approach

	Item	Value
Forward prediction	Learning factor (C_1,C_2)	1.5
	Speed factor (w_{max})	0.9
	Speed factor (w_{min})	0.4
	Velocity maximum	0.5
	Maximum step	100
	Classify number	350
	Particle number	10
	Size of training input data	(3000 × 4)
	Size of training output data	(3000 × 4)
	Size of input centers	(350 × 4×4)
	Size of output centers	(350 × 4)
	Size of weights	(350 × 4)
Backward reasoning	Population size	40
	Generations number	1000
	Objectives number	5
	Variables number	4

3. **Results and analysis**

1. Clustering results

The clustering results are the foundation of forward prediction and backward reasoning. Its accuracy and rationality have directly influences on the bi-directional

Table 9.16 Clustering results of different algorithms

	PSO	K-means	This section			
			EYS1.5	EYSCV	DT	DE
Centers	213	160	343	345	344	344
Fitness	0.140	0.087	0.049	0.048	0.045	0.049

Table 9.17 The center number of different algorithms

Training data					Center no.		
SV	ST	QV	QT	DE	PSO	K-means	This section
0.12	0.06	0.99	0.46	0.62	6	23	153
0.01	0.27	0.99	0.59	0.74	6	23	169
0.04	0.22	0.74	0.46	0.68	6	23	234
0.09	0.11	0.99	0.46	0.66	6	23	305

prediction results. In order to prove that the clustering results are accurate and reasonable, other algorithms are used to compare with it. The clustering results, including the final classify number and the fitness are as shown in Table 9.16.

All the original classify numbers are set to 350. Through the clustering process, some centers will be deleted because they are too alike to other centers. In the proposed algorithm, the clustering process is repeated 4 times because the corresponding output data has 4 kinds. The final center number of the proposed algorithm is twice as much as that of the tradition algorithms. It can be said that the proposed algorithm keeps the variety of the centers. Table 9.17 lists the comparison of clustering results with different algorithms, some data classified in the same center by traditional algorithms, while by the proposed algorithm, it is classified in different centers, where all the data are pre-processed by (9.36).

In addition, the fitness of all algorithms is the total value of the difference of the training data and its corresponding center, so the smaller the fitness is, the well the clustering result is. It can be said that the proposed algorithm in this section has a good clustering result.

2. Forward prediction results

The fiber performances including EYS1.5, EYSCV, DT and DE can be predicted by forward prediction with the production parameters consisted of SV, ST, QV, and QT.

The production parameters as the input data are listed in Table 9.18, in which one parameter is regarded as the variable and the others are the invariant parameters. Figure 9.20 shows the actual values and the prediction results of different parameters respectively. Some prediction results in Fig. 9.20 may have a bigger error than others, but it mostly because of the y-range, so the forward prediction achieves the good results.

Table 9.18 Prediction parameters to input

No.	SV	ST	QV	QT
(a)	Variable	290	134	21
(b)	1050	Variable	130	20
(c)	1140	290	Variable	21
(d)	1050	280	130	Variable

In order to prove that the forward prediction algorithm has a good result generally and it can be adapted in differential fibers, we use more data sets from the experiment above and another experiment data to test it. The training data is about 3000 data sets, and the testing data is about 300 data sets. Table 9.19 is the average error of both fibers. The errors of the differential fibers can be acceptable, and these results can lay a well foundation for the backward reasoning.

3. Backward reasoning results

The production parameters including SV, ST, QV and QT can be predicted by backward reasoning with the fiber performances consisted of EYS15, EYSCV, DT and DE.

From Fig. 9.20, it can be observed that there is a variation tendency in every relationship between production parameters and fiber performance. For example, the increase of ST and QT will increase EYS15 and DE, while reduce EYSCV and DT. The increase of SV and ST will increase DE, while reduce EYS15 and DE. EYSCV will be increased by SV and reduced by QV. This also shows the backward reasoning is a one-to-many problem.

In this section, there are four kinds of fiber performance and there is no special order for each performance. So four objective function values will be averaged. Table 9.20 lists several backward reasoning results of the 1.56dtex cotton-type polyester staple fiber with fully-closed blowing area fiber production line. All of the experiment data is based on the practical fiber production, most of the backward reasoning errors are below 5%, and the average error is around 2%. As mentioned before, the fibers of similar performances can have different production parameters. Table 9.21 lists the backward reasoning of similar performances. The top part of Table 9.21 lists some similar performances and their corresponding parameters. There are differences of output among N1, N2 and N3, which have similar input. After giving the input of N1, because backward reasoning has not only converging speed but also diversity, it will have several solutions of one problem when it is necessary. Finally, the solutions of backward reasoning include S1, S2 and S3. S1 is the best solution to this prediction, however, S2 and S3 are also feasible answers, because following the N2 and N3, these two production parameters can achieve similar fiber performance. Table 9.21 shows that the proposed algorithm does well in this backward reasoning.

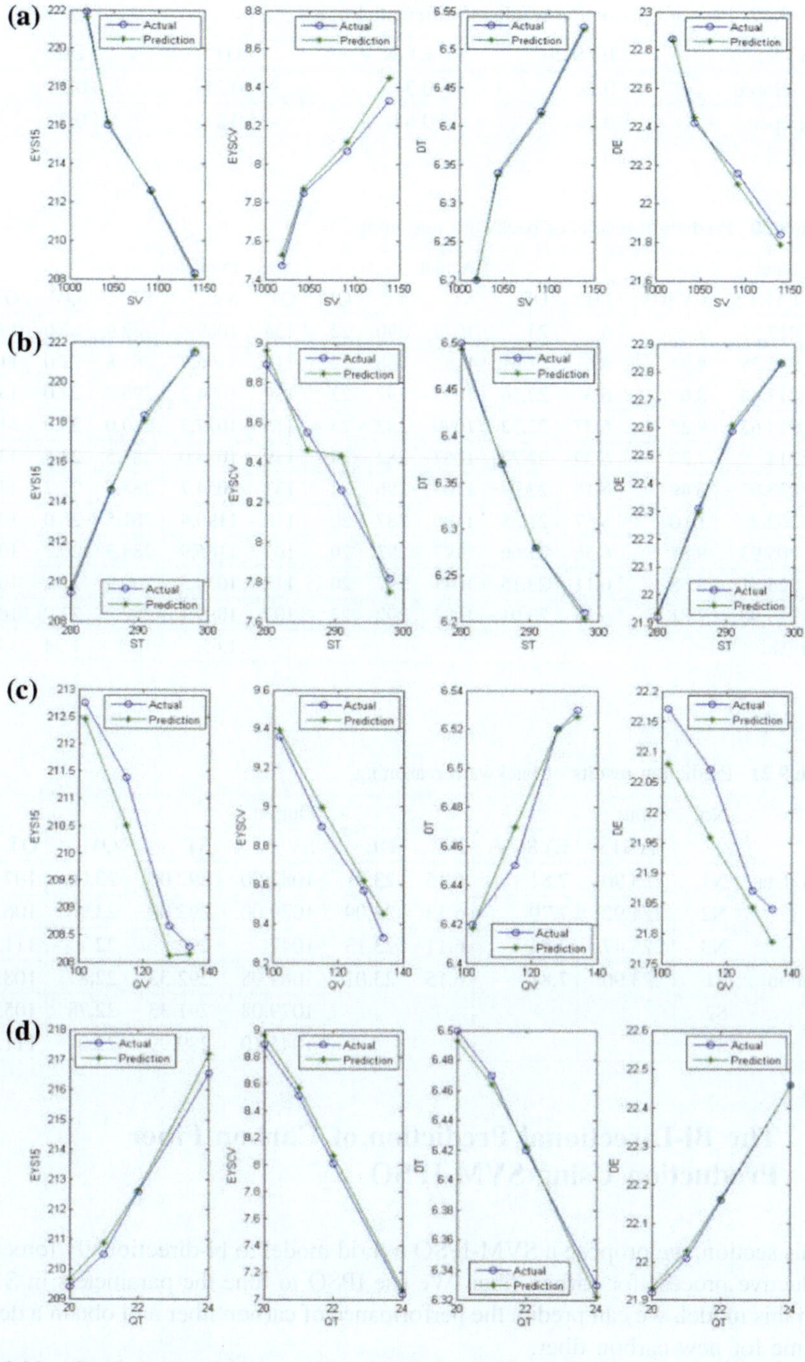

Fig. 9.20 **a** Prediction results rely on SV. **b** Prediction results rely on ST. **c** Prediction results rely on QV. **d** Prediction results rely on QT

Table 9.19 Forward prediction results of differential fibers

Error (%)	EYS15	EYSCV	DT	DE
Fully-closed	0.28	0.74	0.21	0.20
Semi-open	0.26	0.64	0.22	0.20

Table 9.20 Prediction results of backward reasoning

	Input				Actual				Predict			
	EYS1.5	EYSCV	DT	DE	SV	ST	QV	QT	SV	ST	QV	QT
1	217.31	7	6	23	1074	290	22	134	1069.8	289.6	22.0	136.5
2	205.28	8.94	6.6	21.61	1182	286	22	117	1196.7	285.8	22.0	117.2
3	217.84	8.6	6.3	22.56	1095	297	22	139	1094.2	296.8	22.0	138.7
4	214.63	8.25	6.37	22.32	1090	282	23	113	1077.3	280.0	23.0	113.7
5	214	7.23	6.39	22.27	1057	282	23	139	1078.0	283.5	23.5	138.6
6	223.62	6.49	6.16	22.99	1103	296	24	131	1054.7	288.7	23.7	137.0
7	202.2	10.04	6.67	21.38	1190	287	20	110	1190.4	281.5	20.0	106.3
8	205.93	9.87	6.58	21.66	1197	292	20	107	1165.9	284.3	20.2	106.3
9	225.81	8.38	6.11	23.15	1041	297	20	112	1030.5	290.9	21.1	108.8
10	223.90	7.81	6.15	23.01	1087	292	23	107	1089.1	292.3	22.9	108.3
Error (%)									1.65	1.11	1.34	1.85

Table 9.21 Prediction results of backward reasoning

	No.	Input				Output			
		EYS15	EYSCV	DT	DE	SV	ST	QV	QT
Real data	N1	223.90	7.81	6.15	23.01	1087.00	292.00	23.00	107.00
	N2	224.92	7.79	6.13	23.09	1079.00	292.00	23.00	106.00
	N3	225.47	7.86	6.11	23.13	1047	292	22	111
Solution	S1	223.90	7.81	6.15	23.01	1089.08	292.33	22.87	108.34
	S2					1079.08	291.33	22.78	105.34
	S3					1045.10	289.99	22.04	111.93

9.5 The Bi-Directional Prediction of Carbon Fiber Production Using SVM-IPSO

In this section, we propose a SVM-IPSO hybrid model to bi-directionally forecast a productive process for carbon fiber. We use IPSO to tune the parameters in SVM. With this model, we can predict the performance of carbon fiber and obtain a design scheme for new carbon fiber.

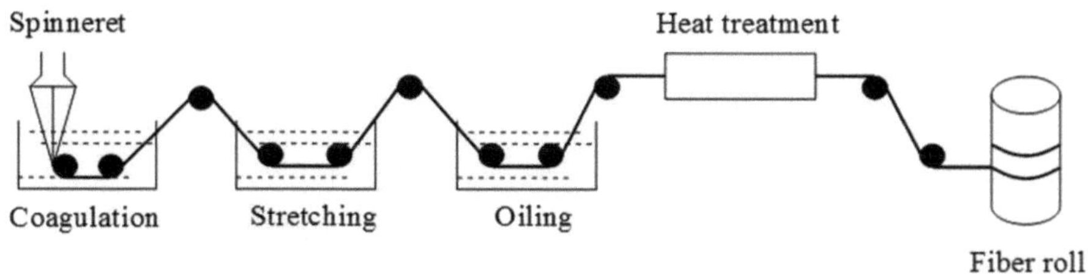

Fig. 9.21 The technological process of carbon fiber production

9.5.1 Carbon Fiber Production and Its Bi-Directional Optimization

9.5.1.1 The Process of Carbon Fiber Production

The technological process for carbon fiber is a sufficiently complex system. It is a very meaningful research subject to control the performance of the carbon fiber process in as small a range as we possibly can. For this purpose, we should observe the parameters of the productive process in real-time, synchronously predict the product properties of the carbon fiber, after which we can determine the defective carbon fiber, timely adjust the parameters of the productive process, avoid reducing the quality of the carbon fiber, and prevent significantly wasting of raw materials. The production of carbon fiber mainly comprises several processes, such as coagulation, stretching, oiling, and heat treatment. Figure 9.21 shows the main technological process of carbon fiber production.

9.5.1.2 The Bi-Directional Prediction Methods

The bi-directional prediction methods include forward prediction and backward prediction. Forward prediction is the process of predicting property indexes from productive parameters. It means that when the productive parameters of carbon fiber are changed, the corresponding performance of carbon fiber will fluctuate accordingly. In other words, if the productive parameters have little differences, there must be the same among the carbon fiber performances. The backward prediction is a reverse process of the forward prediction, so the process can be seen as a method for designing new types of carbon fiber. In the early stage, we design a new kind of carbon fiber through previous experience, then we test this kind of carbon fiber with chemical and physical instruments to verify its expected performance, get inconsistent results, adjust the corresponding parameters in a new scheme, produce carbon fiber products again, repeat this process of production, testing, and adjusting parameters until we get a satisfactory performance of the carbon fiber. From the foregoing, this method wastes a lot of raw materials, energy, and time. The worst thing is that, if we choose inappropriate parameters for the manufacturing

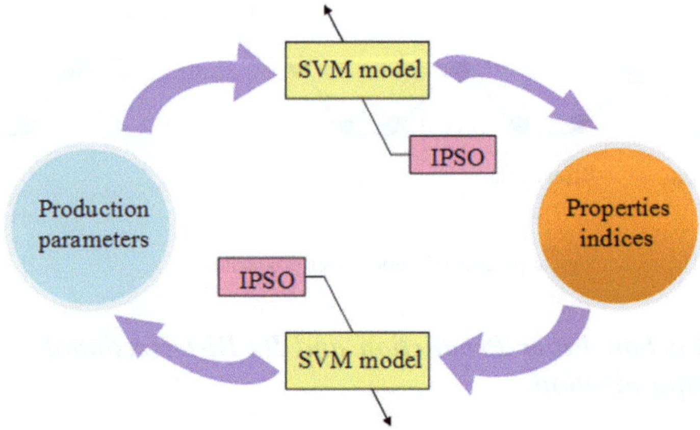

Fig. 9.22 Bidirectional prediction model of carbon fiber production

equipment, it will cause damage to the corresponding equipment. For this reason, a better way to solve this problem is to use computer simulation in the whole productive process, avoid unilateral decisions made by workers with interests and experience, and give out more reasonable control parameters based on the actual production process.

In this section, we propose a bi-directional prediction model of carbon fiber production based on SVM. We adopt IPSO to tune the parameters of the model. The overall chart of bi-directional prediction approaches are as shown in Fig. 9.22.

9.5.2 *SVM-IPSO Model*

9.5.2.1 The SVM Model

The SVM, proposed by Vapnik in 1995 [74], is widely used for many machine learning problems such as bio-informatics, pattern recognition, linear regression analysis and nonlinear regression estimation problems.

The basic principle is as follows [75]: given a training data $\{x_i, y_i\}_{i=1}^{N}$, where $x_T \in R^n$ is the input vector and $y_i \in R$ is the corresponding output of input x_i, and N is the total number of the input vectors. By nonlinear mapping $\varphi(x)$, the data are mapped nonlinearly from the original feature space to a high-dimensional feature space, thus, it can be approximated in a linear way as follows:

$$f(x) = w^T \varphi(x) + b \tag{9.45}$$

where $f(x)$ denotes the regression estimate function; w denotes the weight vector; and b denotes the bias vector.

As mentioned above, the problem of regression approximation is equivalent to minimizing the empirical risk,

$$R(f) = \frac{1}{2}\|w\|^2 + \frac{C}{N}\sum_{i=1}^{N} \Theta_\xi(y_i, f(x)) \tag{9.46}$$

where C and ξ are regularized constants; $\theta_\xi(y_i, f(x))$ denotes the ξ-insensitive loss function and is defined as Eq. (9.47)

$$\Theta_\xi(y_i, f(x)) = \begin{cases} |f(x) - y_i| - \xi, & if\ |f(x) - y_i| \geq \xi \\ 0, & otherwise \end{cases} \tag{9.47}$$

Then, we introduce two positive slack variables ε_i' and ε_i to simplify the risk function R(f) in Eq. (9.46). Thus, Eq. (9.46) is transformed into the objective function shown as follows:

$$Min_{w,b,\varepsilon',\varepsilon} R = \frac{1}{2}\|w\|^2 + C\sum_{i=1}^{N} (\varepsilon_i' + \varepsilon_i) \tag{9.48}$$

with the constraints

$$\begin{cases} y_i - w^T\varphi(x_i) - b \leq \xi + \varepsilon_i', & i = 1, 2\ldots, N \\ -y_i + w^T\varphi(x_i) + b \leq \xi + \varepsilon_i, & i = 1, 2\ldots, N \\ \varepsilon_i' \geq 0, & i = 1, 2\ldots, N \\ \varepsilon_i \geq 0, & i = 1, 2\ldots, N \end{cases} \tag{9.49}$$

where ε_i is the lower training error (ε_i' is the upper) subjected to the ξ-insensitive loss function.

After the convex quadratic programming problem with constraints is solved, the weight vector w in Eq. (9.45) is obtained as Eq. (9.50),

$$w = \sum_{i=1}^{N} \beta'_i - \beta_i)\varphi(x_i) \tag{9.50}$$

where β'_i, β_i are the Lagrangian multipliers and are obtained by solving a convex quadratic optimization.

Finally, the regression function is obtained as the following equation:

$$f(x) = \sum_{i=1}^{N} (\beta'_i - \beta_i)K(x_i, x_j) + b \tag{9.51}$$

where K(x_i, x_j) is called the kernel function, which is the inner product of $\varphi(\mathrm{x}_i)$ and $\varphi(\mathrm{x}_j)$, that is, $\mathrm{K}(\mathrm{x}_i, \mathrm{x}_j) = \varphi(\mathrm{x}_i) \cdot \varphi(\mathrm{x}_j)$. The values of $\varphi(\mathrm{x}_i)$ and $\varphi(\mathrm{x}_j)$ are produced by mapping $x_i$and $x_j$into the higher-dimensional feature space, respectively. It can be shown that any function meeting Mercer's criteria [74] can be used as the kernel function.

However in general, three different types of kernel functions are commonly used, e.g., the sigmoid kernel function, the polynomial kernel function, and the Gaussian RBF kernel function. The Gaussian RBF is a most used for the kernel function because of the fewer free variables that need to be set and an excellent nonlinear forecasting performance [76]. Therefore, in this study, the Gaussian RBF was selected (as shown in Eq. (9.52)) as the kernel function:

$$K(x_i, x_j) = \exp\left(\frac{-\|x_i - x_j\|^2}{2\sigma^2}\right) \tag{9.52}$$

Therefore, there are two variables that need to be selected in the SVM model, which are the constant "C" and the width of the Gaussian RBF kernel "σ". In this section, IPSO is used to optimize the parameters of the SVM model.

9.5.2.2 Overview of Particle Swarm Optimization

The PSO is inspired by simulation of the behavior of birds flocking in nature [64, 77]. It has won high reputation in numerical function optimization due to its precision and efficiency in the fields of stochastic optimization algorithms. It solves a problem by initializing a random population of candidate solutions (particles) and can search for the optimal solution through the search space. Each particle is characterized by a dynamic adjustable velocity, which originates from the momentum of a flying particle. Giving the PSO sufficient numbers of iterations, it is hoped to end up with a global optimal solution. The above-mentioned behavior of the i-th particles can be mathematically expressed as:

$$V_{iD}^{t+1} = \omega \times V_{iD}^{t} + c_1 \times r_1 \times (pbest_{iD}^{t} - X_{iD}^{t}) + c_2 \times r_2 \times (gbest_{iD}^{t} - X_{iD}^{t}) \tag{9.53}$$

$$X_{iD}^{t+1} = X_{iD}^{t} + V_{iD}^{t+1} \tag{9.54}$$

In the above formulas, ω denotes the inertia weight, V_{iD}^{t} denotes its velocity vector of the i-th particle on dimension D; $pbest_{iD}^{t}$ denotes its previously best known position of the i-th particle on dimension D and $gbest_{iD}^{t}$ denotes its best global position, c1 and c2 are social and personal learning parameters, r1 and r2 are random numbers in the range [0,1]; and X_{iD}^{t} denotes its previous position of the i-th particle at time t.

9.5.2.3 Improved Particle Swarm Optimization

1. **The Basic Concept of the Cell Communication**

Cell communication refers to a message through a medium to another cell and interacts with the corresponding receptors of target cells, and then through a series of physiological changes inside the cell. Indeed, cells cannot live in isolation, and survive by processing and receiving information from the external environment, whether that information suits the availability of nutrients and change in temperature. Cell to cell communication is essential for orchestration and coordination of cellular events in multi-cellular systems [78]. Intercellular communication has three main ways: gap junction, cell recognition, and chemical communication.

Gap junction: narrow water-filled channels that connect the cytoplasm of adjacent epithelial cells, as well as of some other cell types.

Cell recognition: mutual identification among cells, usually by particular complementary interaction among their separate membrane glycoproteins or surface molecules.

Chemical communication: cells secrete some chemicals (such as hormones) to outside the cell, as a signal molecule in target cells, adjusts its function, is specific and can be divided into four forms: (a) endocrine; (b) paracrine; (c) synaptic; (d) autocrine.

2. **IPSO Based on the Communication Mechanism of Cells**

Although PSO has been applied to many kinds of practical problems, PSO still suffers to some degree of premature convergence and poor quality of solution. To overcome the shortcomings, we present IPSO by combining the cell communication mechanism. Based on the three main ways of cell communication, three evolutionary strategies are established. Specifically, the gap junction, the cell recognition, the chemical communication in the mode of cell communication can be replaced by evolutionary strategies, Eqs. (9.59)–(9.61) in the iterative and evolutionary process, respectively. IPSO applies modified update formula to maintain population diversity and enhances the convergence velocity and precision, which by incorporating the information of the global best solution into the search strategy improves the exploitation. In particular, the modified update formulas are determined as follows.

A solution of problem P_i with D variables can be expressed as:

$$P_i = [p_{i1}, p_{i2}, \ldots, p_{iD}] \quad i = 1, 2 \ldots m \tag{9.55}$$

An intuitive illustration of how the Euclidean distance is measured is given in Fig. 9.23. As can be seen from Fig. 9.23, the first line represents the Euclidean distance between the 1-th solution to all solutions, the second line represents the Euclidean distance between the 2-th solution to all solutions, and so forth. The Euclidean distance between solutions of problem can be defined as:

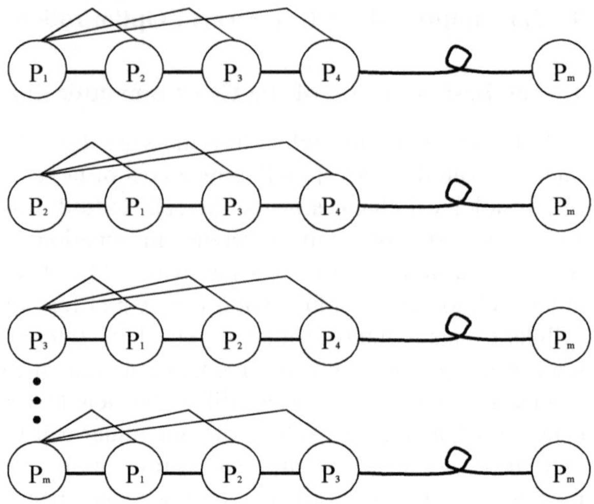

Fig. 9.23 Schematic drawing to show distances with different solution

$$d_{i,j} = \sqrt{\sum_{k=1}^{D} (p_{ik} - p_{jk})^2} \tag{9.56}$$

$$d_{i,j} = \begin{cases} 1, & \text{if } \mathrm{d}_{i,j} \leq V_{\max} \\ 0, & \mathit{otherwise} \end{cases} \tag{9.57}$$

The concentration of the particles can be expressed as:

$$C_k = \frac{N_k}{N} \tag{9.58}$$

where N_k denotes the total number of similar particles, and N denotes the total number of particles.

In this section, two parameters of SVM are optimized, so, each particle's position is checked using the following formula:

If ($C_k > 0.5$):

$$p_{i,1} = x_{b1} \tag{9.59}$$

where p_{i1} denotes the first value of the i-th particle; and xb1 denotes the first value of the best particle.

If ($f_i' = f_i$):

$$p_{i,2} = x_{b2} \tag{9.60}$$

where f_i denotes the fitness function value; f_i' denotes the updated fitness function value; $p_{i,2}$ denotes the second value of the i-th particle, and x_{b2} denotes the second value of the best particle.

The last particle's position is checked using the following formula:

$$P_i = x_b \tag{9.61}$$

where P_i denotes the *i*-th particle; and x_b denotes the best particle of the whole particles.

The values of the aforementioned parameters are usually prescribed to be selected as follows: ω = 0.8 and c1 = c2 = 2. Furthermore, ω denotes the particle's search capability, and defines the balance between local search capabilities and global search capabilities. Once the velocity of the particle is calculated, its new position is updated according to Eq. (9.54). The range of permissible velocity of the particle is usually restricted in order to eliminate fluctuations and obtain better exploration during the iteration process.

9.5.2.4 SVM Based on the IPSO

The diagram of SVM-IPSO forecasting model is illustrated in Fig. 9.24.

The details of IPSO for selecting parameters for the SVM model are as follows:

Step 1: Initialization parameters

The population size is 20, the maximum iteration number is 100, the initial particle swarm location is X, and the random flight velocity is V should first be determined. In the SVM-IPSO program, we set

X = rand(1,2), where rand() is a random function in the range [0,1].

Step 2: Evolution starts

Set gen = 0, and give the random position and the flight velocity for food finding of an individual particle i. In the SVM-IPSO program, we use two variables [x1, x2] to denote the new position for food finding of an individual particle i.

Step 3: Preliminary calculations

Calculate the new position X_i of an individual particle i, and then input X_i into the SVM model for carbon fiber production forecasting; the parameters [C, σ] of the SVM model are replaced by [$X_i(1)$, $X_i(2)$]. According to the carbon fiber production forecasting result, the fi can be calculated. The fi is represented by the root-mean-square error, as shown in Eq. (9.62), which calculates the errors between the actual values and the forecasted values:

$$f_i = \sqrt{\frac{1}{n}\sum_{i=1}^{n}(y_i - \hat{y}_i)^2} \tag{9.62}$$

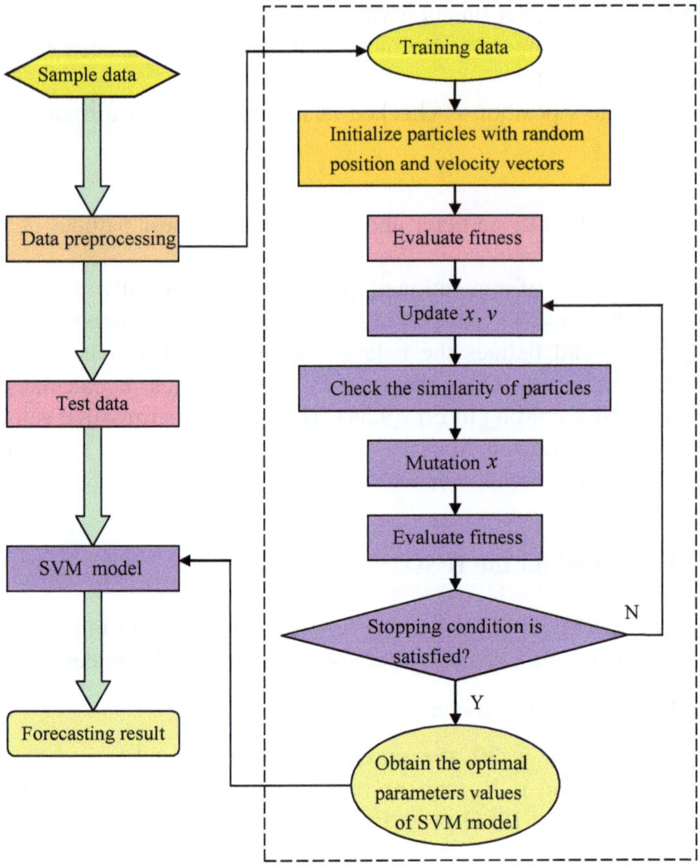

Fig. 9.24 Diagram of the support vector machine and improved particle swarm optimization (SVM-IPSO) forecasting model

where n is the number of training input data; y_i is the i-th actual value; $\hat{y}_i$ denotes the i-th forecasting value.

Step 4: Offspring generation

The global best value is generated according to Eqs. (9.59)–(9.61) after updating the position value according to Eq. (9.10). Then the global best value is inputted into the SVM model and the fitness function value calculated again. Set gen = gen + 1.

Step 5: Circulation stops

When gen is equal to the maximum iteration number, and the stop criterion is met, the optimal parameters of the SVM model are finally obtained. Otherwise, it is necessary to go back to Step 2.

9.5.3 Simulation and Discussion

9.5.3.1 The Preprocessing of Sample Data

In order to examine the performance of the new prediction model, we used the data organized from experiments in Table 9.22 [2]. The errors by the proposed method for the below mentioned experimental data, classified into four different types are given in Table 9.23, where to facilitate comparison the corresponding errors obtained by other methods on the same data are also listed.

In the SVM modeling, the preprocessing steps are implemented before attempting to calculate, to eliminate any bad data or outliers. These steps guarantee that the raw data retrieved from the database is completely suitable for modeling. In order to improve the accuracy of prediction and smooth the training procedure, all the sample data are normalized to fit them in the interval [0,1] using the following linear mapping formula:

$$X' = \{x_i'\} = \frac{x_i - x_{\min}}{x_{\max} - x_{\min}}, \quad i = 1, 2, 3..., N \tag{9.63}$$

where x_i' is the mapped value; x_i is the initial value from the experimental data; N is the total number of input data; xmax and xmin denote the maximum and minimum values of initial data, respectively.

In Table 9.22, the values for the first six columns indicate different parameters, and the values for the last two columns indicate different property indices. These detailed parameters can synthetically represent the whole manufacturing process. Meanwhile, the two kinds of property indices can fully represent the performance of the carbon fiber. In the predictive process, we obtained training sets from the first 45 rows of samples and test sets from the last five rows of samples. Also, to examine the performance of the several algorithms, the mean absolute error, root mean square error, mean relative error and Theil's inequality coefficient were defined according to the following formulations, respectively:

Mean absolute error:

$$MAE = \frac{1}{n}\sum_{i=1}^{n}\left|y_i - \hat{y}_i\right| \tag{9.64}$$

Root mean square error:

$$RMSE = \sqrt{\frac{1}{n}\sum_{i=1}^{n}(y_i - \hat{y}_i)^2} \tag{9.65}$$

Table 9.22 Experimental data of carbon fiber in the productive process

No.	Viscosity average molecular weight (104)	Conversion ratio (%)	Solid content (%)	Spinning jet drawing ratio (%)	Coagulating bath temperature (°C)	Total drawing ratio	Strength (CN/d)	Structure parameter
1	8.9	94.5	20.8	−50.3	14	6.33	4.08	14.82
2	6.3	91.0	20.0	−59.7	15	5.89	3.23	12.63
3	11.6	92.0	20.4	−50.5	14	6.03	3.76	13.24
4	8.8	94.8	21.8	−63.4	13	6.65	4.17	17.24
5	7.0	81.8	17.9	−63.4	15	6.32	3.99	15.14
6	8.2	85.5	21.7	−59.5	15	5.49	4.58	16.61
7	7.2	89.8	19.5	−53.1	13	5.88	3.64	15.49
8	8.9	82.5	17.5	−56.8	19	6.38	4.07	17.57
9	8.0	83.4	18.6	−62.1	17	5.72	3.18	15.48
10	11.7	90.6	17.9	−53.8	16	6.47	3.22	12.10
11	11.5	82.8	18.7	−64.8	17	5.79	3.27	12.73
12	6.3	95.1	19.6	−54.9	16	6.37	4.36	17.18
13	10.4	98.6	20.2	−68.3	17	6.41	3.99	14.91
14	7.6	93.1	19.7	−55.4	16	5.88	3.38	17.07
15	8.5	84.6	22.3	−65.3	18	5.04	3.99	13.26
16	9.3	89.8	20.1	−53.8	16	5.66	3.30	15.31
17	11.7	79.4	22.7	−55.8	19	5.85	3.11	15.78
18	8.5	96.9	20.8	−51.8	14	5.54	4.70	12.19
19	11.9	96.4	22.7	−61.5	13	5.39	4.12	15.69
20	7.8	95.0	18.4	−63.7	13	6.64	4.86	14.17
21	10.2	82.7	21.1	−60.9	13	5.86	4.39	12.30
22	10.0	90.1	18.7	−58.5	15	6.78	4.17	14.94

(continued)

Table 9.22 (continued)

No.	Viscosity average molecular weight (104)	Conversion ratio (%)	Solid content (%)	Spinning jet drawing ratio (%)	Coagulating bath temperature (°C)	Total drawing ratio	Strength (CN/d)	Structure parameter
23	9.2	77.5	21.0	−62.9	16	5.78	4.63	13.16
24	10.2	86.4	21.2	−63.0	15	6.54	4.76	12.74
25	10.0	83.9	17.4	−63.6	18	5.79	4.98	13.23
26	7.1	80.6	18.5	−62.7	17	6.62	3.00	12.88
27	6.8	80.9	18.3	−68.9	18	6.51	4.73	13.13
28	12.0	86.3	21.0	−54.2	19	5.75	4.23	12.26
29	7.0	79.1	22.1	−64.2	19	5.43	4.98	15.81
30	6.2	90.2	19.1	−54.7	14	6.58	4.06	13.69
31	9.4	87.4	21.7	−52.4	13	6.90	3.96	15.23
32	11.3	92.3	21.1	−62.1	17	5.66	4.60	16.17
33	10.0	92.4	17.0	−59.0	13	6.34	3.46	14.99
34	7.1	91.0	20.6	−59.2	16	5.88	4.00	15.21
35	8.2	77.7	19.3	−63.2	16	6.67	4.80	14.67
36	8.8	78.5	22.5	−65.4	19	6.54	4.15	12.74
37	11.9	84.0	17.0	−57.0	16	5.33	4.69	14.94
38	6.9	88.7	19.8	−63.2	15	6.72	4.48	17.12
39	11.1	91.4	19.5	−58.3	17	6.98	4.17	17.24
40	9.9	86.0	19.8	−66.8	18	6.03	3.49	13.62
41	8.3	95.0	21.6	−66.7	16	6.77	4.33	13.25
42	7.1	92.8	18.9	−55.1	15	6.18	3.17	15.39
43	8.6	98.3	21.7	−62.3	14	5.31	4.25	15.84
44	8.9	88.7	19.8	−61.6	17	5.40	4.32	14.50

(continued)

Table 9.22 (continued)

No.	Viscosity average molecular weight (104)	Conversion ratio (%)	Solid content (%)	Spinning jet drawing ratio (%)	Coagulating bath temperature (°C)	Total drawing ratio	Strength (CN/d)	Structure parameter
45	6.7	84.2	17.2	−60.8	14	5.81	4.46	13.24
46	11.0	98.0	22.7	−74.6	12	6.89	3.90	12.82
47	9.5	92.2	20.3	−50.5	17	5.92	3.91	13.20
48	7.7	78.2	17.2	−63.4	19	6.17	3.99	15.19
49	9.5	79.3	18.1	−67.4	13	6.50	4.78	17.69

Note Sample data sources come from reference [9]

Table 9.23 Comparison of errors among the proposed method, the GA-IPSO-RNN, the basic PSO-RNN, and the conventional RNN

Algorithms		Conventional RNN	Basic PSO-RNN	GA-IPSO-RNN	Proposed method
MAE	1	1.1950	0.4818	0.4258	0.3839
	2	3.6827	2.0262	1.9833	1.7821
	Mean	2.4389	1.2540	1.2045	1.0830
MRE (%)	1	28.63	10.65	9.39	8.71
	2	27.96	14.61	14.01	12.41
	Mean	28.30	12.63	11.70	10.56
RMSE	1	1.4843	0.5841	0.5157	0.4076
	2	4.5364	2.2637	2.1177	1.9649
	Mean	3.0104	1.4239	1.3167	1.1863
TIC	1	0.1675	0.0690	0.0609	0.0481
	2	0.1452	0.0766	0.0727	0.0679
	Mean	0.1563	0.0728	0.0668	0.0580

Notes 1. strength, 2. structure parameter

Mean relative error:

$$MRE = \frac{1}{n}\sum_{i=1}^{n}\left|\frac{y_i - \hat{y}_i}{y_i}\right| \tag{9.66}$$

Theil's inequality coefficient:

$$TIC = \frac{\sqrt{\sum_{i=1}^{n}(y_i - \hat{y}_i)^2}}{\sqrt{\sum_{i=1}^{n}(y_i)^2} + \sqrt{\sum_{i=1}^{n}(\hat{y}_i)^2}} \tag{9.67}$$

where y_i is the true value; $\hat{y}_i$ is the predicted value.

9.5.3.2 The Selection of Comparative Models

As mentioned above, the technological process for carbon fiber is a nonlinear system, it involves many process parameters that must be carefully optimized and adjusted. Most of the previous work was focused on analyzing the properties of carbon fibers by means of physical or chemical instruments. However it is a kind of method which wastes time and energy. Thus, we should build the predictive model for the productive process of carbon fiber. In this way, we can effectively predict the properties of carbon fiber in real-time, and thus we can find out inferior carbon fiber

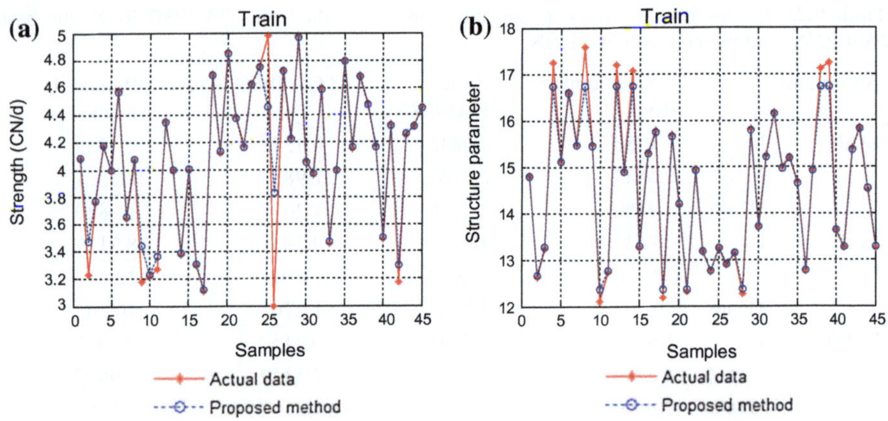

Fig. 9.25 Training results of the proposed SVM-IPSO model. **a** Strength. **b** Structure parameter

with poor quality in advance, timely adjust the process parameters, and avoid wasting raw materials. However, on the contrary, the prediction of productive parameters by the properties of the carbon fiber is a backward reasoning process. Therefore, it is a better idea to analyze all antecedent production schemes through computer simulation. This will avoid partial production schemes created by operational staff who rely on their personal experience and interests.

9.5.3.3 Forward Prediction

Carbon fiber production is a positive process, and the predictable performance of carbon fiber by productive parameters is a forward prediction, too. Thus, in forward prediction, we can take the parameters of carbon fiber production as input and the corresponding performance of carbon fiber as output. By employing IPSO to optimize the parameters of SVM, we gain the best parameters for SVM. The training effectiveness of the proposed method is shown in Fig. 9.25.

Comparing performance of the conventional RNN, the basic PSO-RNN, the GA-IPSO-RNN with the SVM-IPSO is carried out in terms of (a) mean absolute error; (b) root mean square error; (c) mean relative error and (d) Theil's inequality coefficient result. The detailed results are listed in Table 9.23.

From Table 9.23, we can see that on the basis of MAE, the proposed SVM-IPSO achieves roughly a decrease of 55.59, 13.64 and 10.09% compared to the other three methods, respectively. Similar results can be obtained from the other errors such as MRE with a decrease of 62.69, 16.39 and 9.74%, RMSE with a decrease of 60.59, 16.69 and 9.90%, and TIC with a decrease of 62.89, 20.33 and 13.17%. Similar results are obtained on the basis of strength and structure parameters.

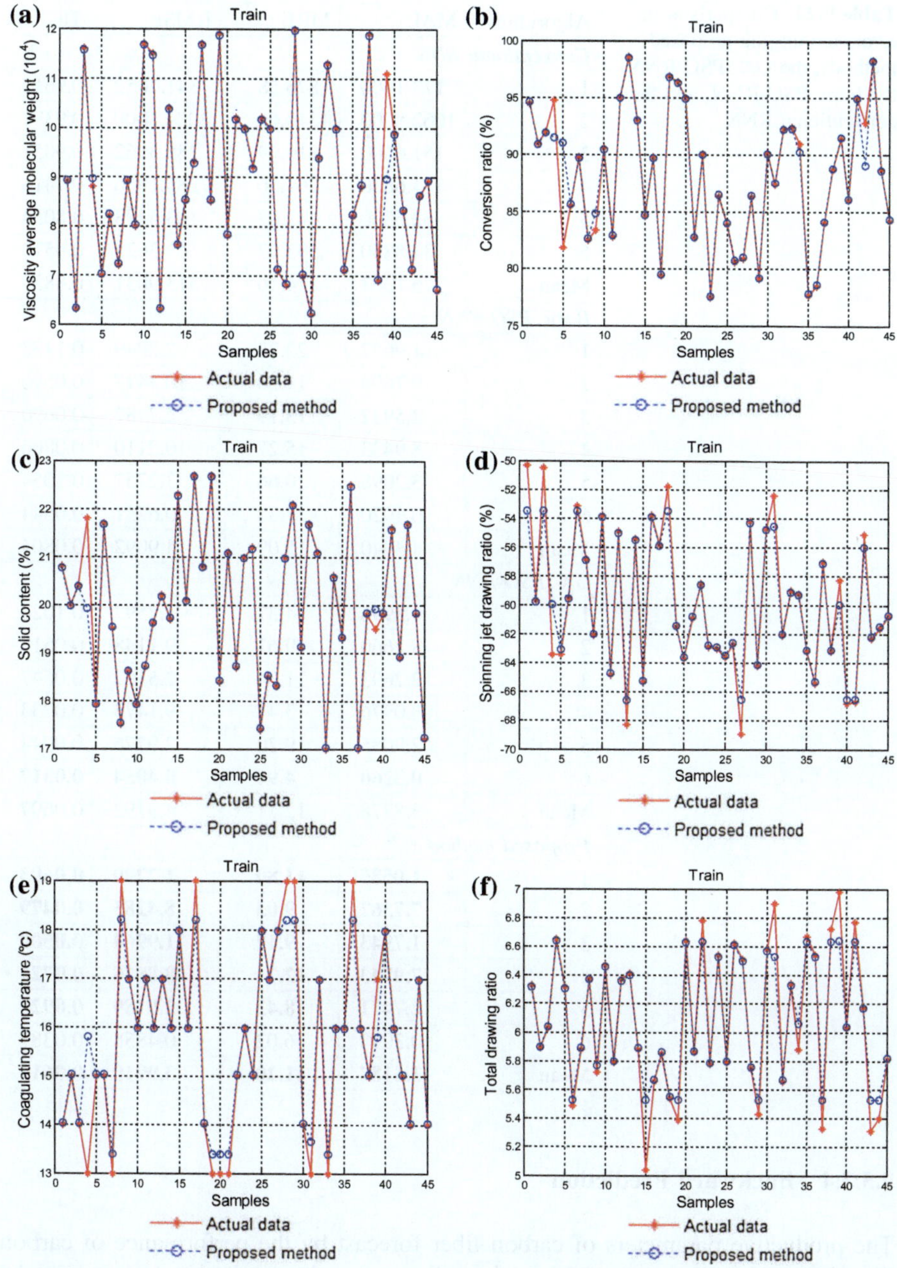

Fig. 9.26 Training results of the proposed SVM-IPSO model. **a** Viscosity average molecular weight. **b** Conversion ratio. **c** Solid content. **d** Spinning jet drawing ratio. **e** Coagulating temperature. **f** Total drawing ratio

Table 9.24 Comparison of errors among the proposed methods, the GA-IPSO-RNN, the basic PSO-RNN, and the conventional RNN

Algorithms	MAE	MRE (%)	RMSE	TIC
Conventional RNN				
1	173.1400	1876.58	341.7052	0.9612
2	1062.8000	1310.66	2004.3000	0.9391
3	151.6600	816.09	285.3062	0.9051
4	171.6800	264.56	285.1956	0.7416
5	62.6800	443.07	103.5493	0.8054
6	34.8800	533.97	63.3625	0.9510
Mean	276.1367	874.16	513.9031	0.8839
Basic PSO-RNN				
1	1.9652	22.73	2.2049	0.1138
2	9.7604	11.30	10.3417	0.0590
3	2.5932	13.19	2.7487	0.0690
4	8.9433	15.27	10.2110	0.0806
5	3.2098	20.66	3.2757	0.1054
6	0.4920	7.38	0.6731	0.0544
Mean	4.4940	15.09	4.9092	0.0804
GA-IPSO-RNN				
1	1.3996	14.37	1.7597	0.1026
2	8.2660	9.61	9.1148	0.0518
3	2.2693	11.04	2.5107	0.0647
4	8.0976	13.47	9.1473	0.0733
5	2.9085	19.20	2.9776	0.0944
6	**0.3260**	**4.94**	**0.4054**	**0.0317**
Mean	3.8778	12.11	4.3192	0.0697
Proposed method				
1	**1.0585**	**11.80**	**1.2290**	**0.0693**
2	**7.7362**	**9.03**	**8.4285**	**0.0479**
3	**1.7943**	**9.17**	**1.9964**	**0.0500**
4	**7.9743**	**12.74**	**8.8876**	**0.0724**
5	**2.7471**	**18.42**	**2.9189**	**0.0920**
6	0.3994	6.02	0.4855	0.0387
Mean	**3.6187**	**11.20**	**3.9910**	**0.0617**

9.5.3.4 Backward Prediction

The productive parameters of carbon fiber forecast by the performance of carbon fiber is a reverse process of forward prediction, so the predictive process can be considered inverse inference. In the backward prediction, we take the performance of carbon fiber as input and the productive parameters as output, which can be considered as a design scheme for a novel style carbon fiber. In order to achieve the above goals, by employing IPSO to optimize the parameters of SVM, we gain the best parameters for SVM. The training effectiveness of the proposed method is shown in Fig. 9.26.

We compared the effectiveness of SVM-IPSO with that of conventional RNN, the basic PSO-RNN, and the GA-IPSO-RNN optimal schemes. The detailed results are listed in Table 9.24. In Table 9.24, (1) is the viscosity average molecular weight; (2) is the conversion ratio; (3) is the solid content; (4) is the spinning jet drawing ratio; (5) is the coagulating bath temperature; (6) is the total drawing ratio.

From Table 9.24, we can see that on the basis of MAE, the proposed SVM-IPSO achieves roughly a decrease of 98.69, 19.48 and 6.68% compared to the other three methods, respectively. Similar results can be obtained from the other errors such as MRE with a decrease of 98.72, 25.78 and 7.51%, RMSE with a decrease of 99.22, 18.70 and 7.60%, and TIC with a decrease of 93.02, 23.26 and 11.48%. Similar results are obtained on the basis of viscosity average molecular weight, conversion ratio, solid content, spinning jet drawing ratio, and coagulating bath temperature, respectively.

9.6 Conclusions

In this chapter, first, a bi-directional optimizing approach based on an immune-enhanced neural network is proposed. The goal of the proposed bi-directional model is not only revealing the internal nonlinear relationship between the process configuration and the quality indices of the fibers in the melting spinning process but also providing a tool for engineers to develop new fiber products with expected quality specifications. A neural network is taken as the foundation of the bi-directional model, and an artificial immune component with algorithm is introduced to enlarge the searching scope of the solution field so that the neural network has a larger possibility to find the appropriate and reasonable solution. The proposed intelligent model can also help to determine what kind of process configuration should be made in order to produce satisfactory fiber products. Simulation results show the proposed model can eliminate the approximation error raised by the neural network-based optimizing model, which is due to the extension of focusing scope by the artificial immune mechanism. Meanwhile, the proposed model with the corresponding software can conduct optimization in two directions, namely, the process optimization and category development, and the corresponding results outperform those with an ordinary neural network-based intelligent model. It is also proved that the proposed model has the potential to act as a valuable tool that the engineers and decision makers of the spinning process can turn to for advice, which actually reaches the goal mentioned above.

And then, we present a model for using an enhanced RNN to bi-directional optimize the production process of PAN-based carbon fiber, which can be viewed as properties prediction and new type fiber designer. We adopt the nearest neighbor-clustering algorithm to simplify the structure of RNN, and investigate the learning method of the RNN based on the PSO, introduce penalty factor to improve the PSO, then combine GA with the improved PSO to train the RNN. Meanwhile, we compare the GA-IPSO-RNN model to the existing basic PSO-RNN and conventional RNN.

Next, a newly bi-directional prediction approach based on hybrid intelligent algorithms, which involved neural networks and multi-objective evolutionary algorithm, is adopted and further developed. The bi-directional prediction approach makes attempt to find the relationship between the production parameters and fiber performance of differential fiber production and optimize the fiber production. In addition, the parameters of the bi-directional prediction approach can be changed to suit differential fibers. The predictions of differential fibers given in this section prove the accuracy of this approach. This approach can give a way to determine the parameters before real production and to meet different fiber performance.

At last, we present the SVM-IPSO hybrid algorithm to bi-directionally forecast the productive process for carbon fiber, which can be considered as predicting the performance of carbon fiber and designing a method for new carbon fiber production. By the proposed model, we can online forecast the performance of the carbon fiber, and adjust the productive parameters in time, while reducing time and energy. In addition, this model can provide a means to determine the productive parameters for production to meet the performance requirements of different fibers.

References

1. Liang, X., Ding, Y.-S., Wang, Z.-D., Hao, K.-R., Hone, K., Wang, H.-P.: Bidirectional optimization of the melting spinning process. IEEE Trans. Cybern. **44**(2), 240–251 (2014)
2. Chen, J.-J., Ding, Y.-S., Hao, K.-R.: The bidirectional optimization of carbon fiber production by neural network with a GA-IPSO hybrid algorithm. Math. Probl. Eng. 16 pages, (2013)
3. Wang, Y., Ding, Y.-S., Hao, K.-R., Wang, T., Liu, X.-Y.: Performance prediction of differential fibers with a bi-directional optimization approach. Mater **6**(12), 5967–5985 (2013)
4. Xiao, C.-C., Hao, K.-R., Ding, Y.-S.: The bi-directional prediction of carbon fiber production using a combination of improved particle swarm optimization and support vector machine. Materials **8**(1), 117–136 (2015)
5. Hsu, H.-M., Hsiung, Y., Chen, Y.-Z., Wu, M.-C.: A GA methodology for the scheduling of yarn-dyed textile production. Expert Syst. Appl. **36**(10), 12095–12103 (2009)
6. Ding, H.-W., Benyoucef, L., Xie, X.-L.: A simulation-based multi-objective genetic algorithm approach for networked enterprises optimization. Eng. Appl. Artif. Intel. **19**(6), 609–623 (2006)
7. Sette, S., Boullart, L., Van langenhove, L.: Using genetic algorithms to design a control strategy of an industrial process. Control. Eng. Pract. **6**(4), 523–527 (1998)
8. Karimi, H.R., Duffie, N.A., Dashkovskiy, S.: Local capacity H-infinity control for production networks of autonomous work systems with time-varying delays. IEEE Trans. Autom. Sci. Eng. **7**(4), 849–857 (2010)
9. Kadi, H.E.: Modeling the mechanical behavior of fiber-reinforced polymeric composite materials using artificial neural networks-A review. Compos. Struct. **73**(1), 1–23 (2006)
10. Arafeh, L., Singh, H., Putatunda, S.K.: A neuro fuzzy logic approach to material processing. IEEE. Trans. Syst. Man. Cybern. **29**(3), 362–370 (1999)
11. Yu, Y., Hui, C.-L., Choi, T.-M., Au, R.: Intelligent fabric hand prediction system with fuzzy neural network. IEEE. Trans. Syst. Man. Cybern. C: Appl. Rev. **40**(6), 619–629 (2010)

12. Liu, M., Dong, M.-Y., Wu, C.: A new ANFIS for parameter prediction with numeric and categorical inputs. IEEE. Trans. Autom. Sci. Eng. **7**(3), 645–653 (2010)
13. Deng, X.-G., Vroman, P., Zeng, X.-Y., Laouisset, B.: Intelligent decision support tools for multicriteria product design. In: Proceedings of the 2010 IEEE International Conference on Systems, Man, and Cybernetics, Istanbul, Turkey, 1223–1230, (2010)
14. Yang, Y.-K., Yang, R.-T., Tzeng, C.-J.: Optimization of mechanical characteristics of short glass fiber and polytetrafluoroethylene reinforced polycarbonate composites using the neural network approach. Expert Syst. Appl. **39**(3), 3783–3792 (2012)
15. Yang, W.-Z., Li, D.-L., Zhu, L.: An improved genetic algorithm for optimal feature subset selection from multi-character feature set. Expert Syst. Appl. **38**(3), 2733–2740 (2011)
16. Wang, L.-Z., Ding, Y.-S., Hao, K.-R., Liang, X.: An immune-enhanced unfalsified controller for a high-speed spinning process, ISKE, 19–24, (2011)
17. Zheng, N.G., Wu, Z.H., Lin, M., Yang, L.T., Pan, G.: Infrastructure and reliability analysis of electric networks for E-textiles. IEEE. Trans. Syst. Man. Cybern. C: Appl. Rev. **40**(1), 36–51 (2010)
18. Du, D., Li, K., Fei, M.: A fast multi-output RBF neural network construction method. Neurocomputing **73**(10–12), 2196–2202 (2010)
19. Roy, A., Govil, S., Miranda, R.: A neural-network learning theory and a polynomial time RBF algorithm. IEEE. Trans. Neural Netw. **8**(6), 1301–1313 (1997)
20. Hong, X., Chen, S.: A new RBF neural network with boundary value constraints. IEEE. Trans. Syst. Cybern. B **39**(1), 298–303 (2009)
21. Huang, G., Saratchandran, P., Sundararajan, N.: A generalized growing and pruning RBF (GGAP-RBF) neural network for function approximation. IEEE. Trans. Neural Netw. **16**(1), 57–67 (2005)
22. Qiao, J., Han, H.: Identification and modeling of nonlinear dynamical systems using a novel self-organizing RBF-based approach. Automatica **48**(8), 1729–1734 (2012)
23. Wang, Y., Liu, G.: A forecasting method based on online self-correcting single model RBF neural network. Procedia Eng. **29**, 2516–2520 (2012)
24. Meireles, M.R.G., Almeida, P.E.M., Simoes, M.G.: A comprehensive review for industrial applicability of artificial neural networks. IEEE. Trans. Ind. Electron. **50**(3), 585–601 (2003)
25. Hippert, H., Pedreira, C., Souza, R.: Neural networks for short-term load forecasting: A review and evaluation. IEEE. Trans. Power. Syst. **16**(1), 44–55 (2002)
26. Khosravi, A., Nahavandi, S., Creighton, D., Atiya, A.F.: Comprehensive review of neural network-based prediction intervals and new advances. IEEE. Trans. Neural. Netw. **22**(9), 1341–1356 (2011)
27. Palnitkar, R. M., Cannady, J.: A review of adaptive neural networks. SoutheastCon, Proceedings. IEEE, 38–47, (2004)
28. Quadus, M.A., Khan, M.S.: Business applications of artificial neural networks an updated review and analysis. Neural Inf. Process **2**, 819–824 (1999)
29. Grond, M.O.W., Luong, N.H., Slootweg, J.G.: Multi-objective optimization techniques and applications in electric power systems. In Proceeding of the Universities Power Engineering Conference, London, UK, 4–7, (2012)
30. Ishibuchi, H., Murata, T.: A multi-objective genetic local search algorithm and its application to flowshop scheduling. IEEE. Trans. Syst. Man. Cybern. C **28**(3), 392–403 (1998)
31. Hu, Y.-F., Ding, Y.-S., Hao, K.-R.: An immune cooperative particle swarm optimization algorithm for fault-tolerant routing optimization in heterogeneous wireless sensor networks. Math. Probl. Eng, 1–19 (2012)
32. Alireza, A.: PSO with adaptive mutation and inertia weight and its application in parameter estimation of dynamic systems. Acta Automatica Sinica **37**(5), 541–549 (2011)
33. Behrang, M.A., Assareh, E., Noghrehabadi, A.R., Ghanbarzadeh, A.: New sunshine-based models for predicting global solar radiation using PSO (particle swarm optimization) technique. Energy **36**(5), 3036–3049 (2011)
34. Mahor, A., Rangnekar, S.: Short term generation scheduling of cascaded hydro electric system using novel self adaptive inertia weight PSO. Int. Electr. Power **34**(1), 1–9 (2012)

35. Tang, Y., Gao, H., Kurths, J., Fang, J.: Evolutionary pinning control and its application in UAV coordination. IEEE. Trans. Ind. Inf. **8**(8), 828–838 (2012)
36. Luitel, B., Venayagamoorthy, G.K.: Quantum inspired PSO for the optimization of simultaneous recurrent neural networks as MIMO learning systems. Neural Netw. **23**(5), 583–586 (2012)
37. Vasumathi, B., Moorthi, S.: Implementation of hybrid ANN-PSO algorithm on FPGA for harmonic estimation. Eng. Appl. Artif. Intel. **25**(3), 476–483 (2012)
38. Oh, S.K., Kim, W.D., Pedrycz, W., Park, B.J.: Polynomial-based radial basis function neural networks (P-RNF NNs) realized with the aid of particle swarm optimization. Fuzzy Sets Syst. **163**(1), 54–77 (2011)
39. Huang, C.M., Wang, F.L.: An RBF network with OLS and EPSO algorithm for real-time power dispatch. IEEE Trans. Power Syst. **22**(1), 96–104 (2007)
40. Li, J., Liu, X.: Melt index prediction by RBF neural network optimized with an MPSO-SA hybrid algorithm. Neurocomputing **74**(5), 735–740 (2011)
41. Hong, W.C., Dong, Y., Zhang, W.Y., Chen, L.Y., Panigrahi, B.K.: Cyclic electric load forecasting by seasonal SVR with chaotic genetic algorithm. Int. Electr. Power Energy Syst. **44**(1), 604–614 (2013)
42. Goh, A.T.C., Goh, S.H.: Support vector machines: Their use in geotechnical engineering as illustrated using seismic liquefaction data. Comput. Geotech. **34**(5), 410–421 (2007)
43. Kordjazi, A., Nejad, F.P., Jaksa, M.B.: Prediction of ultimate axial load-carrying capacity of piles using a support vector machine based on CPT data. Comput. Geotech. **55**(1), 91–102 (2014)
44. Lin, J., Cheng, C., Chau, K.: Using support vector machines for long-term discharge prediction. Hydrol. Sci. **51**(4), 599–612 (2006)
45. Niu, D., Wang, Y., Wu, D.D.: Power load forecasting using support vector machine and ant colony optimization. Expert Syst. Appl. **37**(3), 2531–2539 (2010)
46. Wang, J., Li, L., Niu, D., Tan, Z.: An annual load forecasting model based on support vector regression with differential evolution algorithm. Appl. Energy **94**(6), 65–70 (2012)
47. Gilan, S.S., Jovein, H.B., Ramezanianpour, A.A.: Hybrid support vector regression-Particle swarm optimization for prediction of compressive strength and RCPT of concretes containing metakaolin. Constr. Build. Mater. **34**(3), 321–329 (2012)
48. Hong, W.C.: Chaotic particle swarm optimization algorithm in a support vector regression electric load forecasting model. Energy Convers. Manag. **50**(1), 105–117 (2009)
49. Kang, Q., Zhou, M.C., An, J., Wu, Q.: Swarm intelligence approaches to optimal power flow problem with distributed generator failures in power networks. IEEE Trans. Autom. Sci. Eng. **10**(2), 343–353 (2013)
50. Liang, X., Li, W., Zhang, Y., Zhou, M.C.: An adaptive particle swarm optimization method based on clustering. Soft Comput. **19**(2), 431–448 (2015)
51. Cho, C.S., Chung, B.M., Park, M.J.: Development of real-time vision-based fabric inspection system. IEEE Trans. Ind. Electron. **52**(4), 1073–1079 (2005)
52. Wong, C.C., Dean, T.A., Lin, J.: A review of spinning, shear forming and flow forming processes. Int. Mach. Tool. Manuf. **43**(14), 1419–1435 (2003)
53. Yusof, N., Ismail, A.F.: Post spinning and pyrolysis processes of polyacrylonitrile (pan)-based carbon fiber and activated carbon fiber: a review. Anal. Appl. Pyrolysis **93**(1), 1–13 (2012)
54. Wei, W., Zhang, Y., Zhao, Y., Luo, J., Shao, H., Hu, X.: Bio-inspired capillary dry spinning of regenerated silk fibroin aqueous solution. Mater. Sci. Eng. C **31**(7), 1602–1608 (2011)
55. Sulong, A.B., Park, J., Che, H.A., Jusoff, K.: Process optimization of melt spinning and mechanical strength enhancement of functionalized multi-walled carbon nanotubes reinforcing polyethylene fibers. Compos. Part B-Eng. **42**(1), 11–17 (2011)
56. Antonietti, P.F., Biscari, P., Tavakoli, A., Verani, M., Vianello, M.: Theoretical study and numerical simulation of textiles. Appl. Math. Model. **35**(6), 2669–2681 (2011)

57. Balci, B., Keskinkan, O., Avci, M.: Use of bdst and an ann model for prediction of dye adsorption efficiency of eucalyptus camaldulensis, barks in fixed-bed system. Expert Syst. Appl. **38**(1), 949–956 (2011)
58. Xu, L., Wu, Y., Nawaz, Y.: Numerical study of magnetic electrospinning processes. Comput. Math. Appl. **61**(8), 2116–2119 (2011)
59. Thissen, U., Brakel, R.V., Weijer, A.P.D., Melssen, W.J., Buydens, L.M.C.: Using support vector machines for time series prediction. Chemom. Intell. Lab. **69**(1–2), 35–49 (2003)
60. Kuo, C.J., Chien, C.F., Chen, J.D.: Manufacturing intelligence to exploit the value of production and tool data to reduce cycle time. IEEE. Trans. Autom. Sci. Eng. **8**(1), 103–111 (2011)
61. Horng, S.C., Yang, F.Y., Lin, S.S.: Embedding evolutionary strategy in ordinal optimization for hard optimization problems. Appl. Math. Model. **36**(8), 3753–3763 (2012)
62. Moody, J., Darken, C.J.: Fast learning in networks of locally-tuned processing units. Neural Comput. **1**(2), 281–294 (1989)
63. Jarvis, R.A., Patrick, E.A.: Clustering using a similarity measure based on shared near neighbors. IEEE Trans. Comput. **22**(11), 1025–1034 (1973)
64. Kennedy, J., Eberhart, R.C.: Particle swarm optimization. In: Proceedings of the 1995, IEEE International Conference on Neural Network IV, 4, 1942–1948 (1995)
65. Ditzian, Z.: Relating smoothness to expressions involving Fourier coefficients or to a Fourier transform. Approx. Theory **164**(10), 1369–1389 (2012)
66. Del Valle, Y., Venayagamoorthy, G.K., Mohagheghi, S., Hernandez, J.C., Harley, R.G.: Particle swarm optimization Basic Concepts Variants and applications in power systems. IEEE Trans. Evol. Comput. **12**(2), 171–195 (2008)
67. Paoli, A., Melgani, F., Pasolli, E.: Clustering of hyperspectral images based on multiobjective particle swarm optimization. IEEE Trans. Geosci. Remote Sens. **47**(12), 4175–4188 (2009)
68. Selim, S.Z., Ismail, M.A.: K-means type algorithms: A generalized convergence theorem and characterization of local optimality. IEEE Trans. Pattern Anal. Mach. Intell. **6**(1), 81–87 (1984)
69. Ahmadyfard, A., Modares, H.: Combining PSO and K-means to enhance data clustering. In: International Symposium on Telecommunications, pp. 688–691 (2008)
70. Liu, G., Xiao, X., Mei, C., Ding, Y.: A review of learning algorithm for radius basis function neural network. Paper presented at control and decision conference, vol. 229, pp. 1112–1117, May (2012)
71. Chang, F.J., Liang, J.M., Chen, Y.C.: Flood forecasting using Radial Basis Function Neural Networks. IEEE. Trans. Syst. Man. Cybern. C **31**(4), 530–535 (2001)
72. Wei, J.X., Wang, Y.P.: Multi-objective fuzzy particle swarm optimization based on elite archiving and its convergence. Syst. Eng. Electron. **19**(5), 1035–1040 (2008)
73. Deb, K., Pratap, A., Agarwal, S., Meyarivan, T.: A fast and elitist multiobjective genetic algorithm: nsga-ii. IEEE Trans. Evolut. Comput. **6**(2), 182–197 (2002)
74. Vapnik, V.: The Nature of Statistical Learning Theory. Springer, New York (1995)
75. Li, X., Zheng, A.N., Zhang, X., Li, C., Zhang, L.: Rolling element bearing fault detection using support vector machine with improved ant colony optimization. Measurement **46**(8), 2726–2734 (2013)
76. Keerthi, S.S., Lin, C.J.: Asymptotic behaviors of support vector machines with Gaussian kernel. Neural Comput. **15**(7), 1667–1689 (2003)
77. Zhu, H., Wang, Y., Wang, K., Chen, Y.: Particle Swarm Optimization (PSO) for the constrained portfolio optimization problem. Expert Syst. Appl. **38**(8), 10161–10169 (2011)
78. Ahmed, K.A., Xiang, J.: Mechanisms of cellular communication through intercellular protein transfer. Cell Mol. Med. **15**(7), 1458–1473 (2011)

Index

Y. Ding et al., *Bio-Inspired Collaborative Intelligent Control and Optimization*, Studies in Systems, Decision and Control 118,
https://doi.org/10.1007/978-981-10-6689-4

MIX
Papier aus verantwortungsvollen Quellen
Paper from responsible sources
FSC® C105338

If you have any concerns about our products,
you can contact us on
ProductSafety@springernature.com

In case Publisher is established outside the EU,
the EU authorized representative is:
Springer Nature Customer Service Center GmbH
Europaplatz 3, 69115 Heidelberg, Germany

Printed by Libri Plureos GmbH
in Hamburg, Germany